Introduction to Diagnosis of Active Systems

Gianfranco Lamperti • Marina Zanella
Xiangfu Zhao

Introduction to Diagnosis of Active Systems

Springer

Gianfranco Lamperti
Dipt. di Ingegneria dell'Informazione
Università degli Studi di Brescia
Brescia, Italy

Marina Zanella
Dipt. di Ingegneria dell'Informazione
Università degli Studi di Brescia
Brescia, Italy

Xiangfu Zhao
College of Math, Phys & Inform Engin
Zhejiang Normal University
Jinhua City, Zhejiang, China

ISBN 978-3-030-06503-4 ISBN 978-3-319-92733-6 (eBook)
https://doi.org/10.1007/978-3-319-92733-6

Softcover re-print of the Hardcover 1st edition 2018

Printed on acid-free paper

This Springer imprint is published by the registered company Springer International Publishing AG part of Springer Nature.
The registered company address is: Gewerbestrasse 11, 6330 Cham, Switzerland

To our parents.

In memory of Robert Milne.

Preface

This book is an up-to-date, self-contained compendium of the research carried out for two decades by the authors on model-based diagnosis of a class of discrete-event systems (DESs) called active systems.

In the most traditional view, diagnosis is the task of finding out what is wrong with a system, namely its faults, these being the root causes that make the actual behavior deviate from the expected (normal) behavior. The expression *model-based* refers to the artificial intelligence technology that is adopted in order to carry out the task. Such a technology relies on the availability of specific models, which represent the behavior of the system to be diagnosed, and of a general diagnosis engine, which can process the models of any entity considered.

A DES conceptual model, typically an automaton or a network of communicating automata, describes a dynamic behavior by means of a finite number of states and transitions between them, which are triggered by a finite number of discrete events. The diagnosis engine takes as input an observation relevant to the actual behavior exhibited by the system and exploits models in order to explain the given observation.

The observation is what can be perceived, from the outside, about the inner behavior of the system itself, for instance, a stream of sensor measurements. If an abnormal behavior can explain the observation, the diagnosis engine generates the faults of the system considered.

Several notions of faults can be found in the literature. Roughly, a fault is a broken component, which implies that the system consists of several interconnected components. At a finer granularity, a fault in a DES is an abnormal state transition. Once the diagnosis task has been accomplished, the results produced are exploited to fix the system.

The interest in diagnosis of DESs is remarkably strong and is not likely to fade in the future as, for diagnosis purposes, several existing systems can conveniently be modeled as DESs. Moreover, diagnosis (as well as diagnosability analysis) of DESs is essential in order to deal with hybrid systems, these being conceptual models that integrate discrete and continuous behavior.

An active system is an abstraction that models a DES as a network of automata communicating through buffered links. The input events of active systems are assumed to be unobservable, as they often are in the real world.

This book comes several years after a previous book on the same topic [91], and it records the evolution that the theory of model-based diagnosis of active systems has undergone since then.

The book is structured into twelve chapters, each of which has its own introduction and is concluded by some bibliographic notes and an itemized summary. Concepts and techniques are presented with the help of numerous examples, figures, and tables. When appropriate, these concepts are formalized into propositions and theorems. Detailed algorithms expressed in pseudocode are incorporated.

Chapter 1 briefly surveys the fundamentals of model-based diagnosis as originally conceived in the domain of static systems and its progression toward the realm of dynamic (time-varying) systems. It also defines the central notion of a DES diagnosis problem.

Chapter 2 introduces the essential characteristics of finite automata, which are the basis for the modeling of active systems, along with relevant operations. It also presents a technique, called incremental determinization, which is exploited in the diagnosis methods presented in Chapters 7 and 10.

Chapter 3 defines the notion of an active system, which, unlike the usual DES models in the literature, is assumed to be asynchronous. This means that events generated by components are temporarily stored within connection links before being (asynchronously) consumed.

Chapter 4 is concerned with the notion of a diagnosis problem, along with its solution, specifically relevant to an active system. Such a notion requires in turn the definition of a temporal observation. The dependence of a diagnosis problem on the assumed observability and "fallibility" of the active system considered is highlighted.

Chapter 5 proposes a technique to compute the solution of an a posteriori diagnosis problem for an active system. Diagnosis is a posteriori when its results are computed given a whole observation relevant to the dynamic evolution of the system up to the time it came to a halt. Such a method is monolithic in that the diagnosis problem is neither divided into subproblems nor reduced to a single subproblem of smaller size in order to be solved.

Chapter 6 proposes an alternative technique to solve an a posteriori active-system diagnosis problem. Unlike the monolithic diagnosis described in Chapter 5, modular diagnosis is based on a decomposition of the problem into subproblems. Such a decomposition is relevant to behavior reconstruction, the first step performed by the diagnosis technique. Each separate behavior reconstruction subproblem is focused on a subsystem and can be solved in parallel with the other subproblems. This allows for a more efficient computation.

Chapter 7 presents reactive diagnosis, a diagnosis task that repeatedly computes results on the reception of each new fragment of observation, every time updating the set of results. Reactive diagnosis solves a new diagnosis problem, called the monitoring problem, which is based on the notion of a fragmented observation.

Chapter 8 defines the quite general concept of monotonicity of the solutions of a monitoring problem, and finds some conditions the fragmented observation has to fulfill in order to ensure that the results produced by the method presented in Chapter 7 are monotonic.

Chapter 9 was contributed by Federica Vivenzi, who carried out research with the authors on this topic at the University of Brescia. This chapter deals with the reuse of intermediate data structures produced by a diagnosis engine when it solved a previous diagnosis problem in order to solve a new, similar problem. The aim is to save computation time. The chapter finds some conditions relevant to the similarity between the old and the new problem under which reusability is applicable.

Chapter 10 proposes a pruning technique that can be adopted during the first step of a diagnosis process when reconstructing the behavior of an active system based on a given observation. The aim is to confine the explosion of the reconstruction. The technique is called lazy diagnosis.

Chapter 11 questions the so-called *atomic assumption* of the diagnosis of active systems, namely that faults are invariably associated with component transitions. By contrast, it shows that faults can be conveniently associated with patterns of transitions, possibly involving different components. Consequently, what is considered abnormal is not a single component transition involved in the pattern, but a specific combination of transitions matching such a pattern. In other words, diagnosis becomes *context-sensitive*. A further extension is achieved by conceiving the active system as a hierarchy of subsystems, where faults are defined at different abstraction levels. This opens the way to the diagnosis of *complex* active systems.

Chapter 12 concludes the book by providing a survey of several related publications on the diagnosis and diagnosability of DESs.

Being a research monograph on model-based diagnosis, this book is primarily intended for researchers and academics in the field of artificial intelligence. However, as testified by the references, since the subject of model-based diagnosis of DESs is largely shared with the control theory community, the book is also intended for researchers from that area.

The reader is assumed to be familiar with programming concepts. Since some of the proposed techniques are specified in detail by pseudocode algorithms, the book is also appropriate for engineers and software designers who are charged with developing tools for the supervision and monitoring of complex industrial systems.

The audience addressed includes practitioners and professionals who are interested in knowing about recent research issues in the diagnosis of DESs and in getting new ideas and stimuli for coping with diagnosis in their application domains.

The present work can also be profitably adopted as an additional textbook for graduate courses (such as in artificial intelligence or algorithms and data structures), as well as Ph.D. courses and continuing education initiatives.

Acknowledgments

We owe a debt of gratitude to Federica Vivenzi for her contribution to the book and for her devotion to the subject since she was a talented student at the University of Brescia. We also express our gratitude to the people we met at various scientific symposia, who gave us the opportunity to exchange fruitful ideas, in particular within the DX community. We would like to convey our sincere appreciation to the staff of Springer, especially Ronan Nugent and Annette Hinze, for their professional handling of the manuscript and for the infinite patience and unfailing courtesy that distinguish editors from authors.

Gianfranco Lamperti
Marina Zanella
Xiangfu Zhao

Brescia, Italy
Jinhua, China
April 2018

Contents

Acronyms

BDD	binary decision diagram
BNF	Backus-Naur form
DAG	directed acyclic graph
DFA	deterministic finite automaton
DES	discrete-event system
DNNF	decomposable negation normal form
FA	finite automaton
HFSM	hierarchical finite state machine
HSM	hierarchical state machine
IDEA	incremental determinization of expanding automata
IDMA	incremental determinization and minimization of acyclic automata
ISC	incremental subset construction
ISCA	incremental subset construction with acyclicity
LDE	lazy diagnosis engine
LISCA	lazy incremental subset construction with acyclicity
MFA	minimal deterministic finite automaton
MIN	minimization of DFA
NFA	nondeterministic finite automaton
OBDD	ordered binary decision diagram
SBS	subsumption checking
SC	subset construction

Chapter 1
Introduction

Diagnosis is one of the earliest tasks that were tackled by means of artificial intelligence techniques and it is still an important research topic for the artificial intelligence community, not only for the variety and relevance of its potential applications but also because it is a challenge for several aspects of automated reasoning.

Historically, the first approach to diagnosis was *rule-based* (or *heuristic*): production rules, possibly organized according to decision trees, represented empirical knowledge about the associations between *symptoms* and *faults* in a given system to be diagnosed. Such knowledge embodied rules of thumb elicited from human experts or retrieved from large data samples. The reasoning mechanism was capable of tracing back observed symptoms to their causes. MYCIN [150], which was developed in the 1970s to assist physicians both in determining possible causes of bacteremia and meningitis and in selecting an appropriate therapy, is perhaps the best known of the early rule-based systems. However, rule-based diagnosis suffers from a variety of disadvantages, including the following:

1. It can cope only with a bounded number of known symptoms and faults,
2. Knowledge acquisition from human experts is difficult and time-consuming,
3. Empirical associative knowledge is huge and hard to maintain,
4. The available empirical knowledge is only applicable to malfunctions that have already occurred; thus it is unlikely such a knowledge encompasses rare faults,
5. Empirical knowledge does not exist for newly born artifacts,
6. Empirical knowledge is specific to the system considered and, in general, it cannot be reused for other systems,
7. A large set of probabilities may possibly be needed, as discussed below.

Rule-based diagnosis determines either all the possible sets of faults, each of which is a possible cause of the observed symptoms, or just the most likely set of faults, given the observed symptoms. In order to estimate the likelihood of any set of faults, a large number of probabilities is needed, specifically $2^{n+m} - 1$ distinct probabilities in the worst case, where n is the number of faults and m the number of symptoms.

G. Lamperti et al., *Introduction to Diagnosis of Active Systems*,
https://doi.org/10.1007/978-3-319-92733-6_1

The probability of a fault is a piece of information that is difficult to elicit and is subjective in nature, as it depends on the personal experience of each expert considered and on his/her recalled perception of such experience.

In summary, the knowledge exploited by rule-based systems is *shallow*, as it is subjective and/or derived from recorded experience, and it does not take into account any piece of information about how a system was designed (in the case where it is an artifact) and how it is expected to behave.

In contrast, the chronologically successive approach of *model-based* diagnosis [72, 139] exploits *deep* knowledge, that is, objective knowledge about the structure and behavior of the system considered, which takes into account the physical principles governing the functioning of the system. Such knowledge is easier to elicit than empirical knowledge as it can be drawn from design documents, textbooks, and first principles. Therefore, such knowledge is available also for newly conceived systems that have never been put into use before. Moreover, deep *causal* knowledge enables one to distinguish whether a pair of symptoms are dependent or independent when a certain fault is assumed. Thus, the number of probabilities that need to be considered in order to determine the most likely set of faults is smaller than for rule-based diagnosis. Selected readings about the early years of model-based diagnosis can be found in [60].

In the remainder of the chapter, Sec. 1.1 provides the basics of model-based diagnosis, while Secs. 1.2 and 1.3 address model-based diagnosis of static systems and dynamic systems, respectively. Finally, Sec. 1.4 deals with model-based diagnosis of dynamic systems that are modeled as discrete-event systems.

1.1 Fundamentals of Model-Based Diagnosis

Model-based diagnosis leads to a crisp, unambiguous definition of a *symptom*, according to which a symptom is a discrepancy between the observed behavior and the predicted behavior of a system. The *observed behavior* is the set of actual values of the output variables; the *predicted behavior* is the set of the values of the same output variables computed based on the modeled normal behavior of the system. In both cases, the values of the input variables are assumed to be known.

A solution produced by a diagnostic session is a set of *candidate diagnoses*, or simply *candidates*, with each candidate being, for instance, a set of faulty components or a set of specific faults assigned to components, that *explains* the observation. In a broader perspective, each candidate diagnosis is a set of assumptions relevant to the behavior of the system that explains the observed behavior.

What is considered as a *component* depends on the granularity level of the *structural model* of the system, which describes the topology of the system in terms of constituent parts (components) and interconnections among them. The structure of an artifact is available, at least theoretically, from its design, for example from CAD documents, wherein architectural modules are highlighted.

In general, however, the components to be considered by the diagnosis task are identified based on repair/replacement actions that can be carried out to fix the system. In fact, diagnosis is not a goal in itself, the ultimate goal being instead to increase the availability of the system considered, by repairing it if it is affected by faults. Therefore, for supporting the task of diagnosis, it does not make sense to split a (macro)component C into several components at a finer granularity level if, in the case where any of these components contained in C is faulty, the whole C unit is replaced.

A fundamental assumption underlying most model-based diagnosis approaches is that the modeled structure of the system never changes; that is, the existing interconnections among components cannot be broken and no new interconnections can be added. When this is assumed, model-based diagnosis cannot find any fault that changes the structure of the system; instead, it can find only faults that change the behavior of components.

What is considered as a fault depends on the abstraction level of the *behavioral model* of each component of the system. The *normal behavior* of such a system, that is, how the artifact is intended to work according to the purpose it was designed and built for, is known from design documents.

However, generally speaking, in model-based diagnosis such a normal behavior is not given in a global (system-level) description; rather, it has to implicitly emerge from the composition of the explicit normal behaviors of the components.

Based on the survey in [126], some of the basic principles of model-based diagnosis are the following (where the first two are a consequence of the "no function in structure" principle stated in the literature):

1. The behavior of the system to be diagnosed is expressed by the behavioral models of its components,
2. The behavioral model of a component is independent of any composite device the component may belong to,
3. The interpreter of the behavioral models is independent of the application domain of the system considered,
4. The diagnostic task is independent of the specific representations adopted by the structural and behavioral models.

Hence, the architecture of a model-based diagnostic system is modular: it includes an interpreter of models, which is specific to the modeling language adopted, a diagnostic algorithm, and a library of models.

The above principles determine several advantages of model-based diagnosis over rule-based diagnosis, typically including the reusability of behavioral knowledge, as the same model of a component can be used several times for diagnosing distinct systems the component belongs to. In principle, the same behavioral model of a component can also be used for accomplishing tasks other than diagnosis, provided that the module for the diagnostic algorithm in the architecture is replaced by a module performing another model-based task.

Another asset of model-based diagnosis is that the diagnostic reasoning mechanism is independent of the specific system to be diagnosed: the same diagnostic architecture can be used to diagnose distinct systems, even belonging to different application domains, provided that the models of such systems are contained in the model library.

Whilst the component-oriented ontology contained in the above principles is the one most often used in model-based diagnosis, a different conceptualization of domain knowledge, namely *causal networks*, can also be adopted (see, e.g., [35, 46]). Such networks, however, are not covered in this book.

1.1.1 Modeling Issues

Modeling is defined by Poole [134] as "going from a problem P to a representation R_P of P in the formal language so that the use of the inference relation for the representation will yield a solution to the problem." Models are the heart of model-based diagnosis: they strongly influence the efficiency of the diagnostic process and the effectiveness of its results. Therefore, providing a good model is a crucial central responsibility that involves some issues.

First, model-based diagnosis, while identifying faults, implicitly assumes that the models it exploits are dependable/reliable. However, models differ from the real systems they represent, as every model depends on its own set of simplifying assumptions and approximations.

A model is good for a diagnostic purpose if such simplifications and approximations are acceptable in the case at hand. For example, the models adopted may feature too coarse a granularity for the current purpose. Such a modeling problem can be solved by providing more detailed models, which lead to a better approximation of the real system. This, however, is bound to increase the computational complexity of the task, an effect that is particularly relevant when model-based diagnosis is scaled up to large devices.

Second, experience in applying model-based diagnosis in industrial settings has shown that the modeling cost is a bottleneck for the acceptance of model-based diagnosis.

At ASML (the world's leading lithography machine vendor), a software program has been installed on over 3000 service laptops worldwide, and has successfully been used to diagnose faults in an important electromechanical subsystem that frequently suffered from failures [133]. It was demonstrated that model-based diagnosis could reduce the solution cost from days to minutes for a once-only investment of 25 man-days of modeling effort (approximately 2000 lines of code, comprising sensor modeling, electrical circuits, and some simple mechanics). Despite the obvious financial gains, management discontinued the project once it had become clear that only some 80% of the model could be obtained automatically from the system source code (graphical schematics data for electrical components, and VHDL for the logic circuits) [138, 146].

More recently, a similar experience was obtained with a model-based diagnosis project for the European Space Agency (ESA), which involved modeling the electrical power subsystem of the GOCE satellite [47]. Despite the successful conclusion of this proof-of-concept research project, it appears to be highly uncertain whether the results will ever be applied in a follow-up project by the target ESA satellite operators, now that the ESA has become fully aware of the complexity and human effort involved in the associated diagnostic satellite modeling.

The bottom line is that, in industrial contexts where software and hardware are constantly evolving, there is growing evidence that mainstream adoption of model-based diagnosis is only feasible when modeling can be fully automated.

1.2 Model-Based Diagnosis of Static Systems

The model-based-diagnosis manifesto, which has deeply influenced all research efforts on the topic since 1987, is the content of volume 32 of the journal *Artificial Intelligence*. A fundamental publication there, namely Reiter's theory of diagnosis from first principles [139], provides the logical formalization of a diagnostic problem and of its solutions, as well as the relevant terminology adopted thereafter.

The classical theory of model-based diagnosis addresses a limited class of systems, called *static systems*, whose nominal behavior can be specified as a time-invariant mapping from some input variables to some output variables. In model-based diagnosis of static systems, three diagnostic subtasks can be singled out:

1. *Fault detection*, which consists in finding out *whether* a fault has occurred or not,
2. *Fault isolation* (or *localization*), which ascertains *where* a fault has occurred,
3. *Fault identification*, which determines *what* fault has occurred.

No diagnosis of static systems is performed if no fault has been detected. Fault detection is the only diagnostic subtask that interfaces directly with the system being diagnosed. This subtask is typically carried out by means of a discrepancy detector, which compares the measured outputs from the system with the predicted values from the model, according to some predefined metrics.

If, instead, one or more discrepancies (symptoms) are detected, fault isolation and/or fault identification is performed. The location of the fault depends on the structural description of the plant, that is, the constituent responsible for the fault is part of a process, a component, or simply a (set of) constraint(s).

1.3 Model-Based Diagnosis of Dynamic Systems

In the 1980s it was common to assume (and model) the system to be diagnosed as static, and its observation was given at a single time point [72, 139]. A static behavioral model, possibly encompassing several behavioral modes, does not capture temporal aspects; it is just a set of constraints that restricts the set of states possible in each respective mode. When a system is observed, it is in a state and continues to be in the same state as far as its outputs are measured, even if some additional measurements are performed later in order to discriminate among candidate diagnoses.

Nowadays, research is focused on *dynamic systems*, that is, systems whose behavior at the current time instant depends on what has happened before. A dynamic model characterizes the evolution of a behavior over time, that is, it constrains not only the states but also the relations between states across time. A fault is *permanent* if it has a permanent duration, while it is *transient* or *intermittent* if it causes temporally finite deviations from the normal behavior; in particular, the "intermittent" qualification emphasizes the repeatability of the fault.

In the domain of dynamic systems, the complexity of both the conceptual setting and the diagnosis task increases, since a temporal diagnosis has to account for both the values in the observation and their temporal location. In order to accommodate any distinct approach to temporal diagnosis within a lattice of definitions, the two aspects mentioned above, namely the presence of observed values and their temporal location, can be considered separately [19].

In principle, the approaches adopted for static systems could be applied also to dynamic ones. Such approaches require checking the behavior of the real device over time against the behavior predicted by a dynamic model of the device itself. This amounts to:

1. Tracking the actual behavior of the system over time,
2. Simulating the modeled behavior,
3. Comparing the actual outputs with the results of the simulation.

The simulation task and the comparison of behavior sets can often be computationally prohibitive, particularly when qualitative simulation yields ambiguous results or when fault identification requires the simulation of many fault scenarios. In spite of this, the above idea underlies several approaches [34, 40, 43, 44, 49, 58, 59, 61, 73, 127, 128] from the 1980s and 1990s, sometimes called *simulation-based*, to model-based diagnosis of dynamic systems. In fact, a large group of approaches to the diagnosis of dynamic systems are based mainly on modifications and extensions of Reiter's theory and tend to keep reasoning about temporal behavior separate from diagnostic reasoning.

In these attempts, attention is focused mainly on the problem of fault detection in the context of the dynamic behavior exhibited by the real system, whereas, once a fault has been detected, a variant of Reiter's diagnostic algorithm, normally not including significant temporal reasoning features, is applied.

Therefore reasoning about temporal behavior in these approaches mainly concerns *system monitoring*, this being the activity of looking after the system and detecting symptoms over time, rather than true system diagnosis, that is, the activity of finding explanations for symptoms. When the system is operating correctly, there will be no discrepancy and model-based diagnosis will in effect be operating as a *condition monitor*.

1.4 Model-Based Diagnosis of Discrete-Event Systems

This book is devoted to model-based diagnosis of dynamic systems that are modeled as discrete-event systems. A *discrete-event system* (DES) [25] is a conceptual model to describe a dynamic system as a discrete-state, event-driven entity.

The DESs considered in this book are untimed, as we are interested only in the chronological order of the occurrence of events, disregarding the specific time instants when such events occurred. Moreover, these DESs are only partially observable, in that not all the occurrences of events can be observed outside the system.

Diagnosis of DESs is a task aimed at finding out what is (possibly) wrong with a given dynamic system described as a DES, on account of a given *observation* relevant to the system that has been gathered over time while the system was running. Model-based diagnosis of DESs is the automation of such a task through knowledge-based technology, so as to build a general software engine that can solve diagnosis problem instances relevant to any given DES, by exploiting the DES description (model) of the behavior of the specific system at hand.

Model-based diagnosis of DESs is an important task as a large variety of real-world systems can be conveniently modeled as DESs for diagnosis purposes, including digital hardware, telecommunication networks, and power networks. Automated diagnosis is especially interesting for safety-critical systems (such as chemical or nuclear plants), where, in the case of abnormal behavior, stringent time constraints on recovery actions on the one hand and overwhelming streams of alarms from the protection hardware on the other are equally bound to cause human errors, which, under dangerous circumstances, may have catastrophic consequences.

In the literature, model-based diagnosis of DESs is typically applied to intelligent alarm processing in a remote control room, where the alarms received are the observable events taken as input by the diagnosis process. For instance, in [110] the protection devices of power transmission lines were considered as a case study in monitoring and diagnosis of DESs. Similarly, experimental diagnosis results relevant to a real alarm log from the operation center of an electricity transmission network in Australia were presented in [55, 57]. In addition, a set of event logs recorded on the ground during test flights of an autonomous unmanned helicopter was addressed in [55], while in [131] the alarms received by the supervision center of a telecommunication network were taken into account.

The relevance of the task in the real world explains why model-based diagnosis of DESs has received considerable attention since the 1990s via different approaches, adopting either artificial intelligence [33, 110, 131, 142] or automatic control techniques [24, 28, 39, 124, 136, 143, 144, 145, 147, 153, 167]. However, despite the efforts of the research community, model-based diagnosis of DESs is still a challenging task owing to its computational complexity.

In this book, we are specifically interested in artificial intelligence techniques for model-based diagnosis of DESs. In the following, after having defined, in quite general terms, the DES diagnosis problem (Sec. 1.4.1), we present the various aspects that characterize the task (Secs. 1.4.2-1.4.9), as they have to be taken into account every time the design of a new diagnosis engine is performed.

1.4.1 Problem

A DES diagnosis problem consists of a DES D and a finite observation O, the latter representing what has been observed while D was running during a time interval of interest.

1.4.1.1 System

An (untimed partially observable) DES D is a 4-tuple

$$D = (\Sigma, L, obs, \mathit{flt}), \tag{1.1}$$

where Σ is the finite set of events that can take place in the system, $L \subseteq \Sigma^*$ is the *behavior space*, a language that models all (and only) the possible sequences of events, or *trajectories*, that can take place in the system, and obs is a function that associates each trajectory u with $obs(u) \in \Sigma_o^*$, this being the projection of u on the subset $\Sigma_o \subseteq \Sigma$ of *observable events*, i.e., $obs(u)$ is a copy of u where all unobservable events have been removed. The set of *faulty events*, or *faults*, is denoted as Σ_f, where $\Sigma_f \subseteq \Sigma$. The function *flt* associates each trajectory with its projection on Σ_f.

The model L is assumed to be *complete*, i.e., it contains all possible sequences of events of the system, corresponding to normal and abnormal behaviors. No assumption is made about the number of occurrences of a fault in a trajectory (i.e., faults may be permanent, transient, or intermittent), or about the diagnosability of the system [144]. The language L can be represented by a finite automaton (FA), also called the *behavioral model*.

Example 1.1. Figure 1.1 shows an FA, this being a DES behavioral model, where the alphabet of faulty events is $\Sigma_f = \{a, b, c\}$. Each transition is marked with the relevant event: transitions associated with observable events are drawn as solid arrows, while those associated with unobservable events are represented as dashed arrows. A trajectory in L corresponds to a path in the FA starting from the *initial state*, which in this example is identified by 0.

For instance, $u = [o_2, o_1, b, u, a, o_1, u, a, o_1]$ is a trajectory, while the sequence of observable events relevant to it is $obs(u) = [o_2, o_1, o_1, o_1]$, and the corresponding sequence of faults is $\mathit{flt}(u) = [b, a, a]$.

1.4.1.2 Observation

While system D is evolving along a path of its behavioral model, and hence a trajectory u, the observable events in the totally temporally ordered sequence $obs(u)$ take place. Let us call the device/s in charge of gathering observation O the *observer*, where O is the second parameter of the diagnosis task. In the simplest case, the observer gathers exactly the sequence of observable events $obs(u)$. An observation O

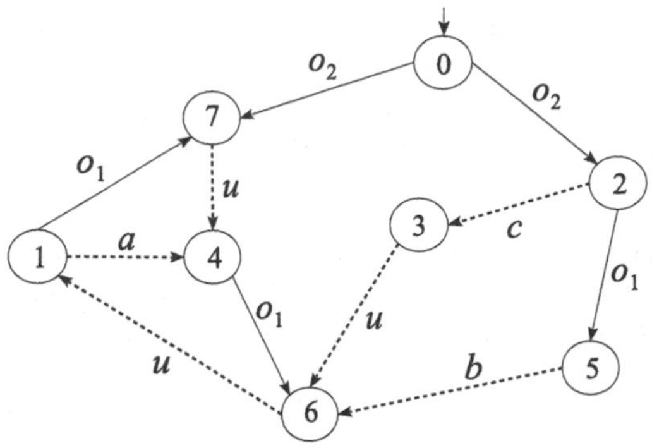

Fig. 1.1 DES model

that equals the (totally temporally ordered) sequence of observable events relevant to the trajectory actually followed by the system is a *certain observation.*

However, while a system is evolving along a path, some temporal constraints may not be perceived by the external observer, that is, what is observed can be a partially temporally ordered observation. Alternatively, or within the same observation, the observer may not perceive some event(s) exactly. Basically, there may coexist several causes for the observer to be in doubt about what it has observed.

An observation that cumulatively represents several sequences of observable events out of which the observer cannot single out the real sequence that occurred in the system is *uncertain*. The set of sequences represented by an observation O is called the *observation extension*, and denoted $\|O\|$. The extension of a certain observation is a singleton.

Example 1.2. Assume that the system D in Fig. 1.1 has followed the trajectory u described in Example 1.1. The (single) relevant certain observation is $O = [o_2, o_1, o_1, o_1]$.

If, instead, the observer has perceived all four observable events but does not know whether event o_2 took place in D before or after the first occurrence of event o_1, the relevant observation O' is uncertain: the observer is in doubt about whether the sequence of observable events that took place in D is $[o_2, o_1, o_1, o_1]$ or $[o_1, o_2, o_1, o_1]$.

If, instead, the observer has perceived all four observable events but does not know whether the first event was o_1 or o_2, the relevant observation O'' is still uncertain, as the observer is in doubt about whether the sequence of observable events that took place in D is $[o_2, o_1, o_1, o_1]$ or $[o_1, o_1, o_1, o_1]$.

So $\|O\| = \{[o_2, o_1, o_1, o_1]\}$, while $\|O'\| = \{[o_2, o_1, o_1, o_1], [o_1, o_2, o_1, o_1]\}$ and $\|O''\| = \{[o_2, o_1, o_1, o_1], [o_1, o_1, o_1, o_1]\}$.

1.4.1.3 Solution

The solution of a diagnosis problem (D,O) is called a *diagnosis*. It consists of a (possibly empty) set of diagnosis *candidates*, where each candidate is an explanation of the given observation.

Definition 1.1. Let (D,O) be a diagnosis problem of a DES $D = (\Sigma, L, obs, flt)$. At a low abstraction level, a *candidate* is a trajectory $u \in L$ such that $obs(u) \in \|O\|$. A *diagnosis* is the set Δ_{id} of all the candidates, that is,

$$\Delta_{\text{id}} = \{ u \in L \mid obs(u) \in \|O\| \} . \tag{1.2}$$

Example 1.3. Let us consider the diagnosis problem (D,O) where the model of the DES D is shown in Fig. 1.1 and the (certain) observation is $O = [o_2, o_1, o_1, o_1]$, that is, the sequence of observable events relevant to trajectory u in the two previous examples. Notice that we do not know which trajectory D has followed; we just know observation O. The solution of this problem is

$$\begin{aligned}\Delta_{\text{id}} = \{ & [o_2, o_1, b, u, a, o_1, u, a, o_1], \\ & [o_2, o_1, b, u, a, o_1, u, a, o_1, u], \\ & [o_2, o_1, b, u, a, o_1, u, a, o_1, u, a], \\ & [o_2, u, o_1, u, o_1, u, o_1], \\ & [o_2, u, o_1, u, o_1, u, o_1, u], \\ & [o_2, u, o_1, u, o_1, u, o_1, u, a], \\ & [o_2, c, u, u, o_1, u, o_1, u, o_1], \\ & [o_2, c, u, u, o_1, u, o_1, u, o_1, u], \\ & [o_2, c, u, u, a, o_1, u, a, o_1, u, a, o_1], \\ & [o_2, c, u, u, a, o_1, u, a, o_1, u, a, o_1, u], \\ & [o_2, c, u, u, a, o_1, u, a, o_1, u, a, o_1, u, a], \ldots \},\end{aligned}$$

where the dots represent (for shortness) many other trajectories.

Each and every trajectory in Δ_{id} represents a distinct evolution of the DES that can explain what has been observed. This means that, given an observation, we do not know which trajectory is the actual evolution the system has followed; however, we know that that evolution is contained in the computed diagnosis.

Different approaches to model-based diagnosis of DESs can adopt different notions of explanation of an observation as provided by a candidate. Definition 1.1 adopts a broad meaning; however, other approaches can be aimed at finding only trajectories ending with an observable event. The rationale behind this constraint is an interest in what has occurred in the DES as far as the latest observed event, not in what may have (silently) occurred after it. Let us now define the set of trajectories ending with an observable event and the refined notion of a candidate (and hence of a diagnosis) relevant to it.

Definition 1.2. Let (D, O) be a DES diagnosis problem. A *refined candidate* is a candidate (that is, an element of Δ_{id}) ending with an observable event, while a *refined diagnosis* $\Delta_{\text{id}}^{+} \subseteq \Delta_{\text{id}}$ is the set of all refined candidates:

$$\Delta_{\text{id}}^{+} = \{ u \in \Delta_{\text{id}} \mid u = po, p \in \Sigma^{*} \wedge o \in \Sigma_{o} \} . \tag{1.3}$$

Example 1.4. The refined diagnosis relevant to the diagnosis problem defined in Example 1.3 is

$$\begin{aligned} \Delta_{\text{id}}^{+} = \{ & [o_2, o_1, b, u, a, o_1, u, a, o_1], \\ & [o_2, u, o_1, u, o_1, u, o_1], \\ & [o_2, c, u, u, o_1, u, o_1, u, o_1], \\ & [o_2, c, u, u, a, o_1, u, a, o_1, u, a, o_1], \ldots \}, \end{aligned}$$

where the dots represent all the refined candidates relevant to the missing trajectories in Δ_{id} (Example 1.3).

Notice that both Δ_{id} and Δ_{id}^{+} may be infinite sets if the behavioral model of the DES considered includes unobservable cycles. In fact, a trajectory that explains an observation can indifferently include zero, one, or several iterations of any unobservable cycle along the path. In order to make Δ_{id} and Δ_{id}^{+} finite, one can agree to collapse the representations of all the (nonnull) iterations of the same unobservable cycle into the representation of a single iteration.

1.4.2 Abstraction

As introduced above, a DES diagnosis problem consists of a system D and an observation O, and its solution is a set of candidates. A candidate has to explain the observation, where the notion of explanation depends on the abstraction level considered. A detailed explanation is a trajectory, hence, at a low abstraction level, a candidate is a trajectory u such that $obs(u) \in \|O\|$. However, usually, when solving a diagnosis problem, we are not interested in all the events in a trajectory but only in the faulty events. Hence, candidates, refined or not, and consequently diagnosis, refined or not, can be provided at different abstraction levels.

At a higher abstraction level than that of trajectories, a candidate is just the sequence of faults relevant to a trajectory in Δ_{id} (and a refined candidate is the sequence of faults relevant to a trajectory in Δ_{id}^{+}).

At a higher abstraction level than that of sequences of faults, a candidate is the multiset of faults relevant to a trajectory in Δ_{id} (and a refined candidate is the multiset of faults relevant to a trajectory in Δ_{id}^{+}).

At a still higher abstraction level, a candidate is the set of faults relevant to a trajectory in Δ_{id} (and a refined candidate is the set of faults relevant to a trajectory in Δ_{id}^{+}).

Still more, at the highest abstraction level, a candidate, refined or not, is just a tag, either *normal* or *abnormal*, corresponding to a trajectory in Δ_{id} for a candidate and to a trajectory in Δ^+_{id} for a refined candidate. The tag is *abnormal* if it corresponds to a trajectory that includes some faulty events; it is *normal* otherwise. Note that, at this abstraction level, we are interested just in fault detection, not in fault isolation or identification.

Example 1.5. With reference to Example 1.4, the refined diagnosis at the abstraction level where a candidate is a sequence of faults is

$$\Delta^+_{\text{seq}} = \{\, [b,a,a], [], [c], [c,a,a,a], \ldots] \,\} \,.$$

Likewise, at the abstraction level where multisets of faults are considered, the refined diagnosis is

$$\Delta^+_{\text{ms}} = \{\, \langle 2,1,0 \rangle, \langle 0,0,0 \rangle, \langle 0,0,1 \rangle, \langle 3,0,1 \rangle, \ldots \},$$

where each triple between angle brackets includes the counters of the occurrences of faults a, b, and c, respectively.

At the set-of-faults abstraction level, the refined diagnosis is

$$\Delta^+_{\text{set}} = \{\, \{a,b\}, \emptyset, \{c\}, \{a,c\}, \ldots \} \,.$$

Finally, at the fault detection abstraction level, the refined diagnosis is

$$\Delta^+_{\text{norm}} = \{\, \textit{normal}, \textit{abnormal} \,\};$$

that is, what has been observed is compatible both with the absence of faults and with the presence of some fault(s) in the DES.

1.4.3 Tasks

Depending on whether the observation is given together in one chunk or whether several observation chunks are provided progressively so as to follow the system while it is evolving, the task is called either *a posteriori* diagnosis or *reactive* diagnosis.

A posteriori diagnosis is a task performed offline, which is meant to find out what has happened to the system in the time interval during which the given observation was gathered. A posteriori diagnosis may be performed long after the observation has been recorded, typically when the system (unexpectedly or even catastrophically) comes to a halt, and one has to find out the reasons for such a halt based on some records.

By contrast, reactive diagnosis is a task performed online, while the system is running (possibly abnormally), typically when the system is being constantly monitored, and sensor values are sampled and sent periodically to a control unit. The output of reactive diagnosis is exploited in order to create a feedback, so as to affect the next evolution of the system, and/or to plan some maintenance/repair action.

One could argue that reactive diagnosis is actually a generalization of a posteriori diagnosis, since it can process incoming observation chunks indefinitely, while a posteriori diagnosis can process just one chunk. Although this is right from a theoretical point of view, that is, the definition of the first task is actually a generalization of the definition of the other, from an operational point of view the tasks are quite different. In fact, it is computationally impractical to implement reactive diagnosis as a task that starts from scratch every time, that is, a task that simply invokes a posteriori diagnosis every time a new observation chunk is received, giving it as input the sequence of all the observation chunks received so far. Rather, in order to update the diagnosis results based on the latest chunk of observation, reactive diagnosis exploits the results of previous computations which are relevant to previous observation chunks. This book deals both with a posteriori diagnosis and reactive diagnosis of DESs.

1.4.4 Initial State

In a diagnosis problem, the state of the DES at the beginning of the diagnosis process is given: such a state may be *certain*, that is, it is a single state, or *uncertain*, that is, it is a *belief state* including several states of the behavioral model of the system, among which we do not know which is the real system state. For instance, in a monitoring context, each belief state can be computed online by iteratively computing the set of states that can be reached from the current belief state through a sequence of transitions relevant to the latest observation chunk. However, the iterative process roughly described above is not the only way to perform reactive diagnosis. In fact, there are distinct approaches to model-based reactive diagnosis of DESs that are quite different from each other. Some of these approaches are briefly presented in Chapter 12.

1.4.5 Models

The aforementioned diagnosis tasks are model-based in nature, as they exploit a behavioral model of the system to be diagnosed. At an untimed abstraction level, the behavior of a DES is modeled as an FA [16] that reacts to input events by state transitions. Since model-based diagnosis of DESs is not usually meant just to find out *whether* something is wrong (*fault detection*) but also to find out *what* is wrong (*fault identification*), the model of the DES encompasses not only the normal behavior but also the abnormal one. This is why a DES behavioral model adopted for diagnosis purposes is said to be *complete*. The need for a complete model is a drawback for the diagnosis of DESs. Building a model representing the normal behavior is burdensome; however, if the system is an artifact, one can rely on design descriptions. Building a model representing also the abnormal behavior is really difficult, especially if one does not know exactly (all) the ways in which the system can become faulty and/or how it behaves when it does so. Hence, modeling is both the hard and the weak part of model-based diagnosis of DESs.

1.4.6 Faults

Describing the complete behavior of a DES demands a description of the *faults*, namely the root causes of malfunctions. Faults can be defined either at the event level or at the state level; both formalisms are equally expressive [67]. In this book, faults are modeled as specific events.

When an FA is used to describe the complete behavior of a DES, it is usually easier and more intuitive to represent it as a nondeterministic finite automaton (NFA). Basically, two alternative modeling rules can be adopted. According to one rule, the modeled state transitions are triggered only by input events coming from the outside world. This means that, given a DES state, the same input event that triggers one transition when the system is behaving normally may trigger another transition if the system is faulty. This indicates that a fault is a label (not an event) associated with a state transition, where such a transition can occur only if the system is affected by the fault itself.

According to the other rule, faults are represented as endogenous events, that is, a transition is triggered either by an exogenous event or by a faulty event. The FA may still be nondeterministic, since a faulty event may trigger several transitions exiting the same state.

In most approaches, faults are modeled as unobservable events. However, this is not strictly needed. In fact, in the formalization of the diagnosis problem provided in Sec. 1.4.1, no such restriction is required. An observable faulty transition does not necessarily imply that the fault, once it has occurred, can be unequivocally identified, as the observed event may be shared with other (normal and/or faulty) transitions.

Traditionally, faults represent abnormal system behaviors, that is, behaviors that the system may exhibit but are undesired; typically, when a fault event takes place in an artifact, that artifact is not behaving as expected according to its specification. However, model-based diagnosis lends itself to a broader notion of a fault: a fault is an event whose occurrence is to be tracked in the system, an event that we are specifically interested in.

1.4.7 Diagnosability

The complete behavioral model of a DES can be processed in order to analyze some important properties, particularly *diagnosability* [32, 122, 130, 144, 164, 165]. Basically, such a property holds if the system can be diagnosed, that is, if the diagnosis task, as applied to that system, can identify each fault within a finite number of observable events following its occurrence.

Example 1.6. The system D whose behavioral model is displayed in Fig. 1.1 is not diagnosable. Let us assume that fault b has occurred while the system was following the trajectory $v = [o_2, o_1, b, u, o_1]$. This trajectory has generated the sequence of observable events $obs(v) = [o_2, o_1, o_1]$. Let us assume that the observer has gathered the relevant certain observation $O = [o_2, o_1, o_1]$.

The refined solution of the diagnosis problem (D, O) at the set-of-faults abstraction level is $\Delta^+_{\text{set}} = \{\{b\}, \emptyset, \ldots\}$, where the (refined) candidate $\{b\}$ is relevant to trajectory v while the empty-set (refined) candidate is relevant to the trajectory $w = [o_2, u, o_1, u, o_1]$. Notice that both trajectories end in state 7; hence, so far, they are indistinguishable based on the observation, and from now on they continue to be indistinguishable since they can continue by following the same path(s).

The former candidate, although updated, will include fault b while the latter will not include it, since, once state 7 has been reached, no occurrence of fault b can follow. Therefore, this ambiguity (according to one candidate, fault b has occurred, while, according to another, it has not occurred) cannot be resolved as there is no finite sequence of observable events following the occurrence of fault b that allows one to identify the fault itself.

Diagnosability analysis is performed offline. When artifacts are considered, it should be carried out at design time, thus being a means to achieve design for diagnosability. The attention paid to diagnosability of DESs preceded that for diagnosis. Significantly, the seminal publication [144] on model-based diagnosis of DESs, presenting the *diagnoser approach*, was indeed focused on diagnosability.

Several subsequent research efforts have been made in this direction in the last twenty years, all of them relying on a completely certain observation. The final chapter of the book includes a presentation of the diagnoser approach, along with a survey of some well-known publications in the literature on DES diagnosability analysis.

1.4.8 Computational Issues

In the traditional approaches to model-based diagnosis of DESs, the diagnosis task is performed by reconstructing (either explicitly [91] or not [149]) all the trajectories in the complete behavioral model of the system that are consistent with the observation received so far.

Regardless of the implementation, every algorithm that can yield such results is an *exact* DES diagnosis algorithm. In other words, an exact algorithm performs a complete search of the behavioral space of the DES considered, and produces a sound and complete diagnosis output, that is, the set of *all* and *only* the candidates that can be drawn from the trajectories that are consistent with the observation.

However, exact algorithms are computationally expensive, and applying model-based diagnosis to DESs is computationally heavy, particularly when the DESs are large and distributed (see Sec. 1.4.9).

Intuitively, the complexity of the diagnosis of dynamic systems cannot be lower than the complexity of the diagnosis of static systems. The problem of diagnosis of static systems, as defined in [135] after [36], can be viewed as falling within the class of incompatibility abduction problems [22], which have been proven to be NP-hard in general, with some special subclasses of them being NP-complete. In particular, in [135] a decision problem consisting in determining whether an instance of a static diagnostic problem has a solution was proven to be NP-complete.

To support our intuition, model-based diagnosis of DESs, first characterized as a reachability analysis problem [144], has recently been reformulated as an exponential number of satisfiability (SAT) problems [52, 140] and as a family of planning problems [152]. According to [111], checking the existence of a candidate (meaning a trajectory) is NP-hard if the DES is distributed, while deciding whether a DES diagnosis problem is solvable is in PSPACE.

1.4.9 Distribution

In the example presented in the previous pages, the DES $D = (\Sigma, L, obs, flt)$ consists of just one component. However, in the general case, a DES is endowed with a structural model that includes several components.

Using the same formalization as in Sec. 1.4.1, a DES is *distributed* when it consists of several interacting *components*, each of which is a DES (Σ_i, L_i, obs_i, flt_i) itself, where $\Sigma_i \subseteq \Sigma$. If an event belongs to the set of events Σ_i of several components, then it occurs only when it can occur simultaneously in all the components that share it, that is, when it triggers a transition starting from all the current states of the FAs of all the components that share it. Therefore, the FA relevant to the whole system is the one (implicitly) resulting from the *parallel composition* (often called the *synchronous composition*) [25] of the FAs relevant to all the components, where such a synchronization is based on transitions relevant to shared events, which are in fact called *synchronous* transitions.

It is well known that in many, perhaps most, real applications, the generation of the FA representing the behavior of a whole DES by composing the automata relevant to its components is prohibitively expensive, even if performed offline (that is, before the diagnosis engine starts operating).

To circumvent this difficulty, since the end of the 1990s, what are now called the classical approaches [11, 12, 131] to model-based diagnosis of DESs have not needed the FA representing the global behavior of the whole system to be generated explicitly. Rather, they exploit the behaviors of the components of the DES in an observation-driven way. In fact, they generate (all or some) sequences of transitions that are both consistent with the (implicit) global behavior of the DES and in agreement with the observation (relevant to the evolution of the DES over time) that is given as input to the diagnosis task and that the diagnosis results have to explain. These sequences may include transitions that occur only if some faults are present, which is the information the diagnosis task is seeking.

However, the models may be very large, and model-based reasoning may yield huge intermediate data structures, thereby causing space allocation problems. The difficulty about intermediate structures can be partially dealt with by means of incremental techniques [53, 113], while that associated with models can be tackled by replacing some large models, such as those for buffered communication among components, with specific concise primitives.

This is the reason why in the *active system* approach [91] components interact "asynchronously," by exchanging (communication) events over finite buffers called *links*. Including links in the model of a DES is actually a matter of expressive power, not of computational power. In fact, a link, whatever its capacity and its policy (FIFO, LIFO, etc.), could be represented as a component itself, which shares state changes with both the components that feed the link and the components that extract events from the link. However, such a model would be exponential in the capacity of the link.

Active systems [11, 91, 93] are distributed DESs described by using a specific primitive for links. They are a subclass of the (synchronously interacting) DESs described above. In fact, while an asynchronous communication can be modeled (at least in theory) through the primitive for synchronous communication, the reverse is not true (unless zero-capacity links are introduced). The core of this book is devoted to model-based diagnosis of active systems.

1.5 Summary of Chapter 1

Rule-Based Diagnosis. A paradigm according to which a general diagnostic reasoner performs a diagnosis task by exploiting rules that represent empirical knowledge (or knowledge extracted from massive data sets) about associations between faults and symptoms within the system to be diagnosed.

Symptom/Malfunction. A discrepancy between the observed behavior (i.e., the actual values of the output variables) and the expected normal behavior of a system. If the system is an artifact, its normal behavior is described by its specifications.

Fault. A cause of a malfunction.

Model-Based Diagnosis. A paradigm according to which a general diagnostic reasoner performs a diagnosis task by exploiting models, typically structural and behavioral, of the system to be diagnosed.

Structural Model. A topological description of a system, in terms of components and interconnections between them.

Behavioral Model. A description that enables one to make predictions about how a system should work; it can also encompass faulty operating modes.

Candidate. A set of assumptions (about the behavior of a system) that explains an observation.

Diagnosis. A set of candidates that represents the solution of an instance of a diagnosis problem.

Predicted Behavior. The values of the output variables of a system computed by simulating its normal behavior.

A Posteriori Diagnosis. A diagnosis task when it is carried out offline, such as, for instance, when the system has come to a halt and the faults that have caused this halt have to be found based on an observation that was gathered when the system was still running.

Reactive Diagnosis. A diagnosis task when it is carried out online, that is, while the system is running, so that the diagnostic results can be exploited in order to make interventions in the system or to make decisions so as to prevent the system from coming to a halt.

Static System. A system described by means of time-independent mappings from (the values of) input variables to (the values of) output variables. The class of static systems is addressed by the classical theory of model-based diagnosis.

Dynamic System. A system whose output values depend not only on the current input values but also on the past values of the inputs.

Discrete-Event System (DES). A conceptual model of a dynamic system, which consists in a set of discrete states and a set of state transitions, where each transition is triggered by an asynchronous instantaneous event from a finite set of events.

Diagnosability. An intrinsic property exhibited by a discrete-event system when each fault can be isolated within a finite number of (observable) events following its occurrence.

Distributed Discrete-Event System. A discrete-event system consisting of several interacting components, each being a discrete-event system itself.

Active System. A distributed discrete-event system where events exchanged between components are temporarily stored in buffers (the communication is said to be asynchronous) and such buffers are described by ad hoc primitives.

Chapter 2
Finite Automata

Automata are a special class of abstract devices for computation, also called *machines*, which have been the subject of a vast amount of research for several decades [64].

In the 1930s, before the advent of computers, Alan Turing invented an abstract machine having the same computational power as current computers, at least from the point of view of what can be computed (irrespective of performance). The goal of Turing was to find the boundaries of the computational power of machines, what they can compute and what they cannot. His findings are applicable not only to his abstract machine but also to real modern computers.

In the 1940s and 1950s, many researchers studied several classes of simpler abstract machines, which are nowadays called *finite automata* or *finite state machines*. These automata, first proposed as a model for brain functions, have become extremely useful in a vast range of different fields, as a model to describe several classes of hardware and software, including:

1. Software for the design and validation of the behavior of digital circuits,
2. Lexical analyzers for compilers,
3. Software for scanning large bodies of text, in order to match words, phrases, or other patterns,
4. Software for checking systems with a finite number of distinct states, such as communication protocols or protocols for secure exchange of information.

Components or systems such as those listed above can be seen to be, at any time instant, in one of a finite set of *states*, with each state being a possible configuration of the system. As such, a state represents a possible evolution (history) of the system. Since the number of states is finite, a finite automaton cannot remember all possible evolutions. Hence, the system is designed to remember only relevant evolutions.

The advantage in having a finite set of states is that the system can be implemented using a fixed set of resources, for example, by wiring the automaton as a digital circuit or, more simply, as a software program that make decisions based on a limited set of data.

G. Lamperti et al., *Introduction to Diagnosis of Active Systems*,
https://doi.org/10.1007/978-3-319-92733-6_2

A finite automaton is characterized by a set of states and a transition function that makes the system evolve, by moving the system from one state to another based on external input.

Different classes of finite automata exist. In this chapter we consider the class of *recognizers*, which generate either *yes* or *no*, depending on whether or not they have recognized the input. Each state of such an automaton is said to be either *final* or *nonfinal*. When all input has been processed, if the current state is a final state, the input is recognized; otherwise, it is rejected. The automaton is associated with a *regular language*, which consists of the strings recognized by the automaton.

An essential distinction relevant to finite automata is concerned with the type of transition function, which can be either of the following:

1. *Deterministic*: the automaton can be in just one state at a time,
2. *Nondeterministic*: the automaton can be in several states at the same time.

Although nondeterminism does not augment the expressive power of finite automata, it may be useful for both formal and practical reasons.

Before formally defining the notion of a finite automaton, we introduce regular languages and regular expressions, which are strictly related to finite automata.

2.1 Languages

Regular languages are a special class of languages. To define a language, we need to introduce the notions of an alphabet and a string.

Definition 2.1 (Alphabet). An *alphabet* is a finite set of symbols.

Example 2.1 (Alphabet). The traditional binary alphabet is $\{0,1\}$. The set $\{a,b,c,d\}$ is an alphabet composed of four characters. Typical alphabets for computers are the *American Standard Code for Information Interchange* (ASCII) and the more extended *Unicode*.

Definition 2.2 (String). A *string* on an alphabet Σ is a finite (possibly empty) sequence of characters on Σ.

Example 2.2 (String). Let $\Sigma = \{a,b,c,d\}$. The strings on Σ include a, $aacbb$, and ε, where the last symbol denotes the *null string* (formally, the empty sequence of symbols on any alphabet).

Strings can be concatenated to generate new strings. If x and y are two strings on a given alphabet, xy is the concatenation of x and y. As such, the null string ε is the identity element of concatenation, as $x\varepsilon = \varepsilon x = x$. If we think of concatenation as a product, we can define the exponentiation operator on strings as follows:

$$x^n = \begin{cases} \varepsilon & \text{if } n = 0, \\ x^{n-1}x & \text{if } n > 0\,. \end{cases} \tag{2.1}$$

Definition 2.3 (Language). A *language* on an alphabet Σ is a (possibly infinite) set of strings on Σ.

Example 2.3 (Language). Formally, both $\emptyset$ (the empty set) and $\{\varepsilon\}$ (the singleton of the null string) are languages. Also, the (infinite) set of syntactically correct programs written in C++ and the (finite) set of bytes (strings of eight bits) are languages.

Languages can be combined by means of operators to generate new languages. In the following, we assume L and M to be languages. Since languages are sets, the usual set-theoretic union can be applied:

$$L \cup M = \{x \mid x \in L \vee x \in M\}\,. \tag{2.2}$$

Like strings, languages can also be concatenated:

$$LM = \{xy \mid x \in L \wedge y \in M\}\,. \tag{2.3}$$

The exponentiation of a language is defined as follows:

$$L^n = \begin{cases} \{\varepsilon\} & \text{if } n = 0, \\ L^{n-1}L & \text{if } n > 0\,. \end{cases} \tag{2.4}$$

Some final useful operations are the *Kleene closure* of a language,

$$L^* = \bigcup_{i=0}^{\infty} L^i, \tag{2.5}$$

and the *positive closure* of a language,

$$L^+ = \bigcup_{i=1}^{\infty} L^i\,. \tag{2.6}$$

Example 2.4 (Language Operations). Let $L = \{A, B, \ldots, Z, a, b, \ldots, z\}$ be the language of letters, and $M = \{0, 1, \ldots, 9\}$ the language of digits.[1] $L \cup M$ is the language whose strings are either letters or digits. $L\,M$ is the language of strings composed of a letter followed by a digit. L^3 is the language of strings of three letters. L^* is the (infinite) language of strings of letters, including ε. $L\,(L \cup M)$ is the language of alphanumeric strings starting with a letter (typically, identifiers in programming languages). M^+ is the (infinite) language of unsigned integer constants.

2.2 Regular Expressions

So far, we have considered languages in the general case. Now, we focus on the special class of regular languages, which are intimately related to finite automata.

[1] As such, in both L and M there is an isomorphism between the alphabet and the language.

However, regular languages can be defined irrespective of finite automata, based on the formalism of *regular expressions*. Intuitively, a regular expression is similar to an arithmetic expression involving integer numbers: instead of numbers, a regular expression involves symbols of the alphabet (plus ε). An arithmetic expression can be represented by a tree, where leaves are numbers and nodes are operators. The same applies to a regular expression: leaves are alphabet symbols (plus ε) and nodes are specific string operators. Each node of the tree of an arithmetic expression is associated with the integer value of the arithmetic subexpression rooted in that node. Likewise, each node of the tree of a regular expression is associated with the language of the regular subexpression rooted in that node. In other words, an arithmetic expression is evaluated as a number (the value of the expression), while a regular expression is evaluated as a set of strings, namely a language.

Definition 2.4 (Regular Expression). Let Σ be an alphabet. A *regular expression* on Σ is inductively defined by the following rules:

1. ε is a regular expression denoting the language $\{\varepsilon\}$.
2. If $a \in \Sigma$, then a is a regular expression denoting the language $\{a\}$.
3. If x and y are regular expressions denoting languages $L(x)$ and $L(y)$, then:
   a. (x) is a regular expression denoting $L(x)$ (use of parentheses, as usual);
   b. $(x) \mid (y)$ is a regular expression denoting the union $L(x) \cup L(y)$;
   c. $(x)\,(y)$ is a regular expression denoting the concatenation $L(x)\,L(y)$;
   d. $(x)^*$ is a regular expression denoting the Kleene closure $(L(x))^*$;
   e. $(x)^+$ is a regular expression denoting the positive closure $(L(x))^+$.

The language denoted by a regular expression is a *regular language*.

Example 2.5 (Regular Expression). Let $\Sigma = \{a, b, c\}$. The expression $(a|c)^*\,b\,(a|c)^*$ denotes the language of strings including exactly one b, while $(a|c)^*\,(b|\varepsilon)\,(a|c)^*$ denotes the language of strings including at most one b.

Regular expressions can only define regular languages; in other words, the expressive power of regular expressions is limited. There exist even simple languages which are not regular and thus cannot be specified by a regular expression.

Example 2.6 (Nonregular Language). Let $\Sigma = \{a, b\}$ and $L = \{\, a^n\, b\, a^n \,\}$, the language of balanced strings, namely *aba*, *aabaa*, *aaabaaa*, and so on. Even if L is very simple, it is not a regular language and, as such, there does not exist a regular expression which defines it. One may argue that the expression $a^+\, b\, a^+$ denotes a language that includes such balanced strings. The point is that, even if this language is complete, it is not sound, as it contains several (actually infinite) other strings which are not balanced, for instance *abaaa*.

2.3 Deterministic Finite Automata

Intuitively, a deterministic finite automaton (DFA) is a finite automaton with a deterministic transition function. As such, for each input, the corresponding state of the DFA is unique.

Definition 2.5 (DFA). A DFA is a 5-tuple $(\Sigma, D, \tau, d_0, F)$, where Σ is a finite set of symbols, called the alphabet, D is a finite set of states, $\tau : D \times \Sigma \mapsto D$ is the transition function mapping state–symbol pairs into states, $d_0 \in D$ is the initial state, and $F \subseteq D$ is the set of final states.

The language of a DFA is the (possibly infinite) set of strings recognized by that DFA. The way the DFA recognizes the input strings, with each string being a sequence of symbols in the alphabet Σ, can be described as follows.

Let $[a_1, a_2, \ldots, a_n]$ be a sequence of symbols (a string) in Σ. Starting from the initial state d_0, according to the transition function τ, $d_1 = \tau(d_0, a_1)$ will be the state reached based on input a_1. Proceeding in a similar manner for the different symbols, we will have $d_i = \tau(d_{i-1}, a_i)$, thereby generating the sequence of states $[d_1, d_2, \ldots, d_n]$: if $d_n \in F$ then the input sequence $[a_1, a_2, \ldots, a_n]$ will be recognized, otherwise it will be rejected.

To make the notation more concise, we can define the notion of an extended transition function $\hat{\tau}$, which computes the state reached from a given state based on an input string, namely $\hat{\tau} : D \times \Sigma^* \mapsto D$ (with Σ^* being the domain of strings on $\Sigma \cup \{\varepsilon\}$), defined as follows:

$$\hat{\tau}(d,x) = \begin{cases} d & \text{if } x = \varepsilon, \\ \tau(\hat{\tau}(d,y),a) & \text{if } x = ya\,. \end{cases} \tag{2.7}$$

Thus, a string x belongs to the language of the DFA if and only if $\hat{\tau}(d_0, x) \in F$. The language L of a DFA $\mathcal{D} = (\Sigma, D, \tau, d_0, F)$ can therefore be defined as follows:

$$L(\mathcal{D}) = \{x \mid x \in \Sigma^*, \hat{\tau}(d_0, x) \in F\}\,. \tag{2.8}$$

Theory shows that the language of any DFA is regular, that is, for each DFA there exists a regular expression denoting the language of the DFA, and that for each regular expression there exists a DFA recognizing the same regular language. Thus, DFAs and regular expressions are different but equivalent formalisms for defining the same class of languages; in other words, they share the same expressive power.

Example 2.7 (DFA). Let $\Sigma = \{a, b\}$ be an alphabet, and $a^*\, b\, (a|b)^*$ a regular expression on Σ. The DFA recognizing the language denoted by the regular expression is defined on Σ as follows. The set of states is $D = \{d_0, d_1, d_2\}$, where d_0 is the initial state. The set of final states is $F = \{d_2\}$. The transition function includes the following transitions: $\tau(d_0, a) = d_1$, $\tau(d_0, b) = d_2$, $\tau(d_1, a) = d_1$, $\tau(d_1, b) = d_2$, $\tau(d_2, a) = d_2$, and $\tau(d_2, b) = d_2$.

2.3.1 *Transition Diagrams*

In Example 2.7 we gave an instance of a DFA based on its formal definition. However, a DFA can be conveniently represented by a transition diagram, which is a graph where:

1. Each state d is represented by a node (single circle) marked by d,
2. Each transition $d' = \tau(d,a)$ is represented by an arc from the node marked by d to the node marked by d', with the arc being marked by the symbol a,
3. The initial state is entered by an unmarked arc (without any origin),
4. Each node denoting a final state is represented by a double circle (instead of a single circle).

Note 2.1. Referring to finite automata as transition diagrams, rather than based on their mathematical definition, may be more intuitive. In a transition diagram, τ is represented by a set of transitions, with each transition being an arc marked by a symbol of the alphabet, connecting two states. For instance, a transition t from state d to state d' and marked by a symbol ℓ is written $d \xrightarrow{\ell} d'$. We say that t exits (leaves) d and enters (reaches) d'.

Example 2.8 (DFA Diagram). The diagrammatic representation corresponding to the DFA defined in Example 2.7 is shown in Fig. 2.1. Notice how the two transitions exiting node d_2 have been represented by one arc marked by two relevant symbols, namely a and b (separated by a comma).

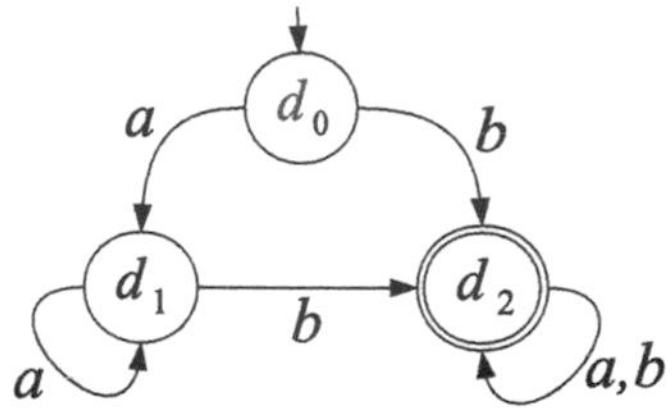

Fig. 2.1 Diagrammatic representation of the DFA defined in Example 2.7

2.4 Nondeterministic Finite Automata

Intuitively, a nondeterministic finite automaton is a finite automaton where it is possible to be in more than one state at the same time. Based on the diagrammatic representation introduced in Sec. 2.3.1, nondeterminism may arise for two reasons:

1. An arc from state n to state n' is marked by ε (namely, an ε-transition),
2. The same state n is exited by two arcs marked by the same symbol a and entering two distinct states n' and n''.

In the first case, when the NFA is in state n, it is also in state n' (as ε is immaterial). In the second case, when the NFA is in state n and the input symbol is a, the NFA is in n' and n'' at the same time.

Definition 2.6 (NFA). An NFA is a 5-tuple $(\Sigma, N, \tau, n_0, F)$, where Σ is a finite set of symbols (the alphabet), N is a finite set of states, $\tau : N \times (\Sigma \cup \{\varepsilon\}) \mapsto 2^N$ is the transition function mapping state–symbol pairs (possibly ε) into a set of states, $n_0 \in N$ is the initial state, and $F \subseteq N$ is the set of final states.

Example 2.9 (NFA). An NFA on the alphabet $\Sigma = \{a, b\}$ is defined as follows. The set of states is $N = \{n_0, n_1, n_2, n_3, n_4\}$, where n_0 is the initial state. The set of final states is $F = \{n_3, n_4\}$. The transition function includes the following transitions: $\tau(n_0, a) = \{n_1\}$, $\tau(n_0, \varepsilon) = \{n_2\}$, $\tau(n_1, a) = \{n_2\}$, $\tau(n_2, a) = \{n_2\}$, $\tau(n_2, b) = \{n_3\}$, $\tau(n_3, \varepsilon) = \{n_4\}$, $\tau(n_4, a) = \{n_3\}$, and $\tau(n_4, b) = \{n_3\}$.

An NFA can also be defined diagrammatically, as illustrated in Sec. 2.3.1. Specifically, a transition function computing several states, namely $\tau(n, a) = \{n_1, \ldots, n_k\}$, is represented by k different arcs, from state n to states $n_1, \ldots, n_k$, respectively.

Example 2.10 (NFA Diagram). The transition diagram corresponding to the NFA defined in Example 2.9 is shown in Fig. 2.2. Notice how the transition function is not total, as there are some pairs for which the mapping is not defined, for instance, (n_1, b). This may also happen for DFAs. In this book, we assume that, generally speaking, transition functions of finite automata are partial.

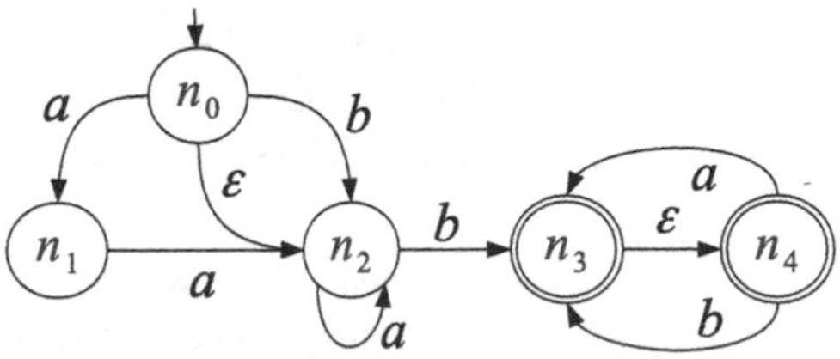

Fig. 2.2 Diagrammatic representation of the NFA defined in Example 2.9

2.4.1 ε-Closure

In order to define the extended transition function for NFAs, we need to introduce the notion of *ε-closure* of an NFA state. Intuitively, the ε-closure of a state n is the set of states reachable from n by means of ε-transitions only, including n.

Definition 2.7 (ε-Closure). Let $\mathcal{N} = (\Sigma, N, \tau, n_0, F)$ be an NFA. The ε-closure of a state $n \in N$ is a function $N \mapsto 2^N$, such that:

$$\varepsilon\text{-closure}(n) = \begin{cases} \{n\} & \text{if } \tau(n,\varepsilon) = \emptyset, \\ \{n\} \cup \bigcup_{n' \in \tau(n,\varepsilon)} \varepsilon\text{-closure}(n') & \text{if } \tau(n,\varepsilon) \neq \emptyset\,. \end{cases} \tag{2.9}$$

Example 2.11 (ε-Closure). With reference to the NFA displayed in Fig. 2.2, we have ε-closure$(n_1) = \{n_1\}$ and ε-closure$(n_3) = \{n_3, n_4\}$.

The ε-closure can also be defined for a set of states, namely ε-closure*.

Definition 2.8 (ε-Closure*). Let $\mathcal{N} = (\Sigma, N, \tau, n_0, F)$ be an NFA. The ε-closure* of a set of states $\mathbb{N} \subseteq N$ is a function $2^N \mapsto 2^N$, such that

$$\varepsilon\text{-closure}^*(\mathbb{N}) = \bigcup_{n \in \mathbb{N}} \varepsilon\text{-closure}(n) \,. \tag{2.10}$$

Example 2.12 (ε-Closure*). With reference to the NFA displayed in Fig. 2.2, we have ε-closure$^*(\{n_1, n_2\}) = \{n_1, n_2\}$ and ε-closure$^*(\{n_0, n_3\}) = \{n_0, n_2, n_3, n_4\}$.

As in the case of DFAs, we can extend the transition function to strings, $\hat{\tau} : N \times \Sigma^* \mapsto 2^N$:

$$\hat{\tau}(n, x) = \begin{cases} \varepsilon\text{-closure}(n) & \text{if } x = \varepsilon, \\ \varepsilon\text{-closure}^*\left(\displaystyle\bigcup_{n' \in \hat{\tau}(n,y)} \tau(n', a) \right) & \text{if } x = ya \,. \end{cases} \tag{2.11}$$

Thus, a string x belongs to the language of the NFA if and only if $\hat{\tau}(n_0, x) \in F$. The language L of an NFA $\mathcal{N} = (\Sigma, N, \tau, n_0, F)$ can therefore be defined as follows:

$$L(\mathcal{N}) = \{ x \mid x \in \Sigma^*, \hat{\tau}(n_0, x) \cap F \neq \emptyset \} \,. \tag{2.12}$$

Theory shows that NFAs and DFAs share the same expressive power: for each NFA, there exists a DFA recognizing the same language. Therefore, from the point of view of expressiveness, DFAs, NFAs, and regular expressions are equivalent notations.

Example 2.13 (NFA Language). Shown in Fig. 2.1 is the diagram of the DFA defined in Example 2.7. The language recognized by the DFA is defined by the regular expression $a^* \, b \, (a|b)^*$. Notice how the language recognized by the NFA displayed in Fig. 2.2 is the same. Intuitively, each string of the language recognized by an NFA can be determined by a path rooted in the initial state and ending at a final state, by generating the sequence of symbols marking the arcs in the path and by removing the ε symbols. For instance, in the NFA displayed in Fig. 2.2, the string *bab* can be generated by the path $n_0 \xrightarrow{b} n_2 \xrightarrow{a} n_2 \xrightarrow{b} n_3$. However, because of nondeterminism, the same string can also be generated by a different path, specifically $n_0 \xrightarrow{\varepsilon} n_2 \xrightarrow{b} n_3 \xrightarrow{\varepsilon} n_4 \xrightarrow{a} n_3 \xrightarrow{\varepsilon} n_4 \xrightarrow{b} n_3$.

2.5 Determinization of Finite Automata

In Sec. 2.4 we pointed out that, in terms of expressive power, NFAs and DFAs are equivalent. However, tasks associated with finite automata, such as string recognition, are generally more efficient when performed on DFAs, because no backtracking is required. Therefore, if we have an NFA $\mathcal{N}$, it may be essential to generate a DFA $\mathcal{D}$ equivalent to (sharing the language of) $\mathcal{N}$. In other words, it may be essential to *determinize* $\mathcal{N}$ into $\mathcal{D}$ [64].

Finite-automata determinization is of pivotal importance in a wide range of applications, from pattern matching based on regular expressions [48] to analysis of protein sequences [5]. The determinization of an NFA is commonly performed by the *Subset Construction* algorithm, which we call *SC*, introduced in the literature several decades ago [137].

In this section we specify the pseudocode of *SC*. Although this specification differs in terminology from that provided in the literature, it has the advantage of being congruent with the terminology adopted in this chapter, in particular, the incremental determinization technique presented in Sec. 2.7.

Definition 2.9 (ℓ-Closure). Let $\mathbb{N}$ be a subset of the states of an NFA $(\Sigma, N, \tau, n_0, F)$, and ℓ a symbol in Σ. The ℓ-closure of $\mathbb{N}$ is

$$\ell\text{-closure}(\mathbb{N}) = \varepsilon\text{-closure}^* \left(\bigcup_{n \in \mathbb{N}} \tau(n, \ell) \right). \tag{2.13}$$

Note 2.2. In the sequel, when $\ell \in \Sigma \cup \{\varepsilon\}$, the actual value of ℓ-closure depends on ℓ: if $\ell = \varepsilon$ then it corresponds to ε-closure*, otherwise it corresponds to Def. 2.9.

Listed below is a specification of *SC*, where $\mathcal{N}$ is the input NFA and $\mathcal{D}$ is the output DFA. By construction, each state in $\mathcal{D}$ is identified by a subset of the states of $\mathcal{N}$ (hence the name of the algorithm). The subset of states of $\mathcal{N}$ identifying a state d of $\mathcal{D}$ is denoted by $\|d\|$. The initial state d_0 of $\mathcal{D}$ is the ε-closure of the initial state n_0 of $\mathcal{N}$.

$\mathcal{D}$ is generated with the support of a *bud-stack* $\mathcal{B}$, where each *bud* is a state of $\mathcal{D}$ to be processed. At the beginning, $\mathcal{B}$ contains the initial state d_0 only. Then, *SC* pops and processes one bud d from $\mathcal{B}$ at a time, by generating the transitions exiting d, until $\mathcal{B}$ becomes empty.

Each transition is generated by considering each symbol $\ell \in \Sigma$ marking a transition exiting a state $n \in \|d\|$. For each symbol ℓ, a transition $d \xrightarrow{\ell} d'$ is created, where $\|d'\| = \ell$-closure$(\|d\|)$. Furthermore, if d' does not exist, then it is created (and possibly qualified as final, if it contains at least one state that is final in the NFA), and pushed into the bud-stack $\mathcal{B}$.

The pseudocode for *SC* is as follows:

algorithm *SC* (**in** $\mathcal{N}$, **out** $\mathcal{D}$)
    $\mathcal{N} = (\Sigma, N, \tau_n, n_0, F_n)$: an NFA,
    $\mathcal{D} = (\Sigma, D, \tau_d, d_0, F_d)$: a DFA equivalent to $\mathcal{N}$;

    auxiliary function *New*($\mathbb{N}$): a new state in $\mathcal{D}$
        $\mathbb{N}$: a subset of states N;
    begin $\langle New \rangle$

7. Create a new state d' with $\|d'\| = \mathbb{N}$ and insert it into D;
8. **if** $\|d'\|$ includes a state in F_n **then** insert d' into F_d **endif**;
9. **return** d'
10. **end** $\langle \mathit{New} \rangle$;

11. **begin** $\langle \mathit{SC} \rangle$
12. $d_0 := \mathit{New}(\varepsilon\text{-closure}(n_0))$;
13. $\mathcal{B} := [d_0]$;
14. **repeat**
15. Pop a bud d from $\mathcal{B}$;
16. **foreach** $\ell \in \Sigma$ such that $n \xrightarrow{\ell} n' \in \tau_\mathrm{n}, n \in \|d\|$ **do**
17. $\mathbb{N} := \ell\text{-closure}(\|d\|)$;
18. **if** D does not include a state d' such that $\|d'\| = \mathbb{N}$ **then**
19. $d' := \mathit{New}(\mathbb{N})$;
20. Push d' into $\mathcal{B}$
21. **else**
22. Let d' be the state in $\mathcal{D}$ such that $\|d'\| = \mathbb{N}$
23. **endif**;
24. Insert transition $d \xrightarrow{\ell} d'$ into τ_d
25. **endfor**
26. **until** $\mathcal{B}$ becomes empty
27. **end** $\langle \mathit{SC} \rangle$.

Example 2.14 (Subset Construction). Consider the NFA displayed in Fig. 2.2. The determinization of such an NFA performed by *SC* is traced in Fig. 2.3 (from left to right). In each step, the configuration of the DFA is displayed, where buds are represented by dashed nodes, with the gray node being the bud on top of the stack (to be processed in the next step). The determinization process involves six steps:

1. The initial state of the DFA is generated, namely $\{n_0, n_2\}$, and inserted into the bud-stack (lines 12 and 13),
2. Bud $\{n_0, n_2\}$ is popped (line 15) and two transitions are generated (lines 16–25), $\{n_0, n_2\} \xrightarrow{a} \{n_1, n_2\}$ and $\{n_0, n_2\} \xrightarrow{b} \{n_2, n_3, n_4\}$, where state $\{n_2, n_3, n_4\}$ is final,
3. Bud $\{n_2, n_3, n_4\}$ is popped and two transitions are generated, marked by a and b, from $\{n_2, n_3, n_4\}$ to the newly generated (final) state $\{n_3, n_4\}$,
4. Bud $\{n_3, n_4\}$ is popped and two auto-transitions[2] are generated, marked by a and b, (no new state is generated),
5. Bud $\{n_1, n_2\}$ is popped and two transitions are generated, one directed to the new state $\{n_2\}$ and marked by a, and the other directed to state $\{n_3, n_4\}$ and marked by b,
6. Bud $\{n_2\}$ is popped and two transitions are generated, one auto-transition marked by a, and the other, marked by b, directed to $\{n_3, n_4\}$.

[2] An auto-transition is a transition where the state reached equals the state from which it comes.

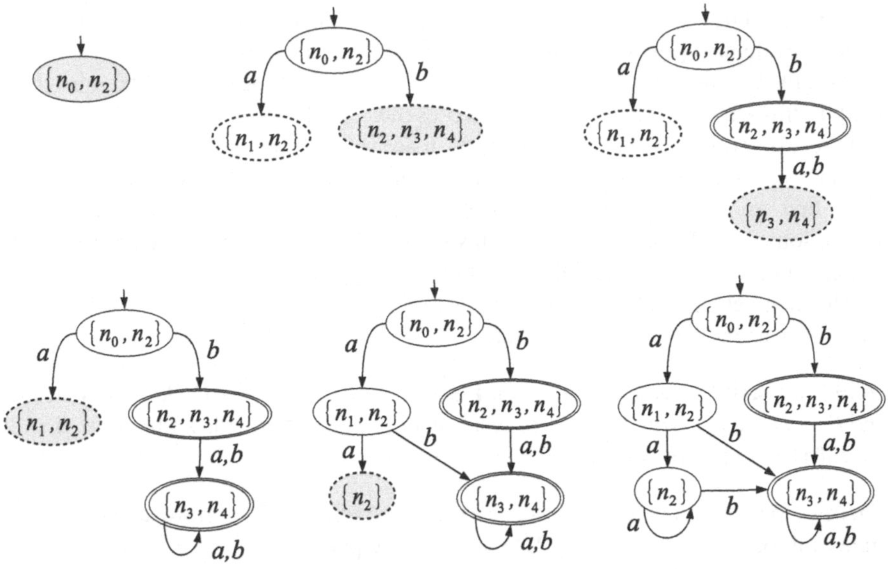

Fig. 2.3 Determinization by *SC* of the NFA displayed in Fig. 2.2: buds are dashed, with the gray node being the bud on top of the stack (to be processed first)

At this point, since the bud-stack is empty (line 26), *SC* terminates. The graph displayed on the lower right of Fig. 2.3 is a DFA equivalent to the NFA displayed in Fig. 2.2. In fact, it recognizes the same language, namely $a^* \, b \, (a|b)^*$.

One may ask why the DFA generated in Example 2.14 (Fig. 2.3) differs from the DFA displayed in Fig. 2.1. After all, in Example 2.9 we showed that the NFA displayed in Fig. 2.2 (which is the input NFA in Example 2.14) is equivalent to the DFA in Fig. 2.1. The point is, generally speaking, there exist several DFAs equivalent to an NFA, and *SC* generates just one of them. In our example, the DFA generated by *SC* is larger than the DFA in Fig. 2.1, as it includes five states (rather than three).

For practical reasons, we might be interested in finding a way to generate a *minimal* equivalent DFA, that is, a DFA with the minimum number of states. Interestingly, theory shows that, among all equivalent DFAs, there always exists one and only one minimal DFA. Furthermore, it provides an algorithm to generate such a minimal DFA.

2.6 Minimization of Finite Automata

In Sec. 2.5 we coped with the problem of determinization of finite automata, which can be performed by the *SC* algorithm. Based on Example 2.14, we also noted that the DFA generated by *SC* may be larger than necessary. In this section we are concerned with DFA minimization, that is, the problem of determining the smallest DFA equivalent to (that is, sharing the same language as) a given DFA.

As with *SC*, algorithms for DFA minimization are provided in the literature [63, 125]. In this chapter, we refer to the one presented in [125], which we call *MIN*.

Often, determinization and minimization are tightly coupled. For instance, lexical analysis based on regular expressions is performed by a DFA generated in three steps: first, an NFA is produced starting from a regular expression by means of the Thompson method [1], then a DFA equivalent to the NFA is generated by means of *SC*, and, finally, the minimal DFA is determined by *MIN*.

The idea behind *MIN* is to make a partition of the set of states of the DFA such that each part includes equivalent states.

Definition 2.10 (State Equivalence). Let $(\Sigma, D, \tau_{\mathrm{d}}, d_0, F)$ be a DFA. Two states d and d' in D are *equivalent* when, for each string $x \in \Sigma^*$, $\hat{\tau}(d,x)$ is a final state if and only if $\hat{\tau}(d',x)$ is a final state.

Theory shows that state equivalence is a transitive relation. Therefore, all equivalent states will be accommodated by *MIN* in the same part.

Definition 2.11 (Transition Signature). Let $(\Sigma, D, \tau, d_0, F)$ be a DFA, and Π a partition of the set of states D. The *transition signature* relevant to Π of a state $d \in D$ is

$$\sigma_\Pi(d) = \{ (\ell, \pi) \mid d' = \tau(d, \ell), d' \in \pi, \pi \in \Pi \}\,. \tag{2.14}$$

Listed below is the pseudocode of *MIN*, which takes as input a DFA $\mathcal{D}$ and generates as output the minimal DFA (MFA), namely $\mathcal{M}$, equivalent to $\mathcal{D}$.

The body of *MIN* starts with the initialization of the partition Π (line 16), which is composed of two parts: the set of final states (F) and the set of nonfinal states ($D - F$). Then, a loop is iterated, where each iteration generates a new partition based on the current partition, by means of the *NewPartition* auxiliary function.

The new partition is either the same partition (in which case the loop of *MIN* terminates) or a refinement of the old partition. If it is a refinement, at least one part of Π is in turn partitioned into several groups, where the states of each group share the same transition signature relevant to the current partition Π.

Once the refinement of Π is completed, the new partition Π_{new} is returned. However, because of the splitting of parts of the old partition, there is no guarantee that states within the same part will still share the same transition signature. This is why the loop continues until no further refinement is possible (line 20).

In the end, the MFA is determined based on the most refined partition Π. Specifically, the initial state m_0 is the part of Π including the initial state d_0 of $\mathcal{D}$. The set of final states F_{m} is composed of all the parts of Π composed of final states of $\mathcal{D}$. Finally, since all states d within the same part π share the same transition signature, all transitions $d \xrightarrow{\ell} d'$ in $\mathcal{D}$ are such that d' belongs to the same part π'. Thus, a transition $\pi \xrightarrow{\ell} \pi'$ is created for $\mathcal{M}$.

The pseudocode for *MIN* is as follows:

1. **algorithm** *MIN* (**in** $\mathcal{D}$, **out** $\mathcal{M}$)
2. $\mathcal{D} = (\Sigma, D, \tau_{\mathrm{d}}, d_0, F_{\mathrm{d}})$: a DFA,
3. $\mathcal{M} = (\Sigma, M, \tau_{\mathrm{m}}, m_0, F_{\mathrm{m}})$: the minimal DFA equivalent to $\mathcal{D}$;

4. **auxiliary function** *NewPartition* (Π): a new partition based on Π
5. Π: base partition;
6. **begin** $\langle$*NewPartition*$\rangle$
7. $\Pi_{\mathrm{new}} := \{\}$;
8. **foreach** part $\pi \in \Pi$ **do**
9. Determine the transition signature $\sigma_\Pi(d)$ of each $d \in \pi$;
10. Group elements of π based on the same transition signature;
11. Create a new part in Π_{new} for each group
12. **endfor**;
13. **return** Π_{new}
14. **end** $\langle$*NewPartition*$\rangle$;

15. **begin** $\langle$*MIN*$\rangle$
16. $\Pi := \{F_{\mathrm{d}}, D - F_{\mathrm{d}}\}$;
17. **repeat**
18. $\Pi_{\mathrm{old}} = \Pi$;
19. $\Pi := NewPartition(\Pi_{\mathrm{old}})$
20. **until** $\Pi = \Pi_{\mathrm{old}}$;
21. Determine $\tau_{\mathrm{m}}, F_{\mathrm{m}}$, and m_0 based on Π
22. **end** $\langle$*MIN*$\rangle$.

Example 2.15 (Minimization). With reference to Example 2.14, consider the DFA displayed on the left of Fig. 2.4. This DFA is the result of the determinization (Fig. 2.3) performed by *SC* on the NFA displayed in Fig. 2.2. Based on the specification of *MIN*, the initial partition is $\Pi_1 = \{\pi_1, \pi_2\}$, where $\pi_1 = \{d_0, d_1, d_3\}$ (nonfinal states) and $\pi_2 = \{d_2, d_4\}$ (final states). We have $\sigma_\Pi(d_0) = \sigma_\Pi(d_1) = \sigma_\Pi(d_3) = \{(a, \pi_1), (b, \pi_2)\}$, while $\sigma_\Pi(d_2) = \sigma_\Pi(d_4) = \{(a, \pi_2), (b, \pi_2)\}$. Since all states in each partition share the same transition signature, the new partition returned by *NewPartition* (line 19) equals the old partition. Therefore, the loop terminates, with $\mathcal{M}$ being composed of two states only, $m_0 = \{d_0, d_1, d_3\}$ and $m_1 = \{d_2, d_4\}$, where m_1 is the initial state and m_2 is the final state. The transition function is determined by the transition signature. The resulting MFA is displayed on the right of Fig. 2.4. As expected, the MFA recognizes the same language as the input DFA, namely $a^* b\,(a|b)^*$.

Note 2.3. The minimization of the DFA in Fig. 2.4 leads to an MFA with two states only. Since the DFA displayed in Fig. 2.1 is equivalent to this MFA, it follows that the former is not minimal, as it includes three states, namely d_0, d_1, and d_2. In fact, if we apply *MIN* to such a DFA then d_0 and d_1 will be in the same part, thereby giving the same two-state MFA displayed in Fig. 2.4 (node identification aside).

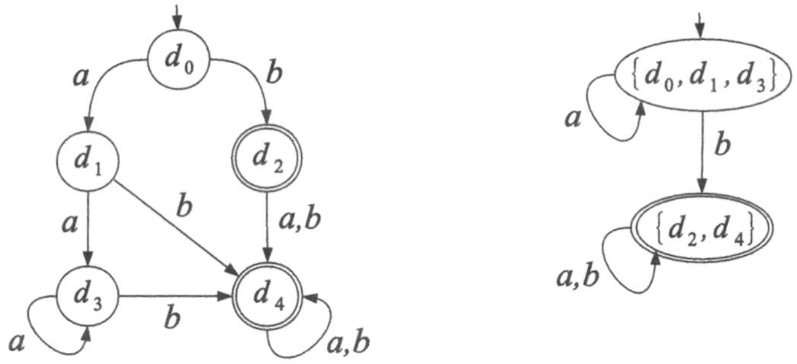

Fig. 2.4 DFA (left) and equivalent MFA (right) generated by *MIN*

2.7 Incremental Determinization of Finite Acyclic Automata

In Sec. 2.5, an algorithm for NFA determinization was introduced, namely *SC*. We say that *SC* is *context free*, that is, the algorithm is based on the input NFA only, without any additional information. This, of course, is perfect when the NFA is given once and for all. As such, when a new NFA $\mathcal{N}'$ is given in the input after an NFA $\mathcal{N}$, *SC* will perform the determinization of $\mathcal{N}'$ irrespective of the previous determinization of $\mathcal{N}$. However, some application domains, such as monitoring of active systems [93, 110, 113] (see Chapter 7) and model-based testing in software engineering [2, 3, 4], require determinization to be carried out incrementally, as stated below.

Definition 2.12 (Expansion). Let $\mathcal{N} = (\Sigma, N, \tau, n_0, F)$ be an NFA. An *expansion* of $\mathcal{N}$ is a quadruple $\Delta\mathcal{N} = (\Delta N, \Delta\tau, \Delta F^+, \Delta F^-)$, where ΔN is the set of additional states, $\Delta\tau$ is the set of additional transitions, $\Delta F^+ \subseteq \Delta N$ is the set of additional final states, and $\Delta F^- \subseteq F_{\mathrm{n}}$ is the set of retracted final states.

In other words, an expansion $\Delta\mathcal{N} = (\Delta N, \Delta\tau, \Delta F^+, \Delta F^-)$ extends $\mathcal{N}$ by $\Delta\mathcal{N}$ with new states (ΔN) and new transitions ($\Delta\tau$); moreover, it may introduce new final states (ΔF^+), while some final states in $\mathcal{N}$ may become nonfinal (ΔF^-). This expansion results in an extended NFA, $\mathcal{N}' = \mathcal{N} \cup \Delta\mathcal{N} = (\Sigma, N', \tau', n_0, F')$, where $N' = N \cup \Delta N$, $\tau' = \tau \cup \Delta\tau$, and $F' = (F - \Delta F^-) \cup \Delta F^+$.

Note 2.4. We assume that both $\mathcal{N}$ and $\mathcal{N}'$ are connected, with each state being reachable from the initial state n_0.

Problem 2.1 (Incremental Determinization). Let $\mathcal{N}$ be an acyclic NFA, and $\mathcal{D}$ the DFA equivalent to $\mathcal{N}$ (as generated by *SC*). Let $\Delta\mathcal{N}$ be an expansion of $\mathcal{N}$ yielding $\mathcal{N}' = \mathcal{N} \cup \Delta\mathcal{N}$, a new acyclic NFA. Generate the DFA $\mathcal{D}'$ equivalent to $\mathcal{N}'$, based on $\mathcal{N}$, $\mathcal{D}$, and $\Delta\mathcal{N}$.

2.7.1 Incremental Determinization Algorithm

Solving the incremental determinization problem naively by means of *SC* is bound to lead to poor performance, as no exploitation of the special nature of $\mathcal{N}' = \mathcal{N} \cup \Delta\mathcal{N}$ is considered. This can be understood by envisioning an extreme scenario, where $\mathcal{N}$ is very large and $\Delta\mathcal{N}$ very small. In this case, *SC* performs the determinization of $\mathcal{N}'$ starting from scratch, disregarding both $\mathcal{D}$ and $\Delta\mathcal{N}$, thereby performing most of the processing already done to yield $\mathcal{D}$.

For the purpose of this book, we focus on incremental determinization of *acyclic* automata only, by an algorithm called *ISCA*. The pseudocoded specification of *ISCA* is given below. *ISCA* takes as input an NFA $\mathcal{N}$, the equivalent DFA $\mathcal{D}$ (as generated by *SC*), and an expansion $\Delta\mathcal{N}$ of $\mathcal{N}$. The algorithm updates $\mathcal{D}$ based on $\Delta\mathcal{N}$ so as to make it equivalent to the extended NFA $\mathcal{N}' = \mathcal{N} \cup \Delta\mathcal{N}$ (as generated by *SC*).

Throughout the pseudocode, like before, we keep a distinction between the *identifier* of a state in $\mathcal{D}$ and its *content*, where the former is a symbol (e.g., d), while the latter is a set of nodes in $\mathcal{N}'$ (e.g., $\mathbb{N}$). The content of a node d is written $\|d\|$. During execution, the content can change, while the identifier cannot.

ISCA is supported by a stack of buds $\mathcal{B}$, with each bud being a triple $(d, \ell, \mathbb{N})$, where d is the identifier of a state in $\mathcal{D}$, ℓ either a symbol of the alphabet or ε, and $\mathbb{N}$ a subset of states in $\mathcal{N}$. Specifically, $\mathbb{N}$ is the ℓ-closure of the NFA states incorporated into d. A bud indicates that further processing needs to be performed to update the transition exiting d, marked by ℓ in $\mathcal{D}$.

Roughly, the bud-stack parallels the stack of DFA states in *SC*. Just as new DFA states are inserted into the *SC* stack and thereafter processed, so are the new buds accumulated into the bud-stack and processed one by one.

In *SC*, the first state pushed into the stack is the initial state of $\mathcal{D}$. In *ISCA*, the bud-stack is initialized by a number of buds relevant to the state exited by the new transitions in $\mathcal{N}$.

The algorithm loops, by popping a bud at each iteration, until the bud-stack becomes empty. While a bud is being processed, new buds may possibly be inserted into the bud-stack. The processing of each bud depends on both the bud and the current configuration of $\mathcal{D}$.

Buds in $\mathcal{B}$ are implicitly grouped by the first two fields: if a new bud $B = (d, \ell, \mathbb{N})$ is pushed into $\mathcal{B}$ and a bud $B' = (d, \ell, \mathbb{N}')$ is already in $\mathcal{B}$, then B' will be absorbed by B, thereby becoming $B = (d, \ell, \mathbb{N} \cup \mathbb{N}')$.

The algorithm *ISCA* makes use of the auxiliary procedure *Expand* (lines 8–25), which takes as input a state d in $\mathcal{D}$ and adds to its content the subset $\mathbb{N}$ of states in $\mathcal{N}'$. The bud-stack $\mathcal{B}$ is extended by the buds relevant to d and the labels exiting nodes in $(\mathbb{N} - \|d\|)$ in $\mathcal{N}'$. If the content of the extended node d equals the content of a node d' already in $\mathcal{D}$, then the two nodes are merged into a single node (lines 17–23): all transitions entering or exiting d are redirected to or from d', respectively. In addition, after the removal of d, all buds relevant to d are renamed using d'. The redirection of transitions exiting d may cause nondeterminism on exiting d', that is, two transitions exiting d' that are marked by the same label ℓ; however, such nondeterminism disappears at the end of the processing (see Theorem 2.1 in Sec. 2.7.2).

Considering the body of *ISCA* (lines 26–55), after determining the subset $\bar{\mathbb{N}}$ of states in $\mathcal{N}$ that are exited by transitions in $\Delta\mathcal{N}$, $\mathcal{N}$ is extended based on $\Delta\mathcal{N}$, while the set of final states in $\mathcal{D}$ may possibly be shrunk based on ΔF_{n}^{-} (line 28).

The bud-stack $\mathcal{B}$ is initialized with buds $(d, \ell, \mathbb{N})$, where $\mathbb{N}$ is the (nonempty) ℓ-closure of $\|d\| \cap \bar{\mathbb{N}}$ (line 29).

A loop is then iterated until $\mathcal{B}$ becomes empty (lines 30–54). At each iteration, a bud $(d, \ell, \mathbb{N})$ is popped from $\mathcal{B}$. Depending on the bud, one of six *action rules*, $\mathcal{R}_1 \cdots \mathcal{R}_6$, is applied, with each rule being defined as a pair $[\,\textit{Condition}\,] \Rightarrow \textit{Action}$, as specified after Def. 2.13.

Definition 2.13 (ℓ-transition). An ℓ-transition is a transition marked by the symbol ℓ.

($\mathcal{R}_1$) $[\ \ell = \varepsilon\ ] \Rightarrow d$ is expanded by $\mathbb{N}$ (line 33).

($\mathcal{R}_2$) $[\ \ell \neq \varepsilon$, no ℓ-transition exits d, $\exists Sd'$ such that $\|Sd'\| = \mathbb{N}\ ] \Rightarrow$ A new transition $d \xrightarrow{\ell} d'$ is inserted into $\mathcal{D}$ (line 36).

($\mathcal{R}_3$) $[\ \ell \neq \varepsilon$, no ℓ-transition exits d, $\nexists d'$ such that $\|d'\| = \mathbb{N}\ ] \Rightarrow$ A new empty state d' and a new transition $d \xrightarrow{\ell} d'$ are created; then, d' is expanded by $\mathbb{N}$ (lines 38 and 39).

($\mathcal{R}_4$) $[\ \ell \neq \varepsilon$, an ℓ-transition exits d to d', no other transition enters $d'\ ] \Rightarrow d'$ is expanded by $\mathbb{N}$ (line 44).

($\mathcal{R}_5$) $[\ \ell \neq \varepsilon$, an ℓ-transition exits d to d', another transition enters d', $\exists d''$ such that $\|d''\| = \|d'\| \cup \mathbb{N}\ ] \Rightarrow$ transition t is redirected toward d'' (line 46).

($\mathcal{R}_6$) $[\ \ell \neq \varepsilon$, an ℓ-transition exits d to d', another transition enters d', $\nexists d''$ such that $\|d''\| = \|d'\| \cup \mathbb{N}\ ] \Rightarrow d'$ is duplicated into d'' (along with exiting transitions and buds), transition t is redirected toward d'', and d'' is expanded by $\mathbb{S}$ (lines 48–50).[3]

Note 2.5. Rules $\mathcal{R}_4$, $\mathcal{R}_5$, and $\mathcal{R}_6$ correspond to a single bud but may be applied several times, depending on the number of ℓ-transitions exiting d (as stated above, a temporary nondeterminism in $\mathcal{D}$ may be caused by merging two states in the function *Expand*).

The pseudocode for *ISCA* is as follows:

```
1.  algorithm ISCA (inout 𝒩, inout 𝒟, in Δ𝒩)
2.     𝒩 = (Σ, N, τ_n, S_0n, F_n): an acyclic NFA,
3.     𝒟 = (Σ, D, τ_d, S_0d, F_d): the DFA equivalent to 𝒩,
4.     Δ𝒩 = (ΔN, Δτ_n, ΔF_n^+, ΔF_n^-): an expansion of 𝒩;

5.     side effects
6.        𝒩 is extended by Δ𝒩,
```

[3] In other words, after the duplication of d' we have $\|d''\| = \|d'\|$ and, for each bud $(d', \ell, \mathbb{N}) \in \mathcal{B}$, a new bud $(d'', \ell, \mathbb{N})$ is created and pushed into $\mathcal{B}$.

7. $\mathcal{D}$ becomes the DFA equivalent to $(\mathcal{N}' = \mathcal{N} \cup \Delta\mathcal{N})$;

8.   **auxiliary procedure** *Expand* (**inout** d, **in** $\mathbb{N}$)
9. d: a state in $\mathcal{D}$,
10. $\mathbb{N}$: a subset of states in $\mathcal{N}$;
11.   **begin** $\langle Expand \rangle$
12.   **if** $\mathbb{N} \not\subseteq \|d\|$ **then**
13. $\mathbb{B}' := \{ (d, \ell, \mathbb{N}') \mid \ell \in \Sigma, \mathbb{N}' = \ell\text{-closure}(\mathbb{N} - \|d\|), \mathbb{N}' \neq \emptyset \}$;
14.   Push buds $\mathbb{B}'$ into $\mathcal{B}$;
15.   Enlarge $\|d\|$ by $\mathbb{N}$;
16.   **if** $d \notin F_\mathrm{d}, \mathbb{N} \cap F_\mathrm{n} \neq \emptyset$ **then** Insert d into F_d **endif**;
17.   **if** $\mathcal{D}$ includes a state d' such that $\|d'\| = \|d\|$ **then**
18.   Redirect to d' all transitions entering d;
19.   Redirect from d' all transitions exiting d;
20.   **if** $d \in F_\mathrm{d}$ **then** Remove d from F_d **endif**;
21.   Remove d from $\mathcal{D}$;
22.   Convert to d' the buds in $\mathcal{B}$ relevant to d
23.   **endif**
24.   **endif**
25.   **end** $\langle Expand \rangle$;

26.   **begin** $\langle ISCA \rangle$
27. $\bar{\mathbb{N}} :=$ the set of states in $\mathcal{N}$ exited by transitions in $\Delta\tau_\mathrm{n}$;
28.   Extend $\mathcal{N}$ based on $\Delta\mathcal{N}$, and update F_d based on ΔF_n^-;
29. $\mathcal{B} := [(d, \ell, \mathbb{N}) \mid d \in D, n \in \|d\| \cap \bar{\mathbb{N}}, n \xrightarrow{\ell} n' \in \Delta\tau_\mathrm{n},$
                                                  $\mathbb{N} = \ell\text{-closure}(\|d\| \cap \bar{\mathbb{N}})]$;
30.   **repeat**
31.   Pop bud $(d, \ell, \mathbb{N})$ from the top of bud-stack $\mathcal{B}$;
32. $(\mathcal{R}_1)$ **if** $\ell = \varepsilon$ **then**
33.   *Expand* $(d, \mathbb{N})$
34.   **elsif** no ℓ-transition exits d **then**
35. $(\mathcal{R}_2)$ **if** $\mathcal{D}$ includes a state d' such that $\|d'\| = \mathbb{N}$ **then**
36.   Insert a new transition $d \xrightarrow{\ell} d'$ into $\mathcal{D}$
37. $(\mathcal{R}_3)$ **else**
38.   Create a new state d' and insert $d \xrightarrow{\ell} d'$ into $\mathcal{D}$;
39.   *Expand* $(d', \mathbb{N})$
40.   **endif**
41.   **else**
42.   **foreach** transition $t = d \xrightarrow{\ell} d'$ where $\mathbb{N} \not\subseteq \|d'\|$ **do**
43. $(\mathcal{R}_4)$ **if** no other transition enters d' **then**
44.   *Expand* $(d', \mathbb{N})$
45. $(\mathcal{R}_5)$ **elsif** $\mathcal{D}$ includes d'' such that $\|d''\| = \|d'\| \cup \mathbb{N}$ **then**
46.   Redirect t toward d''
47. $(\mathcal{R}_6)$ **else**

```
                Create a copy d'' of d', along with all exiting transitions
                 and related buds;
                Redirect t toward d'';
                Expand (d'', N)
              endif
            endfor
          endif
        until bud-stack B becomes empty
      end ⟨ISCA⟩.
```

Example 2.16 (Incremental Subset Construction). Consider the NFA $\mathcal{N}$ and the equivalent DFA $\mathcal{D}$ (generated by *SC*) displayed in Fig. 2.5.

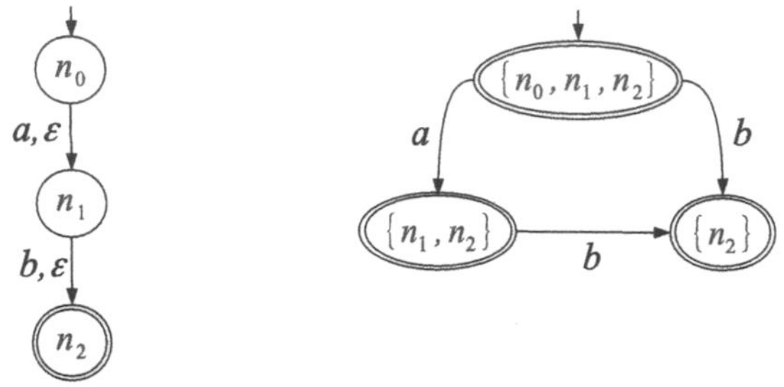

Fig. 2.5 NFA $\mathcal{N}$ (left) and equivalent DFA $\mathcal{D}$ (right) generated by *SC*

Then, assume that $\mathcal{N}$ is expanded to $\mathcal{N}'$ by $\Delta\mathcal{N} = (\Delta N, \Delta F_{\mathrm{n}}^{+}, \Delta F_{\mathrm{n}}^{-})$, where $\Delta N = \{n_3, n_4\}$, $\Delta F_{\mathrm{n}}^{+} = \{n_4\}$, and $\Delta F_{\mathrm{n}}^{-} = \{n_2\}$, as shown in Fig. 2.6 (left). The DFA equivalent to $\mathcal{N}'$ generated by *SC* is displayed next to $\mathcal{N}'$ in the same figure.

The incremental determinization of the expanded NFA $\mathcal{N}'$ performed by *ISCA* is traced in Fig. 2.7 (from left to right). In each step, the configuration of the DFA is

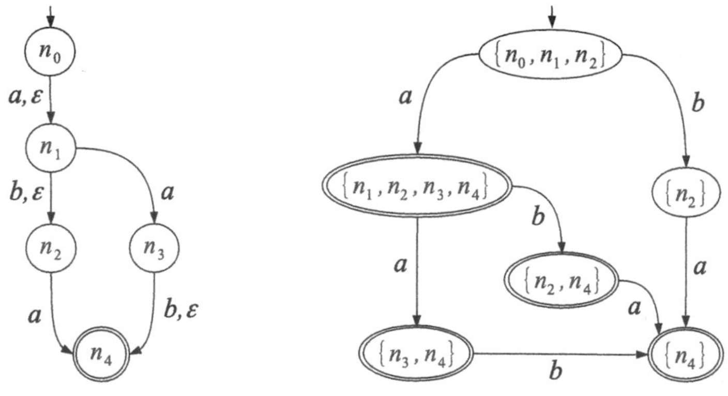

Fig. 2.6 Expanded NFA $\mathcal{N}'$ (left) and equivalent DFA $\mathcal{D}'$ (right) generated by *SC*

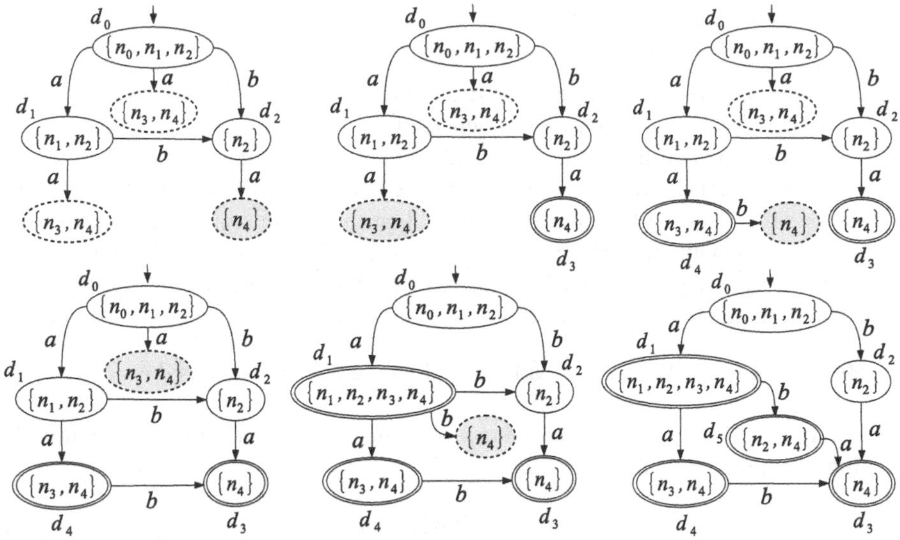

Fig. 2.7 Incremental determinization by *ISCA* of the NFA displayed in Fig. 2.6: buds are dashed, with the gray node being the bud on top of the stack (to be processed first)

displayed, where buds are represented by dashed nodes, with the gray node being the bud on top of the stack (to be processed in the next step).

Initially (line 27 of *ISCA*) we have $\bar{\mathbb{N}} = \{n_1, n_2\}$. At line 28, the set of final states of $\mathcal{D}$ becomes empty, as $\Delta F_{\mathrm{n}}^{-} = \{n_2\}$ means that n_2 is no longer a final state in $\mathcal{N}$. At line 29, the bud-stack is initialized as $\mathcal{B} = \{(d_0, a, \{n_3, n_4\}), (d_1, a, \{n_3, n_4\}), (d_2, a, \{n_4\})\}$. In Fig. 2.7, each bud $(d, \ell, \mathbb{N})$ is represented by an arrow exiting d and entering the dashed node marked by $\mathbb{N}$. A gray dashed node indicates the bud on top of the stack (the next to be processed). Then, the loop (lines 30–54) iterates five times, as detailed below.

1. Bud $(d_2, a, \{n_4\})$ triggers rule $(\mathcal{R}_3)$ of *ISCA*, causing the insertion of state d_3, with $\|d_3\| = \{n_4\}$ and transition $d_2 \xrightarrow{a} d_3$.
2. Bud $(d_1, a, \{n_3, n_4\})$ triggers rule $(\mathcal{R}_3)$, causing the insertion of state d_4, with $\|d_4\| = \{n_3, n_4\}$, transition $d_1 \xrightarrow{a} d_4$, and bud $(d_4, b, \{n_4\})$.
3. Bud $(d_4, b, \{n_4\})$ triggers rule $(\mathcal{R}_2)$, causing the insertion of transition $d_4 \xrightarrow{b} d_3$.
4. Bud $(d_0, a, \{n_3, n_4\})$ triggers rule $(\mathcal{R}_4)$, causing the expansion of state $d_1 = \{n_1, n_2\}$ to $\{n_1, n_2, n_3, n_4\}$, and the creation of bud $(d_1, b, \{n_4\})$.
5. Bud $(d_1, b, \{n_4\})$ triggers rule $(\mathcal{R}_6)$, causing the duplication of state $d_2 = \{n_2\}$ into $d_5 = \{n_2, n_4\}$, and the insertion of transitions $d_1 \xrightarrow{b} d_5$ and $d_5 \xrightarrow{a} d_3$.

$\mathcal{B}$ is now empty and the loop terminates. The final states are those including n_4, namely d_1, d_3, d_4, and d_5. As expected, the DFA displayed on the bottom right of Fig. 2.7 equals the DFA $\mathcal{D}'$ displayed on the right side of Fig. 2.6.

2.7.2 Correctness of ISCA

The correctness of the algorithm *ISCA* is stated in Theorem 2.1 below. Roughly, the proof is by induction on the sequence of automata (and relevant bud-stacks) generated by the consumption of buds. This requires this sequence to be finite, in other words, that the bud-stack eventually becomes empty (Lemma 2.1). Hence, we prove that, starting from the DFA equivalent to $\mathcal{N}$ in the input, the sequence of automata ends with the DFA equivalent to $\mathcal{N}' = \mathcal{N} \cup \Delta\mathcal{N}$, that is, the last automaton in the sequence equals the DFA equivalent to $\mathcal{N}'$ (as generated by *SC*). First we introduce some useful definitions.

Definition 2.14 (ℓ-Bud). An ℓ-bud is a bud where the label involved is ℓ.

Definition 2.15 (Balanced Transition). A transition $d \xrightarrow{\ell} d'$ in the automaton $\mathcal{D}$ being manipulated by *ISCA* is *balanced* iff

$$\|d'\| = \ell\text{-closure}(\|d\|)\,. \tag{2.15}$$

Otherwise, the transition is *unbalanced.*

Definition 2.16 (Pair Sequence). Let $\mathcal{D}_i$ be the configuration of automaton $\mathcal{D}$ after the processing of i buds, and $\mathcal{B}_i$ the configuration of the corresponding bud-stack $\mathcal{B}$. The *pair sequence* $\mathcal{Q}$ is the sequence $[(\mathcal{D}_0,\mathcal{B}_0),(\mathcal{D}_1,\mathcal{B}_1),\ldots,(\mathcal{D}_i,\mathcal{B}_i),\ldots]$ (where $\mathcal{D}_0$ and $\mathcal{B}_0$ are the DFA in the input and the initial configuration of $\mathcal{B}$, respectively).

Definition 2.17 (Viable Pair). A pair $(\mathcal{D}_i,\mathcal{B}_i)$ in $\mathcal{Q}$ is *viable* iff, for each unbalanced $d \xrightarrow{\ell} d'$ in $\mathcal{D}_i$, $\mathcal{B}_i$ includes a set of buds whose processing make the transition balanced.

Theorem 2.1. *Let* $\mathcal{N}$ *be an acyclic NFA,* $\mathcal{D}_{SC}$ *the DFA equivalent to* $\mathcal{N}$ *generated by* SC, $\mathcal{N}'$ *the acyclic NFA obtained by expanding* $\mathcal{N}$ *by* $\Delta\mathcal{N} = (\Delta N, \Delta \mathcal{T}_{\mathrm{n}}, \Delta F_{\mathrm{n}}^{+}, \Delta F_{\mathrm{n}}^{-})$, $\mathcal{D}'_{SC}$ *the DFA equivalent to* $\mathcal{N}'$ *generated by* SC, *and* $\mathcal{D}'_{ISCA}$ *the DFA generated by* ISCA *(based on* $\mathcal{N}$, $\mathcal{D}_{SC}$, *and* $\Delta\mathcal{N}$*). Then*

$$\mathcal{D}'_{ISCA} = \mathcal{D}'_{SC}. \tag{2.16}$$

Proof. The proof comes from various lemmas.

Lemma 2.1. *The algorithm* ISCA *terminates.*

Proof. We have to show that the bud-stack $\mathcal{B}$ becomes empty in a finite number of iterations of the main loop (lines 30–54). To this end, we exploit the acyclicity of the extended NFA $\mathcal{N}'$. Let k' be the number of states in $\mathcal{N}'$. The length of paths in $\mathcal{N}'$ is bounded by k'. Since each state d in $\mathcal{D}$ is such that $\|d\|$ is a subset of states in $\mathcal{N}'$, the number of states in $\mathcal{D}$ is bounded by $2^{k'}$. $\mathcal{B}$ is first instantiated with a finite number of buds (line 29). Although the consumption of each bud $B = (d, \ell, \mathbb{N})$ (line 31) may possibly cause the generation of new buds $B' = (d', \ell', \mathbb{N}')$, $\|d'\|$ necessarily incorporates (at least) one state n which is a successor of (at least) one state in $\|d'\|$. Consequently, after a finite number of iterations, the chain of buds generated by B must stop, notably when no successor exists in $\mathcal{N}'$ for all the states in $\|d'\|$. Applying this consideration to all the buds of the initial configuration of $\mathcal{B}$ leads to the conclusion that $\mathcal{B}$ becomes empty in a finite number of iterations.

Lemma 2.2. *Each transition of $\mathcal{D}'_{ISCA}$ is balanced.*

Proof. We prove viability by induction on $\mathcal{Q} = [(\mathcal{D}_0, \mathcal{B}_0), (\mathcal{D}_1, \mathcal{B}_1), \ldots, (\mathcal{D}_m, \mathcal{B}_m)]$.

(*Basis*) $(\mathcal{D}_0, \mathcal{B}_0)$ *is a viable pair.* Before $\mathcal{N}$ is extended by $\Delta\mathcal{N}$, based on the definition of *SC*, each transition in $\mathcal{D}_0 = \mathcal{D}_{SC}$ is balanced. After the extension by $\Delta\mathcal{N}$, some transitions may become unbalanced because of $\Delta\mathcal{N}$. For a newly unbalanced transition $t = d \xrightarrow{\ell} d'$, at most two cases are possible for the imbalance:

(a) New ε-transitions in $\Delta\mathcal{N}$ exiting some states in $\|d'\|$. In this case, a bud $(d', \varepsilon, \mathbb{N}')$ is inserted into $\mathcal{B}_0$, where $\mathbb{N}' = \varepsilon\text{-closure}^*(\bar{\mathbb{N}})$, with $\bar{\mathbb{N}}$ being the set of states in $\|d'\|$ exited by new ε-transitions in $\Delta\tau_{\mathrm{n}}$.

(b) New ℓ-transitions in $\Delta\mathcal{N}$ exiting some states in $\|d\|$. In this case, a bud $(d, \ell, \mathbb{N})$ is inserted into $\mathcal{B}_0$, where $\mathbb{N} = \ell\text{-closure}(\bar{\mathbb{N}})$, with $\bar{\mathbb{N}}$ being the set of states in $\|d\|$ exited by new ℓ-transitions in $\Delta\tau_{\mathrm{n}}$.

We show that processing the new (at most two) buds makes transition t balanced anew. For case (a), processing bud $(d', \varepsilon, \mathbb{N}')$ corresponds to rule $\mathcal{R}_1$ of *ISCA*, where $\|d'\|$ is expanded by $\mathbb{N}'$, thereby making $\|d'\|$ include $\varepsilon\text{-closure}^*(\bar{\mathbb{N}})$. Since in $\mathcal{D}_0$ the transition t is balanced, d becomes balanced anew. A possible merge of d' with another state $\bar{d}$ maintains balanced nature of the ℓ-transition redirected toward $\bar{d}$. Furthermore, the conversion of buds relevant to d' into buds relevant to $\bar{d}$ allows possible newly unbalanced transitions $\bar{d} \xrightarrow{\ell} d''$ to become balanced anew.

Now, consider case (b), by processing bud $(d, \ell, \mathbb{N})$. Since $t = d \xrightarrow{\ell} d'$ is unbalanced, this means that t exists. Hence, only rules $\mathcal{R}_4$, $\mathcal{R}_5$, and $\mathcal{R}_6$ are applicable. Considering $\mathcal{R}_4$, *Expand*$(d', \mathbb{N})$ is called, thereby leading to the same conclusion as that discussed for case (a). Considering $\mathcal{R}_5$, since in $\mathcal{D}_{SC}$ the transition $d \xrightarrow{\ell} d'$ is balanced, after the extension of $\mathcal{N}$ by $\Delta\mathcal{N}$ the ℓ-closure of $\|S_d\|$ becomes $\|S'_d\| \cup \mathbb{S}$, in fact, $\|d''\|$; in other words, $d \xrightarrow{\ell} d''$ is balanced. Finally, considering $\mathcal{R}_6$, the relevant actions of duplication, redirection, and extension cause the removal of the unbalanced transition t_d by creating a new unbalanced transition $d \xrightarrow{\ell} d''$, which again corresponds to rule $\mathcal{R}_4$.

(*Induction*) *If* $(\mathcal{D}_i, \mathcal{B}_i)$ *is a viable pair then* $(\mathcal{D}_{i+1}, \mathcal{B}_{i+1})$ *is a viable pair.* We have to analyze all action rules relevant to a generic bud $B = (d, \ell, \mathbb{N})$. Considering $\mathcal{R}_1$, that is $\ell = \varepsilon$, the extension of $\|d\|$ by $\mathbb{N}$ is accompanied by the creation of buds $(d, \ell', \mathbb{N}')$, where $\mathbb{N}' = \ell'\text{-closure}(\mathbb{N} - \|d\|)$, which, following the same reasoning as for case (b) of the basis, allow possibly unbalanced transitions $d \xrightarrow{\ell'} d'$ to become balanced anew. Considering $\mathcal{R}_2$, the newly created transition $t' = d \xrightarrow{\ell} d'$ cannot be unbalanced. In fact, bud B can be created either initially or by means of the function *Expand*. In either case, $\mathbb{N} = \ell\text{-closure}(\|d\|)$; in other words, t' is balanced. Considering $\mathcal{R}_3$, the newly created transition $d \xrightarrow{\ell} d'$ is balanced as, after the creation of the new state d' and the relevant extension, similarly to rule $\mathcal{R}_2$, we have $\|d'\| = \ell\text{-closure}(\|d\|)$. Considering $\mathcal{R}_4$, similarly to $\mathcal{R}_1$, the extension of $\|d'\|$ by $\mathbb{N}$ is accompanied by the creation of buds $(d', \ell', \mathbb{N}')$, where $\mathbb{N}' = \ell'\text{-closure}(\mathbb{N} - \|d'\|)$, which, following the same reasoning as for case (b) of the basis, allow possibly unbalanced transitions $d \xrightarrow{\ell'} d'$ to become balanced anew. Considering $\mathcal{R}_5$, notice how $\mathbb{N}$ is the ℓ-closure of a subset of $\|d\|$ for which $\|d'\|$ may possibly miss some states that make $d \xrightarrow{\ell} d'$ unbalanced. However, since $\|d''\| = \|d'\| \cup \mathbb{N}$, the new (redirected) transition $d \xrightarrow{\ell} d''$ is balanced. Considering $\mathcal{R}_6$, the relevant actions of duplication, redirection, and extension cause the removal of the unbalanced transition t by creating a new unbalanced transition $S_d \xrightarrow{\ell} d''$, which leads to the same reasoning as that applied to rule $\mathcal{R}_4$.

Note 2.6. Since rules $\mathcal{R}_4$, $\mathcal{R}_5$, and $\mathcal{R}_6$ are associated with a single bud, the induction step corresponds to the total effect of them.

Therefore, in either case, $(\mathcal{D}_{i+1}, \mathcal{B}_{i+1})$ is viable. The actual conclusion of the proof of Lemma 2.2 follows from Lemma 2.1, by virtue of which $\mathcal{B}_m = \emptyset$.

Lemma 2.3. *If there exists an ℓ-transition exiting a state d of $\mathcal{D}_i$ in $\mathcal{Q}$, for each automaton $\mathcal{D}_j$ of $\mathcal{Q}$, $j > i$, if d is still in $\mathcal{D}_j$ then an ℓ-transition exits d in $\mathcal{D}_j$.*

Proof. Considering the six action rules in *ISCA*, the only possibility for an existing transition exiting state d to be removed is by merging d with another state by means of the function *Expand* (lines 29–35). In this case, however, d is removed too.

Lemma 2.4. *Let $t = n \xrightarrow{\ell} n'$, $\ell \neq \varepsilon$, be a transition in $\mathcal{N}'$. Then, each state d in $\mathcal{D}'_{ISCA}$ where $n \in \|d\|$ is exited by an ℓ-transition.*

Proof. Notice that $\mathcal{D}'_{ISCA}$ is the last automaton $\mathcal{D}_m$ in $\mathcal{Q}$. Consider a state d in $\mathcal{D}'_{ISCA}$ such that $n \in \|d\|$. Two cases are possible for the genesis of d:

(a) d is in $\mathcal{D}_{SC}$ already (possibly including a different subset of states). When $n \in \|d\|$, if $t \notin \Delta\tau_n$ then, based on *SC*, an ℓ-transition exits d; otherwise (if $t \in \Delta\tau_n$), $\mathcal{B}_0$ includes an ℓ-bud relevant to d, whose consumption will create (if it does not already exist) an ℓ-transition exiting d. When, instead, $n \notin \|d\|$, the insertion of n into $\|d\|$ will be caused by the function *Expand* in either rule $\mathcal{R}_1$ or rule $\mathcal{R}_4$, along with the generation of an ℓ-bud relevant to d, which will create (if it does not already exist) an ℓ-transition exiting d.

(b) d is created by *ISCA*. Two cases are possible: either by rule $\mathcal{R}_3$ or by rule $\mathcal{R}_6$. On the one hand, $\mathcal{R}_3$ instantiates $\|d\|$ by means of *Expand*, thereby generating an ℓ-bud relevant to d, which will create (if it does not already exist) an ℓ-transition exiting d. On the other, $\mathcal{R}_6$ creates the new state as a copy of an existing state, along with its exiting transitions. The subsequent call to *Expand* generates an ℓ-bud relevant to the newly created state, which will create (if it does not already exist) an exiting ℓ-transition.

Hence, in either case, (a) or (b), an ℓ-transition exiting d exists for an automaton $\mathcal{D}_i$ in the pair sequence $\mathcal{Q}$ (see Def. 2.16). By virtue of Lemma 2.3, since we have assumed d to be in $\mathcal{D}'_{ISCA}$, an ℓ-transition exits d in $\mathcal{D}'_{ISCA}$.

Lemma 2.5. $\|d_0^{\mathcal{D}'_{ISCA}}\| = \|d_0^{\mathcal{D}'_{SC}}\|$.

Proof. Let n_0 and $d_0^{\mathcal{D}_{SC}}$ be the initial states of $\mathcal{N}$ and $\mathcal{D}_{SC}$, respectively. If there does not exist in $\Delta\mathcal{T}_{\mathrm{n}}$ any ε-transition exiting a state in $\|d_0^{\mathcal{D}_{SC}}\|$, then the equality holds true, as $d_0^{\mathcal{D}_{SC}}$ can be extended by initial ε-buds only (because of acyclicity). Otherwise, processing the bud $(d_0^{\mathcal{D}_{SC}}, \varepsilon, \mathrm{N})$ makes $\|d_0^{\mathcal{D}_{SC}}\| = \varepsilon\text{-closure}(n_0)$ in $\mathcal{N}'$, which corresponds to the definition of the initial state of $\mathcal{D}'_{SC}$.

Lemma 2.6. *$\mathcal{D}'_{ISCA}$ is deterministic.*

Proof. The proof is by contradiction. Assume $\mathcal{D}'_{ISCA}$ is nondeterministic. Since $\mathcal{D}'_{ISCA}$ is the result of the manipulation of $\mathcal{D}_{SC}$ by rules $\mathcal{R}_1 \cdots \mathcal{R}_6$, which do not create any ε-transitions, no ε-transition is included in $\mathcal{D}'_{ISCA}$. Hence, the only way for nondeterminism to exist in $\mathcal{D}'_{ISCA}$ is for there to be a state d exited by (at least) two ℓ-transitions reaching two different states, namely $d \xrightarrow{\ell} d'$ and $d \xrightarrow{\ell} d''$, with $d' \neq d''$, where $\|d'\| \neq \|d''\|$ (otherwise the two states would be merged by *Expand*).

Note 2.7. Since *ISCA* starts by modifying a deterministic automaton, each possible extension of a state is monitored by the function *Expand*, which, if the expanded state equals another existing state in content, merges the two states into one state (see lines 17–23 of *ISCA*).

However, since the two transitions exit the same state d, by virtue of Lemma 2.2, $\|d'\| = \|d''\| = \ell\text{-closure}(\|d\|)$, a contradiction.

Lemma 2.7. *The transition function of $\mathcal{D}'_{ISCA}$ equals the transition function of $\mathcal{D}'_{SC}$.*

Proof. According to Lemma 2.5, $\|d_0^{\mathcal{D}'_{ISCA}}\| = \|d_0^{\mathcal{D}'_{SC}}\|$. Consider two states, $d^{\mathcal{D}'_{ISCA}}$ in $\mathcal{D}'_{ISCA}$ and $d^{\mathcal{D}'_{SC}}$ in $\mathcal{D}'_{SC}$, such that $\|d^{\mathcal{D}'_{ISCA}}\| = \|d^{\mathcal{D}'_{SC}}\|$. On the one hand, assume $d^{\mathcal{D}'_{SC}} \xrightarrow{\ell} d'^{\mathcal{D}'_{SC}}$ in $\mathcal{D}'_{SC}$. According to *SC*, there exists a transition $n \xrightarrow{\ell} n'$ in $\mathcal{N}'$ such that $n \in \|d^{\mathcal{D}'_{SC}}\|$ and $\|d'^{\mathcal{D}'_{SC}}\| = \ell\text{-closure}(\|d^{\mathcal{D}'_{SC}}\|)$. Based on Lemma 2.4, this entails that there exists a transition $d^{\mathcal{D}'_{ISCA}} \xrightarrow{\ell} d'^{\mathcal{D}'_{ISCA}}$ in $\mathcal{D}'_{ISCA}$ such that, because of Lemma 2.2, $\|d'^{\mathcal{D}'_{ISCA}}\| = \|d'^{\mathcal{D}'_{SC}}\|$. Lemma 2.6 ensures that this transition is unique in $\mathcal{D}'_{ISCA}$. On the other hand, assume $d^{\mathcal{D}'_{ISCA}} \xrightarrow{\ell} d'^{\mathcal{D}'_{ISCA}}$ in $\mathcal{D}'_{ISCA}$. According to Lemma 2.2 and *SC*, there exists a transition $d^{\mathcal{D}'_{SC}} \xrightarrow{\ell} d'^{\mathcal{D}'_{SC}}$ in $\mathcal{D}'_{SC}$ such that $\|d'^{\mathcal{D}'_{SC}}\| = \|d'^{\mathcal{D}'_{ISCA}}\|$. In summary, the set of transitions exiting $d^{\mathcal{D}'_{ISCA}}$ equals the set of transitions exiting $d^{\mathcal{D}'_{SC}}$. The conclusion follows from induction based on the (common) initial state (Lemma 2.5).

Corollary 2.1. *The set of states of $\mathcal{D}'_{ISCA}$ equals the set of states of $\mathcal{D}'_{SC}$.*

Lemma 2.8. *The set of final states of $\mathcal{D}'_{ISCA}$ equals the set of final states of $\mathcal{D}'_{SC}$.*

Proof. At line 28, the set of final states F_{d} of $\mathcal{D}_0 = \mathcal{D}_{SC}$ is updated based on $\Delta F_{\mathrm{n}}^{-}$. This way, $\mathcal{D}_0$ becomes the DFA generated by *SC* based on $\mathcal{N}$, with the set of final states being $F_{\mathrm{n}} - \Delta F_{\mathrm{n}}^{-}$. Then, the maintenance of F_{d} is performed by the function *Expand*, specifically:

1. After expanding $\|d\|$ at line 16, by possibly inserting a new final state into F_{d}.
2. After merging d and d' at line 20, by possibly removing a final state from F_{d}.

Therefore, since $\mathcal{D}_m = \mathcal{D}'_{ISCA}$, the latter will include the same final states as $\mathcal{D}_{SC}$.

Finally, the proof of Theorem 2.1 follows from Lemma 2.1, Corollary 2.1, Lemma 2.7, Lemma 2.5, and Lemma 2.8. □

2.8 Bibliographic Notes

The need for incremental determinization of finite automata originated in the context of monitoring-based diagnosis of active systems (called *reactive diagnosis* in Chapter 7), which requires the incremental generation of the *index space* (introduced in Chapter 5). This incremental technique was then generalized in [112] into an algorithm for incremental determinization of cyclic automata called *ISC*. At first sight, *ISC* looks more powerful than *ISCA*, inasmuch as it deals with cyclic automata. However, the algorithm was then discovered to generate DFAs with spurious states, which are disconnected from the initial state (see [78] for details). Since spurious states are irrelevant to the DFA, the practical problem is that the disconnected part may require considerable processing during the execution of incremental determinization, which is eventually wasted. To cope with the disconnection problem of cyclic automata, an algorithm called *RISC* (*Revisited ISC*) was proposed in [6], which may operate either in *lazy* or in *greedy* mode. In lazy mode, it behaves like *ISCA* and, at the end of the processing, garbage collection is applied, where the disconnected part of the DFA is removed. However, similarly to *ISC*, this approach does not eliminate the problem of wasteful processing of spurious states. To avoid such a waste of processing, *RISC* may operate in greedy mode, where the goal is to avoid a priori the disconnection of the DFA. In so doing, it exploits the distance of states from the initial state, as detailed in [7]. However, despite guaranteeing the connection of the DFA, no proof of correctness has been provided, with termination of the algorithm being purely conjectural. Hence, a new algorithm, called *IDEA* (*Incremental Determinization of Expanding Automata*), was proposed in [18], which guarantees not only the connection of the DFA but also termination. An algorithm for incremental determinization and minimization of finite automata, called *IDMA*, was presented in [79] and refined in [80].

2.9 Summary of Chapter 2

Language. A set of strings (sequences of symbols) on a given alphabet.

Regular Expression. A formalism to define a language on an alphabet. Starting from symbols of the alphabet and ε, operators of concatenation, alternative, and repetition are applied to subexpressions (just like arithmetic expressions are built upon numbers using arithmetic operators).

Regular Language. A language defined by a regular expression.

Finite Automaton (FA). An abstract machine characterized by a mathematical model of computation, defined by an alphabet, a (finite) set of states, a transition function, an initial state, and a set of final states.

Deterministic Finite Automaton (DFA). A finite automaton with a deterministic transition function, mapping each state–symbol pair to just one state.

Nondeterministic Finite Automaton (NFA). A finite automaton with a nondeterministic transition function. Nondeterminism is caused either by extending the domain of symbols with the null symbol ε (giving rise to ε-transitions) or by mapping a state–symbol pair into several states.

Transition Diagram. A graph-based representation of a finite automaton, where circles (marked with identifiers) denote states, while arcs (marked with symbols) denote transitions. The initial state is entered by a dangling arrow. Final states are denoted by double circles.

Determinization. The process of generating a DFA equivalent to (sharing the same language as) a given NFA.

Minimization. The process of generating the minimal DFA equivalent to (sharing the same language as) a given DFA.

NFA Expansion. The extension of an NFA by new (possibly final) states and transitions, where some final states in the original NFA may be qualified as nonfinal in the expanded NFA.

Incremental Determinization. The determinization of an expanded NFA based on the original NFA, the expansion, and the DFA equivalent to the original NFA.

Chapter 3
Active Systems

In this chapter we are concerned with the definition of a specific class of DESs called active systems. The notion of an active system is given in terms of formal properties rather than concrete, physical characteristics. Depending on the context and the nature of the task to be performed (simulation, supervision, diagnosis, etc.), a physical system can be modeled in different ways. For example, the protection apparatus of a power transmission network can be modeled as either a continuous system or a DES.

In fact, a physical system is not continuous or discrete in itself, although practical reasons may suggest a particular modeling choice. However, choosing between continuous and discrete modeling is only the first problem, as several abstraction levels can be considered in either case. Roughly, the more detailed the model, the more complicated the reasoning task of the diagnostic technique, which, in the worst case, is bound to become overwhelming in space and time complexity.

Unlike other approaches in the literature where DESs are synchronous, active systems are assumed to be *asynchronous*; this means that events generated by components are stored within connection links before being (asynchronously) consumed.

Thus, the modeling ontology involves two classes of elements: components and links. Topologically, an active system is a network of components that are connected to each other by means of links. The behavior of each component is specified by a finite automaton whose transitions between states are triggered by occurrences of events. Such events may come either from the external world or from neighboring components through links. The component model is assumed to be complete, thereby including both normal and faulty behaviors.

Unlike components, which exhibit active behavior, links are passive elements that incorporate the events exchanged between components.

Active systems are compositionally modeled in terms of the topology (the way components are connected to one another through links) and the behavior of each component (the way the component reacts to events) and of each link. Links may differ in the number of storable events and in the mode in which they behave when the buffer of events is full.

G. Lamperti et al., *Introduction to Diagnosis of Active Systems*,
https://doi.org/10.1007/978-3-319-92733-6_3

Therefore, the actual (global) model of an active system is implicitly defined by its topology, and by component and link models. Such a model is in turn a finite automaton, where a state encompasses component and link states (where a link state is represented by the queue of events stored in it), while events are component transitions.

3.1 Components

Components are the basic elements of active systems. Each component is characterized by a structure, called the *topological model*, and a mode in which it reacts to incoming events, called the *behavioral model.*

Topologically, a component is described by a set of input terminals and a set of output terminals, which can be thought of as pins through which events are fetched and generated, respectively.

Both normal and faulty behaviors of a component are uniformly described by a single finite automaton, including a set of states and a transition function. A state transition is triggered by an input event being ready on one input terminal. During the transition, before the new state is reached, a (possibly empty) set of output events are generated on output terminals.

The specification of the topological model and the behavioral model of a component can be abstracted from the specific component and unified into a single component model.

Definition 3.1 (Component Model). A *component model* is a communicating automaton

$$M_c = (S, E_{\text{in}}, I, E_{\text{out}}, O, \tau), \tag{3.1}$$

where S is the set of states, E_{in} is the set of inputs, I is the set of input terminals, E_{out} is the set of outputs, O is the set of output terminals, and τ is the (nondeterministic) transition function:

$$\tau : S \times (E_{\text{in}} \times I) \times 2^{(E_{\text{out}} \times O)} \mapsto 2^S . \tag{3.2}$$

An input or output event is a pair (e, ϑ), where e is an input or output and ϑ an input or output terminal, respectively. A *component* is the instantiation of a component model.

A transition t from a state s to a state s' is triggered by an input event (e,x) at an input terminal x and generates the (possibly empty) set of output events $\{(e_1,y_1),\ldots,(e_n,y_n)\}$ at output terminals $y_1,\ldots,y_n$, respectively, denoted by

$$t = s \xrightarrow{(e,x) \Rightarrow (e_1,y_1),\ldots,(e_n,y_n)} s' . \tag{3.3}$$

When (e,x) is consumed, s' is reached and the set of output events are generated at the relevant output terminals.

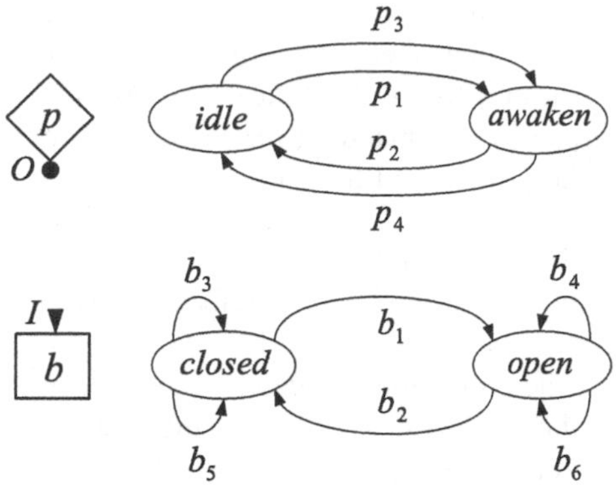

Fig. 3.1 Component models of protection device p (top) and breaker b (bottom); topological models are on the left-hand side; behavioral models are on the right-hand side; input and output terminals are denoted by a triangle and a bullet, respectively; details of component transitions are given in Table 3.1

Note 3.1. A transition is always triggered by an input event, that is, the input event cannot be missing. However, the set of output events can be empty. In such a case, the transition is simply denoted by $s \xrightarrow{(e,x)} s'$.

An input *virtual terminal* called *In* is implicitly defined, through which the component can collect events coming from the external world.

Table 3.1 Details of behavioral models outlined on the right-hand side of Fig. 3.1.

Component	*Transition*	*Action of component*
	$p_1 = idle \xrightarrow{(sh,In)\Rightarrow(op,O)} awaken$	Detects short circuit and outputs *op* event
p	$p_2 = awaken \xrightarrow{(ok,In)\Rightarrow(cl,O)} idle$	Detects end of short circuit and outputs *cl* event
	$p_3 = idle \xrightarrow{(sh,In)\Rightarrow(cl,O)} awaken$	Detects short circuit, yet outputs *cl* event
	$p_4 = awaken \xrightarrow{(ok,In)\Rightarrow(op,O)} idle$	Detects end of short circuit, yet outputs *op* event
	$b_1 = closed \xrightarrow{(op,I)} open$	Consumes *op* event and opens
	$b_2 = open \xrightarrow{(cl,I)} closed$	Consumes *cl* event and closes
b	$b_3 = closed \xrightarrow{(op,I)} closed$	Consumes *op* event, yet remains closed
	$b_4 = open \xrightarrow{(cl,I)} open$	Consumes *cl* event, yet remains open
	$b_5 = closed \xrightarrow{(cl,I)} closed$	Consumes *cl* event
	$b_6 = open \xrightarrow{(op,I)} open$	Consumes *op* event

Example 3.1 (Component). Displayed in Fig. 3.1 are the models of a protection device p (top) and a breaker b (bottom). Considering the topological models (left), protection device p has no input terminals and one output terminal O, while breaker b has one input terminal I and no output terminals.[1]

[1] In the protection system, a link will connect output terminal O of p to input terminal I of b.

We assume that these components are devoted to the protection of a transmission line in a power network when a short circuit occurs in the line (for instance, as a result of a lightning strike). The protection device is designed to react to a short circuit by commanding the breaker to open, so that the line can be isolated. This isolation will end the short circuit, and then the protection device commands the breaker to close, so as to reconnect the line to the network.

The behavioral models of p and b are outlined on the right-hand side of Fig. 3.1. Details of the component transitions are listed in Table 3.1. Both behavioral models include two states.

The protection device can be either in the *idle* state (no short circuit detected) or the *awaken* state (short circuit detected). When a short circuit is detected, p can perform either transition p_1 or transition p_3. In the case of p_1, an *op* (open) output event is generated at terminal O. In the case of p_3, p (abnormally) generates a *cl* (close) event. When the short circuit ends, p can perform either transition p_2 or transition p_4. In the case of p_2, a *cl* (close) event is generated. In the case of p_4, p (abnormally) generates an *op* (open) event.

The breaker can be either *closed* or *open*. When it is closed and an *op* event is ready at input terminal I, the breaker b can perform either transition b_1 (it opens) or transition b_3 (it remains closed). Dually, when it is open and a *cl* event is ready, b can perform either transition b_2 (it closes) or transition b_4 (it remains open). Furthermore, b can receive either a *cl* event when it is already closed (transition b_5) or an *op* event when it is already open (transition b_6).

3.2 Links

In active systems, components are connected to each other by means of links. Each link exits an output terminal o of a component c and enters an input terminal of another component c'. This way, when an output is generated at o by c, it is queued within the link and, if it is at the head of the queue, it becomes a triggering event for component c'.

Similarly to components, links are characterized by a relevant model, which is an abstraction of the actual link. As such, several links may share the same model.

Definition 3.2 (Link Model). A *link model* is a quadruple

$$M_l = (x, y, z, w), \tag{3.4}$$

where x is the *input terminal*, y is the *output terminal*, z is the *size*, and w is the *saturation policy*.

A *link* l is an instantiation of a link model, that is, a directed communication channel between two different components c and c', where an output terminal y of c and an input terminal x of c' coincide with the input and output terminals, respectively, of

l. Specifically, $y(c) \Rightarrow x(c')$ denotes a link through which each output e generated at output terminal y of c is available (after a certain time[2]) at input terminal x of c'.

When it is within link l, event e is said to be a *queued event* of l. The size z of l is the maximum number of events that can be queued in l.

The actual sequence of queued events of l is the *configuration* of l, denoted by $\|l\|$. The cardinality of $\|l\|$ is denoted by $|l|$. If $|l|$ equals the size z then l is *saturated*.

The first consumable event in $\|l\|$ is the *ready event*. When a ready event is consumed, it is dequeued from $\|l\|$. Events are consumed one at a time.

When l is saturated, the semantics for the triggering of a transition t of a component c that generates a new output event (e, y) on l is dictated by the saturation policy w of l, which can be:

1. *LOSE*: e is lost,
2. *OVERRIDE*: e replaces the last event in $\|l\|$,
3. *WAIT*: transition t cannot be triggered until l becomes unsaturated, that is, until at least one event in $\|l\|$ is consumed.

Let $Head(\|l\|)$ denote the first consumable element in $\|l\|$. Let $Tail(\|l\|)$ be the queue of events in $\|l\|$ following the first event. Let $App(\|l\|, e)$ denote the queue of events obtained by appending e to $\|l\|$. Let $Repl(\|l\|, e)$ denote the queue of events obtained by replacing the last event in $\|l\|$ by e. Based on these functions, the queue resulting from the insertion of event e into $\|l\|$ is defined as follows:

$$Ins(\|l\|, e) = \begin{cases} App(\|l\|, e) & \text{if } |l| < z, \\ \|l\| & \text{if } |l| = z, w = LOSE, \\ Repl(\|l\|, e) & \text{if } |l| = z, w = OVERRIDE \,. \end{cases} \tag{3.5}$$

In other words, if l is not saturated, e is queued into l. If, instead, l is saturated, two cases are possible, depending on the saturation policy. If the policy is *LOSE*, event e is lost and the configuration of l does not change. If it is *OVERRIDE*, the last event in l is replaced by e.

A transition $s \xrightarrow{(e,x) \Rightarrow (e_1,y_1),\ldots,(e_n,y_n)} s'$ is said to be *triggerable* iff event e is ready at input terminal x (e is the first event within the link entering x), and for each output terminal y_i, $i \in [1 \cdots n]$, if the saturation policy of the link l_i exiting y_i is *WAIT* then $\|l\|$ is empty.

Example 3.2 (Link). With reference to the component models displayed in Fig. 3.1, Fig. 3.2 shows a link pb connecting output terminal O of protection device p to input terminal I of breaker b, where we assume a size $z = 1$ and a saturation policy $w = WAIT$.

[2] No assumption is made about the length of this time.

Fig. 3.2 Active system $\mathcal{A}_1$

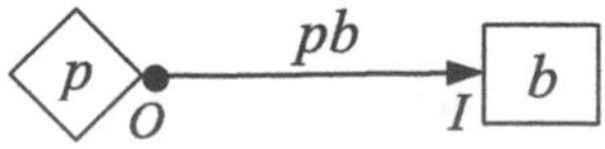

3.3 Active System

Roughly, an *active system* (or simply a *system*) is a network of components which are connected to one another by means of links. Each component and each link in the system are characterized by a relevant model. Several components, and similarly several links, may share the same model. Component input terminals cannot be overloaded, that is, at most one link is connected to each of the component input terminals in the system.

A system may incorporate a number of component terminals that are not connected to any link in the system. These are called the *dangling terminals* of the system.

Definition 3.3 (Active System). An active system $\mathcal{A}$ is a triple

$$\mathcal{A} = (C, L, D), \tag{3.6}$$

where C is the set of components, L is the set of links among the terminals of the components in C, and D is the set of dangling terminals, with D being the union of two disjoint sets:

$$D = D_{\text{on}} \uplus D_{\text{off}}, \tag{3.7}$$

where D_{on} is the set of *on-terminals* and D_{off} is the set of *off-terminals.*

If $D_{\text{on}} \neq \emptyset$, the system $\mathcal{A}$ is *open*, otherwise $\mathcal{A}$ is *closed*. If $\mathcal{A}$ is closed ($D_{\text{on}} = \emptyset$), no events are available at the input terminals in D_{off}, while events generated at output terminals in D_{off} are lost.

If, instead, $\mathcal{A}$ is open, we assume that the dangling terminals in D_{on} are connected to links outside $\mathcal{A}$; in other words, $\mathcal{A}$ is supposed to be incorporated within another (larger, virtually unknown) system. Therefore, events generated at output terminals in D_{on} are buffered within the relevant (unknown) link external to $\mathcal{A}$. Similarly, a component corresponding to an input terminal in D_{on} is assumed to be sensitive to events ready at this terminal. In other words, state transitions may be triggered by events buffered in the (external) link connected to this terminal.

Example 3.3 (Active System). With reference to the components p (protection device) and b (breaker) displayed in Fig. 3.1, and the link pb defined in Example 3.2, Fig. 3.2 shows an active system $\mathcal{A}_1 = (\{p, b\}, \{pb\}, \emptyset)$ with no dangling terminals.

3.4 Trajectory

An active system $\mathcal{A}$ can be thought of as a machine that can be either *quiescent* or *reacting*. If it is quiescent, components in $\mathcal{A}$ do not perform any transitions, as no consumable event is ready on the input terminals. Upon the occurrence of an event, coming either from the external world or from a dangling input terminal in D_{on}, $\mathcal{A}$ becomes reacting. Since the behavior of $\mathcal{A}$ is asynchronous, such a reaction, called a *trajectory*, consists of a sequence of transitions performed by components in $\mathcal{A}$.

Each component transition moves $\mathcal{A}$ from one state to another, where a state of $\mathcal{A}$ is identified by component states and link configurations. In other words, a state of $\mathcal{A}$ is a pair (S, Q), where S is an n-tuple of states of components in $\mathcal{A}$, while $\mathcal{Q}$ is an m-tuple of configurations of links in $\mathcal{A}$.

Therefore, a transition T of $\mathcal{A}$, from state (S, Q) to state (S', Q'), triggered by the occurrence of a transition t of component c, can be written as

$$T = (S, Q) \xrightarrow{t(c)} (S', Q') \, . \tag{3.8}$$

Note 3.2. Since the identifier c of a component is unique within $\mathcal{A}$ and t is unique within the model of c, it follows that $t(c)$ is the unique identifier of a component transition.

Assuming that $a_0 = (S_0, Q_0)$ is the initial (quiescent) state of $\mathcal{A}$, the trajectory of $\mathcal{A}$ starting at a_0 is the sequence of states of $\mathcal{A}$ determined by the occurrences of triggerable component transitions $t_1, \ldots, t_k$, namely:

$$h = a_0 \xrightarrow{t_1(c_1)} a_1 \xrightarrow{t_2(c_2)} a_2 \cdots \xrightarrow{t_k(c_k)} a_k \, . \tag{3.9}$$

Since each $t_i(c_i)$, $i \in [1 \cdots k]$, identifies a component transition, state a_i uniquely depends on the pair $(a_{i-1}, t_i(c_i))$. Assuming the initial state a_0, this allows us to identify the trajectory defined in (3.9) by the sequence of component transitions only:

$$h = [t_1(c_1), t_2(c_2), \ldots, t_k(c_k)] \, . \tag{3.10}$$

Example 3.4 (Trajectory). With reference to the active system $\mathcal{A}_1$ defined in Example 3.3 and displayed in Fig. 3.2, assume that the initial state of $\mathcal{A}_1$ is the triple $(i, c, [\,])$, where the first two elements are the states[3] of p and b, respectively, while the third element denotes the configuration of the link pb (where $[\,]$ stands for the empty sequence of events). A possible trajectory for $\mathcal{A}_1$ is

$$h = (i, c, [\,]) \xrightarrow{p_1(p)} (a, c, [op]) \xrightarrow{b_1(b)} (a, o, [\,]) \xrightarrow{p_2(p)} (i, o, [cl]) \xrightarrow{b_2(b)} (i, c, [\,]),$$

which corresponds to the following evolution:

[3] Identifiers of states are denoted by the initial letter of the state: i stands for *idle* and c stands for *closed*.

1. The protection device detects a short circuit and commands the breaker to open,
2. The breaker opens,
3. The protection device detects the end of the short circuit and commands the breaker to close,
4. The breaker closes.

Note 3.3. The final state of the trajectory h equals the initial state, where the protection device p is *idle*, the breaker b is *closed*, and the link pb is empty. Such cyclicity entails that the active system $\mathcal{A}_1$ can have an infinite number of trajectories (each one obtained by performing the cycle a different number of times).

3.5 Behavior Space

Given an active system $\mathcal{A}$ and its initial state a_0, several (even an infinite number of) trajectories are possible for $\mathcal{A}$. This proliferation of possible trajectories is determined by the different modes in which events can be (asynchronously) consumed at input terminals. Moreover, because of nondeterminism in component models, different component transitions are equally triggerable by the same event.

Just as the set of possible trajectories of a single component c (sequences of transitions of c) is captured by the model of c, so the set of possible trajectories of $\mathcal{A}$ can be captured by a finite automaton, called a *behavior space* of $\mathcal{A}$, whose alphabet is the whole set of component transitions.

Definition 3.4 (Behavior Space). Let $\mathcal{A} = (C, L, D)$ be an active system, where $C = \{c_1, \ldots, c_n\}$ and $L = \{l_1, \ldots, l_m\}$. A *behavior space* of $\mathcal{A}$ is a DFA

$$Bsp(\mathcal{A}) = (\Sigma, A, \tau, a_0), \tag{3.11}$$

where:

1. Σ is the alphabet, consisting of the union of the transitions of components in C,
2. A is the set of states (S, Q), with $S = (s_1, \ldots, s_n)$ being a tuple of states of all components in C, and $Q = (q_1, \ldots, q_m)$ being a tuple of configurations of all links in L,
3. $a_0 = (S_0, Q_0)$ is the initial state,
4. τ is the transition function $\tau : A \times \Sigma \mapsto A$, such that $(S, Q) \xrightarrow{t(c)} (S', Q') \in \tau$, where $S = (s_1, \ldots, s_n)$, $Q = (q_1, \ldots, q_m)$, $S' = (s'_1, \ldots, s'_n)$, $Q' = (q'_1, \ldots, q'_m)$, and

$$t(c) = s \xrightarrow{(e,x) \Rightarrow (e_1,y_1),\ldots,(e_p,y_p)} s' \tag{3.12}$$

iff:

(4a) Either $x = In$, or $x \in D_{\text{on}}$, or e is ready at terminal x within a link denoted l^x,

(4b) For each $i \in [1 \cdots n]$, we have

$$s'_i = \begin{cases} s' & \text{if } c_i = c, \\ s_i & \text{otherwise}. \end{cases} \tag{3.13}$$

(4c) Based on (3.5), for each $j \in [1 \cdots m]$, we have

$$q'_j = \begin{cases} Tail(q_j) & \text{if } l_j = l^x, \\ Ins(q_j, e_k) & \text{if } l_j \text{ exits terminal } y_k, k \in [1 \cdots p], \\ q_j & \text{otherwise}. \end{cases} \tag{3.14}$$

Example 3.5 (Behavior Space). The behavior space of the active system $\mathcal{A}_1$ introduced in Example 3.3 is displayed in Fig. 3.3, where names of states are shortened to initial letters ($i = idle$, $a = awaken$). Each state a_i, $i \in [1 \cdots 11]$, is marked by a triple indicating the state of the protection device p, the state of the breaker b, and the configuration of the link pb.[4] The initial state a_0 is assumed to be $(idle, closed, [\,])$. Based on the component models outlined in Fig. 3.1 and detailed in Table 3.1, starting from the initial state, the active space is generated by considering the set of component transitions that are triggerable at the state considered. For instance, the triggerable transitions at state a_0 are p_1 and p_3. With p_1, the system $\mathcal{A}_1$ performs the transition $a_0 \xrightarrow{p_1} a_1$, with $a_1 = (awaken, closed, [op])$. With p_3, $\mathcal{A}_1$ performs the transition $a_0 \xrightarrow{p_3} a_3$, with $a_3 = (awaken, closed, [cl])$. When new states are generated (e.g., a_1 and a_3), these must be considered in turn until no new state is generated. Since the saturation policy of link pb is *WAIT*, if pb is not empty at a given system state a_i then no transition of p is triggerable at a_i. For instance, at state a_1, transitions p_3 and p_4 of the protection device are not triggerable, because link pb is not empty (it incorporates the event op).

Note 3.4. The behavior space $Bsp(\mathcal{A}_1)$ displayed in Fig. 3.3 includes the path

$$a_0 \xrightarrow{p_1(p)} a_1 \xrightarrow{b_1(b)} a_4 \xrightarrow{p_2(p)} a_6 \xrightarrow{b_2(b)} a_0,$$

which equals the trajectory h introduced in Example 3.4. This is no coincidence, as stated by Proposition 3.1.

Proposition 3.1. *The language of the behavior space $Bsp(\mathcal{A})$ with initial state a_0 equals the set of possible trajectories of $\mathcal{A}$ starting at a_0.*

Proof. First we show that $h \in Lang(Bsp(\mathcal{A})) \implies h$ is a trajectory of $\mathcal{A}$, by induction on the sequence of system transitions in h.

(*Basis*) *The string h starts at the initial state a_0, the same state from which a trajectory of $\mathcal{A}$ starts.* This is true by assumption.

(*Induction*) *If $a \xrightarrow{t(c)} a'$ is a transition in $Bsp(\mathcal{A})$, then it is also a system transition of a trajectory of $\mathcal{A}$.* Based on the definition of a transition function in a behavior

[4] Since, based on Example 3.2, the size of the link pb is 1, the configuration of pb can be either the empty sequence $[\,]$ or a single event generated by p (either op or cl).

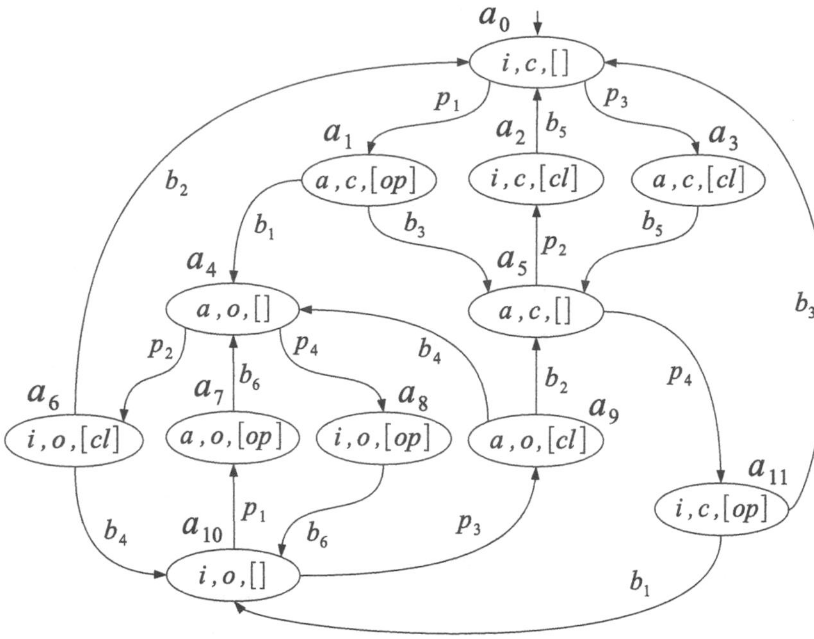

Fig. 3.3 Behavior space of active system $\mathcal{A}_1$

space, $t(c)$ must be a triggerable transition of component c at system state $a = (S, Q)$, with $a' = (S', Q')$ being the new state of $\mathcal{A}$, where S' differs from S in the state reached by $t(c)$ for component c, and Q' is the new configuration of links obtained by the consumption of input events of $t(c)$ and the possible generation of output events. Therefore, $a \xrightarrow{t(c)} a'$ is also a transition of a trajectory of $\mathcal{A}$.

Then, we have to show that h is a trajectory of $\mathcal{A} \implies h \in Lang(Bsp(\mathcal{A}))$. This can be proven by induction on the sequence of system transitions in h in a way similar to that presented for the first entailment. □

Note 3.5. The notion of a behavior space is presented here for formal reasons only. In Example 3.5 the behavior space of system $\mathcal{A}_1$ is actually generated. After all, $\mathcal{A}_1$ is very small, incorporating just two components and one link. For real active systems, the generation of the active space may become impossible because of the huge number of states. Therefore, a strong assumption in the diagnosis techniques presented in this book is the unavailability of the behavior space.

3.6 Bibliographic Notes

Active systems stem from the application domain of model-based diagnosis of power networks [8, 9, 75]. Specifically, the term *active system* was first introduced in [10]. Active systems with both synchronous and asynchronous behavior were proposed in [84], and subsequently analyzed in the context of monitoring-based diagnosis in [93].

3.7 Summary of Chapter 3

Component. An artifact equipped with a set of input terminals and a set of output terminals, whose behavior is defined by a finite automaton in which transitions are triggered by events that are ready at input terminals. The change of state following a transition is possibly accompanied by the generation of events at output terminals.

Virtual Terminal. An implicit input terminal, named *In*, devoted to the reception of events coming from the external world.

Link. A channel connecting the output terminal of a component to the input terminal of another component. Events generated at the output terminal are queued within the relevant link, and thereafter consumed in sequential order. The configuration of a link is the sequence of events stored in the link. A link is characterized by a size, the maximum number of storable events, and a saturation policy, establishing what happens when one attempts to insert a new event into a saturated (full) link.

Active System. A set of components connected by a set of links. Possibly, a number of component terminals may be dangling, that is, they are not connected to any link of the active system. However, a dangling terminal may be connected to a link outside the active system (in such a case, the active system is a subpart of a larger system). Starting from the initial state, an active system may change state in a manner corresponding to its component transitions. A state of an active system is defined by the states of its components and the configurations of its links.

Trajectory. A sequence of transitions of an active system starting from the initial state.

Behavior Space. A finite automaton whose language equals the set of all possible trajectories of an active system starting from the initial state.

Chapter 4
Diagnosis Problems

In this chapter we are concerned with the notion of a diagnosis problem. Roughly, a diagnosis problem relevant to an active system $\mathcal{A}$ involves a temporal observation of $\mathcal{A}$, relevant information on the observability of $\mathcal{A}$, and a discrimination between normal and faulty behavior of components in $\mathcal{A}$.

According to Chapter 3, $\mathcal{A}$ is made up of components which are connected to one another through links. The complete behavior of each component is modeled by a communicating automaton whose state transitions are triggered by events coming either from the external environment or from links.

Assuming that $\mathcal{A}$ is quiescent (empty links) in the initial state, it may become reacting on the occurrence of an external event. When it is reacting, $\mathcal{A}$ performs a sequence of system transitions which form a trajectory, where each system transition is triggered by a component transition. In other words, the occurrence of a transition of a component causes the occurrence of a transition of the system $\mathcal{A}$. This way, the trajectory of $\mathcal{A}$ is identified by the sequence of component transitions occurring in $\mathcal{A}$. Since each component is modeled by its complete behavior, a trajectory of $\mathcal{A}$ carries information on how its components have behaved, either normally or abnormally.

Example 4.1. With reference to the active system $\mathcal{A}_1$ introduced in Example 3.3, based on the behavior space $Bsp(\mathcal{A}_1)$ displayed in Fig. 3.3, we can consider two possible trajectories, namely $h_1 = [p_1, b_1, p_2, b_2]$ and $h_2 = [p_1, b_1, p_4, b_6]$. According to the component models of the protection device p and breaker b displayed in Fig. 3.1 and detailed in Table 3.1, the behavior of both components in trajectory h_1 is normal, while in h_2 the protection device is faulty because of transition p_4, which sends the wrong command to the breaker, namely *op* rather than *cl*.

Since behavioral models do not make any explicit distinction between normal and abnormal behavior, in order to find out whether a trajectory of $\mathcal{A}$ is normal or faulty, additional specification is required.

A further problem is related to the observability of $\mathcal{A}$, as, generally speaking, a trajectory is not perceived as it occurs, but only as a projection on a domain of observable labels. Each component transition is either observable or unobservable: if it

G. Lamperti et al., *Introduction to Diagnosis of Active Systems*,
https://doi.org/10.1007/978-3-319-92733-6_4

is unobservable, no sign of its occurrence is left for the observer; if it is observable, the component transition is perceived as a specific (visible) label. Consequently, a trajectory h of $\mathcal{A}$ is perceived as a sequence of observable labels, the *trace* of h.

Note 4.1. Several (possibly infinite) trajectories may be projected onto the same trace. Consequently, a trace does not identify one trajectory of $\mathcal{A}$. Rather, it identifies a (possibly infinite) set of candidate trajectories.

Example 4.2. With reference to the trajectories h_1 and h_2 in Example 4.1, assuming that transitions b_1 and b_2 are observable in terms of observable labels *opens* and *closes*, respectively, while p_1 and p_3 are both observable in terms of an observable label *awakes*, with all other transitions being unobservable, the trace of h_1 will be $[awakes, opens, closes]$, while for h_2 the trace will be $[awakes, opens]$.

Note 4.2. Generally speaking, the observable label associated with an observable component transition does not identify the transition, as different component transitions may be associated with the same observable label.

Example 4.3. Based on the observable labels defined in Example 4.2 and the behavior space displayed in Fig. 3.3, the trace $[awakes]$ is consistent with several candidate trajectories, including (among others) $[p_1, b_3, p_2, b_5]$, $[p_3, b_5, p_2, b_5]$, and $[p_3, b_5, p_4, b_3]$.

Since behavioral models do not make any distinction between observable and unobservable transitions, additional specification is required, establishing which transitions are observable, along with relevant observable labels.

To further complicate the task of linking what is observed to the actual trajectory of $\mathcal{A}$, there is the possible alteration of a trace, typically because of noise in the transmission channel and the distribution of sensors. As a result, rather being a temporally ordered sequence of observable labels, the trace may be perceived by the observer as a *temporal observation*, a directed acyclic graph (DAG) where nodes are marked by sets of candidate observable labels, while arcs denote partial (rather than total) temporal ordering between nodes: the temporal observation is a *relaxation* of the trace.

Example 4.4. Consider the trace $\mathcal{T} = [awakes, opens, closes]$ introduced in Example 4.2. Displayed in Fig. 4.1 is a temporal observation $\mathcal{O}$ obtained as a relaxation of $\mathcal{T}$, which is composed of nodes n_1, n_2, and n_3. Each node is marked by a set of candidate labels and represents a relaxation of an observable label in $\mathcal{T}$. Among this set is the actual label.[1] This is called *label relaxation.*

Within a node, labels other than the actual labels are *spurious*, resulting from noise in the transmission of the trace. Specifically, n_1 is a relaxation of the first label *awakes*, n_2 is a relaxation of the second label *closes*, and n_3 is a relaxation of the third label *closes*. The special symbol ε indicates the *null* label. When a node

[1] The actual label is known only when the set of candidate labels is a singleton. When several candidate labels are involved, the actual label remains unknown to the observer.

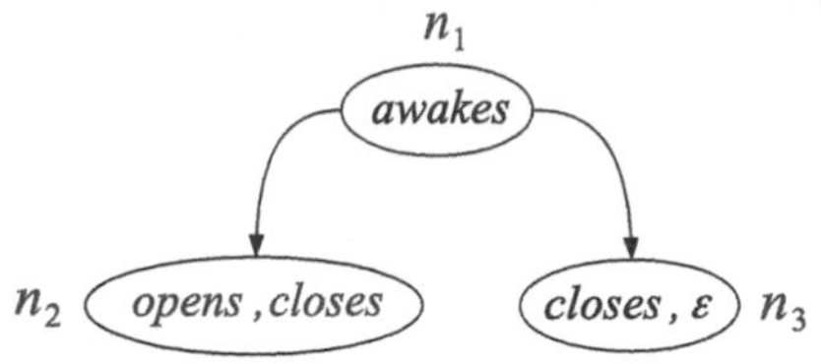

Fig. 4.1 Temporal observation $\mathcal{O}$, which is a relaxation of trace $\mathcal{T} = [awakes, opens, closes]$

includes ε as the actual label, this means that there is no corresponding observable label in the trace. This is called *node relaxation*, where the node incorporating the actual null label is *spurious*.[2] Therefore, there is an isomorphism between $\mathcal{T}$ and the set of nodes involving an actual label different from ε. In our example, $[awakes, opens, closes]$ is isomorphic to $[n_1, n_2, n_3]$.

Relaxation applies to temporal constraints too, namely in the form of *temporal relaxation*. In trace $\mathcal{T}$, the observable labels are totally ordered: first *awakes*, then *opens*, and finally *closes*. Instead, within temporal observation $\mathcal{O}$, the nodes corresponding to labels in $\mathcal{T}$ are related by a partial temporal ordering determined by arcs. Specifically, although n_1 is the first node, the temporal ordering between n_2 and n_3 is unknown. In other words, the total temporal ordering between labels in $\mathcal{T}$ is relaxed to a partial temporal ordering between the corresponding nodes in $\mathcal{O}$. Rather being totally ordered, as in $[n_1, n_2, n_3]$, the nodes in $\mathcal{O}$ are only partially ordered. The arcs in $\mathcal{O}$ indicate that n_1 comes before both n_2 and n_3. However, the temporal precedence of n_2 over n_3 is missing.

Despite the relaxation of a trace $\mathcal{T}$ into a temporal observation $\mathcal{O}$, the former is still present in $\mathcal{O}$ as a *candidate trace*. A candidate trace can be distilled from $\mathcal{O}$ by selecting a candidate label in each node of $\mathcal{O}$ based on the temporal constraints imposed by arcs. Among such candidates is the actual trace $\mathcal{T}$.

Example 4.5. With reference to the observation $\mathcal{O}$ defined in Example 4.4 and displayed in Fig. 4.1, the trace $\mathcal{T}$ can be distilled from $\mathcal{O}$ by selecting the label *awakes* from n_1 (the only possible label), *opens* from n_2, and *closes* from n_3. Notice how the chosen sequence $[n_1, n_2, n_3]$ fulfills the temporal constraints in $\mathcal{O}$. However, four additional candidate traces can equally be distilled from $\mathcal{O}$, namely $[awakes, opens, \varepsilon] = [awakes, opens]$, $[awakes, closes, closes]$, $[awakes, closes, \varepsilon] = [awakes, closes]$, and $[awakes, closes, opens]$.[3]

In the rest of the chapter we formalize the concepts we have informally introduced so far. Eventually, we provide a formal definition of a diagnosis problem and a relevant solution. To this end, we refer to the active system $\mathcal{A}_2$ introduced in the following example.

Example 4.6. Displayed in Fig. 4.2 is an active system called $\mathcal{A}_2$, which is obtained as an extension of the system $\mathcal{A}_1$ in Fig. 3.2. Specifically, the protection device p is

[2] Node relaxation does not occur in our example, as the symbol ε in node n_3 is not the actual label.

[3] The same candidate trace may be distilled in different ways, for example $[awakes, closes]$.

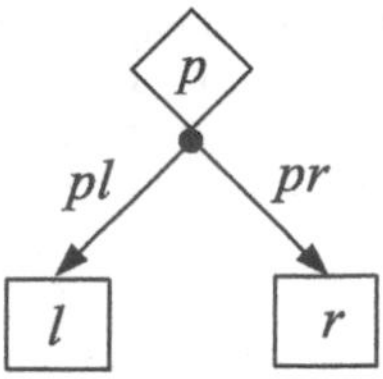

Fig. 4.2 Active system $\mathcal{A}_2$

connected to two breakers, namely l (left) and r (right), by means of links pl and pr, respectively. This may be a good solution for improving the reliability of the protection apparatus: if one breaker fails to open and the other opens correctly, then the corresponding side of the line is isolated, despite the faulty breaker.[4] Based on the behavioral models of the protection device and breaker outlined in Fig. 3.1 and detailed in Table 3.1, when the protection device p generates an output event (either op or cl), the latter is queued in both of the links pl and pr, which are assumed to have size 1 and a saturation policy *WAIT*.[5]

4.1 Viewer

As pointed out above, the modeling of an active system $\mathcal{A}$ does not provide any information on the observability of $\mathcal{A}$. Providing information about observability means defining, for each component in $\mathcal{A}$, which transitions are observable and which ones are unobservable. A transition is observable if it generates an observable label when it occurs, otherwise the transition is unobservable. Therefore, specifying the observability of $\mathcal{A}$ amounts to defining a mapping from observable transitions to corresponding observable labels. This is captured by the notion of a viewer introduced in Def. 4.1.

Definition 4.1 (Viewer). Let $\mathcal{A}$ be an active system and Ω a domain of observable labels. A *viewer* $\mathcal{V}$ for $\mathcal{A}$ is a surjective mapping from a subset of component transitions in $\mathcal{A}$ to Ω.

Note 4.3. Since the mapping from observable transitions to Ω is surjective, each of the observable labels in Ω is associated in $\mathcal{V}$ with at least one of these transitions.

Example 4.7. Displayed in the first line of Table 4.1 is the specification of a viewer $\mathcal{V}$ for the active system $\mathcal{A}_2$ introduced in Example 4.6. Notice how only two breaker transitions are observable, namely b_1 (opening) and b_2 (closing), where each observable label in fact identifies the corresponding component transition. By contrast, all

[4] This redundancy may bring a problem when the breakers are commanded to close, as reconnection of the line occurs only if both breakers close.

[5] This means that the transition of the protection device can occur only if both links are empty.

Table 4.1 Viewer (observable labels in italic) and ruler $\mathcal{R}$ (fault labels in bold) for system $\mathcal{A}_2$

	$b_1(l)$	$b_2(l)$	$b_3(l)$	$b_4(l)$	$b_5(l)$	$b_6(l)$	p_1	p_2	p_3	p_4	$b_1(r)$	$b_2(r)$	$b_3(r)$	$b_4(r)$	$b_5(r)$	$b_6(r)$
$\mathcal{V}$	*opl*	*cll*					*awk*	*ide*	*awk*	*ide*	*opr*	*clr*				
$\mathcal{R}$			**nol**	**ncl**					**fop**	**fcp**			**nor**	**ncr**		

four transitions of the protection device p are observable. However, only two observable labels are involved, namely *awk* (for p_1 and p_3) and *ide* (for p_2 and p_4). In other words, two different transitions of p are associated with the same label.

Definition 4.2 (Trace). Let h be a trajectory of an active system $\mathcal{A}$, and $\mathcal{V}$ a viewer for $\mathcal{A}$. The *trace* of h based on $\mathcal{V}$, written $h_{[\mathcal{V}]}$, is the sequence of observable labels

$$h_{[\mathcal{V}]} = [\ell \mid t \in h, (t,\ell) \in \mathcal{V}] . \tag{4.1}$$

In other words, a trace of a trajectory h of an active system $\mathcal{A}$ is a projection of the observable transitions of h into the corresponding observable labels defined in a viewer $\mathcal{V}$ for $\mathcal{A}$.

4.2 Temporal Observation

As stated above, a trajectory of an active system $\mathcal{A}$ generates a trace, that is, a sequence of observable labels. However, what is actually perceived by the observer is a temporal observation, that is, a relaxation of the trace. The precise mode in which a trace may be relaxed into a temporal observation is presented in Def. 4.3. In the definition, the notion of *precedence* "$\prec$" between two nodes of a DAG is used. This is defined as the smallest relation satisfying the following conditions (where n, n', and n'' denote nodes, while $n \to n'$ denotes an arc from n to n'):

1. If $n \to n'$ is an arc, then $n \prec n'$,
2. If $n \prec n'$ and $n' \prec n''$, then $n \prec n''$.

Definition 4.3 (Temporal Observation). Let $\mathcal{A}$ be an active system, $\mathcal{V}$ a viewer with a domain Ω of observable labels, and $\mathcal{T} = [\ell_1, \ell_2, \ldots, \ell_k]$ a trace of $\mathcal{A}$. A *temporal observation* $\mathcal{O}$ of $\mathcal{A}$, derived from $\mathcal{T}$, is a directed acyclic graph (L, N, A), where $L \subseteq \Omega \cup \{\varepsilon\}$ is the set of labels, with ε being the *null* label; N is the set of nodes, with each node n being marked by a nonempty subset of L (excluding the singleton $\{\varepsilon\}$) denoted by $\|n\|$; and $A \subseteq N \times N$ is the set of arcs, inductively defined as follows.

(*Basis*) $\mathcal{O}_\mathcal{T} = (L_\mathcal{T}, N_\mathcal{T}, A_\mathcal{T})$ is a temporal observation, where

$L_\mathcal{T} = \{\ell_1, \ell_2, \ldots, \ell_k\}$,
$N_\mathcal{T} = \{n_1, n_2, \ldots, n_k\}$, where each n_i, $i \in [1 \cdots k]$, is marked by $\{\ell_i\}$,

$A_{\mathcal{T}} = \{n_1 \rightarrow n_2, n_2 \rightarrow n_3, \ldots, n_{k-1} \rightarrow n_k\}$.

(*Induction*) If $\mathcal{O} = (N, L, A)$ is a temporal observation, then the result of each of the following three *relaxation* operations (λ, τ, ν) is a temporal observation:

λ (*Label relaxation*): the content of one node is extended by labels in $\Omega \cup \{\varepsilon\}$,
τ (*Temporal relaxation*): this performs the following actions:

1. An arc $n \rightarrow n'$ is removed,
2. For each parent node n_p of n such that $n_p \nprec n'$, an arc $n_p \rightarrow n'$ is inserted,
3. For each child node n_c of n' such that $n \nprec n_c$, an arc $n \rightarrow n_c$ is inserted.

ν (*Node relaxation*): a new node n is inserted, marked by ε and one or more other labels in L, and possibly connected to other nodes so that:

1. For each pair of nodes n_1, n_2, with $n_1 \neq n$, $n_2 \neq n$, if $n_1 \prec n_2$ in the extended graph then $n_1 \prec n_2$ in the original graph,
2. Each arc $n_1 \rightarrow n_2$ in the extended graph, such that $n_1 \prec n'' \prec n_2$, is removed.

According to Def. 4.3, $\mathcal{O}_{\mathcal{T}}$ is a temporal observation isomorphic to trace $\mathcal{T}$, called a *plain* temporal observation; any application of a relaxation operator (λ, τ, or ν) to a temporal observation generates a temporal observation.

Example 4.8. Shown on the right side of Fig. 4.3 is a temporal observation for the active system $\mathcal{A}_2$ (introduced in Example 4.6). $\mathcal{O}_2$ is obtained as a relaxation of trace $\mathcal{T}_2 = [awk, opl, ide]$. Starting from the plain observation relevant to $\mathcal{T}_2$ (displayed on the left of the figure), $\mathcal{O}_2$ can be generated by three relaxation operations:

1. λ-relaxation: the content of node n_1 is extended by ε,
2. ν-relaxation: node n_4 is inserted between n_2 and n_3, with content $\{cll, \varepsilon\}$,
3. τ-relaxation: arc $n_4 \rightarrow n_3$ is removed, and arc $n_2 \rightarrow n_3$ is inserted.

This results in the temporal observation $\mathcal{O}_2$ displayed on the right of Fig. 4.3.

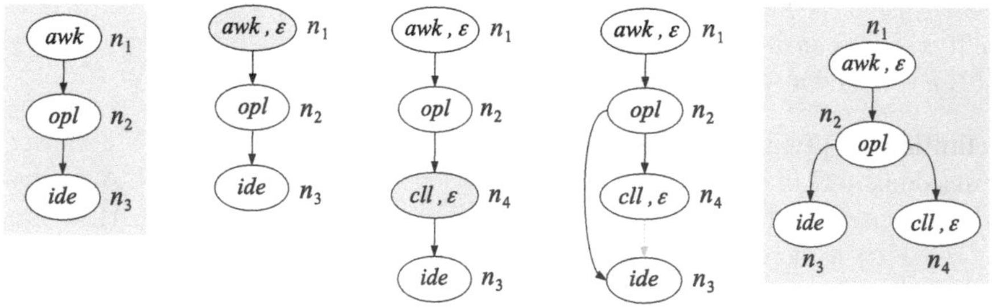

Fig. 4.3 Relaxation of trace $\mathcal{T}_2 = [awk, opl, ide]$ of system $\mathcal{A}_2$ into temporal observation $\mathcal{O}_2$

Even though the mode in which a trace is relaxed into a temporal observation is unknown to the observer, two properties hold for temporal observations, namely *canonicity* (Proposition 4.1) and *preservation* (Proposition 4.2).

Proposition 4.1. *Let $\mathcal{O} = (L, N, A)$ be a temporal observation. $\mathcal{O}$ is in* canonical form, *that is, if $n \rightarrow n' \in A$ then $\nexists n'' \in N$ such that $n \prec n'' \prec n'$.*

Proof. The proof is by induction on the sequence of relaxations applied to the trace $\mathcal{T}$ relevant to $\mathcal{O}$.

(*Basis*) If $\mathcal{O} = \mathcal{O}_{\mathcal{T}}$, then canonicity trivially holds.

(*Induction*) Assume that O_i is in canonical form, the temporal observation derived from $\mathcal{T}$ by i relaxations. We show that the application of a relaxation to $\mathcal{O}_i$ generates a temporal observation $\mathcal{O}_{i+1}$ in canonical form. In fact, in the case of λ-relaxation, the set of arcs is not affected. In the case of τ-relaxation, the removal of the arc $n \rightarrow n'$ does not affect canonicity. Besides, since the insertion of each arc $n_{\mathrm{p}} \rightarrow n'$ occurs only if $n_{\mathrm{p}} \not\prec n'$, canonicity is preserved. The same applies for arcs $n \rightarrow n_{\mathrm{c}}$. In the case of ν-relaxation, by definition, after the new node is connected to other nodes, each arc that violates canonicity is removed. Thus, in all cases, $\mathcal{O}_{i+1}$ is in canonical form. □

Definition 4.4 (Candidate Trace). Let $\mathcal{O} = (L, N, A)$ be a temporal observation. A *candidate trace* $\mathcal{T}_{\mathrm{c}}$ of $\mathcal{O}$ is a sequence of labels

$$\mathcal{T}_{\mathrm{c}} = [\ell \mid \ell \in \|n\|, n \in N], \tag{4.2}$$

where the nodes n are chosen to fulfill the partial order defined by arcs, and labels ε are removed. The *extension* of $\mathcal{O}$ is the set of candidate traces of $\mathcal{O}$, written $\|\mathcal{O}\|$.

Example 4.9. The temporal observation $\mathcal{O}_2$ displayed on the right of Fig. 4.3 involves six candidate traces, namely $[awk, opl, ide, cll]$, $[awk, opl, ide]$, $[opl, ide, cll]$. $[opl, ide]$, $[awk, opl, cll, ide]$, and $[opl, cll, ide]$.

Proposition 4.2. *If $\mathcal{O}$ is a temporal observation derived from trace $\mathcal{T}$, then $\mathcal{T} \in \|\mathcal{O}\|$.*

Proof. The proof is by induction on the sequence of relaxations applied to the trace $\mathcal{T}$ relevant to $\mathcal{O}$.

(*Basis*) If $\mathcal{O} = \mathcal{O}_{\mathcal{T}}$, then $\mathcal{T} \in \|\mathcal{O}\|$, as $\|\mathcal{O}\| = \{\mathcal{T}\}$.

(*Induction*) Assume $\mathcal{T} \in \|O_i\|$, where $\mathcal{O}_i$ is the temporal observation derived from $\mathcal{T}$ by i relaxations. We show that the application of a relaxation to $\mathcal{O}_i$ generates $\mathcal{O}_{i+1}$, where $\mathcal{T} \in \|O_{i+1}\|$. In the case of λ-relaxation, the mode in which nodes are chosen in $\mathcal{O}_i$ is not affected. Hence, the sequence of labels that make up $\mathcal{T}$ can still be generated. In the case of τ-relaxation, since the total effect of removing the arc $n \rightarrow n'$ and inserting new arcs $n_{\mathrm{p}} \rightarrow n'$ and $n \rightarrow n_{\mathrm{c}}$ is the deletion of one precedence only, namely $n \prec n'$, the mode in which nodes are chosen in $\mathcal{O}_i$ is not affected, and hence $\mathcal{T}$ can still be generated in $\mathcal{O}_{i+1}$. In the case of ν-relaxation, what is relevant is condition 1, which ensures that no new precedence between nodes in $\mathcal{O}_i$ is created in $\mathcal{O}_{i+1}$. Hence, the chosen sequence of nodes in $\mathcal{O}_i$ for generating $\mathcal{T}$ is still valid in $\mathcal{O}_{i+1}$. Provided that we select the null label from the new node, $\mathcal{T}$ can be generated in $\mathcal{O}_{i+1}$ too. Thus, in all cases, $\mathcal{T} \in \|\mathcal{O}_{i+1}\|$. □

4.3 Ruler

The modeling of an active system $\mathcal{A}$ does not make any distinction between normal and faulty behavior. Providing a specification for such a distinction means defining, for each component in $\mathcal{A}$, which transitions are normal and which ones are faulty. In other words, faults are associated with component transitions. This way, a transition $a \xrightarrow{t(c)} a'$ of $\mathcal{A}$ is faulty if and only if transition t of component c is faulty.

Therefore, specifying the abnormal behavior of $\mathcal{A}$ amounts to defining a mapping from faulty component transitions to corresponding fault labels.[6] This is captured by the notion of a ruler introduced in Def. 4.5.

Definition 4.5 (Ruler). Let $\mathcal{A}$ be an active system and Φ a domain of fault labels. A *ruler* $\mathcal{R}$ for $\mathcal{A}$ is a bijective mapping from a subset of component transitions in $\mathcal{A}$ to Φ.

Note 4.4. Since the mapping from faulty transitions to Φ is bijective, each of the fault labels in Φ is associated in $\mathcal{R}$ with one and only one of these transitions.

Example 4.10. Displayed in the second line of Table 4.1 is the specification of a ruler $\mathcal{R}$ for the active system $\mathcal{A}_2$ introduced in Example 4.6. Accordingly, each component involves two fault labels (printed in bold). In both breakers, the faulty transitions are b_3 (not open) and b_4 (not closed). In the protection device, the faulty transitions are p_3 (failure to command breakers to open) and p_4 (failure to command breakers to close).

Note 4.5. One may ask why we do not bind the ruler to system $\mathcal{A}$ at system-modeling time, rather than as a subsequent specification. After all, this would fix once and for all the abnormal behavior of $\mathcal{A}$. The point is, we prefer binding the ruler to each individual diagnosis problem because, generally speaking, different diagnosis problems may be characterized by different diagnosis ontologies, where each such ontology is a particular way to specify faulty transitions and corresponding faults. A hybrid approach might be to define a default ruler at modeling time, which can then be overridden at diagnosis-problem time.

Definition 4.6 (Diagnosis). Let h be a trajectory of an active system $\mathcal{A}$, and $\mathcal{R}$ a ruler for $\mathcal{A}$. The *diagnosis* of h based on $\mathcal{R}$, written $h_{[\mathcal{R}]}$, is the set of fault labels

$$h_{[\mathcal{R}]} = \{ f \mid t \in h, (t, f) \in \mathcal{R} \} . \tag{4.3}$$

Example 4.11. Consider the system $\mathcal{A}_2$ introduced in Example 4.6. Assuming an initial state with the protection device *idle*, the breakers *closed*, and the links empty, Table 4.2 shows a tabular representation of the transition function of the behavior space $Bsp(\mathcal{A}_2)$, the latter being depicted in Fig. 4.4. Each of the 57 lines defines the

[6] Just as specifying the observability of $\mathcal{A}$ amounts to defining a mapping from observable component transitions to corresponding observable labels.

Table 4.2 Transition function of $Bsp(\mathcal{A}_2)$ outlined in Fig. 4.4 (details of nodes are given in Table 4.3)

State	$b_1(l)$	$b_2(l)$	$b_3(l)$	$b_4(l)$	$b_5(l)$	$b_6(l)$	p_1	p_2	p_3	p_4	$b_1(r)$	$b_2(r)$	$b_3(r)$	$b_4(r)$	$b_5(r)$	$b_6(r)$
0							1		53							
1	2		4								3		5			
2											6		8			
3	6		7													
4											7		9			
5	8		9													
6								10		30						
7								25		34						
8								26		33						
9								29		49						
10		11										12		13		
11												0		15		
12		0		16												
13		15		17												
14												16		17		
15							18		39							
16							19		40							
17							52		54							
18	23		21													3
19						20					24		22			
20											6		8			
21																7
22						8										
23																6
24						6										
25					11							27		47		
26		28		48											12	
27					0											
28															0	
29					28										27	
30						31										32
31																17
32						17										
33						36					32		35			
34	31		37													38
35						16										
36											17		16			
37																15
38	17		15													
39					41							43		42		
40		44		45											46	
41												9		7		
42					7											
43					9											
44															9	
45															8	
46		9		8												
47					15											
48															16	
49	36		50								38		51			
50											15		0			
51	16		0													
52						23										24
53					44										43	
54		41		55								46		56		
55												8		6		
56		7		6												

Table 4.3 Details of states in $Bsp(\mathcal{A}_2)$ (see Table 4.2)

State	l	p	r	$\|pl\|$	$\|pr\|$
0	*closed*	*idle*	*closed*	[]	[]
1	*closed*	*awaken*	*closed*	[op]	[op]
2	*open*	*awaken*	*closed*	[]	[op]
3	*closed*	*awaken*	*open*	[op]	[]
4	*closed*	*awaken*	*closed*	[]	[op]
5	*closed*	*awaken*	*closed*	[op]	[]
6	*open*	*awaken*	*open*	[]	[]
7	*closed*	*awaken*	*open*	[]	[]
8	*open*	*awaken*	*closed*	[]	[]
9	*closed*	*awaken*	*closed*	[]	[]
10	*open*	*idle*	*open*	[cl]	[cl]
11	*closed*	*idle*	*open*	[]	[cl]
12	*open*	*idle*	*closed*	[cl]	[]
13	*open*	*idle*	*open*	[cl]	[]
14	*open*	*idle*	*open*	[]	[cl]
15	*closed*	*idle*	*open*	[]	[]
16	*open*	*idle*	*closed*	[]	[]
17	*open*	*idle*	*open*	[]	[]
18	*closed*	*awaken*	*open*	[op]	[op]
19	*open*	*awaken*	*closed*	[op]	[op]
20	*open*	*awaken*	*closed*	[]	[op]
21	*closed*	*awaken*	*open*	[]	[op]
22	*open*	*awaken*	*closed*	[op]	[]
23	*open*	*awaken*	*open*	[]	[op]
24	*open*	*awaken*	*open*	[op]	[]
25	*closed*	*idle*	*open*	[cl]	[cl]
26	*open*	*idle*	*closed*	[cl]	[cl]
27	*closed*	*idle*	*closed*	[cl]	[]
28	*closed*	*idle*	*closed*	[]	[cl]
29	*closed*	*idle*	*closed*	[cl]	[cl]
30	*open*	*idle*	*open*	[op]	[op]
31	*open*	*idle*	*open*	[]	[op]
32	*open*	*idle*	*open*	[op]	[]
33	*open*	*idle*	*closed*	[op]	[op]
34	*closed*	*idle*	*open*	[op]	[op]
35	*open*	*idle*	*closed*	[op]	[]
36	*open*	*idle*	*closed*	[]	[op]
37	*closed*	*idle*	*open*	[]	[op]
38	*closed*	*idle*	*open*	[op]	[]
39	*closed*	*awaken*	*open*	[cl]	[cl]
40	*open*	*awaken*	*closed*	[cl]	[cl]
41	*closed*	*awaken*	*open*	[]	[cl]
42	*closed*	*awaken*	*open*	[cl]	[]
43	*closed*	*awaken*	*closed*	[cl]	[]
44	*closed*	*awaken*	*closed*	[]	[cl]
45	*open*	*awaken*	*closed*	[]	[cl]
46	*open*	*awaken*	*closed*	[cl]	[]
47	*closed*	*idle*	*open*	[cl]	[]
48	*open*	*idle*	*closed*	[]	[cl]
49	*closed*	*idle*	*closed*	[op]	[op]
50	*closed*	*idle*	*closed*	[]	[op]
51	*closed*	*idle*	*closed*	[op]	[]
52	*open*	*awaken*	*open*	[op]	[op]
53	*closed*	*awaken*	*closed*	[cl]	[cl]
54	*open*	*awaken*	*open*	[cl]	[cl]
55	*open*	*awaken*	*open*	[]	[cl]
56	*open*	*awaken*	*open*	[cl]	[]

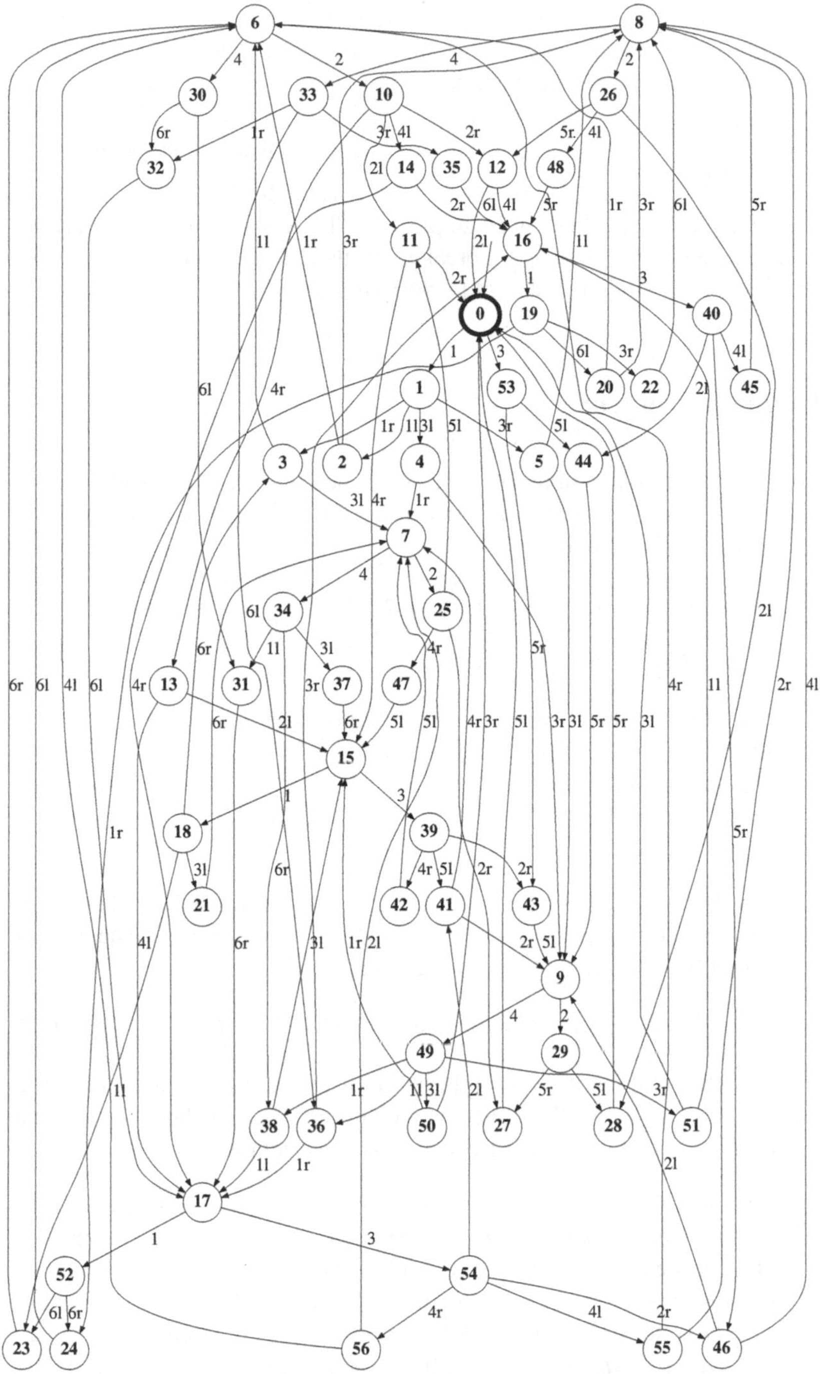

Fig. 4.4 Behavior space of active system $\mathcal{A}_2$, where arcs are marked by indices of component transitions. A transition of the protection device is indicated by a single digit (its number), while a transition of a breaker is indicated by two characters, the number of the transition and the identifier of the breaker (either l or r); for example, 1 stands for p_1 and $3r$ stands for $b_3(r)$

transitions relevant to the corresponding state identified by a member in the range $[0 \cdots 56]$, with 0 being the initial state. Details of the states of $Bsp(\mathcal{A}_2)$ are given in Table 4.3. For instance, at state 0, two transitions are possible, one marked by p_1 and entering state 1, and another one marked by p_3 and entering state 53, in other words $0 \xrightarrow{p_1} 1$ and $0 \xrightarrow{p_3} 53$. Based on the transition function, a possible trajectory of $\mathcal{A}_2$ is $h = 0 \xrightarrow{p_1} 1 \xrightarrow{b_1(l)} 2 \xrightarrow{b_3(r)} 8 \xrightarrow{p_4} 33 \xrightarrow{b_2(r)} 36$. According to Def. 4.6, the diagnosis of h based on the ruler $\mathcal{R}$ defined in Example 4.10 (Table 4.1) is $h_{[\mathcal{R}]} = \{\mathbf{nor}, \mathbf{fcp}\}$. That is, in trajectory h of system $\mathcal{A}_2$, breaker r fails to open (**nor**) and protection device p fails to command the breakers to close (**fcp**).

4.4 Diagnosis Problem

Definition 4.7 (Diagnosis Problem). Let $\mathcal{A}$ be an active system, $\mathcal{V}$ a viewer for $\mathcal{A}$, $\mathcal{O}$ a temporal observation of $\mathcal{A}$ relevant to a trajectory starting at an initial state a_0, and $\mathcal{R}$ a ruler for $\mathcal{A}$. The quadruple

$$\wp(\mathcal{A}) = (a_0, \mathcal{V}, \mathcal{O}, \mathcal{R}) \tag{4.4}$$

is a *diagnosis problem* for $\mathcal{A}$.

Example 4.12. Consider the active system $\mathcal{A}_2$ introduced in Example 4.6. A diagnosis problem for $\mathcal{A}_2$ is $\wp(\mathcal{A}_2) = (a_0, \mathcal{V}, \mathcal{O}_2, \mathcal{R})$, where $a_0 = 0$ as defined in Table 4.3, $\mathcal{V}$ is the viewer defined in the first line of Table 4.1, $\mathcal{O}_2$ is the temporal observation displayed on the right of Fig. 4.3, and $\mathcal{R}$ is the ruler defined in the second line of Table 4.1.

Definition 4.8 (Solution). Let $\wp(\mathcal{A}) = (a_0, \mathcal{V}, \mathcal{O}, \mathcal{R})$ be a diagnosis problem for an active system $\mathcal{A}$, and $Bsp(\mathcal{A})$ a behavior space of $\mathcal{A}$ with initial state a_0. The *solution* of $\wp(\mathcal{A})$, written $\Delta(\wp(\mathcal{A}))$, is the set of diagnoses

$$\Delta(\wp(\mathcal{A})) = \{\, \delta \mid \delta = h_{[\mathcal{R}]}, h \in Bsp(\mathcal{A}), h_{[\mathcal{V}]} \in \|\mathcal{O}\| \,\}\,. \tag{4.5}$$

In other words, the solution of $\wp(\mathcal{A})$ is the set of diagnoses of the trajectories of $\mathcal{A}$ based on the ruler $\mathcal{R}$ such that their trace on the viewer $\mathcal{V}$ is a candidate trace of the temporal observation $\mathcal{O}$, that is, their trace on $\mathcal{V}$ is consistent with $\mathcal{O}$.

Example 4.13. Consider the diagnosis problem $\wp(\mathcal{A}_2)$ defined in Example 4.12. According to the behavior space $Bsp(\mathcal{A}_2)$ specified in Table 4.2 and depicted in Fig. 4.4, 26 trajectories generate a trace which is a candidate in $\|\mathcal{O}_2\|$, as detailed in Table 4.4. For each trajectory h_i, $i \in [1 \cdots 26]$, both the trace $h_{[\mathcal{V}]}$ and the diagnosis $h_{[\mathcal{R}]}$ are listed. Despite the 26 trajectories, only two traces are generated, namely $[awk, opl, ide]$ and $[awk, opl, ide, cll]$, which are candidate traces in $\|\mathcal{O}_2\|$, and three diagnoses, namely $\delta_r = \{\mathbf{nor}\}$, $\delta_{lr} = \{\mathbf{ncl}, \mathbf{nor}\}$, and $\delta_{rp} = \{\mathbf{nor}, \mathbf{fcp}\}$. Therefore, according to Def. 4.8, the solution of $\wp(\mathcal{A}_2)$ will be

Table 4.4 Trajectories in $Bsp(\mathcal{A}_2)$ generating traces in $\|\mathcal{O}_2\|$

h	*Trajectory*	$h_{[\mathcal{V}]}$	$h_{[\mathcal{R}]}$
h_1	$0 \xrightarrow{p_1} 1 \xrightarrow{b_1(l)} 2 \xrightarrow{b_3(r)} 8 \xrightarrow{p_4} 33$	$[awk, opl, ide]$	$\{\mathbf{nor}, \mathbf{fcp}\}$
h_2	$0 \xrightarrow{p_1} 1 \xrightarrow{b_1(l)} 2 \xrightarrow{b_3(r)} 8 \xrightarrow{p_4} 33 \xrightarrow{b_3(r)} 35$	$[awk, opl, ide]$	$\{\mathbf{nor}, \mathbf{fcp}\}$
h_3	$0 \xrightarrow{p_1} 1 \xrightarrow{b_1(l)} 2 \xrightarrow{b_3(r)} 8 \xrightarrow{p_4} 33 \xrightarrow{b_3(r)} 35 \xrightarrow{b_6(l)} 16$	$[awk, opl, ide]$	$\{\mathbf{nor}, \mathbf{fcp}\}$
h_4	$0 \xrightarrow{p_1} 1 \xrightarrow{b_1(l)} 2 \xrightarrow{b_3(r)} 8 \xrightarrow{p_4} 33 \xrightarrow{b_6(l)} 36$	$[awk, opl, ide]$	$\{\mathbf{nor}, \mathbf{fcp}\}$
h_5	$0 \xrightarrow{p_1} 1 \xrightarrow{b_1(l)} 2 \xrightarrow{b_3(r)} 8 \xrightarrow{p_4} 33 \xrightarrow{b_6(l)} 36 \xrightarrow{b_3(r)} 16$	$[awk, opl, ide]$	$\{\mathbf{nor}, \mathbf{fcp}\}$
h_6	$0 \xrightarrow{p_1} 1 \xrightarrow{b_1(l)} 2 \xrightarrow{b_3(r)} 8 \xrightarrow{p_2} 26$	$[awk, opl, ide]$	$\{\mathbf{nor}\}$
h_7	$0 \xrightarrow{p_1} 1 \xrightarrow{b_1(l)} 2 \xrightarrow{b_3(r)} 8 \xrightarrow{p_2} 26 \xrightarrow{b_2(l)} 28$	$[awk, opl, ide, cll]$	$\{\mathbf{nor}\}$
h_8	$0 \xrightarrow{p_1} 1 \xrightarrow{b_1(l)} 2 \xrightarrow{b_3(r)} 8 \xrightarrow{p_2} 26 \xrightarrow{b_2(l)} 28 \xrightarrow{b_5(r)} 0$	$[awk, opl, ide, cll]$	$\{\mathbf{nor}\}$
h_9	$0 \xrightarrow{p_1} 1 \xrightarrow{b_1(l)} 2 \xrightarrow{b_3(r)} 8 \xrightarrow{p_2} 26 \xrightarrow{b_4(l)} 48$	$[awk, opl, ide]$	$\{\mathbf{ncl}, \mathbf{nor}\}$
h_{10}	$0 \xrightarrow{p_1} 1 \xrightarrow{b_1(l)} 2 \xrightarrow{b_3(r)} 8 \xrightarrow{p_2} 26 \xrightarrow{b_4(l)} 48 \xrightarrow{b_5(r)} 16$	$[awk, opl, ide]$	$\{\mathbf{ncl}, \mathbf{nor}\}$
h_{11}	$0 \xrightarrow{p_1} 1 \xrightarrow{b_1(l)} 2 \xrightarrow{b_3(r)} 8 \xrightarrow{p_2} 26 \xrightarrow{b_5(r)} 12$	$[awk, opl, ide]$	$\{\mathbf{nor}\}$
h_{12}	$0 \xrightarrow{p_1} 1 \xrightarrow{b_1(l)} 2 \xrightarrow{b_3(r)} 8 \xrightarrow{p_2} 26 \xrightarrow{b_5(r)} 12 \xrightarrow{b_4(l)} 16$	$[awk, opl, ide]$	$\{\mathbf{ncl}, \mathbf{nor}\}$
h_{13}	$0 \xrightarrow{p_1} 1 \xrightarrow{b_1(l)} 2 \xrightarrow{b_3(r)} 8 \xrightarrow{p_2} 26 \xrightarrow{b_5(r)} 12 \xrightarrow{b_2(l)} 0$	$[awk, opl, ide, cll]$	$\{\mathbf{nor}\}$
h_{14}	$0 \xrightarrow{p_1} 1 \xrightarrow{b_3(r)} 5 \xrightarrow{b_1(l)} 8 \xrightarrow{p_4} 33$	$[awk, opl, ide]$	$\{\mathbf{nor}, \mathbf{fcp}\}$
h_{15}	$0 \xrightarrow{p_1} 1 \xrightarrow{b_3(r)} 5 \xrightarrow{b_1(l)} 8 \xrightarrow{p_4} 33 \xrightarrow{b_3(r)} 35$	$[awk, opl, ide]$	$\{\mathbf{nor}, \mathbf{fcp}\}$
h_{16}	$0 \xrightarrow{p_1} 1 \xrightarrow{b_3(r)} 5 \xrightarrow{b_1(l)} 8 \xrightarrow{p_4} 33 \xrightarrow{b_3(r)} 35 \xrightarrow{b_6(l)} 16$	$[awk, opl, ide]$	$\{\mathbf{nor}, \mathbf{fcp}\}$
h_{17}	$0 \xrightarrow{p_1} 1 \xrightarrow{b_3(r)} 5 \xrightarrow{b_1(l)} 8 \xrightarrow{p_4} 33 \xrightarrow{b_6(l)} 36$	$[awk, opl, ide]$	$\{\mathbf{nor}, \mathbf{fcp}\}$
h_{18}	$0 \xrightarrow{p_1} 1 \xrightarrow{b_3(r)} 5 \xrightarrow{b_1(l)} 8 \xrightarrow{p_4} 33 \xrightarrow{b_6(l)} 36 \xrightarrow{b_3(r)} 16$	$[awk, opl, ide]$	$\{\mathbf{nor}, \mathbf{fcp}\}$
h_{19}	$0 \xrightarrow{p_1} 1 \xrightarrow{b_3(r)} 5 \xrightarrow{b_1(l)} 8 \xrightarrow{p_2} 26$	$[awk, opl, ide]$	$\{\mathbf{nor}\}$
h_{20}	$0 \xrightarrow{p_1} 1 \xrightarrow{b_3(r)} 2 \xrightarrow{b_1(l)} 8 \xrightarrow{p_2} 26 \xrightarrow{b_2(l)} 28$	$[awk, opl, ide, cll]$	$\{\mathbf{nor}\}$
h_{21}	$0 \xrightarrow{p_1} 1 \xrightarrow{b_3(r)} 2 \xrightarrow{b_1(l)} 8 \xrightarrow{p_2} 26 \xrightarrow{b_2(l)} 28 \xrightarrow{b_5(r)} 0$	$[awk, opl, ide, cll]$	$\{\mathbf{nor}\}$
h_{22}	$0 \xrightarrow{p_1} 1 \xrightarrow{b_3(r)} 5 \xrightarrow{b_1(l)} 8 \xrightarrow{p_2} 26 \xrightarrow{b_4(l)} 48$	$[awk, opl, ide]$	$\{\mathbf{ncl}, \mathbf{nor}\}$
h_{23}	$0 \xrightarrow{p_1} 1 \xrightarrow{b_3(r)} 5 \xrightarrow{b_1(l)} 8 \xrightarrow{p_2} 26 \xrightarrow{b_4(l)} 48 \xrightarrow{b_5(r)} 16$	$[awk, opl, ide]$	$\{\mathbf{ncl}, \mathbf{nor}\}$
h_{24}	$0 \xrightarrow{p_1} 1 \xrightarrow{b_3(r)} 5 \xrightarrow{b_1(l)} 8 \xrightarrow{p_2} 26 \xrightarrow{b_5(r)} 12$	$[awk, opl, ide]$	$\{\mathbf{nor}\}$
h_{25}	$0 \xrightarrow{p_1} 1 \xrightarrow{b_3(r)} 5 \xrightarrow{b_1(l)} 8 \xrightarrow{p_2} 26 \xrightarrow{b_5(r)} 12 \xrightarrow{b_4(l)} 16$	$[awk, opl, ide]$	$\{\mathbf{ncl}, \mathbf{nor}\}$
h_{26}	$0 \xrightarrow{p_1} 1 \xrightarrow{b_3(r)} 5 \xrightarrow{b_1(l)} 8 \xrightarrow{p_2} 26 \xrightarrow{b_5(r)} 12 \xrightarrow{b_2(l)} 0$	$[awk, opl, ide, cll]$	$\{\mathbf{nor}\}$

$$\Delta(\wp(\mathcal{A}_2)) = \{\, \delta_r, \delta_{lr}, \delta_{rp} \,\} \,. \tag{4.6}$$

In other words, one of the following three abnormal scenarios has occurred:

1. $\delta_r = \{\mathbf{nor}\}$: Breaker r did not open,
2. $\delta_{lr} = \{\mathbf{ncl}, \mathbf{nor}\}$: Breaker r did not open and breaker l did not close,
3. $\delta_{rp} = \{\mathbf{nor}, \mathbf{fcp}\}$: Breaker r did not open and protection device p did not command the breakers to close.

Despite the uncertainty caused by the different candidate diagnoses, since all candidates include the fault **nor**, breaker r certainly did not open. In addition, possibly either breaker l did not close (**ncl**) or protection device p did not command the breakers to close (**fcp**).

Note 4.6. If a trajectory is consistent with a candidate trace of a temporal observation $\mathcal{O}$ and includes a cycle with no observable component transitions, then there is an unbounded number of trajectories consistent with $\mathcal{O}$, each one obtained by iterating the cycle a different number of times. However, since the diagnosis of a trajectory is a set (without duplicates), additional iterations of the same cycle do not extend the diagnosis with new additional faults. More generally, even if there exists an infinite set of trajectories consistent with a temporal observation, the set of corresponding diagnoses is always finite, this being bounded by $2^{|\Phi|}$, where $|\Phi|$ is the cardinality of the set Φ of fault labels defined in the ruler.

4.5 Bibliographic Notes

The notion of an (uncertain) temporal observation was introduced and studied in [85, 87, 88]. An uncertain temporal observation was then generalized to a *complex observation* in [89], where additional uncertainty about the component emitting the observable label is considered.

4.6 Summary of Chapter 4

Viewer. A mapping from a subset of transitions of components in an active system to a domain of observable labels. If a component transition is involved in the viewer, it is observable, otherwise it is unobservable.

Trace. The sequence of observable labels obtained by replacing each observable transition of a trajectory by the corresponding observable label in the viewer.

Relaxation. The transformation of a trace into a temporal observation by means of a sequence of operations, namely λ (label relaxation), τ (temporal relaxation), and ν (node relaxation).

Temporal Observation. A directed acyclic graph obtained by relaxation of a trace, where each node is marked by a set of candidate observable labels and each arc denotes partial temporal ordering between nodes.

Plain Temporal Observation. A temporal observation isomorphic to a trace, where the nodes are totally temporally ordered and marked by a single observable label.

Candidate Trace. A sequence of observable labels obtained by selecting one label from each node of a temporal observation fulfilling the partial temporal ordering imposed by arcs.

Ruler. A mapping from a subset of transitions of components in an active system to a domain of fault labels. If a component transition is involved in the ruler, it is faulty, otherwise it is normal.

Diagnosis. The set of fault labels obtained by replacing each faulty transition of a trajectory by the corresponding fault label in the ruler.

Diagnosis Problem. A quadruple $\wp(\mathcal{A}) = (a_0, \mathcal{V}, \mathcal{O}, \mathcal{R})$, where a_0 is the initial state of active system $\mathcal{A}$, $\mathcal{V}$ a viewer for $\mathcal{A}$, $\mathcal{O}$ a temporal observation of $\mathcal{A}$, and $\mathcal{R}$ a ruler for $\mathcal{A}$. The solution of $\wp(A)$ is the set of diagnoses of the trajectories of $\mathcal{A}$ whose trace is a candidate trace in $\mathcal{O}$.

Chapter 5
Monolithic Diagnosis

In Chapter 4 we defined the notion of a diagnosis problem for an active system. This is a quadruple involving the initial state of the system, a viewer that maps observable component transitions to observable labels, a temporal observation obtained as a relaxation of the trace of the system, and a ruler that maps faulty component transitions to fault labels. We also defined the solution of a diagnosis problem (Def. 4.8) as the set of diagnoses of the trajectories, in the behavior space of the system, that are consistent with the temporal observation. Consistency means that the trace of the trajectory is a candidate trace of the temporal observation.

However, the definition of the solution is in fact only a definition, just like saying that the solution of a mathematical equation is the set of numbers that make the equality true. From a practical viewpoint, we need a (sound and complete) technique that generates the actual solution of the equation.

In the context of the diagnosis problem, the declarative nature of the definition of the solution is exacerbated by the explicit reference to the behavior space, which is assumed to be missing when we are solving the problem. Therefore, similarly to mathematical equations, we need a sound and complete technique that generates the solution of the diagnosis problem, with the additional constraint that the behavior space is missing.

According to Def. 4.8, what is essential to generate a candidate diagnosis is a trajectory of the system that is consistent with the temporal observation. These trajectories are a subset of all possible trajectories. In other words, the trajectories consistent with the temporal observation form a subset of the trajectories of the behavior space. If the language (set of trajectories) of the behavior space is infinite, then the subset of trajectories consistent with the temporal observation can be infinite in turn.

The point is, since the behavior space is a DFA embodying (in its language) a possibly infinite set of trajectories, that the chances are we can devise a technique for generating a DFA whose language is exactly the subset of the language of the behavior space that includes all the trajectories consistent with the temporal observation. We call this DFA the *behavior* of the diagnosis problem $\wp(\mathcal{A})$, written $Bhv(\wp(\mathcal{A}))$.

G. Lamperti et al., *Introduction to Diagnosis of Active Systems*,
https://doi.org/10.1007/978-3-319-92733-6_5

Building the behavior of the diagnosis problem is the first step of a monolithic diagnosis engine (an algorithm performing monolithic diagnosis), and is called *behavior reconstruction*.

Then, we need to associate with each trajectory of the behavior the corresponding diagnosis. At first glance, when the language of the behavior (the set of trajectories) is infinite, this might seem an endless task, as we have to consider each trajectory one by one in order to generate the associated diagnosis.

However, as pointed out in Note 4.6, the set of diagnoses is always finite because a diagnosis is a set (rather than a multiset) of faults, so that duplicated faults in the same trajectory are removed. In practice, this means that, for the purpose of diagnosis generation, only a finite set of trajectories need to be considered.

Specifically, this finite set is composed of all the trajectories in which cycles are traversed at most once. In fact, if a trajectory h_1 contains a string s corresponding to a cycle in the behavior and a trajectory h_2 differs from h_1 just because it includes a string ss corresponding to the same cycle in h_1 traversed twice, then the set of faults involved in the component transitions of s equals the set of faults involved in ss. In other words, the diagnosis of h_1 equals the diagnosis of h_2. More generally, the diagnosis of each trajectory h_k involving the repetition $k \geq 1$ times of the same string s equals the diagnosis of h_1.

This is good news because the chances are we can associate with each node β of the behavior a (finite) set of diagnoses, with each diagnosis corresponding to a trajectory h ending at β, such that h does not traverse a cycle more than once. That is, we can generate the set of diagnoses associated with node β by considering only a finite subset of the (possibly infinite) set of trajectories ending at β.

Generating the set of diagnoses associated with each node of the behavior is the second step of a monolithic diagnosis engine, and is called *behavior decoration*.

Once the behavior has been reconstructed and decorated for each node, there remains only one last step for the monolithic diagnosis engine, called *diagnosis distillation*. Distilling the candidate diagnoses from the decorated behavior means selecting only the diagnoses of the trajectories consistent with the temporal observation.

This sounds like a contradiction of asserting that the set of trajectories in the behavior is the subset of the trajectories in the behavior space that are consistent with the temporal observation. In other words, where do possibly inconsistent trajectories of the behavior come from?

The point is, in order for a trajectory h of the behavior to be consistent with a temporal observation $\mathcal{O}$, the trace of h based on the relevant viewer should be a candidate trace of $\mathcal{O}$.

This requires the diagnosis engine to make a distinction between nonfinal and final nodes in the behavior, where each final node β_f is such that all trajectories ending at β_f have a trace which is a candidate trace of $\mathcal{O}$. The solution of $\wp(\mathcal{A})$ is eventually distilled by collecting the diagnoses decorating the final nodes of $Bhv(\wp(\mathcal{A}))$ only.

5.1 Observation Indexing

Behavior reconstruction means generating a DFA similar in nature to the behavior space, whose language is the set of system trajectories consistent with the temporal observation. Both the reconstructed behavior and the behavior space are represented by a DFA rooted in the initial state of the system, with each arc being marked by a component transition.

In a way, thinking of the behavior space as a container of trajectories leads to the idea that the reconstructed behavior is a restricted container resulting from selecting from the behavior space all the trajectories fulfilling the temporal observation. The possibly infinite number of trajectories is not a problem, as they can be represented by a DFA, which is finite in all cases (like the DFA representing the behavior space).

Briefly, the reconstructed behavior is the subgraph of the behavior space consistent with the temporal observation.

The essential question now is: *how can behavior reconstruction be performed?* Intuitively, since the reconstructed behavior is a subgraph of the behavior space, the former might be generated in a way similar to the way in which we generated the behavior space of Example 3.5, displayed in Fig. 3.3. That is, starting from the initial state, we consider the triggerable component transitions to generate the neighboring states and continue in this way until no new state is generated.

However, the generation of the reconstructed behavior differs from the (virtual) generation of the behavior space in two ways:

1. A triggerable transition can be applied only if it is also consistent with the temporal observation,
2. A state contains not only the tuple S of component states and the tuple Q of link configurations: it also contains additional information concerning the consistency of relevant trajectories with the temporal observation.

The two points above are strictly related: to check the consistency of a triggerable transition (point 1), we need to keep relevant information in the nodes (point 2).

How can we track the consistency of trajectories with the temporal observation? In other words, what kind of information should we include in the nodes of the reconstructed behavior?

To answer this question, it is convenient to consider the problem in the special case of a plain temporal observation (Sec. 4.2), that is, when the observation coincides with the trace (there is total temporal ordering between nodes, which are marked by a single observable label). If we succeed in coping with the problem based on plain observations, then we can extend the approach to general temporal observations.

Assuming a plain observation $\mathcal{O}$ (a sequence of observable labels), the consistency of a trajectory with $\mathcal{O}$ can be tracked by maintaining, within each node β of the behavior, the prefix of $\mathcal{O}$ already generated by the trajectories up to β.

Starting from the empty prefix associated with the initial state, when considering a triggerable component transition $t(c)$ at state β of the behavior, we must check the consistency of $t(c)$ with $\mathcal{O}$ based on the prefix of $\mathcal{O}$ stored in β. Specifically, $t(c)$

is consistent if either it is unobservable or the associated observable label equals the next label in $\mathcal{O}$: if $t(c)$ is observable, the prefix associated with the new state β' is extended by one label (giving rise to a new prefix), otherwise the prefix remains unchanged. In either case, a new transition $\beta \xrightarrow{t(c)} \beta'$ will be created in the reconstructed behavior.

For space reasons, instead of storing the prefix of $\mathcal{O}$ in each node, it is better if we store a surrogate of it, namely the *index* of that prefix, which might be the identifier[1] of the last node of the prefix or, even better, an integer in the range $[0 \cdots k]$, where k is the length of $\mathcal{O}$.

Thus, a node of the reconstructed behavior can be represented by a triple (S, Q, i), where the additional field $i \in [0 \cdots k]$ is the index of the plain observation. A node (S, Q, i) is final if and only if $i = k$.

Example 5.1. Consider the active space $\mathcal{A}_2$ defined in Example 4.6 (Fig. 4.2) and the relevant viewer $\mathcal{V}$ defined in Table 4.1. Assume for $\mathcal{A}_2$ the trace $\mathcal{T}_2 = [awk, opl, ide]$ introduced in Example 4.8, with the associated plain observation $\mathcal{O}_{\mathcal{T}_2}$ displayed on the left side of Fig. 4.3. Since $\mathcal{T}_2$ includes three observable labels, the index i of $\mathcal{O}_{\mathcal{T}_2}$ is in the range $[0 \cdots 3]$.

Shown in Fig. 5.1 is the reconstructed behavior for $\mathcal{A}_2$ based on $\mathcal{O}_{\mathcal{T}_2}$, involving 12 states, named $\beta_0, \ldots, \beta_{11}$. The subpart of the graph with dotted lines and gray nodes is not consistent with the observation (as explained shortly).

Each state is marked by the triple (S, Q, i), where:

1. S is the triple of states of components l (left breaker), p (protection device), and r (right breaker) (with each state being identified by the initial letter),
2. Q is the pair of configurations of links pl and pr,
3. i is the index (integer value) for $\mathcal{O}_{\mathcal{T}_2}$.

Consider the initial state β_0. Since both links are empty, the only triggerable transitions are p_1 and p_3. Considering p_1, since this transition is observable and associated with the label *awk*, which is the successive label in $\mathcal{O}_{\mathcal{T}_2}$ (actually the first label, as the index in β_0 is $i = 0$, denoting the empty prefix), p_1 is also consistent with the observation. Therefore, transition $\beta_0 \xrightarrow{p_1} \beta_1$ is generated, where β_1 involves an index $i = 1$, denoting the prefix $[awk]$.

At state β_1, four component transitions are triggerable, namely $b_1(l)$, $b_1(r)$, $b_3(l)$, and $b_3(r)$. However, unlike the others, the observable transition $b_1(r)$ is not consistent with $\mathcal{O}_{\mathcal{T}_2}$, because the associated observable label *opr* is not the next label in the sequence (namely *opl*). Thus, three new transitions are generated in the behavior: $\beta_1 \xrightarrow{b_1(l)} \beta_2$, $\beta_1 \xrightarrow{b_3(r)} \beta_3$, and the transition marked by $b_3(l)$ and entering the gray state. Since $b_1(l)$ is associated with the observable label *opl*, the index in β_2 is $i = 2$. By contrast, since both $b_3(l)$ and $b_3(r)$ are unobservable, the index in β_3 and in the gray state is $i = 1$, the same as in β_1.

[1] This identifier cannot be the observable label marking the node, as several occurrences of the same label may occur within $\mathcal{O}$.

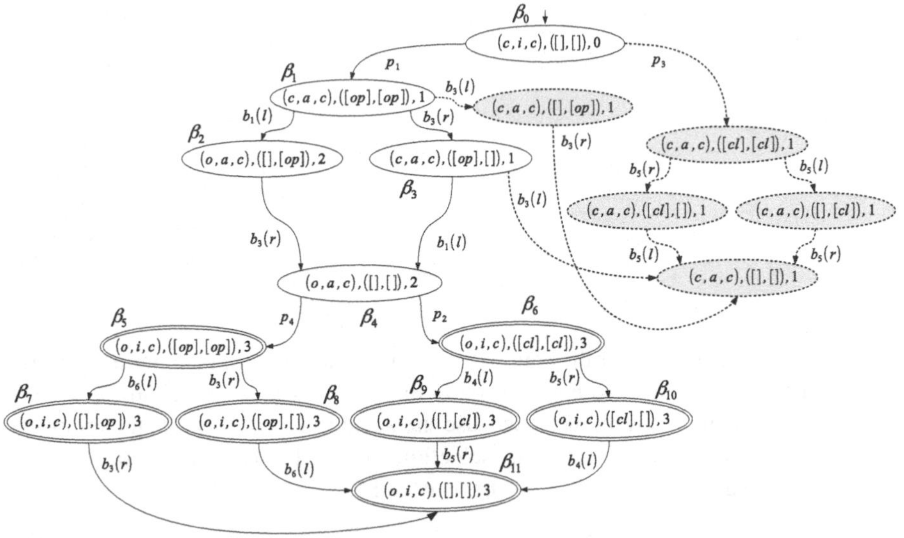

Fig. 5.1 Reconstructed behavior based on plain observation $\mathcal{O}_{\mathcal{T}_2}$ (Example 5.1)

The rest of the behavior is generated by applying the same technique as described above. Notice that states $\beta_5, \ldots, \beta_{11}$ are final (double circled), as the index $i = 3$ equals the length of $\mathcal{O}_{\mathcal{T}_2}$.

We can also understand why the gray part of the graph is inconsistent with $\mathcal{O}_{\mathcal{T}_2}$. In both of the gray states with both links empty, the only triggerable transitions are p_1 and p_3, which are both observable via the label *awk*. However, since the index in both states is $i = 1$, the successive expected observable label is *opl*, not *awk*. Hence, no consistent transitions exit these two states.

Eventually, the part of the graph that is not connected to a final state will be discarded because the trajectories ending at such states are inconsistent with the temporal observation (in our example, six states and eight transitions).

So far, we have considered the problem of indexing plain observations for checking the consistency of trajectories. Now we can extend the indexing technique to cope with general temporal observations. Specifically, we need to generalize the notions of an observation prefix and an observation index.

More generally, a prefix of a plain observation $\mathcal{O}$ can be defined as the (possibly empty) set of nodes that can be chosen based on the total temporal ordering between nodes. This means that, starting from the empty set, if P is a prefix of $\mathcal{O}$, then for each node $n \in P$, each node $n' \prec n$ is in P.

Interestingly, the same definition can be applied to any temporal observation: a prefix of a temporal observation $\mathcal{O}$ is the (possibly empty) set of nodes that can be chosen based on a partial temporal ordering between nodes.

Example 5.2. Consider the temporal observation $\mathcal{O}_2$ introduced in Example 4.8 and reported in Fig. 5.2. The set $P = \{n_1, n_2, n_3\}$ is a prefix of $\mathcal{O}$ because these nodes

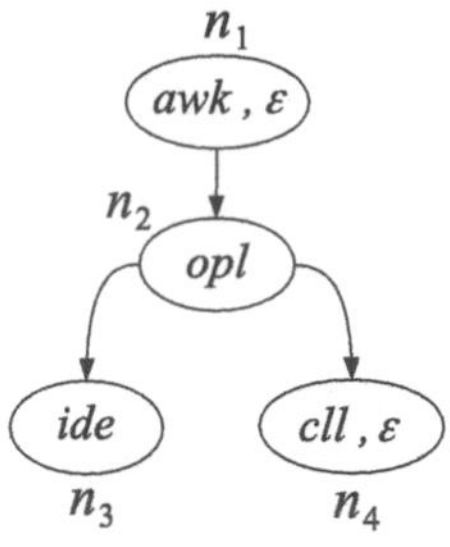

Fig. 5.2 Temporal observation $\mathcal{O}_2$

can be selected (in the order they are written) based on the partial temporal ordering. Equivalently, for each node $n \in P$, each node $n' \prec n$ is in P. By contrast, the set $P' = \{n_1, n_3\}$ is not a prefix of $\mathcal{O}$, because the selection of n_3 violates the temporal ordering (which requires a previous selection of n_2). Equivalently, there exists a node $n_3 \in P'$ such that node n_2, $n_2 \prec n_3$, is not in P'.

A temporal-observation prefix can be generated inductively as specified in Def. 5.1.

Definition 5.1 (Prefix). Let $\mathcal{O} = (L, N, A)$ be a temporal observation. A *prefix* of $\mathcal{O}$ is a subset of nodes N inductively defined as follows:

1. The empty set $\emptyset$ is a prefix of $\mathcal{O}$,
2. If P is a prefix of $\mathcal{O}$ and $n' \in N - P$, where $\forall n \in N$ such that $n \prec n'$ we have $n \in P$, then $P' = P \cup \{n'\}$ is a prefix of $\mathcal{O}$.

In other words, starting from the empty prefix, if P is a prefix of $\mathcal{O}$, then a new prefix P' can be generated by inserting into P a node $n' \notin P$ such that all nodes n preceding n' are in P.

Example 5.3. With reference to Fig. 5.2, the possible prefixes of the temporal observation $\mathcal{O}_2$ are $\emptyset$, $\{n_1\}$, $\{n_1, n_2\}$, $\{n_1, n_2, n_3\}$, $\{n_1, n_2, n_4\}$, and $\{n_1, n_2, n_3, n_4, n_5\}$. For instance, $\{n_1, n_2, n_4\}$ is generated by inserting node n_4 into the prefix $\{n_1, n_2\}$, which includes all preceding nodes of n_4.

Now we have to generalize the notion of a prefix index. In abstract terms, for a plain observation $\mathcal{O}$, the index of a prefix P is the node $n \in P$ such that all other nodes $n' \in P$ precede n in $\mathcal{O}$, that is, $n' \prec n$.

For a general temporal observation $\mathcal{O}$, the index of a prefix P is the minimal set of nodes I such that all nodes $n' \in P - I$ precede a node in I, that is, $n' \prec n$, $n \in I$. In other words, all nodes in $P - I$ precede a node in I and each node in I does not precede any node in I, as formalized in Def. 5.2.

Definition 5.2 (Index). Let P be a prefix of a temporal observation $\mathcal{O} = (L, N, A)$. The *index* of P, denoted $Idx(P)$, is the subset of P defined as follows:

$$Idx(P) = \{n \mid n \in P, \forall n', n' \prec n \, (n' \in P), \forall n'', n \prec n'' \, (n'' \notin P)\} . \quad (5.1)$$

Given an index $I = Idx(P)$, the set of nodes in P can be denoted by the inverse function Idx^{-1} applied to I:

$$P = Idx^{-1}(I). \tag{5.2}$$

An index I is *final* when $Idx^{-1}(I) = N$.

Example 5.4. With reference to Fig. 5.2, the index of the prefix $P = \{n_1, n_2, n_3, n_4\}$ is $I = \{n_3, n_4\}$. Since prefix P involves all nodes of $\mathcal{O}_2$, the index I is final.

Based on the generalized notion of a temporal-observation index (Def. 5.2), we can now extend the technique of behavior reconstruction to temporal observations with partial temporal ordering. For the sake of clarity, we first limit the technique to temporal observations with nodes marked by just one observable (nonnull) label.

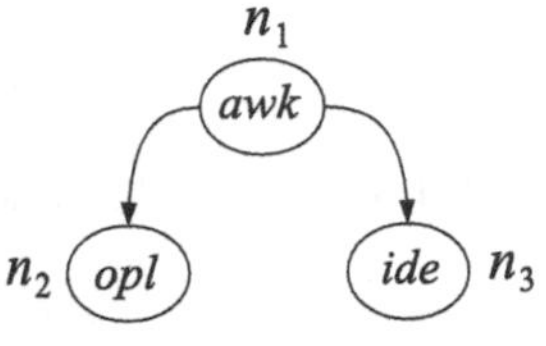

Fig. 5.3 Temporal observation $\mathcal{O}'_2$; nodes are marked by just one (nonnull) observable label

Example 5.5. Shown in Fig. 5.3 is a temporal observation $\mathcal{O}'_2$ obtained by temporal relaxation of the trace $\mathcal{T}_2 = [awk, opl, ide]$. As a result, the labels *opl* and *ide* are no longer temporally related, and the set of candidate traces for $\mathcal{O}'_2$ is

$$\|\mathcal{O}'_2\| = \{[awk, opl, ide], [awk, ide, opl]\}\ . \tag{5.3}$$

The temporal relaxation of $\mathcal{T}_2$ produces the additional trace $[awk, ide, opl]$. Since the reconstructed behavior should be consistent with either trace, we expect the behavior to be an extension of the behavior displayed in Fig. 5.1 (see Example 5.1).

What differentiates behavior reconstruction with partially ordered observable labels from behavior reconstruction with totally ordered observable labels (a plain observation) is the check of the consistency of a triggerable (and observable) component transition.

When the observation is linear, the observable label ℓ associated with the triggerable component transitions should equal the next label in the plain observation. Instead, for a temporal observation $\mathcal{O}$ with partially ordered labels, given the current prefix P, ℓ should equal the label of a node $n \notin P$ such that, for each $n' \prec n$, we have $n' \in P$. Besides, the index of $\mathcal{O}$ in the new node of the behavior should identify the new prefix $P \cup \{n\}$.

Example 5.6. Consider a variation of the behavior reconstruction in Example 5.1, where the plain observation $\mathcal{O}_{\mathcal{T}_2}$ is replaced by the temporal observation $\mathcal{O}'_2$ shown in Fig. 5.3. The corresponding reconstructed behavior is displayed in Fig. 5.4 (where the spurious part is omitted).

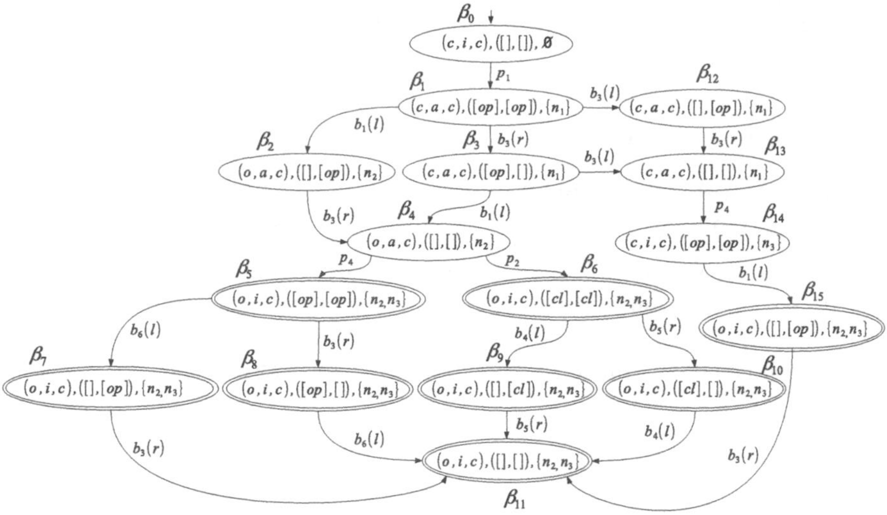

Fig. 5.4 Reconstructed behavior based on temporal observation $\mathcal{O}'_2$ (Fig. 5.3)

Notice how nodes $\beta_0 \cdots \beta_{11}$ correspond to the homonymous nodes in Fig. 5.1, as this part of the behavior is consistent with trace $\mathcal{T}_2$. However, the subgraph composed of nodes $\beta_{12} \cdots \beta_{15}$ is consistent with the additional candidate trace $[awk, ide, opl]$.

Since the nodes in the temporal observation $\mathcal{O}'_2$ (Fig. 5.3) are partially ordered, the third field in each node of the reconstructed behavior is the index (Def. 5.2) of the current prefix P of $\mathcal{O}'_2$, namely $Idx(P)$, that is, the (minimal) subset of nodes in P that identify P based on the temporal ordering of nodes.

In state β_0, the index is the empty set (identifying the empty prefix of $\mathcal{O}'_2$). In this state, the component transition p_1 is consistent because the associated observable label *awk* marks node n_1 of $\mathcal{O}'_2$. Hence, the index in β_1 is the singleton $\{n_1\}$.

The rest of the behavior is built by applying the same rules. Final states (double circled) are characterized by the final index, namely $\{n_2, n_3\}$, which identifies the whole set of nodes $\{n_1, n_2, n_3\}$.

Now, we extend the scenario further by assuming that the partially ordered nodes of the temporal observation are marked by a set of (nonnull) observable labels (rather than a single label).

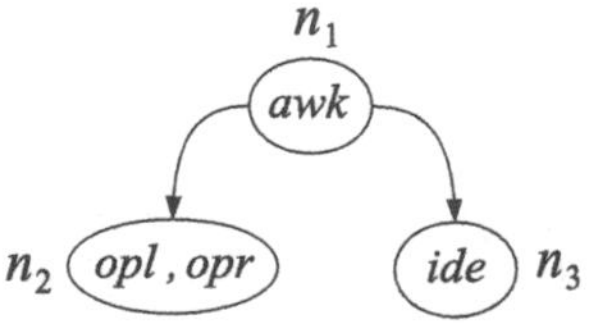

Fig. 5.5 Temporal observation $\mathcal{O}''_2$: node n_2 is marked by two observable labels

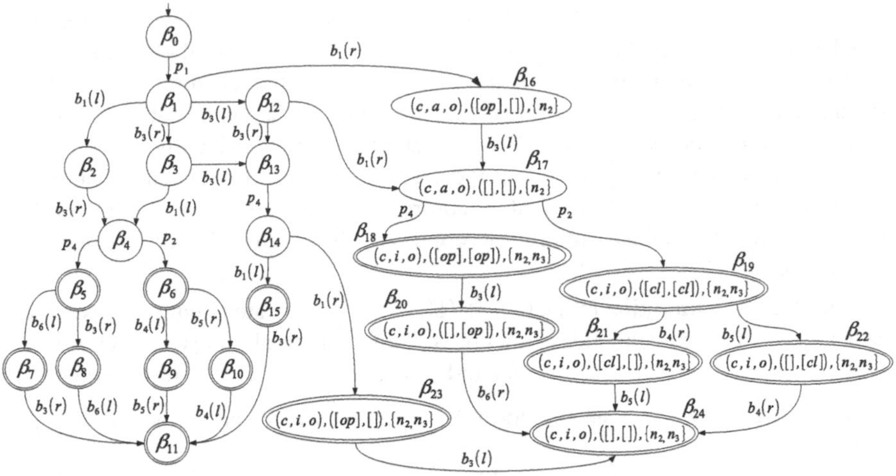

Fig. 5.6 Reconstructed behavior based on temporal observation $\mathcal{O}''_2$ (Fig. 5.5)

Example 5.7. Shown in Fig. 5.5 is a temporal observation $\mathcal{O}''_2$ obtained by label relaxation of the temporal observation $\mathcal{O}'_2$ (Fig. 5.3). Specifically, node n_2 is extended by the label *opr*. As a result, the set of candidate traces for $\mathcal{O}''_2$ is

$$\|\mathcal{O}''_2\| = \{[awk, opl, ide], [awk, opr, ide], [awk, ide, opl], [awk, ide, opr]\}\,. \quad (5.4)$$

The label relaxation of $\mathcal{O}'_2$ produces two additional traces, $[awk, opr, ide]$ and $[awk, ide, opr]$. Consequently, we expect the behavior to be an extension of the behavior displayed in Fig. 5.4 (see Example 5.6).

Let $\beta = (S, Q, P)$ be a node of the reconstructed behavior, and $t(c)$ an observable component transition triggerable in β. Checking the consistency of $t(c)$ requires that the associated observable label ℓ be included in the set of labels marking a node $n \notin P$ such that, for each $n' \prec n$, we have $n' \in P$. In other words, $\ell \in \|n\|$.

Example 5.8. Consider a variation of the behavior reconstruction in Example 5.6, where the temporal observation $\mathcal{O}'_2$ is replaced by the temporal observation $\mathcal{O}''_2$ shown in Fig. 5.5. The corresponding reconstructed behavior is displayed in Fig. 5.6.

Nodes $\beta_0 \cdots \beta_{15}$ coincide with the homonymous nodes in Fig. 5.4, as this part of the behavior is consistent with the candidate traces of $\mathcal{O}'_2$, namely $[awk, opl, ide]$ and $[awk, ide, opl]$. However, the subgraph composed of nodes $\beta_{16} \cdots \beta_{24}$ is consistent with the additional candidate traces $[awk, opr, ide]$ and $[awk, ide, opr]$.

For instance, in state β_1, the index is $\{n_1\}$. In this state, the component transition $b_1(r)$ is consistent because the associated observable label *opr* is a candidate label of node n_2 in $\mathcal{O}''_2$. Hence, the index in β_{16} is $\{n_2\}$, identifying the prefix $\{n_1, n_2\}$. As with the behavior consistent with $\mathcal{O}'_2$ (Fig. 5.4), final states are marked by the index $\{n_2, n_3\}$.

To complete the scenario, we have to consider the possibility of having ε (the null label) in the set of candidate labels marking a node of the temporal observation.

Unlike other (observable) labels, the null label can be matched regardless of the occurrence of component transitions. In effect, if ε is the actual label among all candidates for a node n, it means that no observable label was actually generated for n. Hence, n can be consumed without the triggering of any component transition.[2]

The specific nature of the null label requires the technique for indexing temporal observations to be extended in order to cope with the case in which ε is in fact the actual label.

Assume that a transition $(S,Q,I) \xrightarrow{t(c)} (S',Q',I')$ is generated in the reconstructed behavior, with I' being the index of a prefix P' of a temporal observation $\mathcal{O}$. If there exists a node $n \notin P'$ such that $\varepsilon \in \|n\|$ and, for each $n' \prec n$, $n' \in P'$, then a transition $(S,Q,I) \xrightarrow{t(c)} (S',Q',I'')$ is generated in the behavior, with I'' being the index of the prefix $P'' = P' \cup \{n\}$.

The result is that (S,Q,I) is exited by two transitions marked by the same component transition $t(c)$, with the states reached differing just in the index (I' vs. I'').

More generally, the generation of additional transitions marked by the same component transition $t(c)$ continues as long as a consumable node includes the null label. For instance, if there exists a node $n'' \notin P''$ such that $\varepsilon \in \|n''\|$ and, for each $n' \prec n''$, $n' \in P''$, then a transition $(S,Q,I) \xrightarrow{t(c)} (S',Q',I''')$ is generated in the behavior, with I''' being the index of $P''' = P'' \cup \{n''\}$.

Note 5.1. The creation of several transitions marked by the same symbol $t(c)$ causes the reconstructed behavior to be nondeterministic.

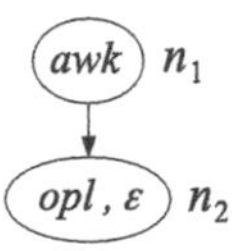

Fig. 5.7 Temporal observation $\mathcal{O}_\varepsilon$: node n_2 includes a null candidate label ε

Example 5.9. Shown in Fig. 5.7 is a temporal observation $\mathcal{O}_\varepsilon$, with $\varepsilon \in \|n_2\|$. The set of candidate traces is $\|\mathcal{O}_\varepsilon\| = \{[awk, opl], [awk]\}$.

Shown in Fig. 5.8 is the reconstructed behavior consistent with $\mathcal{O}_\varepsilon$. At state β_0, the component transitions consistent with the observable label *awk* are p_1 and p_3.

Considering p_1, a transition $\beta_0 \xrightarrow{p_1} \beta_1$ is generated in the behavior, with the index in β_1 being the singleton $\{n_1\}$ (which, in this case, equals the corresponding prefix). Besides, since $n_2 \notin \{n_1\}$, with $\varepsilon \in \|n_2\|$, and the preceding node n_1 is in the prefix, an additional transition $\beta_0 \xrightarrow{p_1} \beta_2$ is generated, with the index of β_2 being $\{n_2\}$.

As such, $[p_1]$ is also one of the 10 trajectories consistent with the candidate trace $[awk]$. In contrast, $[p_1, b_3(r), b_1(l)]$ and $[p_1, b_1(l), b_3(r)]$ are the only trajectories consistent with the candidate trace $[awk, opl]$.

[2] Similarly to the way in which an ε-transition is performed in an NFA (Chapter 2).

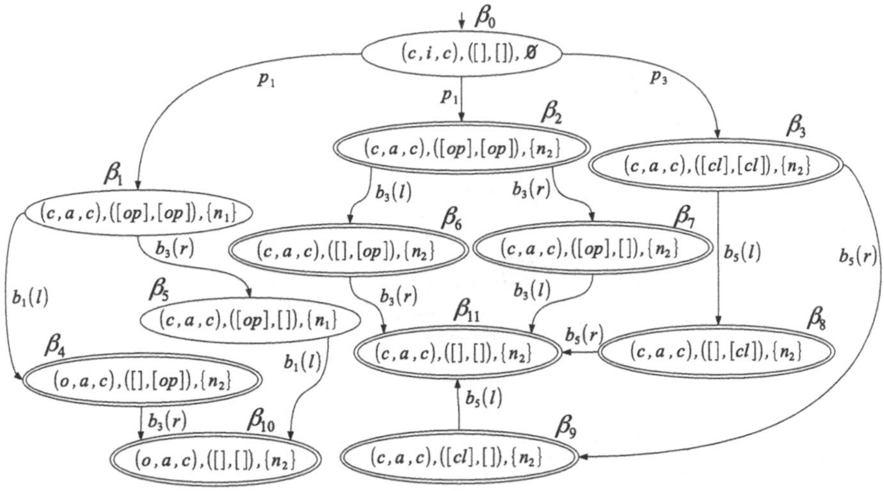

Fig. 5.8 Reconstructed behavior based on temporal observation $\mathcal{O}_\varepsilon$ (Fig. 5.7)

For the same reason, we have two transitions that exit β_0 and are marked by p_3. However, only one of them, at the end of the reconstruction, turns to be consistent with $\mathcal{O}_\varepsilon$, namely $\beta_0 \xrightarrow{p_3} \beta_3$, with β_3 being marked by the index $\{n_2\}$. This means that all trajectories starting with p_3 are consistent with the trace $[awk]$, but not with $[awk, opl]$.

In behavior reconstruction, the extended technique for indexing observations involving nodes marked by the null label ε is not completely satisfactory for one main reason: as pointed out in Note 5.1, the reconstructed behavior is nondeterministic.

From a practical viewpoint, nondeterminism in the reconstructed behavior is a problem because of the replication of subpaths of trajectories, leading to enlargement of the set of states and transitions.

One way to avoid nondeterminism might be to merge states (of the reconstructed behavior) which differ in the index field only. This is accomplished by generalizing the single index to a set $\mathbb{I}$ of indices. Given a state $\beta = (S, Q, \mathbb{I})$, if a new state (S, Q, I') is to be generated based on an index $I \in \mathbb{I}$, then β is updated to $(S, Q, \mathbb{I} \cup \{I'\})$.

Example 5.10. Displayed in Fig. 5.9 is the reconstructed behavior, with generalized indexing, consistent with $\mathcal{O}_\varepsilon$ (Fig. 5.7). Starting from the initial state β_0 marked by the index set $\mathbb{I} = \{\emptyset\}$, considering p_1, a single transition $\beta_0 \xrightarrow{p_1} \beta_1$ is generated, where the index set of β_1 is $\{\{n_1\}, \{n_2\}\}$. Notice how this set results from grouping the indices $\{n_1\}$ of β_1 and $\{n_2\}$ of β_2 in Fig. 5.8.

Considering β_1, since the component transition $b_1(l)$ (associated with the observable label *opl*) is consistent with the index $\{n_1\}$, the reconstructed behavior is extended by a transition $\beta_1 \xrightarrow{b_1(l)} \beta_3$, where the index set of β_3 becomes $\{\{n_2\}\}$, as

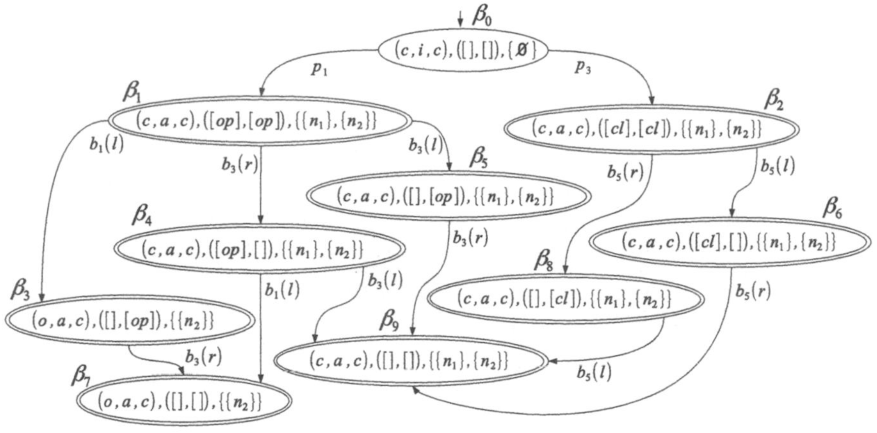

Fig. 5.9 Reconstructed behavior, with generalized indexing, based on $\mathcal{O}_\varepsilon$ (Fig. 5.7)

the result of replacing $\{n_1\}$ by $\{n_2\}$ within the index of β_1 (the duplicate index $\{n_2\}$ is removed). The same applies to the creation of the transition $\beta_4 \xrightarrow{b_1(l)} \beta_7$.

Eventually, the reconstructed behavior shown in Fig. 5.9 includes 10 states and 13 transitions, while the reconstructed behavior in Fig. 5.8 includes 12 states and 15 transitions. This is caused by the merging of states that differ in the index field only.

Although the technique of generalized indexing exemplified in Example 5.10 sounds appropriate for coping with temporal observations involving the null label, storing the index set in each state of the reconstructed behavior is not the optimal solution.

Ideally, the index field should be a scalar value, similarly to the integer index for linear observations (Fig. 5.1). This ideal solution presents both memory and computational advantages:

1. The contribution of the index field to the size of each state is minimized,
2. Owing to the simple (scalar) structure of the index field, the comparison between indices is efficient, thereby making the comparison between states simpler.

To this end, the notion of an index space is introduced in Def. 5.3.

Definition 5.3 (Index Space). Let $\mathcal{O} = (L, N, A)$ be a temporal observation. The *nondeterministic index space* of $\mathcal{O}$, denoted $Nsp(\mathcal{O})$, is an NFA

$$Nsp(\mathcal{O}) = (\Sigma, S, \tau, s_0, S_f), \tag{5.5}$$

where:

1. $\Sigma = L$ is the alphabet, with each symbol being a label in L,
2. S is the set of states, with each state being the index of a prefix of $\mathcal{O}$,
3. $s_0 = \emptyset$ is the initial state,
4. S_f is the singleton $\{s_f\}$, where $s_f \in S$, $Idx^{-1}(s_f) = N$,

5. $\tau : S \times \Sigma \mapsto 2^S$ is the nondeterministic transition function, where $s \xrightarrow{\ell} s' \in \tau$ iff

$$s' = Idx(s \cup \{n\}), \ell \in \|n\|, n \in N - s, \forall n' \to n \in A\,(n' \in s)\,. \qquad (5.6)$$

The *index space* of $\mathcal{O}$, denoted $Isp(\mathcal{O})$, is the DFA obtained by determinization of $Nsp(\mathcal{O})$. The language of $Isp(\mathcal{O})$ is denoted by $\|Isp(\mathcal{O})\|$.

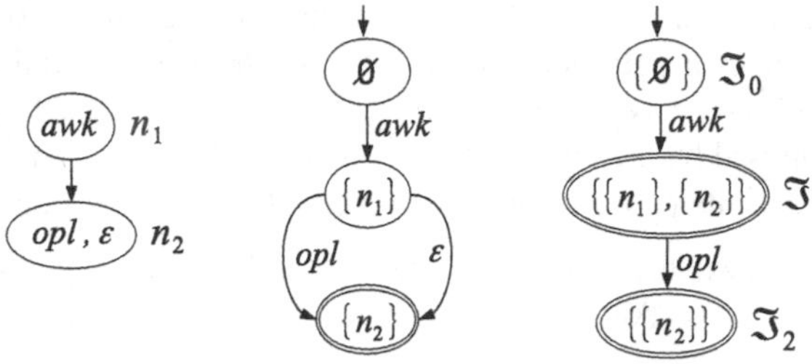

Fig. 5.10 Observation $\mathcal{O}_\varepsilon$ (left), $Nsp(\mathcal{O}_\varepsilon)$ (center), and $Isp(\mathcal{O}_\varepsilon)$ (right)

Example 5.11. Depicted on the left side of Fig. 5.10 is the temporal observation $\mathcal{O}_\varepsilon$ introduced in Example 5.9. The relevant nondeterministic index space is displayed in the center of the figure. $Nsp(\mathcal{O}_\varepsilon)$ can be generated starting from the initial state identified by the empty index.

Based on (5.6), a transition $s \xrightarrow{\ell} s'$ is generated, where s' is the index of the prefix obtained by extending the prefix P of s by a node n (not in P) including the label ℓ and such that all preceding nodes of n are in P.

Hence, starting from the initial state, a transition $\emptyset \xrightarrow{awk} \{n_1\}$ is generated. Then, since n_2 is the next (and last) node, where $\|n_2\| = \{opl, \varepsilon\}$, two transitions are inserted, namely $\{n_1\} \xrightarrow{opl} \{n_2\}$ and $\{n_1\} \xrightarrow{\varepsilon} \{n_2\}$, where $\{n_2\}$ is the final state, the index identifying all states of $\mathcal{O}_\varepsilon$.

Finally, the index space of $Isp(\mathcal{O}_\varepsilon)$ is obtained by determinization of $Nsp(\mathcal{O}_\varepsilon)$, as displayed on the right side of Fig. 5.10, where nodes are marked by the identifiers $\Im_0$, $\Im_1$, and $\Im_2$, with $\Im_0$ being the initial state, and $\Im_1$ and $\Im_2$ the final states.

The essential property of the index space of a temporal observation $\mathcal{O}$ is that its language equals the set of candidate traces of $\mathcal{O}$, as proven in Proposition 5.1.

Proposition 5.1. *The language of $Isp(\mathcal{O})$ equals $\|\mathcal{O}\|$.*

Proof. From Def. 5.3, since the language of $Isp(\mathcal{O})$ equals the language of $Nsp(\mathcal{O})$, it suffices to show that the language of $Nsp(\mathcal{O})$ equals $\|\mathcal{O}\|$.

Lemma 5.1. *If $\mathcal{T}$ is a sequence of labels chosen from a set $N_{\mathcal{T}}$ of nodes of $\mathcal{O}$ based on the precedence constraints, then $Nsp(\mathcal{O})$ includes a path generating $\mathcal{T}$, from the initial state to a state s, such that $Idx^{-1}(s) = N_{\mathcal{T}}$.*

Proof. The proof is by induction on $\mathcal{T}$. In the following, $\mathcal{T}(k)$ and $N_{\mathcal{T}(k)}$ denote the prefix of $\mathcal{T}$ up to the k-th label and the corresponding set of nodes in $\mathcal{O}$ from which labels in $\mathcal{T}(k)$ are chosen, respectively.

(*Basis*) For $k = 0$, we have $\mathcal{T}(0) = \varepsilon$, where $N_{\mathcal{T}(0)} = \emptyset$. On the other hand, $\mathcal{T}(0)$ is generated in $Nsp(\mathcal{O})$ by the null path rooted in the initial state $s_0 = \{\emptyset\}$, where $Idx^{-1}(s_0) = N_{\mathcal{T}(0)} = \emptyset$.

(*Induction*) If $\mathcal{T}(k)$ is a sequence of labels chosen from a set $N_{\mathcal{T}(k)}$ of nodes of $\mathcal{O}$ (based on the precedence constraints) and $Nsp(\mathcal{O})$ includes a path generating $\mathcal{T}(k)$, from the initial state to a state s, such that $Idx^{-1}(s) = N_{\mathcal{T}(k)}$, then $Nsp(\mathcal{O})$ includes a path generating $\mathcal{T}(k+1)$, from the initial state to a state s', such that $Idx^{-1}(s') = N_{\mathcal{T}(k+1)}$. In fact, on the one hand, based on the precedence constraints imposed by arcs in $\mathcal{O}$, the last label ℓ in $\mathcal{T}(k+1)$ can be chosen from a node n such that all its parent nodes are in $N_{\mathcal{T}(k)}$. On the other hand, based on (5.6), a transition $s \xrightarrow{\ell} s'$ exists in $Nsp(\mathcal{O})$. Hence, $\mathcal{T}(k+1)$ can be generated by a path in $Nsp(\mathcal{O})$, with $Idx^{-1}(s') = N_{\mathcal{T}(k+1)}$.

Lemma 5.2. *If $Nsp(\mathcal{O})$ includes a path, from the initial state to a state s, which generates a sequence $\mathcal{T}$ of labels, then $\mathcal{T}$ can be generated by choosing one label from each node of $\mathcal{O}$ in $Idx^{-1}(s)$ based on the precedence constraints.*

Proof. The proof is by induction on $\mathcal{T}$. In the following, $\mathcal{T}(k)$ and $N_{\mathcal{T}(k)}$ denote the prefix of $\mathcal{T}$ up to the k-th label and the corresponding set of nodes in $\mathcal{O}$ from which labels in $\mathcal{T}(k)$ are chosen, respectively.

(*Basis*) For $k = 0$, we have $\mathcal{T}(0) = \varepsilon$, which is generated in $Nsp(\mathcal{O})$ by the null path rooted in the initial state $s_0 = \{\emptyset\}$. On the other hand, since $Idx^{-1}(s_0) = \emptyset$, $\mathcal{T}(0)$ can be trivially generated by choosing no label, as $Idx^{-1}(s_0) = \emptyset$.

(*Induction*) If $Nsp(\mathcal{O})$ includes a path, from the initial state to a state s, which generates a sequence $\mathcal{T}(k)$ of labels, and $\mathcal{T}(k)$ can be generated by choosing one label from each node of $\mathcal{O}$ in $Idx^{-1}(s)$ based on the precedence constraints, then $\mathcal{T}(k+1)$ can be generated by choosing one label from each node in $Idx^{-1}(s')$ based on the precedence constraints, where $s \xrightarrow{\ell} s'$ is a transition in $Nsp(\mathcal{O})$ and ℓ is the last label in $\mathcal{T}(k+1)$. In fact, based on (5.6), $s' = Idx(s \cup \{n\})$, where n is a node not in s such that all its parent nodes are in s. As such, n can be selected as the next node in $Idx^{-1}(s')$, from which we choose label ℓ, thereby generating $\mathcal{T}(k+1)$.

Therefore, on the one hand, based on Lemma 5.1, if a string belongs to $\|\mathcal{O}\|$, then it belongs to the language of $Nsp(\mathcal{O})$. On the other hand, if a string belongs to $Nsp(\mathcal{O})$, then it belongs to $\|\mathcal{O}\|$. In other words, the language of $Nsp(\mathcal{O})$ equals $\|\mathcal{O}\|$. Hence, the proof of Proposition 5.1 is finally obtained from the equivalence of $Nsp(\mathcal{O})$ and $Isp(\mathcal{O})$. □

Proposition 5.1 opens the way to the ideal solution of indexing any temporal observation by a scalar value. Since the language of $Isp(\mathcal{O})$ coincides with the set of candidate traces of $\mathcal{O}$, each state of $Isp(\mathcal{O})$ is in fact the index of at least one candidate trace of $\mathcal{O}$. In other words, each state $\mathfrak{I}$ of $Isp(\mathcal{O})$ is the surrogate of one or several indices of $\mathcal{O}$, precisely, the surrogate of the indices identifying $\mathfrak{I}$.

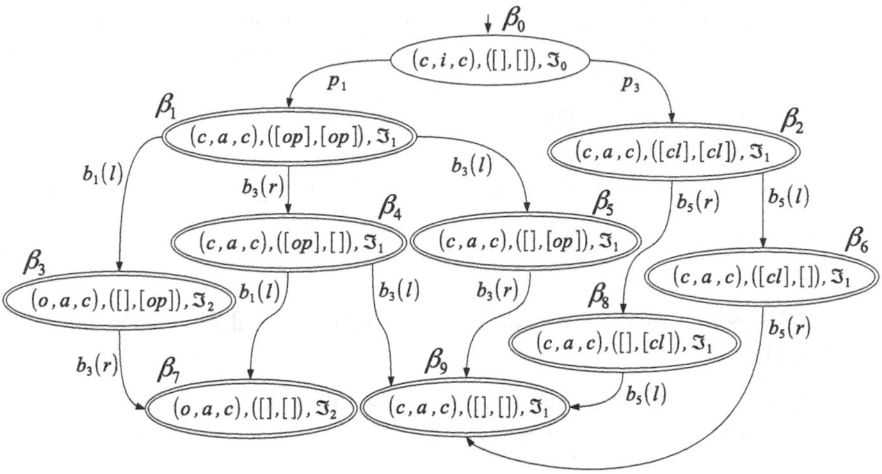

Fig. 5.11 Reconstructed behavior based on index space $Isp(\mathcal{O}_\varepsilon)$ (Fig. 5.10)

Example 5.12. With reference to the index space $Isp(\mathcal{O}_\varepsilon)$ displayed on the right side of Fig. 5.10, we have $\Im_0$ as the surrogate of index $\emptyset$, $\Im_1$ as the surrogate of indices $\{n_1\}$ and $\{n_2\}$, and $\Im_2$ as the surrogate of index $\{n_2\}$. As a result, the reconstructed behavior in Fig. 5.9 can be replaced by the behavior displayed in Fig. 5.11, where each set $\mathbb{I}$ of indices (corresponding to the third field of each state in Fig. 5.9) is substituted by the index $\Im$ identifying node $\mathbb{I}$ in $Isp(\mathcal{O}_\varepsilon)$.

The precise way in which the behavior is reconstructed based on the index space is specified in the next section.

5.2 Behavior Reconstruction

Before introducing the algorithm for behavior reconstruction, we provide a formal definition of a behavior.

Definition 5.4 (Behavior). Let $\wp(\mathcal{A}) = (a_0, \mathcal{V}, \mathcal{O}, \mathcal{R})$ be a diagnosis problem for an active system $\mathcal{A} = (C, L, D)$, where $C = \{c_1, \ldots, c_n\}$ and $L = \{l_1, \ldots, l_m\}$.

Let $Isp(\mathcal{O}) = (\Sigma', I, \tau', \Im_0, I_f)$ be the index space of a temporal observation $\mathcal{O}$, with Σ' being the set of observable labels, I the set of states, τ' the (deterministic) transition function, $\Im_0$ the initial state, and I_f the set of final states.

The *spurious behavior* of $\wp(\mathcal{A})$, written $Bhv^s(\wp(\mathcal{A}))$, is a DFA

$$Bhv^s(\wp(\mathcal{A})) = (\Sigma, B^s, \tau^s, \beta_0, B_f), \tag{5.7}$$

where:

1. Σ is the alphabet, consisting of the union of transitions of components in C,

2. B^s is the set of states $(S, Q, \Im)$, with $S = (s_1, \ldots, s_n)$ being a tuple of states of all components in C, $Q = (q_1, \ldots, q_m)$ a tuple of configurations of all links in L, and $\Im \in I$ a state of $Isp(\mathcal{O})$,
3. $\beta_0 = (S_0, Q_0, \Im_0)$ is the initial state, where $a_0 = (S_0, Q_0)$,
4. τ^s is the transition function $\tau^s : B^s \times \Sigma \mapsto B^s$, where $(S, Q, \Im) \xrightarrow{t(c)} (S', Q', \Im') \in \tau^s$, $S = (s_1, \ldots, s_n)$, $Q = (q_1, \ldots, q_m)$, $S' = (s'_1, \ldots, s'_n)$, $Q' = (q'_1, \ldots, q'_m)$, and

$$t(c) = s \xrightarrow{(e,x) \Rightarrow (e_1,y_1),\ldots,(e_p,y_p)} s', \tag{5.8}$$

with l^x denoting either the link in L entering input terminal x of component c, or **nil**, in the latter case when $x \in (D_{\text{on}} \cup \{In\})$, iff:

(4a) Either $l^x = \textbf{nil}$ or e is ready at terminal x within link l^x,
(4b) For each $i \in [1 \cdots n]$, we have

$$s'_i = \begin{cases} s' & \text{if } c_i = c, \\ s_i & \text{otherwise}. \end{cases} \tag{5.9}$$

(4c) Based on (3.5), for each $j \in [1 \cdots m]$, we have

$$q'_j = \begin{cases} Tail(q_j) & \text{if } l_j = l^x, \\ Ins(q_j, e_k) & \text{if } l_j \text{ exits terminal } y_k, k \in [1 \cdots p], \\ q_j & \text{otherwise}. \end{cases} \tag{5.10}$$

(4d) Either $t(c)$ is observable in viewer $\mathcal{V}$ via label ℓ, $\Im \xrightarrow{\ell} \bar{\Im}$ is a transition in $Isp(\mathcal{O})$, and $\Im' = \bar{\Im}$, or $t(c)$ is unobservable in $\mathcal{V}$ and $\Im' = \Im$,

5. B_f is the set of final states $(S_f, Q_f, \Im_f)$, where $\Im_f$ is final in $Isp(\mathcal{O})$.

The *behavior* of $\wp(\mathcal{A})$, written $Bhv(\wp(\mathcal{A}))$, is the DFA

$$Bhv(\wp(\mathcal{A})) = (\Sigma, B, \tau, \beta_0, B_f) \tag{5.11}$$

which results from removing from the spurious behavior $Bsp^s(\mathcal{A})$ all states and transitions that are not included in any path from the initial state to a final state.

The definition of $Bhv(\wp(\mathcal{A}))$ provided in Def. 5.4 is an extension of the definition of the behavior space $Bsp(\mathcal{A})$ provided in Def. 3.4. Essentially, Def. 5.4 differs from Def. 3.4 in four ways:

1. Each state in $Bhv(\wp(\mathcal{A}))$ includes an additional (third) field, the index $\Im$ of $\mathcal{O}$, which is a state of the index space $Isp(\mathcal{O})$.
2. The transition function requires an additional consistency check of each triggerable component transition, as detailed in point (4d).
3. The DFA characterizing $Bhv(\wp(\mathcal{A}))$ involves the additional set $B_f \subseteq B$ of final states, which are characterized by the final index $\Im_f$, as detailed in point 5.
4. Only states and transitions involved in paths from the initial state to a final state are retained.

Intuitively, the introduction of the index $\Im$ in the states of $Bhv(\wp(\mathcal{A}))$ allows for the checking of consistency of trajectories with the temporal observation $\mathcal{O}$. This way, the language of $Bhv(\wp(\mathcal{A}))$ is the subset of the language of $Bsp(\mathcal{A})$ composed of all the trajectories which are consistent with $\mathcal{O}$.

5.2.1 Reconstruction Algorithm

Listed below is the specification of the algorithm *Build*, which reconstructs the behavior relevant to a given diagnosis problem $\wp(\mathcal{A})$.

To this end, the index space of the temporal observation $\mathcal{O}$, $Isp(\mathcal{O})$, is generated (line 5). The initial state β_0 is set to $(S_0, Q_0, \Im_0)$, where S_0 is the tuple of initial states of components, Q_0 is the tuple of initial configurations of links, and $\Im_0$ is the initial state of $Isp(\mathcal{O})$ (line 6). The set B of states is initialized as the singleton $\{\beta_0\}$, while the set τ of transitions is set empty (line 7).

Within the main loop of the algorithm (lines 8–25), at each iteration, an unmarked state $\beta = (S, Q, \Im)$ is selected (line 9). The goal is to generate all transitions $\beta \xrightarrow{t(c)} \beta'$, with $\beta' = (S', Q', \Im')$.

Thus, each triggerable transition $t(c)$ in β is checked for consistency with $\mathcal{O}$ (lines 11–17). Specifically, if $t(c)$ is not observable, then it is consistent with $\mathcal{O}$ and $\Im'$ equals $\Im$ (lines 11 and 12). If, instead, $t(c)$ is observable via label ℓ and a transition $\Im \xrightarrow{\ell} \bar{\Im}$ is included in the index space, then $\Im'$ is set to $\bar{\Im}$ (lines 13 and 14). Otherwise, $t(c)$ is not consistent with $\mathcal{O}$, and lines 18–22 are skipped (line 16). If $t(c)$ is consistent, then tuples S' and Q' are computed by updating copies of S and Q, respectively (lines 18 and 19).

If the state $\beta' = (S', Q', \Im')$ is new, then it is inserted into B (line 21), while the new transition $\beta \xrightarrow{t(c)} \beta'$ is inserted into τ (line 22).

Once all triggerable transitions have been considered, state β is marked, as no further exiting transition exists (line 24).

When all states in B are marked, no further states or transitions can be generated. Hence, the set B_f of final states is determined (line 26).

At this point, B and τ are the states and transitions of the spurious behavior $Bhv^s(\wp(\mathcal{A}))$. To obtain the behavior $Bhv(\wp(\mathcal{A}))$ it suffices to remove the spurious part of the DFA, that is, the states and transitions which are not included in any path connecting the initial state to a final state (line 27).

The specification of the algorithm is as follows:

1. **algorithm** *Build* (**in** $\wp(A)$, **out** $Bhv(\wp(\mathcal{A}))$)
2. $\quad\wp(\mathcal{A}) = (a_0, \mathcal{V}, \mathcal{O}, \mathcal{R})$: a diagnosis problem for active system $\mathcal{A}$,
3. $\quad Bhv(\wp(\mathcal{A})) = (\Sigma, B, \tau, \beta_0, B_f)$: the behavior of $\wp(\mathcal{A})$;

4. $\quad$**begin** $\langle$*Build*$\rangle$
5. $\quad\quad$Generate the index space $Isp(\mathcal{O}) = (\Sigma, I, \tau', \Im_0, I_f)$;

$\beta_0 := (S_0, Q_0, \Im_0)$, where $a_0 = (S_0, Q_0)$;
$B := \{\beta_0\}$; $\tau := \emptyset$;
repeat
  Choose an unmarked state $\beta = (S, Q, \Im)$ in B;
  foreach component transition $t(c)$ triggerable in β **do**
    if $t(c)$ is unobservable in $\mathcal{V}$ **then**
      $\Im' := \Im$
    elsif $t(c)$ is observable in $\mathcal{V}$ via label ℓ **and** $\Im \xrightarrow{\ell} \bar{\Im}$ is in $Isp(\mathcal{O})$ **then**
      $\Im' := \bar{\Im}$
    else
      continue
    endif;
    Set S' as a copy of S and update state of component c based on $t(c)$;
    Set Q' as a copy of Q and update link configurations based on $t(c)$;
    $\beta' := (S', Q', \Im')$;
    if $\beta' \notin B$ **then** insert β' into B **endif**;
    Insert transition $\beta \xrightarrow{t(c)} \beta'$ into τ
  endfor;
  Mark β
until all states in B are marked;
$B_f := \{\beta_f \mid \beta_f \in B, \beta_f = (S, Q, \Im_f), \Im_f \in I_f\}$;
Remove from B and τ all states and transitions, respectively, which are not included in any path from β_0 to a final state
end $\langle$*Build*$\rangle$.

Example 5.13. Consider the diagnosis problem $\wp(\mathcal{A}_2) = (a_0, \mathcal{V}, \mathcal{O}_2, \mathcal{R})$ defined in Example 4.12. Reported on the left side of Fig. 5.12 is temporal observation $\mathcal{O}_2$. The corresponding index space is displayed on the right side of the figure (details of its computation are omitted). The spurious behavior of $\wp(\mathcal{A}_2)$ generated by *Build* is shown in Fig. 5.13, where the inconsistent part is indicated by dashed lines and gray states (named $\beta_1^s \cdots \beta_5^s$).

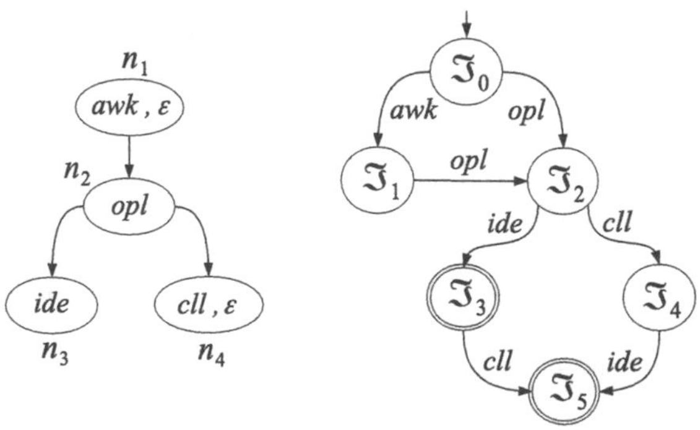

Fig. 5.12 Observation $\mathcal{O}_2$ (left) and $Isp(\mathcal{O}_2)$ (right)

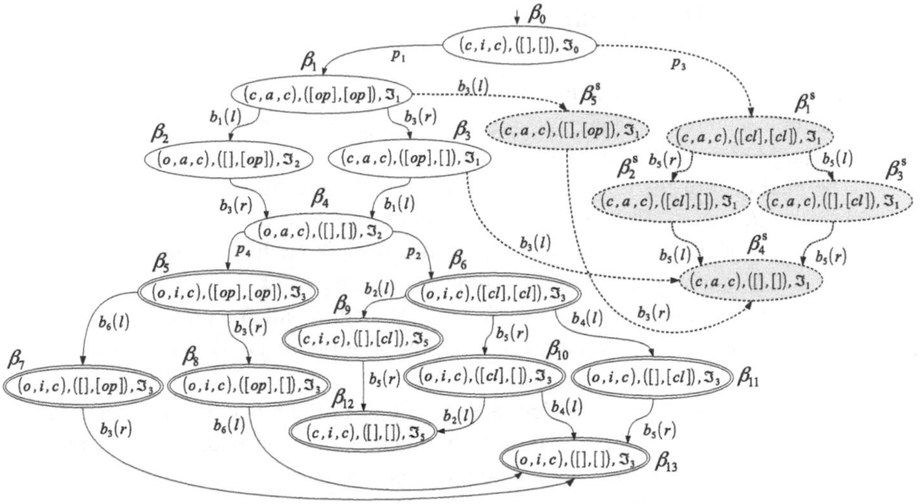

Fig. 5.13 Reconstructed behavior $Bhv(\wp(\mathcal{A}_2))$ based on temporal observation $\mathcal{O}_2$ (Fig. 5.12)

After the initialization of the set B of states with the singleton $\{\beta_0\}$, the main loop is iterated until all states in B have been processed. At the first iteration, the only state to be processed is the initial state β_0, in which the protection device p is idle, both of the breakers l and r are closed, and both links are empty.

In β_0, two component transitions are triggerable, p_1 and p_3, both relevant to the awakening of p. Since both of them are observable via the label *awk* and $Isp(\mathcal{O})$ includes a transition $\Im_0 \xrightarrow{awk} \Im_1$, two states are generated, β_1 and β_1^s, both marked by the index $\Im_1$, as well as two transitions, $\beta_0 \xrightarrow{p_1} \beta_1$ and $\beta_0 \xrightarrow{p_3} \beta_1^s$.

At this point, state β_0 is marked and the processing continues at the next iteration by considering one of the two newly created (unmarked) states.

At the end of the main loop, when all states have been processed (marked), the set B_f of final nodes is determined based on the associated index, which is required to be final in $Isp(\mathcal{O}_2)$. Hence, B_f is composed of states $\beta_5 \cdots \beta_{13}$, which are associated with either $\Im_3$ or $\Im_5$.

Eventually, the actual behavior $Bhv(\wp(\mathcal{A}_2))$ is obtained by removing all states and transitions that are not involved in any path from β_0 to a final state, thereby resulting in states $\beta_0 \cdots \beta_{12}$ and the relevant transitions.

According to Fig. 5.12, out of the six candidate traces in $\mathcal{O}_2$, in fact the language of $Isp(\mathcal{O}_2)$, only two traces are consistent with the reconstructed behavior, namely $[awk, opl, ide]$ and $[awk, opl, ide, cll]$.

Note 5.2. The behavior contains the trajectories (of the behavior space) that conform to the temporal observation. Conversely, a subset of the candidate traces of the temporal observation is possibly inconsistent with the reconstructed behavior.

5.3 Behavior Decoration

Behavior reconstruction is only the first (and primary) step of monolithic diagnosis. The result is a DFA where arcs are marked by component transitions. Each string in the language of the behavior is a trajectory consistent with the temporal observation.

The final goal of the diagnosis process is to generate the solution of a diagnosis problem $\wp(\mathcal{A}) = (a_0, \mathcal{V}, \mathcal{O}, \mathcal{R})$. According to (4.5), the solution $\Delta(\wp(\mathcal{A}))$ is the set of diagnoses of the trajectories based on the ruler $\mathcal{R}$ such that their trace on the viewer $\mathcal{V}$ is a candidate trace of the temporal observation $\mathcal{O}$.

Since the reconstructed behavior $Bhv(\wp(\mathcal{A}))$ includes all and only the trajectories whose trace on $\mathcal{V}$ is a candidate trace of $\mathcal{O}$, the solution of $\wp(A)$ is in fact the set of diagnoses relevant to all such trajectories.

In theory, it suffices to consider each trajectory in the behavior and generate the corresponding diagnosis based on (4.3). In practice, this approach is naive in nature because the set of trajectories in the behavior can be infinite.

When the set of trajectories is infinite, the behavior necessarily includes cycles. However, based on Note 4.6, if a trajectory h includes a cycle then there is an unbounded number of trajectories with the same diagnosis as h, with each trajectory being obtained by traversing the cycle a different number of times.

That is, several (possibly an infinite number of) trajectories may share the same diagnosis. After all, the set of diagnoses is always finite, this being bounded by $2^{|\Phi|}$, where $|\Phi|$ is the cardinality of the set Φ of fault labels defined in the ruler $\mathcal{R}$.

Now, the idea is to mark each state β of the behavior by the set of the diagnoses relevant to *all* the trajectories ending at β. We denote this set by $\Delta(\beta)$. The process of marking the behavior states with relevant diagnoses is called *behavior decoration*, giving rise to the *decorated behavior*, formally specified in Def. 5.5.

Definition 5.5 (Decorated Behavior). Let $Bhv(\wp(\mathcal{A}))$ be the behavior of a diagnosis problem $\wp(\mathcal{A})$. The decorated behavior $Bhv^*(\wp(\mathcal{A}))$ is the DFA obtained from $Bhv(\wp(\mathcal{A}))$ by marking each state β of the latter with a *diagnosis set* $\Delta(\beta)$ based on the application of the following two rules:

1. For the initial state β_0, $\Delta(\beta_0) = \{\emptyset\}$,
2. For each transition $\beta \xrightarrow{t(c)} \beta'$, for each $\delta \in \Delta(\beta)$, if $(t(c), f) \in \mathcal{R}$ then $\delta \cup \{f\} \in \Delta(\beta')$, otherwise $\delta \in \Delta(\beta')$.

The rationale for the above rules is as follows. Based on the first rule, the empty diagnosis corresponds to the empty trajectory. Based on the second rule, if the decoration of state β includes a diagnosis δ then there exists at least one trajectory h, ending at β, whose diagnosis is δ. Consequently, there exists a trajectory $h \cup [t(c)]$ ending at β' whose diagnosis is either the extension of δ by the faulty label f associated with $t(c)$ in $\mathcal{R}$, when $(t(c), f) \in \mathcal{R}$, or δ, when $(t(c), f) \notin \mathcal{R}$.

Unlike the first rule, which represents the base case and, as such, is applied only once, the second rule is inductive in nature. This means that, for the sake of completeness of the decoration, the second rule must be continuously applied until the decoration of the behavior cannot be changed.

5.3.1 Decoration Algorithm

Listed below is the specification of the algorithm *Decorate*, which, based on Def. 5.5, decorates the behavior relevant to a diagnosis problem $\wp(\mathcal{A})$ by means of the recursive auxiliary procedure *Dec*.

Considering the body of *Decorate* (lines 18–22), first the initial state β_0 is marked by the singleton $\{\emptyset\}$ (line 19), while all other states are marked by the empty set (line 20). Then, *Dec* is called with parameters β_0 and $\{\emptyset\}$.

Consider the specification of *Dec* (lines 5–17), which takes as input a state β and a set of diagnoses $\mathcal{D}$, the latter being an extension of the decoration of β. The aim of *Dec* is to propagate $\mathcal{D}$ to each neighboring state β' of β based on the second decoration rule in Def. 5.5. To this end, if the decoration of β' is extended by a nonempty set $\mathcal{D}^+$ of diagnoses, then *Dec* is recursively called on β' and $\mathcal{D}^+$. *Dec* stops when no further diagnosis can be generated.

The specification of the algorithm *Decorate* is as follows:

1. **algorithm** *Decorate* (**in** $Bhv(\wp(A))$)
2. $Bhv(\wp(A))$: the behavior of $\wp(\mathcal{A}) = (a_0, \mathcal{V}, \mathcal{O}, \mathcal{R})$;

3.   **side effects**
4. $Bhv(\wp(\mathcal{A}))$ is decorated, thereby becoming $Bhv^*(\wp(\mathcal{A}))$;

5.   **auxiliary procedure** *Dec* $(\beta, \mathcal{D})$
6. β: a state of $Bhv(\wp(\mathcal{A}))$,
7. $\mathcal{D}$: a set of diagnoses;
8.   **begin** $\langle$*Dec*$\rangle$
9.   **foreach** transition $\beta \xrightarrow{t(c)} \beta'$ **do**
10. $\mathcal{D}^+ := \emptyset$;
11.   **foreach** $\delta \in \mathcal{D}$ **do**
12. δ' := **if** $(t(c), f) \in \mathcal{R})$ **then** $\delta \cup \{f\}$ **else** δ **endif**;
13.   **if** $\delta' \notin \Delta(\beta')$ **then** insert δ' into both $\Delta(\beta')$ and $\mathcal{D}^+$ **endif**
14.   **endfor**;
15.   **if** $\mathcal{D}^+ \neq \emptyset$ **then** *Dec* $(\beta', \mathcal{D}^+)$ **endif**
16.   **endfor**
17.   **end** $\langle$*Dec*$\rangle$;

18.   **begin** $\langle$*Decorate*$\rangle$
19.   Mark the initial state β_0 with the singleton $\{\emptyset\}$;
20.   Mark all other (noninitial) states with the empty set $\emptyset$;
21.   *Dec* $(\beta_0, \{\emptyset\})$
22.   **end** $\langle$*Decorate*$\rangle$.

Example 5.14. Consider the behavior $Bhv(\wp(\mathcal{A}_2))$ displayed in Fig. 5.13. Reported in Fig. 5.14 is the same behavior (with the spurious part omitted), where faulty tran-

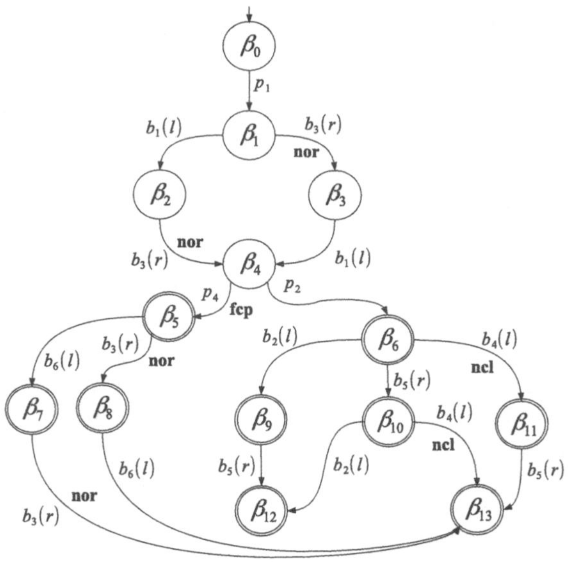

Fig. 5.14 Consistent part of the reconstructed behavior $Bhv(\wp(\mathcal{A}_2))$ displayed in Fig. 5.13, with faulty transitions being marked by corresponding fault labels, as defined in ruler $\mathcal{R}$ (Table 4.1)

sitions are marked by corresponding fault labels, as defined in the ruler $\mathcal{R}$ specified in Table 4.1.

Based on the specification of the *Decorate* algorithm, Table 5.1 traces the sequence of calls to the recursive procedure *Dec*. Specifically, each line indicates the number of the call to *Dec* with the relevant parameters $(\beta, \mathcal{D}^+)$, along with the diagnosis sets $\Delta(\beta_i)$, $i \in [0 \cdots 12]$. For space reasons, diagnoses are indicated as in Example 4.13, namely $\delta_r = \{\textbf{nor}\}$, $\delta_{lr} = \{\textbf{nol}, \textbf{nor}\}$, and $\delta_{rp} = \{\textbf{nor}, \textbf{fcp}\}$.

Table 5.1 Trace of *Decorate* applied to the behavior displayed in Fig. 5.14 (Example 5.14)

N	*Call*	$\Delta(\beta_0)$	$\Delta(\beta_1)$	$\Delta(\beta_2)$	$\Delta(\beta_3)$	$\Delta(\beta_4)$	$\Delta(\beta_5)$	$\Delta(\beta_6)$	$\Delta(\beta_7)$	$\Delta(\beta_8)$	$\Delta(\beta_9)$	$\Delta(\beta_{10})$	$\Delta(\beta_{11})$	$\Delta(\beta_{12})$	$\Delta(\beta_{13})$
1	$(\beta_0, \{\emptyset\})$	$\{\emptyset\}$	$\emptyset$	$\emptyset$	$\emptyset$	$\emptyset$	$\emptyset$	$\emptyset$	$\emptyset$	$\emptyset$	$\emptyset$	$\emptyset$	$\emptyset$	$\emptyset$	$\emptyset$
2	$(\beta_1, \{\emptyset\})$	$\{\emptyset\}$	$\{\emptyset\}$	$\emptyset$	$\emptyset$	$\emptyset$	$\emptyset$	$\emptyset$	$\emptyset$	$\emptyset$	$\emptyset$	$\emptyset$	$\emptyset$	$\emptyset$	$\emptyset$
3	$(\beta_2, \{\emptyset\})$	$\{\emptyset\}$	$\{\emptyset\}$	$\{\emptyset\}$	$\emptyset$	$\emptyset$	$\emptyset$	$\emptyset$	$\emptyset$	$\emptyset$	$\emptyset$	$\emptyset$	$\emptyset$	$\emptyset$	$\emptyset$
4	$(\beta_4, \{\delta_r\})$	$\{\emptyset\}$	$\{\emptyset\}$	$\{\emptyset\}$	$\emptyset$	$\{\delta_r\}$	$\emptyset$	$\emptyset$	$\emptyset$	$\emptyset$	$\emptyset$	$\emptyset$	$\emptyset$	$\emptyset$	$\emptyset$
5	$(\beta_5, \{\delta_{rp}\})$	$\{\emptyset\}$	$\{\emptyset\}$	$\{\emptyset\}$	$\emptyset$	$\{\delta_r\}$	$\{\delta_{rp}\}$	$\emptyset$	$\emptyset$	$\emptyset$	$\emptyset$	$\emptyset$	$\emptyset$	$\emptyset$	$\emptyset$
6	$(\beta_7, \{\delta_{rp}\})$	$\{\emptyset\}$	$\{\emptyset\}$	$\{\emptyset\}$	$\emptyset$	$\{\delta_r\}$	$\{\delta_{rp}\}$	$\emptyset$	$\{\delta_{rp}\}$	$\emptyset$	$\emptyset$	$\emptyset$	$\emptyset$	$\emptyset$	$\emptyset$
7	$(\beta_{13}, \{\delta_{rp}\})$	$\{\emptyset\}$	$\{\emptyset\}$	$\{\emptyset\}$	$\emptyset$	$\{\delta_r\}$	$\{\delta_{rp}\}$	$\emptyset$	$\{\delta_{rp}\}$	$\emptyset$	$\emptyset$	$\emptyset$	$\emptyset$	$\emptyset$	$\{\delta_{rp}\}$
8	$(\beta_8, \{\delta_{rp}\})$	$\{\emptyset\}$	$\{\emptyset\}$	$\{\emptyset\}$	$\emptyset$	$\{\delta_r\}$	$\{\delta_{rp}\}$	$\emptyset$	$\{\delta_{rp}\}$	$\{\delta_{rp}\}$	$\emptyset$	$\emptyset$	$\emptyset$	$\emptyset$	$\{\delta_{rp}\}$
9	$(\beta_{13}, \{\delta_{rp}\})$	$\{\emptyset\}$	$\{\emptyset\}$	$\{\emptyset\}$	$\emptyset$	$\{\delta_r\}$	$\{\delta_{rp}\}$	$\emptyset$	$\{\delta_{rp}\}$	$\{\delta_{rp}\}$	$\emptyset$	$\emptyset$	$\emptyset$	$\emptyset$	$\{\delta_{rp}\}$
10	$(\beta_6, \{\delta_r\})$	$\{\emptyset\}$	$\{\emptyset\}$	$\{\emptyset\}$	$\emptyset$	$\{\delta_r\}$	$\{\delta_{rp}\}$	$\{\delta_r\}$	$\{\delta_{rp}\}$	$\{\delta_{rp}\}$	$\emptyset$	$\emptyset$	$\emptyset$	$\emptyset$	$\{\delta_{rp}\}$
11	$(\beta_9, \{\delta_r\})$	$\{\emptyset\}$	$\{\emptyset\}$	$\{\emptyset\}$	$\emptyset$	$\{\delta_r\}$	$\{\delta_{rp}\}$	$\{\delta_r\}$	$\{\delta_{rp}\}$	$\{\delta_{rp}\}$	$\{\delta_r\}$	$\emptyset$	$\emptyset$	$\emptyset$	$\{\delta_{rp}\}$
12	$(\beta_{12}, \{\delta_r\})$	$\{\emptyset\}$	$\{\emptyset\}$	$\{\emptyset\}$	$\emptyset$	$\{\delta_r\}$	$\{\delta_{rp}\}$	$\{\delta_r\}$	$\{\delta_{rp}\}$	$\{\delta_{rp}\}$	$\{\delta_r\}$	$\emptyset$	$\emptyset$	$\{\delta_r\}$	$\{\delta_{rp}\}$
13	$(\beta_{10}, \{\delta_r\})$	$\{\emptyset\}$	$\{\emptyset\}$	$\{\emptyset\}$	$\emptyset$	$\{\delta_r\}$	$\{\delta_{rp}\}$	$\{\delta_r\}$	$\{\delta_{rp}\}$	$\{\delta_{rp}\}$	$\{\delta_r\}$	$\{\delta_r\}$	$\emptyset$	$\{\delta_r\}$	$\{\delta_{rp}\}$
14	$(\beta_{12}, \{\delta_r\})$	$\{\emptyset\}$	$\{\emptyset\}$	$\{\emptyset\}$	$\emptyset$	$\{\delta_r\}$	$\{\delta_{rp}\}$	$\{\delta_r\}$	$\{\delta_{rp}\}$	$\{\delta_{rp}\}$	$\{\delta_r\}$	$\{\delta_r\}$	$\emptyset$	$\{\delta_r\}$	$\{\delta_{rp}\}$
15	$(\beta_{13}, \{\delta_{lr}\})$	$\{\emptyset\}$	$\{\emptyset\}$	$\{\emptyset\}$	$\emptyset$	$\{\delta_r\}$	$\{\delta_{rp}\}$	$\{\delta_r\}$	$\{\delta_{rp}\}$	$\{\delta_{rp}\}$	$\{\delta_r\}$	$\{\delta_r\}$	$\emptyset$	$\{\delta_r\}$	$\{\delta_{rp}, \delta_{lr}\}$
16	$(\beta_{11}, \{\delta_{lr}\})$	$\{\emptyset\}$	$\{\emptyset\}$	$\{\emptyset\}$	$\emptyset$	$\{\delta_r\}$	$\{\delta_{rp}\}$	$\{\delta_r\}$	$\{\delta_{rp}\}$	$\{\delta_{rp}\}$	$\{\delta_r\}$	$\{\delta_r\}$	$\{\delta_{lr}\}$	$\{\delta_r\}$	$\{\delta_{rp}, \delta_{lr}\}$
17	$(\beta_{13}, \{\delta_{lr}\})$	$\{\emptyset\}$	$\{\emptyset\}$	$\{\emptyset\}$	$\emptyset$	$\{\delta_r\}$	$\{\delta_{rp}\}$	$\{\delta_r\}$	$\{\delta_{rp}\}$	$\{\delta_{rp}\}$	$\{\delta_r\}$	$\{\delta_r\}$	$\{\delta_{lr}\}$	$\{\delta_r\}$	$\{\delta_{rp}, \delta_{lr}\}$
18	$(\beta_3, \{\delta_r\})$	$\{\emptyset\}$	$\{\emptyset\}$	$\{\emptyset\}$	$\{\delta_r\}$	$\{\delta_r\}$	$\{\delta_{rp}\}$	$\{\delta_r\}$	$\{\delta_{rp}\}$	$\{\delta_{rp}\}$	$\{\delta_r\}$	$\{\delta_r\}$	$\{\delta_{lr}\}$	$\{\delta_r\}$	$\{\delta_{rp}, \delta_{lr}\}$

The first call to $Dec(\beta_0, \{\emptyset\})$ by the *Decorate* algorithm (line 21) is activated with a decoration in which all states but β_0 are marked by the empty diagnosis set, while $\Delta(\beta_0) = \{\emptyset\}$. The subsequent calls are generated assuming that the transitions within the loop at line 9 are considered from left to right in Fig. 5.14, in other words, considering the numbers of entered states in ascending order. For instance, since β_1 is exited by two transitions, entering β_2 and β_3, the transition entering β_2 is considered before the transition entering β_3. After the decoration of β_3 has been updated, the procedure *Dec* is called for the last time ($N = 18$) with parameters $(\beta_3, \{\delta_r\})$, as the propagation of δ_r does not change the current decoration of β_4.

Example 5.15. In Example 5.14, the decorated behavior is acyclic. To show how *Decorate* works in the general case, we consider a further example with cyclic behavior, as displayed in Fig. 5.15, where transitions are marked by faults only. Despite the fact that the behavior is very simple and abstract in nature, it nonetheless includes a cycle between nodes β_2 and β_3. The sequence of calls to *Dec* is listed in Table 5.2, where the diagnoses involved are $\delta_l = \{\mathbf{nol}\}$, $\delta_r = \{\mathbf{nor}\}$, $\delta_{lr} = \{\mathbf{nol}, \mathbf{nor}\}$, and $\delta_{rp} = \{\mathbf{nor}, \mathbf{fcp}\}$.

The cycle is traversed the first time by three calls, namely 3, 4, and 5, and a second time by four calls, namely 6, 7, 8, and 9. In either case, the cyclic propagation stops when no change in the decoration occurs, specifically, when the list of diagnosis sets $\Delta(\beta_0), \ldots, \Delta(\beta_3)$ does not change (calls 5 and 9, respectively).

Table 5.2 Trace of *Decorate* applied to the behavior displayed in Fig. 5.15 (Example 5.15)

N	*Call*	$\Delta(\beta_0)$	$\Delta(\beta_1)$	$\Delta(\beta_2)$	$\Delta(\beta_3)$
1	$(\beta_0, \{\emptyset\})$	$\{\emptyset\}$	$\emptyset$	$\emptyset$	$\emptyset$
2	$(\beta_1, \{\emptyset\})$	$\{\emptyset\}$	$\{\delta_r\}$	$\emptyset$	$\emptyset$
3	$(\beta_3, \{\delta_r\})$	$\{\emptyset\}$	$\{\delta_r\}$	$\emptyset$	$\{\delta_{rp}\}$
4	$(\beta_2, \{\delta_{rp}\})$	$\{\emptyset\}$	$\{\delta_r\}$	$\{\delta_{rp}\}$	$\{\delta_{rp}\}$
5	$(\beta_3, \{\delta_{rp}\})$	$\{\emptyset\}$	$\{\delta_r\}$	$\{\delta_{rp}\}$	$\{\delta_{rp}\}$
6	$(\beta_2, \{\emptyset\})$	$\{\emptyset\}$	$\{\delta_r\}$	$\{\delta_{rp}, \delta_l\}$	$\{\delta_{rp}\}$
7	$(\beta_3, \{\delta_l\})$	$\{\emptyset\}$	$\{\delta_r\}$	$\{\delta_{rp}, \delta_l\}$	$\{\delta_{rp}, \delta_{lr}\}$
8	$(\beta_2, \{\delta_{lr}\})$	$\{\emptyset\}$	$\{\delta_r\}$	$\{\delta_{rp}, \delta_l, \delta_{lr}\}$	$\{\delta_{rp}, \delta_{lr}\}$
9	$(\beta_3, \{\delta_{lr}\})$	$\{\emptyset\}$	$\{\delta_r\}$	$\{\delta_{rp}, \delta_l, \delta_{lr}\}$	$\{\delta_{rp}, \delta_{lr}\}$

5.4 Diagnosis Distillation

Once the reconstructed behavior $Bhv(\wp(\mathcal{A}))$ has been decorated, the solution of the diagnosis problem $\wp(\mathcal{A})$ can be straightforwardly determined as the union of the diagnostic sets associated with the final states, namely

$$\Delta(\wp(\mathcal{A})) = \{\, \delta \mid \delta \in \Delta(\beta_f), \beta_f \text{ is final in } Bhv^*(\wp(\mathcal{A}))\,\}\,. \tag{5.12}$$

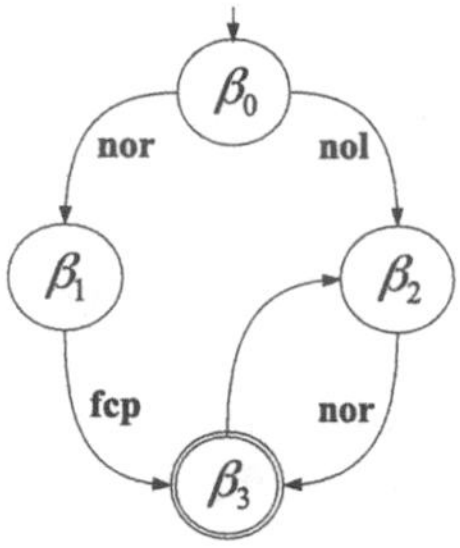

Fig. 5.15 Cyclic reconstructed behavior, with faulty transitions being marked by corresponding fault labels only, as defined in ruler $\mathcal{R}$ (Table 4.1)

This property is formally proven in Theorem 5.1. First, we apply it to our reference example.

Example 5.16. Consider the behavior $Bhv(\wp(\mathcal{A}_2))$ displayed in Fig. 5.14, whose decoration is listed in the last line of Table 5.1 ($N = 18$). Since the final states are $\beta_5, \ldots, \beta_{13}$, the solution of $\wp(\mathcal{A}_2)$ will be

$$\Delta(\wp(\mathcal{A}_2)) = \bigcup_{i \in [5 \cdots 13]} \Delta(\beta_i) = \{\, \delta_r, \delta_{lr}, \delta_{rp} \,\} = \{\, \{\mathbf{nor}\}, \{\mathbf{ncl}, \mathbf{nor}\}, \{\mathbf{nor}, \mathbf{fcp}\} \,\}\,.$$

As expected, this set of candidate diagnoses equals the solution of $\wp(\mathcal{A}_2)$ determined in Example 4.13 based on the formal definition of a diagnosis-problem solution stated in Def. 4.8.

Theorem 5.1. *Let $\wp(\mathcal{A}) = (a_0, \mathcal{V}, \mathcal{O}, \mathcal{R})$ be a diagnosis problem and $Bhv^*(\wp(\mathcal{A}))$ the corresponding decorated behavior. The union of the diagnosis sets associated with the final states of $Bhv^*(\wp(\mathcal{A}))$ equals the solution of $\wp(\mathcal{A})$.*

Proof. The proof is grounded on Lemmas 5.3–5.7, where $\mathcal{B}^{\mathrm{s}}$ and $\mathcal{B}^*$ denote $Bsp(\mathcal{A})$ and $Bhv^*(\wp(\mathcal{A}))$, respectively.

Lemma 5.3. *If h is a trajectory in $\mathcal{B}^*$, then h is a trajectory in $\mathcal{B}^{\mathrm{s}}$.*

Proof. This derives from the fact that B^* differs from $\mathcal{B}^{\mathrm{s}}$ in the additional index $\mathfrak{I}$, which is irrelevant to the triggering of transitions. By induction on h, starting from the initial state, each new transition applicable to $\mathcal{B}^*$ is applicable to $\mathcal{B}^{\mathrm{s}}$ too.

Lemma 5.4. *If h is a trajectory in $\mathcal{B}^*$, then $h_{[\mathcal{V}]} \in \|\mathcal{O}\|$.*

Proof. Recall that $h_{[\mathcal{V}]}$ is the sequence of labels associated with observable transitions in the viewer $\mathcal{V}$. Based on the definition of $\mathcal{B}^*$, $h_{[\mathcal{V}]}$ belongs to the language of $Isp(\mathcal{O})$, which, based on Proposition 5.1, equals $\|\mathcal{O}\|$. Hence, $h_{[\mathcal{V}]} \in \|\mathcal{O}\|$.

Lemma 5.5. *If h is a trajectory in $\mathcal{B}^{\mathrm{s}}$ and $h_{[\mathcal{V}]} \in \|\mathcal{O}\|$, then h is a trajectory in $\mathcal{B}^*$.*

Proof. By induction on h, starting from the initial state, each new transition t applicable to $\mathcal{B}^{\mathrm{s}}$ is applicable to $\mathcal{B}^*$ too. In fact, if t is unobservable then no further condition is required. If t is observable, based on the assumption $h_{[\mathcal{V}]} \in \|\mathcal{O}\|$ and on the fact that the language of $Isp(\mathcal{O})$ equals $\|\mathcal{O}\|$, the label associated with t in the viewer $\mathcal{V}$ matches a transition in $Isp(\mathcal{O})$.

Lemma 5.6. *If h is a trajectory in $\mathcal{B}^*$ ending at a final state β_f, then $h_{[\mathcal{R}]} \in \Delta(\beta_f)$.*

Proof. Based on the two decoration rules in Def. 5.5, the proof is by induction on h. In the following, $h(k)$ denotes the prefix of h up to the k-th component transition, with β_k being the state in $\mathcal{B}^*$ reached by $h(k)$.

(*Basis*) $h(0)_{[\mathcal{R}]} \in \Delta(\beta_0)$. In fact, $h(0)_{[\mathcal{R}]} = \emptyset$, which, based on the first decoration rule, belongs to $\Delta(\beta_0)$.

(*Induction*) If $h(k)_{[\mathcal{R}]} \in \Delta(\beta_k)$, then $h(k+1)_{[\mathcal{R}]} \in \Delta(\beta_{k+1})$. In fact, let t_{k+1} be the last transition in $h(k+1)$. Based on the second decoration rule, for $\beta_k \xrightarrow{t_{k+1}} \beta_{k+1}$, if $\delta \in \Delta(\beta_k)$ then $\delta' \in \Delta(\beta_{k+1})$, where $\delta' = \delta \cup \{f\}$ if $(t_{k+1}, f) \in \mathcal{R}$, otherwise $\delta' = \delta$. Hence, in either case, based on (4.3), $h(k+1)_{[\mathcal{R}]} \in \Delta(\beta_{k+1})$.

Lemma 5.7. *If β_f is a final state in $\mathcal{B}^*$ and $\delta \in \Delta(\beta_f)$, then there exists a trajectory h in $\mathcal{B}^*$ ending at β_f such that $h_{[\mathcal{R}]} = \delta$.*

Proof. Based on the decoration rules for $\mathcal{B}^*$, the diagnosis δ is incrementally generated by a path h starting from the empty diagnosis initially associated with β_0. On the one hand, in order for h to be a trajectory, it must be finite. If h is infinite, then it must include (at least) a cycle in $\mathcal{B}^*$ traversed an infinite number of times. On the other hand, once a cycle is traversed, all associated fault labels are inserted into δ: successive iterations of the cycle do not extend δ, because of duplicate removals caused by set-theoretic union in the second decoration rule. In other words, δ can always be generated by a finite trajectory h, namely $h_{[\mathcal{R}]} = \delta$.

Finally, Theorem 5.1 is proven by showing that $\delta \in \Delta(\mathcal{B}^*) \Leftrightarrow \delta \in \Delta(\wp(\mathcal{A}))$. On the one hand, if $\delta \in \Delta(\mathcal{B}^*)$ then, based on Lemmas 5.3, 5.4, and 5.7, there exists a trajectory $h \in \mathcal{B}^s$ such that $h_{[\mathcal{V}]} \in \|\mathcal{O}\|$ and $h_{[\mathcal{R}]} = \delta$; in other words, based on (4.5), $\delta \in \Delta(\wp(\mathcal{A}))$. On the other hand, if $\delta \in \Delta(\wp(\mathcal{A}))$ then, according to (4.5) and based on Lemmas 5.5 and 5.6, there exists a trajectory $h \in \mathcal{B}^*$ ending at a final state β_f such that $\delta = h_{[\mathcal{R}]}$ and $\delta \in \Delta(\beta_f)$, in other words, $\delta \in \Delta(\mathcal{B}^*)$. □

5.5 Bibliographic Notes

Monolithic diagnosis of active systems with plain observations was introduced in [10], and extended in [84] to systems integrating synchronous and asynchronous behavior. Monolithic diagnosis with (uncertain) temporal observations was proposed and analyzed in [85, 87, 88].

5.6 Summary of Chapter 5

Observation Prefix. A (possibly empty) set of nodes that can be chosen from a temporal observation based on the partial temporal ordering imposed by arcs.

Observation Indexing. The task of maintaining information on the prefixes of a temporal observation.

Prefix Index. Given a prefix P of a temporal observation, the index I of P is the minimal set of nodes identifying P, denoted $I = Idx(P)$; conversely, $P = Idx^{-1}(I)$.

Index Space. Given a temporal observation $\mathcal{O}$, the index space of $\mathcal{O}$, denoted $Isp(\mathcal{O})$, is a DFA whose language equals $\|\mathcal{O}\|$.

Behavior. Given a diagnosis problem $\wp(A) = (a_0, \mathcal{V}, \mathcal{O}, \mathcal{R})$, the behavior of $\wp(A)$, denoted $Bhv(\wp(A))$, is a DFA whose language is the subset of trajectories in the behavior space $Bsp(\mathcal{A})$ that are consistent with the temporal observation $\mathcal{O}$.

Behavior Reconstruction. The task of generating the behavior of a given diagnosis problem $\wp(A) = (a_0, \mathcal{V}, \mathcal{O}, \mathcal{R})$. Each state of the reconstructed behavior is a triple $(S, Q, \mathfrak{I})$, where S is a tuple of component states, Q a tuple of link configurations, and $\mathfrak{I}$ a state of the index space of the temporal observation $\mathcal{O}$.

Behavior Decoration. The process of marking each state β of a reconstructed behavior $Bhv(\wp(\mathcal{A}))$ with the set of diagnoses relevant to all trajectories up to β, denoted $\Delta(\beta)$. The resulting decorated behavior is denoted $Bhv^*(\wp(\mathcal{A}))$.

Diagnosis Distillation. The task of generating the solution of a diagnosis problem $\wp(\mathcal{A})$ based on the decorated behavior $Bhv^*(\wp(\mathcal{A}))$, by merging the diagnosis sets marking its final nodes.

Chapter 6
Modular Diagnosis

In Chapter 5 we presented a technique for generating the solution of a diagnosis problem $\wp(\mathcal{A}) = (a_0, \mathcal{V}, \mathcal{O}, \mathcal{R})$, where a_0 is the initial state, $\mathcal{V}$ the viewer, $\mathcal{O}$ the temporal observation, and $\mathcal{R}$ the ruler.

The technique is composed of three steps: behavior reconstruction, behavior decoration, and diagnosis distillation. In the first step, which is also the most complex one, the behavior of system $\mathcal{A}$ is generated as a DFA, denoted $Bhv(\wp(\mathcal{A}))$, where each arc is marked by a component transition.

The peculiarity of the reconstructed behavior is that the sequence of component transitions generated by a path connecting the initial state to a final state is a trajectory in the behavior space of $\mathcal{A}$ that is consistent with temporal observation $\mathcal{O}$. In other words, the language of $Bhv(\wp(\mathcal{A}))$ is the subset of trajectories in $Bsp(\mathcal{A})$ that conform to $\mathcal{O}$.

In order to get the set of candidate diagnoses, two subsequent steps are required. First, each state of the reconstructed behavior is marked by the set of diagnoses relevant to all the trajectories ending at such a state (behavior decoration). Then, the set of candidate diagnoses is generated by collecting the diagnoses associated with the final states of the behavior (diagnosis distillation).

This diagnosis technique can be described as *monolithic* insofar as the reconstruction of the behavior (first step) is carried out in one shot, based on the topology of system $\mathcal{A}$, on relevant component and link models, and on the temporal observation.

However, monolithic reconstruction is not the only possible way to yield the behavior of the system based on $\wp(\mathcal{A})$. Interestingly, behavior reconstruction can also be performed in a modular way, hence the name *modular* diagnosis.

The idea is simple: instead of reconstructing the behavior of $\mathcal{A}$ in one shot (by means of the *Build* algorithm), system $\mathcal{A}$ is decomposed into a set of subsystems $\{\mathcal{A}_1, \ldots, \mathcal{A}_n\}$. Accordingly, the diagnosis problem $\wp(\mathcal{A})$ is decomposed into a set of subproblems $\{\wp(\mathcal{A}_1), \ldots, \wp(\mathcal{A}_n)\}$, where each subproblem $\wp(\mathcal{A}_i)$ is relevant to subsystem $\mathcal{A}_i$. This way, the generation of $Bhv(\wp(\mathcal{A}))$ is performed in a modular way, by means of two steps:

1. Reconstruction of the set of subsystem behaviors $\{Bhv(\wp(\mathcal{A}_1), \ldots, Bhv(\wp(\mathcal{A}_n)\}$,

G. Lamperti et al., *Introduction to Diagnosis of Active Systems*,
https://doi.org/10.1007/978-3-319-92733-6_6

2. Merging of the behaviors generated in step 1 into the behavior of $\wp(\mathcal{A})$.

A critical point to be addressed is how a subproblem should be defined for each subsystem $\mathcal{A}_i$, $i \in [1 \cdots n]$, given the decomposition of $\mathcal{A}$. Since $\wp(\mathcal{A}) = (a_0, \mathcal{V}, \mathcal{O}, \mathcal{R})$, we expect each element in $\wp(A_i) = (a_{0i}, \mathcal{V}_i, \mathcal{O}_i, \mathcal{R}_i)$ to be somehow a restriction, on subsystem $\mathcal{A}_i$, of the corresponding element in $\wp(\mathcal{A})$.

Specifically, $a_{0i} = (S_i, Q_i)$ should include only the states and configurations relevant to components and links within $\mathcal{A}_i$. Similarly, $\mathcal{V}_i$ and $\mathcal{R}_i$ should include only the subset of pairs in $\mathcal{V}$ and $\mathcal{R}$, respectively, which are relevant to transitions of components in $\mathcal{A}_i$. Finally, we expect the nodes of $\mathcal{O}_i$ to include labels relevant to $\mathcal{V}_i$ only.[1]

Besides, since each subproblem $\wp(\mathcal{A}_i)$ does not account for the constraints imposed by the other subproblems, the resulting behavior $Bhv(\wp(\mathcal{A}_i))$ is expected to be complete, yet unsound with respect to $\wp(\mathcal{A})$.

Completeness means that, for each trajectory h in $Bhv(\wp(\mathcal{A}))$, the restriction of h on $\mathcal{A}_i$ (the subsequence of h relevant to transitions of components in $\mathcal{A}_i$) is also a trajectory in $Bhv(\wp(\mathcal{A}_i))$. Unsoundness means that a trajectory h_i in $Bhv(\wp(\mathcal{A}_i))$ may not be the restriction of any trajectory in $Bhv(\wp(\mathcal{A}))$.

Completeness and unsoundness of $Bhv(\wp(\mathcal{A}_i))$ translate into completeness and unsoundness of candidate diagnoses in the solution $\Delta(\wp(\mathcal{A}_i))$. In fact, once we have reconstructed the behavior $Bhv(\wp(\mathcal{A}_i))$, nothing prevents us from decorating the behavior and eventually distilling the solution of $\wp(\mathcal{A}_i)$. However, as with the subsystem behavior, we expect $\Delta(\wp(\mathcal{A}_i))$ to be complete but unsound.

In this case, completeness means that, for each candidate diagnosis $\delta \in \Delta(\wp(\mathcal{A}))$, the restriction of δ on $\mathcal{A}_i$ (the subset of faults in δ relevant to transitions of components in $\mathcal{A}_i$) is also a candidate diagnosis in $\Delta(\wp(\mathcal{A}_i))$. Unsoundness means that a candidate diagnosis δ_i in $\Delta(\wp(\mathcal{A}_i))$ may not be the restriction of any candidate diagnosis in $\Delta(\wp(\mathcal{A}))$.

While, intuitively, the rationale for unsoundness comes from the relaxation of both topological and observation constraints caused by restriction of the problem, one may ask why completeness holds. In other words, why not incompleteness?

In the case of the reconstructed behavior, incompleteness means that there exists a trajectory h in $Bhv(\wp(\mathcal{A}))$ such that the restriction of h on $\mathcal{A}_i$ is not a trajectory in $Bhv(\wp(\mathcal{A}_i))$. Intuitively, this sounds odd, as the constraints under which $Bhv(\wp(\mathcal{A}_i))$ is generated constitute a subset of the constraints imposed on the generation of $Bhv(\wp(\mathcal{A}))$.

For instance, on the one hand, if h conforms to a candidate trace $\mathcal{T} \in \|\mathcal{O}\|$, then the restriction of h on h_i is expected to conform to the restriction of $\mathcal{T}$ on viewer $\mathcal{V}_i$. On the other hand, since $\mathcal{O}_i$ is a restriction of $\mathcal{O}$, we expect that $\|\mathcal{O}_i\|$ includes the restriction of $\mathcal{T}$ on $\mathcal{V}_i$. The same reasoning applies to the constraints imposed by the system topology (input/output events and link configurations).

Despite the unsoundness of the reconstructed behaviors of the subsystems, the eventual behavior obtained by merging such unsound behaviors (step 2) is expected to be sound as well as complete. This expectation is grounded on the constraints

[1] This poses a problem when the subset of labels relevant to $\mathcal{V}_i$ in a node is empty.

imposed by the temporal observation $\mathcal{O}$ during merging, as well as the topological constraints relevant to the links connecting subsystems to one another (interface constraints).

The interface of a system decomposition is the set of links that do not belong to any subsystem. In other words, a link belongs to the interface when the two components connected by the link belong to different subsystems.

Based on these considerations, we do not have theoretical reasons for not adopting modular diagnosis. However, we need to highlight the practical reasons for such an approach. What is the *raison d'être* of modular diagnosis?

There are two main answers to this question:

1. In some instances, performing the reconstruction of the system behavior in a modular way is more efficient than doing so in a monolithic way,
2. In some application domains, it is better to generate in real time a larger set of candidate diagnoses (including unsound candidates) than to wait for a sound and complete solution of the problem.

The first answer sounds counterintuitive: why might a two-step approach (modular reconstruction) be more efficient than a one-step approach (monolithic reconstruction)? After all, if the final result (the reconstructed system behavior) is expected to be the same (language equivalence), in the modular approach we have to consider the additional computation of the first step (reconstruction of subsystem behaviors).

Even if this argument is reasonable, it nonetheless ignores three subtle facts:

1. Behavior reconstruction of subsystems can be performed in parallel, provided parallelism is supported by the implementation of the diagnosis engine,
2. Both reconstruction and merging require first generating the spurious behavior (which is eventually pruned),
3. In the merging step, the elicitation of triggerable component transitions is facilitated, as no check for link configurations is required as long as the links are within subsystems.

Considering the first fact, on average, the time for reconstructing the behavior of n subsystems in parallel equals the time for reconstructing one subsystem behavior.

Considering the second fact (under the assumption of parallelism), since both reconstruction and merging generate a spurious subgraph, the advantage of modular reconstruction over monolithic reconstruction (in terms of computational time) holds when the time for generating the spurious subgraph by monolithic reconstruction is larger than the sum of the time for generating one subsystem behavior and the time for generating the spurious subgraph by merging.

Intuitively, if the interactions among subsystem components are tight, the constraints imposed on the subsystem behaviors are stringent and, consequently, the reconstructed behaviors are bound to be relatively small. In such a case, the spurious subgraph generated by the merging may be significantly smaller than the spurious subgraph generated by monolithic reconstruction.

Considering the third fact, what makes merging less complex than monolithic reconstruction is that triggerability of component transitions (marking arcs of the reconstructed behaviors) is not required to be checked, provided that such transitions involve input/output events relevant to links within subsystems (this check is instead required when the relevant links are part of the interface of the system decomposition).

With reference to the second answer (real-time constraints), consider the scenario in which the nature of the active system under diagnosis (for instance, an aircraft or a nuclear power plant) makes the completion of the reconstruction of the system behavior either inherently infeasible or beyond the time constraints imposed by the application domain. Instead of waiting for a sound and complete set of candidate diagnoses, the reconstruction of subsystem behaviors generates a complete set of candidate diagnoses in real time. Despite the fact that, generally speaking, several of these candidate diagnoses are unsound, useful information can nevertheless be derived.

For instance, if a specific fault relevant to a component c does not appear in any candidate diagnosis, that fault will not be part of any sound candidate diagnosis (as sound candidate diagnoses form a subset of the set of complete diagnoses generated in real time). In other words, we can be sure in real time that component c is not faulty.

By contrast, if f is in all candidate diagnoses, we can be sure in real time that component c is faulty. This is a consequence of the fact that the solution of the diagnosis problem cannot be empty (it is at least the singleton $\{\delta\}$, where δ is possibly $\emptyset$, indicating no fault occurrence).

The idea of modular diagnosis is not limited to a single decomposition of the system, as has been implicitly assumed up to now. In fact, a system $\mathcal{A}$ can be recursively decomposed into a tree of subsystems, where the root is $\mathcal{A}$, the internal nodes are subsystems that are decomposed in turn, and the leaf nodes are components. In particular, the set of children $\mathcal{A}'_1, \ldots, \mathcal{A}'_{n'}$ of a node $\mathcal{A}'$ is the decomposition of $\mathcal{A}'$.

Given a recursive decomposition, the modular reconstruction of the system behavior can be performed bottom-up, based on the recursive-decomposition tree. Parallel behavior reconstruction can be applied not only to the set of children of one subsystem; more generally, it can be applied to several internal nodes (and, therefore, to several sets of children), provided that those nodes are independent of one another in the decomposition tree (each node is neither an ancestor nor a descendant of the other nodes).

For example, if system $\mathcal{A}$ is decomposed into $\{\mathcal{A}_1, \mathcal{A}_2\}$, $\mathcal{A}_1$ into $\{\mathcal{A}_{11}, \mathcal{A}_{12}\}$, and $\mathcal{A}_2$ into $\{\mathcal{A}_{21}, \mathcal{A}_{22}\}$, then the reconstruction of the behaviors of $\mathcal{A}_{11}$, $\mathcal{A}_{12}$, $\mathcal{A}_{21}$, and $\mathcal{A}_{22}$ may be performed in parallel. Moreover, once those behaviors have been reconstructed, the merging of behaviors relevant to the children of $\mathcal{A}_1$ can be performed in parallel with the merging of behaviors relevant to the children of $\mathcal{A}_2$.

6.1 System Decomposition

Modular diagnosis requires a system $\mathcal{A}$ under diagnosis to be decomposed into a set $\{\mathcal{A}_1, \ldots, \mathcal{A}_n\}$ of subsystems. Thus, before introducing the notion of a decomposition, we first need to formally define a subsystem.

Definition 6.1 (Subsystem). Let $\mathcal{A} = (C, L, D)$ be an active system. A *subsystem* $\mathcal{A}'$ of $\mathcal{A}$, denoted $\mathcal{A}' \sqsubseteq \mathcal{A}$, is an active system (C', L', D') where $C' \subseteq C$, L' is the whole set of links in L between components in C', $D' = D'' \cup D''_{\text{on}}$, such that D'' is the set of dangling terminals in D relevant to components in C', and D''_{on} is the set of input terminals of components in $\mathcal{A}'$ which are connected to links in $L - L'$.

Example 6.1. Consider the active system $\mathcal{A}_2$ displayed in Fig. 4.2 and shown again in Fig. 6.1. One subsystem of $\mathcal{A}_2$ is $\mathcal{A}_{pr}$, where the set of components is $\{p, r\}$, the set of links is $\{pr\}$, and the set of dangling terminals is empty (as the output terminal of the protection device p is still connected to one link, namely pr).

Intuitively, a decomposition of a system $\mathcal{A}$ is a set of subsystems that form a partition of $\mathcal{A}$. However, such a partition applies only to the components and links between components within the same subsystem. Links between components belonging to different subsystems are not part of any partition. Hence, the union of the components of all subsystems equals the set of components of $\mathcal{A}$, while the union of the links of all subsystems is only a subset of the links of $\mathcal{A}$.

Definition 6.2 (Decomposition). Let $\mathcal{A} = (C, L, D)$ be an active system. A *decomposition* $\boldsymbol{\mathcal{A}}$ of $\mathcal{A}$ is a set $\{\mathcal{A}_1, \ldots, \mathcal{A}_n\}$ of disjoint subsystems of $\mathcal{A}$, with $\mathcal{A}_i = (C_i, L_i, D_i)$, $i \in [1 \cdots n]$, such that $\{C_1, \ldots, C_n\}$ is a partition of C.

Example 6.2. A decomposition of the system $\mathcal{A}_2$ is displayed in Fig. 6.1, namely $\boldsymbol{\mathcal{A}}_2 = \{\mathcal{A}_{pr}, \mathcal{A}_l\}$, where $\mathcal{A}_{pr}$ includes the protection device p and breaker r, while $\mathcal{A}_l$ includes just breaker l.

In Example 6.2, link pl does not belong to any subsystem of $\boldsymbol{\mathcal{A}}_2$ (neither $\mathcal{A}_l$ nor $\mathcal{A}_{pr}$). More generally, given a decomposition of a system $\mathcal{A}$, there exists a subset of links in $\mathcal{A}$ that do not belong to any subsystem of $\mathcal{A}$. This is the interface of the decomposition (Def. 6.3).

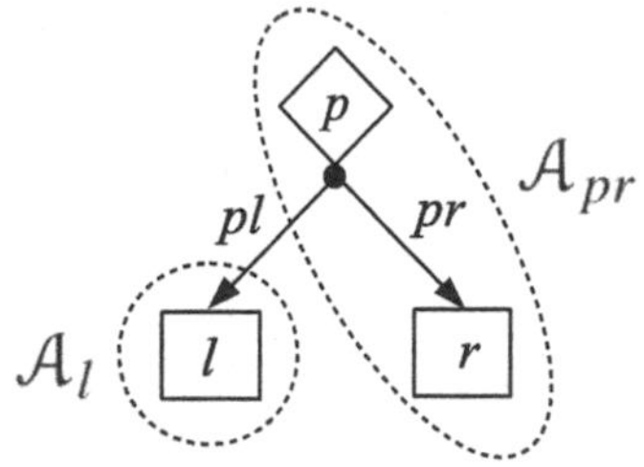

Fig. 6.1 Decomposition $\boldsymbol{\mathcal{A}}_2 = \{\mathcal{A}_{pr}, \mathcal{A}_l\}$ of active system $\mathcal{A}_2$

Definition 6.3 (Interface). Let $\mathbf{\mathcal{A}}$ be a decomposition of an active system. The *interface* of $\mathbf{\mathcal{A}}$, denoted $Int(\mathbf{\mathcal{A}})$, is the subset of links within $\mathcal{A}$ that connect two components in two different subsystems in $\mathbf{\mathcal{A}}$.

Example 6.3. The interface of the decomposition $\mathbf{\mathcal{A}}_2 = \{\mathcal{A}_{pr}, \mathcal{A}_l\}$ displayed in Fig. 6.1 is $Int(\mathbf{\mathcal{A}}_2) = \{pl\}$, as link pl connects the protection device p, belonging to subsystem $\mathcal{A}_{pr}$, to breaker l, belonging to subsystem $\mathcal{A}_l$. By contrast, link pr is not included in the interface, as it connects two components in the same subsystem, $\mathcal{A}_{pr}$.

6.2 Problem Restriction

In its simplest form, the modular diagnosis of a system $\mathcal{A}$ is based on a (nonrecursive) decomposition $\mathbf{\mathcal{A}} = \{\mathcal{A}_1, \dots, \mathcal{A}_n\}$ of $\mathcal{A}$. The first step consists in the (possibly parallel) reconstruction of the behaviors of the subsystems in $\mathbf{\mathcal{A}}$. Such behaviors are generated based on specific diagnosis problems relevant to the subsystems.

The question is, given a diagnosis problem $\wp(\mathcal{A})$ and a subsystem $\mathcal{A}_i \in \mathbf{\mathcal{A}}$, what is the diagnosis problem $\wp(\mathcal{A}_i)$ relevant to $\mathcal{A}_i$?

Intuitively, since $\wp(\mathcal{A}) = (a_0, \mathcal{V}, \mathcal{O}, \mathcal{R})$ and $\wp(\mathcal{A}_i) = (a_{0i}, \mathcal{V}_i, \mathcal{O}_i, \mathcal{R}_i)$, each element in the quadruple $\wp(\mathcal{A}_i)$ is expected to be a restriction of each corresponding element in the quadruple $\wp(\mathcal{A})$, namely $a_{0i} = a_{0[\mathcal{A}_i]}$, $\mathcal{V}_i = \mathcal{V}_{[\mathcal{A}_i]}$, $\mathcal{O}_i = \mathcal{O}_{[\mathcal{A}_i]}$, and $\mathcal{R}_i = \mathcal{R}_{[\mathcal{A}_i]}$, with the subscript $[\mathcal{A}_i]$ denoting the restriction of the corresponding element on subsystem $\mathcal{A}_i$. Therefore, we need to formalize these different restrictions.

Definition 6.4 (State Restriction). Let $a = (S, Q)$ be a state of an active system $\mathcal{A}$, and $\mathcal{A}' = (C', L', D')$ a subsystem of $\mathcal{A}$. The *restriction* of a on $\mathcal{A}'$, denoted $a_{[\mathcal{A}']}$, is the pair (S', Q') such that S' is the subtuple of states in S relevant to components in C' and Q' is the subtuple of configurations in Q relevant to links in L'.

Example 6.4. With reference to the states of system $\mathcal{A}_2$ defined in Table 4.3, consider state $1 = ((closed, awaken, closed), ([op], [op]))$. The restriction of this state on subsystem $\mathcal{A}_{pr}$ defined in Example 6.1 is $((awaken, closed), ([op]))$.

Definition 6.5 (Viewer Restriction). Let $\mathcal{V}$ be a viewer for an active system $\mathcal{A}$, and $\mathcal{A}' = (C', L', D')$ a subsystem of $\mathcal{A}$. The *restriction* of $\mathcal{V}$ on $\mathcal{A}'$, denoted $\mathcal{V}_{[\mathcal{A}']}$ is the subset of pairs $(t(c), \ell) \in \mathcal{V}$ such that $c \in C'$.

Example 6.5. With reference to the viewer $\mathcal{V}$ defined in Table 4.1, the restriction of $\mathcal{V}$ on the subsystem $\mathcal{A}_{pr} \sqsubseteq \mathcal{A}_2$ defined in Example 6.1 includes the following set of associations:

$$\mathcal{V}_{[\mathcal{A}_{pr}]} = \{(p_1, awk), (p_2, ide), (p_3, awk), (p_4, ide), (b_1(r), opr), (b_2(r), clr)\} .$$

The restriction of a temporal observation $\mathcal{O}$ is not as straightforward as the restriction of the initial state or the restriction of the viewer. Since $\mathcal{O}$ is a DAG where each node is marked by a set of candidate labels, each node in the restriction of $\mathcal{O}$ on a subsystem $\mathcal{A}' \sqsubseteq \mathcal{A}$, namely $\mathcal{O}_{[\mathcal{A}']}$, will retain any label which is involved in $\mathcal{V}_{[\mathcal{A}']}$, the restriction of viewer $\mathcal{V}$ on $\mathcal{A}'$. By contrast, labels not involved in $\mathcal{V}_{[\mathcal{A}']}$ will not be included in any node of $\mathcal{O}_{[\mathcal{A}']}$. However, we have to consider two further cases:

First, if the label ε belongs to a node in $\mathcal{O}$ then it also belongs to the restricted node in $\mathcal{O}_{[\mathcal{A}']}$. This is reasonable in that ε means that maybe no observable label was generated for this node, so the same applies to the corresponding node in $\mathcal{O}_{[\mathcal{A}']}$.

Second, assume that a label ℓ is associated in $\mathcal{V}$ with a transition of a component c, where c does not belong to $\mathcal{A}'$, and that the same label is associated in $\mathcal{V}_{[\mathcal{A}']}$ with a transition of a component c' (belonging to $\mathcal{A}'$). In other words, ℓ is relevant to both a component c' in $\mathcal{A}'$ and a component c outside $\mathcal{A}'$.

Now, assume that ℓ is a candidate label in a node of $\mathcal{O}$. Since ℓ is associated with c' in $\mathcal{A}'$, it will be included in the corresponding restricted node. However, and this is the point, since ℓ is also associated with a component c outside $\mathcal{A}'$, we cannot know whether ℓ was generated by c or c'. Consequently, the corresponding restricted node in $\mathcal{O}_{[\mathcal{A}']}$ not only will contain ℓ (in this case ℓ is assumed to be generated by c'), but will also contain ε (in this case ℓ is assumed to be generated by c).

Once each node of $\mathcal{O}$ has been mapped to a corresponding restricted node, we obtain a DAG isomorphic to $\mathcal{O}$, where each node is marked by a new set of candidate labels. Disturbingly, a number of such restricted nodes can be marked by the singleton $\{\varepsilon\}$ (empty nodes), which, based on Def. 4.3, is not allowed in a temporal observation.[2] Therefore, we must remove such empty nodes by retaining the precedence constraints between the other nodes.

Definition 6.6 (Observation Restriction). Let $\mathcal{A}$ be an active system embodying components C, let $\mathcal{O} = (L, N, A)$ be a temporal observation of $\mathcal{A}$ based on a viewer $\mathcal{V}$, and let $\mathcal{A}'$ be a subsystem of $\mathcal{A}$ embodying components C'. Let Ω and Ω' be the domains of observable labels involved in $\mathcal{V}$ and $\mathcal{V}_{[\mathcal{A}']}$, respectively. Let $\Omega_{(C-C')}$ denote the set of observable labels defined as follows:

$$\Omega_{(C-C')} = \{\ell \mid (t(c), \ell) \in \mathcal{V}, c \in (C - C')\}\,. \tag{6.1}$$

The *restriction* of a node $n \in N$ on $\mathcal{A}'$, denoted $n_{[\mathcal{A}']}$, is a node n' whose extension is defined as follows.

For each $\ell \in \|n\|$, the following rules are applied:

1. If $\ell \in \Omega'$, then $\ell \in \|n'\|$,
2. If $\ell \in \Omega' \cap \Omega_{(C-C')}$, then $\varepsilon \in \|n'\|$,
3. If $\ell \notin \Omega'$, then $\varepsilon \in \|n'\|$,
4. If $\ell = \varepsilon$, then $\varepsilon \in \|n'\|$.

The *restriction* of the temporal observation $\mathcal{O}$ on $\mathcal{A}'$, denoted $\mathcal{O}_{[\mathcal{A}']}$, is a temporal observation $\mathcal{O}' = (L', N', A')$, where the set of nodes is

$$N' = \{n' \mid n \in N, n' = n_{[\mathcal{A}']}, \|n'\| \neq \{\varepsilon\}\}, \tag{6.2}$$

[2] For example, if a node in $\mathcal{O}$ is marked by the singleton $\{\ell\}$ and ℓ is not involved in $\mathcal{V}_{[\mathcal{A}']}$, then the corresponding restricted node will be marked by the singleton $\{\varepsilon\}$.

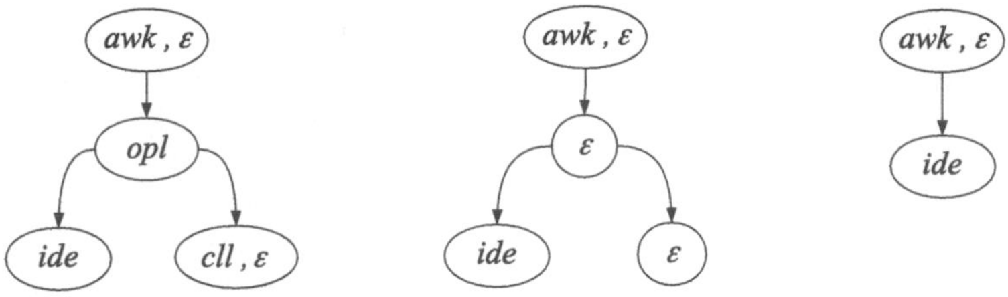

Fig. 6.2 Temporal observation $\mathcal{O}_2$ (left), node restrictions on $\mathcal{A}_{pr}$ (center), and observation restriction $\mathcal{O}_{2[\mathcal{A}_{pr}]}$ (right)

while the set of arcs is defined as follows: $n'_1 \to n'_2 \in \mathcal{A}'$, where $n'_1 = n_{1[\mathcal{A}']}$ and $n'_2 = n_{2[\mathcal{A}']}$, iff:

1. (*Precedence*) $n_1 \prec n_2$ in $\mathcal{O}$,
2. (*Canonicity*) $\nexists n'_3 \in N', n'_3 = n_{3[\mathcal{A}']}$, such that $n_1 \prec n_3 \prec n_2$ in $\mathcal{O}$.

Example 6.6. Consider the temporal observation $\mathcal{O}_2$ (for system $\mathcal{A}_2$) displayed on the left side of Fig. 6.2, and the subsystem $\mathcal{A}_{pr} \sqsubseteq \mathcal{A}_2$ defined in Example 6.1. Based on Def. 6.6, the center of Fig. 6.2 shows the graph resulting from the restriction of the nodes of $\mathcal{O}$ on $\mathcal{A}_{pr}$, where the arcs are retained. The actual restriction of $\mathcal{O}_2$ on $\mathcal{A}_{pr}$ is displayed on the right side of Fig. 6.2, where the nodes marked by the singleton $\{\varepsilon\}$ have been removed, while the remaining nodes have been linked by corresponding precedence arcs.

As will be clarified shortly, for practical reasons, it is also convenient to define the restriction of the index space of an observation $\mathcal{O}$ on a subsystem $\mathcal{A}'$, namely $Isp_{[\mathcal{A}']}(\mathcal{O})$.

According to Proposition 5.1, the language of $Isp(\mathcal{O})$ equals $\|\mathcal{O}\|$ (the set of candidate traces in $\mathcal{O}$). In other words, each sequence of labels generated by a path from the initial state to a final state in $Isp(\mathcal{O})$ is a candidate trace in $\mathcal{O}$, and vice versa. As such, each transition in $Isp(\mathcal{O})$ is marked by a visible label relevant to $\mathcal{O}$. Since $Isp(\mathcal{O})$ is a DFA, such a label cannot be ε.

Intuitively, the restriction of $Isp(\mathcal{O})$ on $\mathcal{A}'$ is obtained by restricting each of these labels on $\mathcal{A}'$ and by creating transitions marked by the new labels. The rules for mapping a label ℓ marking a transition in $Isp(\mathcal{O})$ to a label ℓ' marking a corresponding transition in the restricted index space parallel the first three rules in Def. 6.6.

For each transition $\mathfrak{I} \xrightarrow{\ell} \mathfrak{I}'$ in $Isp(\mathcal{O})$, the following actions are performed:

1. If ℓ is involved in $\mathcal{V}_{[\mathcal{A}']}$, then $\mathfrak{I} \xrightarrow{\ell} \mathfrak{I}'$ is created,
2. If ℓ is both involved in $\mathcal{V}_{[\mathcal{A}']}$ and associated in $\mathcal{V}$ with a component not in $\mathcal{A}'$, then $\mathfrak{I} \xrightarrow{\varepsilon} \mathfrak{I}'$ is created,
3. If ℓ is not involved in $\mathcal{V}_{[\mathcal{A}']}$, then $\mathfrak{I} \xrightarrow{\varepsilon} \mathfrak{I}'$ is created.

Generally speaking, owing to possible ε-transitions, the resulting automaton, namely $\mathcal{N}$, is nondeterministic. Therefore, the actual restriction $Isp_{[\mathcal{A}']}(\mathcal{O})$ is obtained by determinization of $\mathcal{N}$.

Definition 6.7 (Index-Space Restriction). Let $\mathcal{A}$ be an active system embodying components C, let $\mathcal{O} = (L, N, A)$ be a temporal observation of $\mathcal{A}$ based on a viewer $\mathcal{V}$, and let $\mathcal{A}'$ be a subsystem of $\mathcal{A}$ embodying components C'. Let Ω and Ω' be the domains of observable labels involved in $\mathcal{V}$ and $\mathcal{V}_{[\mathcal{A}']}$, respectively. Let $\Omega_{(C-C')}$ denote the set of observable labels defined in (6.1).

The *restriction* of the index space $Isp(\mathcal{O})$ on $\mathcal{A}'$, denoted $Isp_{[\mathcal{A}']}(\mathcal{O})$, is the DFA equivalent to an NFA $\mathcal{N}$ defined by the following rules.

For each $\mathfrak{I} \xrightarrow{\ell} \mathfrak{I}'$ in $Isp(\mathcal{O})$:

1. If $\ell \in \Omega'$, then $\mathfrak{I} \xrightarrow{\ell} \mathfrak{I}'$ is in $\mathcal{N}$,
2. If $\ell \in \Omega' \cap \Omega_{(C-C')}$, then $\mathfrak{I} \xrightarrow{\varepsilon} \mathfrak{I}'$ is in $\mathcal{N}$,
3. If $\ell \notin \Omega'$, then $\mathfrak{I} \xrightarrow{\varepsilon} \mathfrak{I}'$ is in $\mathcal{N}$.

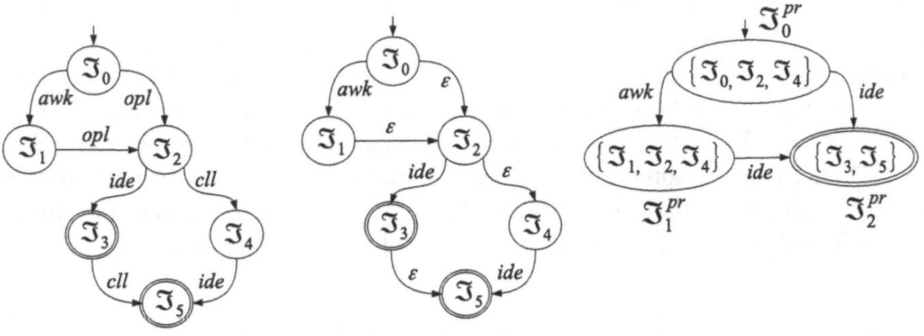

Fig. 6.3 Index space $Isp(\mathcal{O}_2)$ (left), the corresponding nondeterministic restriction on $\mathcal{A}_{pr}$ (center), and index-space restriction $Isp_{[\mathcal{A}_{pr}]}(\mathcal{O}_2)$ (right)

Example 6.7. Consider the index space of $\mathcal{O}_2$ displayed on the left side of Fig. 6.3, and the subsystem $\mathcal{A}_{pr} \sqsubseteq \mathcal{A}_2$ defined in Example 6.1. Based on Def. 6.7, the center of Fig. 6.3 shows the nondeterministic automaton $\mathcal{N}$ resulting from the restrictions of labels marking arcs of $Isp(\mathcal{O}_2)$. The restriction of $Isp(\mathcal{O}_2)$ on $\mathcal{A}_{pr}$ is displayed on the right side of Fig. 6.3, which is obtained by determinization of $\mathcal{N}$.

At this point, a natural question arises: is there any relationship between the restriction of the index space $Isp(\mathcal{O})$ on $\mathcal{A}'$ and the index space of the restriction of observation $\mathcal{O}$ on $\mathcal{A}'$? The answer is given by Proposition 6.1.

Proposition 6.1. *Let $Isp(\mathcal{O})$ be the index space of a temporal observation $\mathcal{O}$ of an active system $\mathcal{A}$, and let $\mathcal{A}'$ be a subsystem of $\mathcal{A}$. Then, the language of $Isp(\mathcal{O}_{[\mathcal{A}']})$ equals the language of $Isp_{[\mathcal{A}']}(\mathcal{O})$.*

Proof. Based on Proposition 5.1, since the language of the index space equals the extension of the temporal observation, it suffices to show that $\|\mathcal{O}_{[\mathcal{A}']}\|$ equals the language of $Isp_{[\mathcal{A}']}(\mathcal{O})$. Moreover, based on the definition of the index-space restriction, since the language of $Isp_{[\mathcal{A}']}(\mathcal{O})$ equals the language of the NFA named $\mathcal{N}$, it suffices to show that $\|\mathcal{O}_{[\mathcal{A}']}\|$ equals the language of $\mathcal{N}$. Let $\mathcal{O}^{\varepsilon}_{[\mathcal{A}']}$ denote the restriction of $\mathcal{O}$ on $\mathcal{A}'$ prior to the removal of the projected nodes marked by the singleton $\{\varepsilon\}$. As such, $\mathcal{O}^{\varepsilon}_{[\mathcal{A}']}$ is isomorphic to $\mathcal{O}$, while $\|\mathcal{O}^{\varepsilon}_{[\mathcal{A}']}\| = \|\mathcal{O}_{[\mathcal{A}']}\|$. Hence, it suffices to show that $\|\mathcal{O}^{\varepsilon}_{[\mathcal{A}']}\|$ equals the language of $\mathcal{N}$. Let $\mathcal{T}_{\varepsilon} = [\ell'_1, \ldots, \ell'_{k'}]$ be the sequence of labels (possibly ε) chosen in each node of $\mathcal{O}$ to generate a candidate trace $\mathcal{T} = [\ell_1, \ldots, \ell_k]$. Let $\mathbf{L} = [L_0, \ldots, L_k]$ be the sequence of substrings of $\mathcal{T}_{\varepsilon}$ defined as follows:

1. L_0 is the (possibly empty) longest prefix of $\mathcal{T}_{\varepsilon}$ composed of ε labels only,
2. Let $\mathcal{T}^i_{\varepsilon}$ be the suffix of $\mathcal{T}_{\varepsilon}$ following the last label in L_{i-1}. For each $i \in [1 \cdots k]$, L_i is the nonempty prefix of $\mathcal{T}^i_{\varepsilon}$ composed of the first label and the (possibly empty) following sequence of ε labels.

In other words, $\mathbf{L}$ is a sequence-based partition of $\mathcal{T}_{\varepsilon}$ where L_0 is the (possibly empty) string of ε labels preceding ℓ_1 in $\mathcal{T}_{\varepsilon}$, while each L_i, $i \in [1 \cdots k]$, is the string starting with observable label ℓ_i and continuing with all ε labels following ℓ_i in $\mathcal{T}_{\varepsilon}$. According to Proposition 5.1, $\mathcal{T}$ also belongs to the language of $Isp(\mathcal{O})$. Let $\mathbf{I} = [\mathfrak{I}_0, \mathfrak{I}_1, \ldots, \mathfrak{I}_k]$ be the sequence of states in $Isp(\mathcal{O})$ involved in the generation of $\mathcal{T}$. There exists an isomorphism between $\mathbf{L}$ and $\mathbf{I}$, where L_0 corresponds to the initial state $\mathfrak{I}_0$, while the first label ℓ_i of L_i, $i \in [1 \cdots k]$, marks the transition from $\mathfrak{I}_{i-1}$ to $\mathfrak{I}_i$ in $Isp(\mathcal{O})$. The proof of Proposition 6.1 is grounded on Lemmas 6.1 and 6.2.

Lemma 6.1. *If $\mathcal{T}$ is a candidate trace in $\|\mathcal{O}^{\varepsilon}_{[\mathcal{A}']}\|$, then $\mathcal{T}$ is in the language of $\mathcal{N}$.*

Proof. Let $\mathcal{T}_{\varepsilon}$ be the sequence of labels chosen in each node of $\mathcal{O}^{\varepsilon}_{[\mathcal{A}']}$ generating $\mathcal{T}$. Let $\mathbf{L}$ be the partition of $\mathcal{T}_{\varepsilon}$ defined above. Assume that each label in $\mathcal{T}_{\varepsilon}$ chosen from $\|n_{[\mathcal{A}']}\|$ is also in $\|n\|$. In this case, $\mathcal{T}$ continues to belong to the language of $\mathcal{N}$, as $\mathcal{N}$ includes a sequence $\mathbf{I}$ of states isomorphic to $\mathbf{L}$ and generating $\mathcal{T}$ (based on the first rule of the definition of the index-space restriction, all relevant transitions are maintained in $\mathcal{N}$). Now, assume that the above condition holds for the first j labels only, $j \in [0 \cdots (k'-1)]$, while $\ell'_{j+1} = \varepsilon$ is chosen in $\|n_{[\mathcal{A}']}\|$, with $\varepsilon \notin \|n\|$. Let $\mathbf{L}_i$ be the partition of $\mathcal{T}_{\varepsilon}$ up to ℓ'_j, including the first i substrings of $\mathcal{T}_{\varepsilon}$. Hence, there exists an isomorphism between $\mathbf{L}_i$ and the sequence of states $\mathbf{I}_i = [\mathfrak{I}_0, \mathfrak{I}_1, \ldots, \mathfrak{I}_i]$ in $\mathcal{N}$ generating the prefix $[\ell_1, \ldots, \ell_i]$ of $\mathcal{T}$. Let $\mathbf{L}_{+1} = \mathbf{L}_i \cup [L_{i+1}]$, where L_{i+1} is the string starting with $\ell'_{j+1} = \varepsilon$ and continuing with all the ε labels in $\mathcal{T}_{\varepsilon}$ that are also in the extension of the corresponding original node (before the restriction). On the one hand, since $\ell'_{j+1} = \varepsilon$ is chosen in $\|n_{[\mathcal{A}']}\|$, with $\varepsilon \notin \|n\|$, either the second or the third rule for node restriction is the cause of the generation of ℓ'_{j+1}. On the other hand, based on either the second or the third rule for index-space restriction, a transition $\mathfrak{I}_i \xrightarrow{\varepsilon} \mathfrak{I}_{i+1}$ exists in $\mathcal{N}$. Hence, an isomorphism still exists between $\mathbf{L}_{i+1}$ and $\mathbf{I}_{i+1} = \mathbf{I}_i \cup [\mathfrak{I}_{i+1}]$, with the latter generating the string $[\ell_1, \ldots, \ell_i]$. By induction on $\mathcal{T}_{\varepsilon}$, we eventually come to the conclusion that $\mathcal{T}$ also belongs to the language of $\mathcal{N}$.

Lemma 6.2. *If $\mathcal{T}$ is a string in the language of $\mathcal{N}$, then $\mathcal{T} \in \|\mathcal{O}^{\varepsilon}_{[\mathcal{A}']}\|$.*

Proof. Let $\mathbf{I} = [\mathfrak{I}_0, \mathfrak{I}_1, \ldots, \mathfrak{I}_k]$ be a sequence of states in $\mathcal{N}$ relevant to the generation of $\mathcal{T}$. Assume that, for each transition $\mathfrak{I}_{i-1} \xrightarrow{\ell_i} \mathfrak{I}_i$ in $\mathcal{N}$, $i \in [1 \cdots k]$, we have $\ell_i \neq \varepsilon$. In this case, based on the definition of $\mathcal{N}$, only the first restriction rule is applied and, hence, the same transitions are also in $Isp(\mathcal{O})$ and $\ell_i \in \Omega'$ for each $i \in [1 \cdots k]$. Therefore, $\mathcal{T} \in \|\mathcal{O}\|$, with $\mathcal{T}_\varepsilon$ being the sequence of labels chosen in $\mathcal{O}$ to generate $\mathcal{T}$ and $\mathbf{L} = [L_0, L_1, \ldots, L_k]$ being the corresponding partition of $\mathcal{T}_\varepsilon$ isomorphic to $\mathbf{I}$. Since all labels in $\mathcal{T}$ are in Ω', the first rule for node restriction is applied for each observable label ℓ_i, while for ε labels the fourth rule is applied. In either case, $\mathcal{T}_\varepsilon$ does not change, and therefore $\mathcal{T} \in \|\mathcal{O}^{\varepsilon}_{[\mathcal{A}']}\|$. Now, assume that $\mathbf{I}_i = [\mathfrak{I}_0, \mathfrak{I}_1, \ldots, \mathfrak{I}_i]$, $i \in [0 \cdots (k-1)]$, is a prefix of $\mathbf{I}$ where all labels involved differ from ε, while $\mathfrak{I}_i \xrightarrow{\ell_{i+1}} \mathfrak{I}_{i+1}$ is such that $\ell_{i+1} = \varepsilon$. In this case, there exists an isomorphism between $\mathbf{I}_i$ and $\mathbf{L}_i = [L_0, L_1, \ldots, L_i]$, with the latter being a partition of a prefix of a string in $\|\mathcal{O}\|$ such that ℓ_i is the first label in L_i, $i \in [1 \cdots i]$. Since $\ell_{i+1} = \varepsilon$, either the second or the third rule for index-space restriction is applied for the transition connecting $\mathfrak{I}_i$ with $\mathfrak{I}_{i+1}$ and marked by ε. Since $Isp(\mathcal{O})$ includes a transition $\mathfrak{I}_i \xrightarrow{\ell} \mathfrak{I}_{i+1}$ where $\ell \neq \varepsilon$, it is possible to select the next node in n in $\mathcal{O}$ such that $\ell \in \|n\|$. Based on either the second or the third rule for node restriction (whose conditions are consistent with the conditions of either the second or the third rule, respectively, for index-space restriction), $\varepsilon \in \|n_{\mathcal{A}'}\|$ and, as such, ε can be chosen as the next label in $\mathcal{T}_\varepsilon$, along with all the ε labels of the successive nodes in which ε is included before the observation restriction, thereby giving rise to the next substring L_{i+1} (composed of ε labels only). Hence, an isomorphism still exists between $\mathbf{L}_{i+1} = [L_0, L_1, \ldots, L_{i+1}]$ and $\mathbf{I}_{i+1} = [\mathfrak{I}_0 . \mathfrak{I}_1, \ldots, \mathfrak{I}_{i+1}]$. By induction on $\mathbf{I}$, we conclude that $\mathcal{T} \in \|\mathcal{O}^{\varepsilon}_{[\mathcal{A}']}\|$. □

Fig. 6.4 Restricted observation $\mathcal{O}2_{[\mathcal{A}_{pr}]}$ (left), nondeterministic index space $Nsp(\mathcal{O}2_{[\mathcal{A}_{pr}]})$ (center), and index space $Isp(\mathcal{O}2_{[\mathcal{A}_{pr}]})$ (right)

Example 6.8. Consider the temporal observation $\mathcal{O}_2$ (for system $\mathcal{A}_2$) displayed on the left side of Fig. 6.2, and the subsystem $\mathcal{A}_{pr} \sqsubseteq \mathcal{A}_2$ defined in Example 6.1. The restriction of $\mathcal{O}_2$ on $\mathcal{A}_{pr}$ generated in Example 6.6 is reported on the left side of

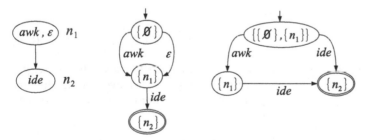

Fig. 6.4. Displayed in the center of the same figure is the corresponding nondeterministic index space $Nsp(\mathcal{O}_{2[\mathcal{A}_{pr}]})$, while the index space $Isp(\mathcal{O}_{2[\mathcal{A}_{pr}]})$ is depicted on the right side of the figure. Apart from node identification, $Isp(\mathcal{O}_{2[\mathcal{A}_{pr}]})$ equals the restriction of $Isp(\mathcal{O}_2)$ on $\mathcal{A}_{pr}$, which is displayed on the right side of Fig. 6.3. Hence, as expected from Proposition 6.1, they share the same language.

Definition 6.8 (Ruler Restriction). Let $\mathcal{R}$ be a ruler for an active system $\mathcal{A}$, and let $\mathcal{A}' = (C', L', D')$ be a subsystem of $\mathcal{A}$. The *restriction* of $\mathcal{R}$ on $\mathcal{A}'$, denoted $\mathcal{R}_{[\mathcal{A}']}$, is the subset of pairs $(t(c), f) \in \mathcal{R}$ such that $c \in C'$.

Example 6.9. With reference to the ruler $\mathcal{R}$ defined in Table 4.1, the restriction of $\mathcal{R}$ on the subsystem $\mathcal{A}_{pr} \sqsubseteq \mathcal{A}_2$ defined in Example 6.1 includes the following set of associations:

$$\mathcal{R}_{[\mathcal{A}_{pr}]} = \{(p_3, \mathbf{fop}), (p_4, \mathbf{fcp}), (b_3(r), \mathbf{nor}), (b_4(r), \mathbf{ncr})\}\ .$$

Based on the definitions of restrictions for the four elements of a diagnosis problem $\wp(\mathcal{A})$, namely Defs. 6.4, 6.5, 6.6, and 6.8, the restriction of $\wp(\mathcal{A})$ is simply defined as the quadruple of these restrictions.

Definition 6.9 (Problem Restriction). Let $\wp(\mathcal{A}) = (a_0, \mathcal{V}, \mathcal{O}, \mathcal{R})$ be a diagnosis problem for an active system $\mathcal{A}$, and let $\mathcal{A}'$ be a subsystem of $\mathcal{A}$. The *restriction* of $\wp(\mathcal{A})$ on $\mathcal{A}'$, denoted $\wp_{[\mathcal{A}']}(\mathcal{A})$, is a diagnosis problem $\wp(\mathcal{A}')$ for $\mathcal{A}'$

$$\wp(\mathcal{A}') = (a'_0, \mathcal{V}', \mathcal{O}', \mathcal{R}'), \tag{6.3}$$

where $a'_0 = a_{0[\mathcal{A}']}$, $\mathcal{V}' = \mathcal{V}_{[\mathcal{A}']}$, $\mathcal{O}' = \mathcal{O}_{[\mathcal{A}']}$, and $\mathcal{R}' = \mathcal{R}_{[\mathcal{A}']}$.

Example 6.10. Consider the diagnosis problem $\wp(\mathcal{A}_2) = (a_0, \mathcal{V}, \mathcal{O}_2, \mathcal{R})$ defined in Example 4.12, and the subsystem $\mathcal{A}_{pr} \sqsubseteq \mathcal{A}_2$ defined in Example 6.1. The restriction of $\wp(\mathcal{A}_2)$ on $\mathcal{A}_{pr}$ is

$$\wp_{[\mathcal{A}_{pr}]}(\mathcal{A}_2) = (a_{0[\mathcal{A}_{pr}]}, \mathcal{V}_{[\mathcal{A}_{pr}]}, \mathcal{O}_{2[\mathcal{A}_{pr}]}, \mathcal{R}_{[\mathcal{A}_{pr}]}), \tag{6.4}$$

where $a_{0[\mathcal{A}_{pr}]} = ((idle, closed), ([]))$, $\mathcal{V}_{[\mathcal{A}_{pr}]} = \{(p_1, awk), (p_2, ide), (p_3, awk), (p_4, ide), (b_1(r), opr), (b_2(r), clr)\}$ (Example 6.5), $\mathcal{O}_{2[\mathcal{A}_{pr}]}$ is displayed in the left side of Fig. 6.4, and $\mathcal{R}_{[\mathcal{A}_{pr}]} = \{(p_3, \mathbf{fop}), (p_4, \mathbf{fcp}), (b_3(r), \mathbf{nor}), (b_4(r), \mathbf{ncr})\}$ (Example 6.9).

Intuitively, the property of diagnosis-problem restriction sounds transitive. That is, if $\mathcal{A}'' \sqsubseteq \mathcal{A}' \sqsubseteq \mathcal{A}$ then we expect the restriction on $\mathcal{A}''$ of the restriction on $\mathcal{A}'$ of $\wp(\mathcal{A})$ to be equal to the restriction of $\wp(\mathcal{A})$ on $\mathcal{A}''$. This claim is supported by Proposition 6.2.

Proposition 6.2. *Let $\mathcal{A}$ be an active system, $\mathcal{A}'$ a subsystem of $\mathcal{A}$, and $\mathcal{A}''$ a subsystem of $\mathcal{A}'$. Let $\wp(\mathcal{A})$ be a diagnosis problem for $\mathcal{A}$, $\wp(\mathcal{A}')$ the restriction of $\wp(\mathcal{A})$ on $\mathcal{A}'$, and $\wp(\mathcal{A}'')$ the restriction of $\wp(\mathcal{A}')$ on $\mathcal{A}''$. Then, $\wp(\mathcal{A}'')$ equals the restriction of $\wp(\mathcal{A})$ on $\mathcal{A}''$.*

Proof. Let $\wp(\mathcal{A}) = (a_0, \mathcal{V}, \mathcal{O}, \mathcal{R})$. Based on Def. 6.9, we show how the transitivity property asserted by Proposition 6.2 holds for all elements in the quadruple $\wp(\mathcal{A})$.

Lemma 6.3. *Transitivity holds for the restriction of the initial state:*

$$\left(a_{0[\mathcal{A}']}\right)_{[\mathcal{A}'']} = a_{0[\mathcal{A}'']} \,. \tag{6.5}$$

Proof. Based on Def. 6.4, the restriction of a state (S, Q) on $\mathcal{A}'$ is the pair (S', Q'), where S' is the subtuple of component states in S that are in $\mathcal{A}'$; similarly, Q' is the subtuple of link configurations in Q relevant to links that are in $\mathcal{A}'$. Thus, (6.5) comes from the transitivity property of tuple containment.

Lemma 6.4. *Transitivity holds for the restriction of the viewer:*

$$\left(\mathcal{V}_{[\mathcal{A}']}\right)_{[\mathcal{A}'']} = \mathcal{V}_{[\mathcal{A}'']} \,. \tag{6.6}$$

Proof. Based on Def. 6.5, the restriction of a viewer $\mathcal{V}$ on $\mathcal{A}'$ is the subset of pairs $(t(c), \ell)$ in $\mathcal{V}$ such that c is in $\mathcal{A}'$. Thus, (6.6) comes from the transitivity property of set containment.

Lemma 6.5. *Transitivity holds for the restriction of the ruler:*

$$\left(\mathcal{R}_{[\mathcal{A}']}\right)_{[\mathcal{A}'']} = \mathcal{R}_{[\mathcal{A}'']} \,. \tag{6.7}$$

Proof. Based on Def. 6.8, the restriction of a ruler $\mathcal{R}$ on $\mathcal{A}'$ is the subset of pairs $(t(c), f)$ in $\mathcal{R}$ such that c is in $\mathcal{A}'$. Thus, (6.7) comes from the transitivity property of set containment.

Lemma 6.6. *Transitivity holds for any node n of the temporal observation:*

$$\left(n_{[\mathcal{A}']}\right)_{[\mathcal{A}'']} = n_{[\mathcal{A}'']} \,. \tag{6.8}$$

Proof. Let n', $\bar{n}'$, and n'' denote $n_{[\mathcal{A}']}$, $n'_{[\mathcal{A}'']}$ and $n_{[\mathcal{A}'']}$, respectively. Let $\Omega_{(C-C')}$, $\Omega_{(C'-C'')}$, and $\Omega_{(C-C'')}$ denote the sets of labels defined as follows:

$$\begin{aligned}
\Omega_{(C-C')} &= \{\ell \mid (t(c), \ell) \in \mathcal{V}, c \in (C - C')\}, \\
\Omega_{(C'-C'')} &= \{\ell \mid (t(c), \ell) \in \mathcal{V}, c \in (C' - C'')\}, \\
\Omega_{(C-C'')} &= \{\ell \mid (t(c), \ell) \in \mathcal{V}, c \in (C - C'')\}.
\end{aligned}$$

First, we show that $\bar{\ell} \in \|\bar{n}'\| \implies \bar{\ell} \in \|n''\|$. To this end, based on the four rules for node restriction in Def. 6.6, we consider each possible way in which a label $\ell \in \|n\|$ may be mapped to a label $\bar{\ell} \in \bar{n}'$, while showing that $\bar{\ell}$ is also in $\|n''\|$:

1. If $\ell \in \Omega'$, then (rule 1) $\ell \in \|n'\|$. Two cases are possible: either $\ell \in \Omega''$ or $\ell \notin \Omega''$. If $\ell \in \Omega''$ then, on the one hand (rule 1), $\ell \in \|\bar{n}'\|$; on the other hand (rule 1), $\ell \in \|n''\|$. If, instead, $\ell \notin \Omega''$, then, on the one hand (rule 3), $\varepsilon \in \|\bar{n}'\|$; on the other hand (rule 3), $\varepsilon \in \|n''\|$.

2. If $\ell \in \Omega' \cap \Omega_{(C-C')}$, then (rule 2) $\varepsilon \in \|n'\|$ and (rule 4) $\varepsilon \in \|\bar{n}'\|$. Since we assume $\ell \in \Omega_{(C-C')}$, we also have $\ell \in \Omega_{(C-C'')}$ (as $C'' \subset C'$). Thus, two cases are possible: either $\ell \in \Omega''$ or $\ell \notin \Omega''$. If $\ell \in \Omega''$, then $\ell \in \Omega'' \cap \Omega_{(C-C'')}$, and hence (rule 2) $\varepsilon \in \|n''\|$. If, instead, $\ell \notin \Omega''$, then (rule 3) $\varepsilon \in \|n''\|$.
3. If $\ell \notin \Omega'$, then (rule 3) $\varepsilon \in \|n'\|$ and (rule 4) $\varepsilon \in \|\bar{n}'\|$. From the assumption that $\ell \notin \Omega'$, it follows that $\ell \notin \Omega''$. Hence (rule 3) $\varepsilon \in \|n''\|$.
4. If $\ell = \varepsilon$, then (rule 4) $\varepsilon \in \|n'\|$ and (rule 4) $\varepsilon \in \|\bar{n}'\|$. Besides (rule 4), $\varepsilon \in \|n''\|$.

Then, we show that $\bar{\ell} \in \|n''\| \implies \bar{\ell} \in \|\bar{n}'\|$. To this end, based on the four rules for node restriction in Def. 6.6, we consider each possible way in which a label $\ell \in \|n\|$ may be mapped to a label $\bar{\ell} \in \|n''\|$, while showing that $\bar{\ell}$ is also in $\bar{n}'$:

1. If $\ell \in \Omega''$, then $\ell \in \Omega'$. Hence (rule 1), $\ell \in \|n'\|$ and (rule 1) $\ell \in \|\bar{n}'\|$.
2. If $\ell \in \Omega' \cap \Omega_{(C-C'')}$, then (rule 2) $\varepsilon \in \|n''\|$. Two cases are possible: either $\ell \in \Omega'' \cap \Omega_{(C'-C'')}$ or $\ell \in \Omega'' \cap \Omega_{(C-C')}$. If $\ell \in \Omega'' \cap \Omega_{(C'-C'')}$, then $\ell \in \Omega'$, and hence (rule 1) $\ell \in \|n'\|$ and (rule 2) $\varepsilon \in \|\bar{n}'\|$. If, instead, $\ell \in \Omega'' \cap \Omega_{(C-C')}$, then (rule 2) $\varepsilon \in \|n'\|$ and (rule 4) $\varepsilon \in \|\bar{n}'\|$.
3. If $\ell \notin \Omega''$, then (rule 3) $\varepsilon \in \|n''\|$. Two cases are possible: either $\ell \in \Omega'$ or $\ell \notin \Omega'$. If $\ell \in \Omega'$, then (rule 1) $\ell \in \|n'\|$ and (rule 3) $\varepsilon \in \|\bar{n}'\|$. If, instead, $\ell \notin \Omega'$, then (rule 3) $\varepsilon \in \|n'\|$ and (rule 4) $\varepsilon \in \|\bar{n}'\|$.
4. If $\ell = \varepsilon$, then (rule 4) $\varepsilon \in \|n''\|$. Besides (rule 4), $\varepsilon \in \|n'\|$ and (rule 4) $\varepsilon \in \|\bar{n}'\|$.

Corollary 6.1. *Let $\mathcal{O}^{\varepsilon}_{[\mathcal{A}']}$ and $\mathcal{O}^{\varepsilon}_{[\mathcal{A}'']}$ denote the graphs obtained by restricting each node in $\mathcal{O}$ on $\mathcal{A}'$ and $\mathcal{A}''$, respectively. Let $\left(\mathcal{O}^{\varepsilon}_{[\mathcal{A}']}\right)^{\varepsilon}_{[\mathcal{A}'']}$ denote the graph obtained by restricting each node in $\mathcal{O}^{\varepsilon}_{[\mathcal{A}']}$ on $\mathcal{A}''$.*[3] *Then,*

$$\left(\mathcal{O}^{\varepsilon}_{[\mathcal{A}']}\right)^{\varepsilon}_{[\mathcal{A}'']} = \mathcal{O}^{\varepsilon}_{[\mathcal{A}'']} \,. \tag{6.9}$$

Proof. Equation (6.9) comes directly from the notion of a restricted graph and (6.8) in Lemma 6.6.

Lemma 6.7. *Let $\mathcal{O}^{\varepsilon}_{[\mathcal{A}']}$ denote the observation graph obtained by restricting each node in $\mathcal{O}$ on $\mathcal{A}'$. Then,*

$$\left(\mathcal{O}^{\varepsilon}_{[\mathcal{A}']}\right)_{[\mathcal{A}'']} = \left(\mathcal{O}_{[\mathcal{A}']}\right)_{[\mathcal{A}'']} \,. \tag{6.10}$$

Proof. On the one hand, since the restricted graph $\mathcal{O}^{\varepsilon}_{[\mathcal{A}']}$ differs from $\mathcal{O}_{[\mathcal{A}']}$ only in the possible additional nodes marked by the singleton $\{\varepsilon\}$ (empty nodes), the restriction of each nonempty node in $\mathcal{O}^{\varepsilon}_{[\mathcal{A}']}$ on $\mathcal{A}''$ yields the same result as that obtained from the restriction on $\mathcal{A}''$ of the identical node in $\mathcal{O}_{[\mathcal{A}']}$. On the other hand, owing to the preservation of the arcs connecting empty nodes in $\mathcal{O}^{\varepsilon}_{[\mathcal{A}']}$, the eventual removal of empty nodes in $\mathcal{O}^{\varepsilon}_{[\mathcal{A}']}$ (after node restriction on $\mathcal{A}''$) produces

[3] As such, these restricted graphs may possibly contain nodes marked by the singleton $\{\varepsilon\}$.

the same result as that obtained by the removal of nonempty nodes in $\mathcal{O}_{[\mathcal{A}']}$ after node restriction on $\mathcal{A}''$.

Lemma 6.8. *Transitivity holds for the temporal observation:*

$$\left(\mathcal{O}_{[\mathcal{A}']}\right)_{[\mathcal{A}'']} = \mathcal{O}_{[\mathcal{A}'']} \, . \tag{6.11}$$

Proof. Based on (6.9), by applying the removal of nodes marked by the singleton $\{\varepsilon\}$ as specified in Def. 6.6, we obtain

$$\left(\mathcal{O}^{\varepsilon}_{[\mathcal{A}']}\right)_{[\mathcal{A}'']} = \mathcal{O}_{[\mathcal{A}'']} \, . \tag{6.12}$$

Finally, (6.11) is obtained by substituting the left-hand side of (6.12) by the right-hand side of (6.10).

The proof of Proposition 6.2 is grounded on Lemmas 6.3, 6.4, 6.5, and 6.8. □

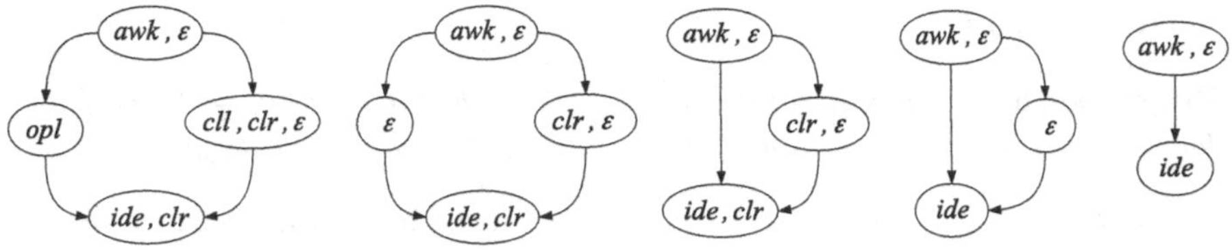

Fig. 6.5 From left to right, observations $\mathcal{O}_3$, $\mathcal{O}^{\varepsilon}_{3[\mathcal{A}_{pr}]}$, $\mathcal{O}_{3[\mathcal{A}_{pr}]}$, $\left(\mathcal{O}_{3[\mathcal{A}_{pr}]}\right)^{\varepsilon}_{[\mathcal{A}_p]}$, and $\left(\mathcal{O}_{3[\mathcal{A}_{pr}]}\right)_{[\mathcal{A}_p]}$

Example 6.11. Consider the active system $\mathcal{A}_2$ displayed in Fig. 4.2. Let $\mathcal{A}_{pr}$ be the subsystem of $\mathcal{A}_2$ involving components p and r, and $\mathcal{A}_p$ the subsystem of $\mathcal{A}_{pr}$ involving just component p. Shown on the left side of Fig. 6.5 is a temporal observation $\mathcal{O}_3$ for $\mathcal{A}_2$. Next, from left to right, are $\mathcal{O}^{\varepsilon}_{3[\mathcal{A}_{pr}]}$, the restricted observation $\mathcal{O}_{3[\mathcal{A}_{pr}]}$, $\left(\mathcal{O}_{3[\mathcal{A}_{pr}]}\right)^{\varepsilon}_{[\mathcal{A}_p]}$, and the restricted observation $\left(\mathcal{O}_{3[\mathcal{A}_{pr}]}\right)_{[\mathcal{A}_p]}$. Shown in Fig. 6.6 are the graphs relevant to the restriction of $\mathcal{O}_3$ on $\mathcal{A}_p$. As expected from Lemma 6.8 in Proposition 6.2, the final result is the same.

The notion of a restriction can be defined for trajectories as well. The restriction of a trajectory h on a subsystem $\mathcal{A}'$ is the subsequence of transitions in h that are relevant to components in $\mathcal{A}'$, as formalized in Def. 6.10.

Definition 6.10 (Trajectory Restriction). Let h be a trajectory of a system $\mathcal{A}$, and let $\mathcal{A}' = (C', L', D')$ be a subsystem of $\mathcal{A}$. The *restriction* of h on $\mathcal{A}'$, denoted $h_{[\mathcal{A}']}$, is the subsequence of h defined as follows:

$$h_{[\mathcal{A}']} = \left\{ t(c) \mid t(c) \in h, c \in C' \right\} . \tag{6.13}$$

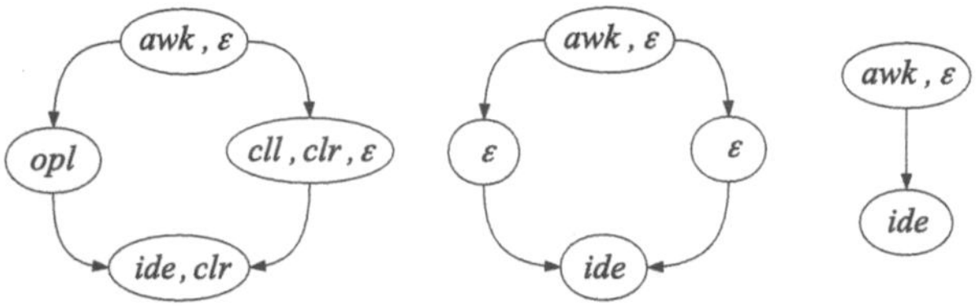

Fig. 6.6 From left to right, observations $\mathcal{O}_3$, $\mathcal{O}^{\varepsilon}_{3[\mathcal{A}_p]}$, and $\mathcal{O}_{3[\mathcal{A}_p]}$

Example 6.12. Consider the diagnosis problem $\wp(\mathcal{A}_2) = (a_0, \mathcal{V}, \mathcal{O}_2, \mathcal{R})$ defined in Example 4.12, and the subsystem $\mathcal{A}_{pr} \sqsubseteq \mathcal{A}_2$ defined in Example 6.1. According to Table 4.4, $h_{10} = [\, p_1, b_1(l), b_3(r), p_2, b_5(r), b_4(l) \,]$ is a trajectory in $Bsp(\mathcal{A}_2)$. Based on Def. 6.10, we have $h_{10[\mathcal{A}']} = [\, p_1, b_3(r), p_2, b_5(r) \,]$.

As we know, a trajectory h of a system $\mathcal{A}$ is a string in the language of the behavior space of $\mathcal{A}$, rooted at an initial state a_0. Interestingly, the restriction of h on $\mathcal{A}' \sqsubseteq \mathcal{A}$ is a trajectory of the behavior space of $\mathcal{A}'$ rooted at the initial state $a_{0[\mathcal{A}']}$, as claimed in Proposition 6.3.

Proposition 6.3. *Let h be a trajectory in $Bsp(\mathcal{A})$ with initial state $a_0 = (S_0, Q_0)$, and let $\mathcal{A}' \sqsubseteq \mathcal{A}$. Then, $h_{[\mathcal{A}']}$ is a trajectory in $Bsp(\mathcal{A}')$ with initial state $a_{0[\mathcal{A}']}$.*

Proof. Let $h' = h_{[\mathcal{A}']} = [\, t_1, \ldots, t_k \,]$. Let $H = [\, T_1, \ldots, T_k \,]$ be the sequence of strings of transitions in h, where $\forall i \in [1 \cdots k]$, T_i is the subsequence of transitions in h starting at t_i and continuing with all successive transitions preceding t_{i+1} (or the end of h). The proof is by induction on h' (disregarding the trivial case in which h' is empty).

(*Basis*) *Transition t_1 is triggerable in h'.*

Based on Def. 6.4, $a_{0[\mathcal{A}']} = (S'_0, Q'_0)$, where S'_0 is the subtuple of states in S_0 relevant to components in $\mathcal{A}'$, and Q'_0 is the subtuple of configurations in Q_0 relevant to links in $\mathcal{A}'$. Let a^*_0 be the state in $Bsp(\mathcal{A})$ before the triggering of t_1. Since all transitions preceding t_1 in h are relevant to components not in $\mathcal{A}'$, they cannot change the configurations of links in $\mathcal{A}'$. Therefore, $a^*_{0[\mathcal{A}']} = a_{0[\mathcal{A}']}$. Since t_1 is triggerable in h, any conditions on input event and link saturation hold for t_1 in h' too; in other words, t_1 is triggerable in h', with the state reached being the restriction on $\mathcal{A}'$ of the state reached by t_1 in h.

(*Induction*) *If t_i is triggerable in h', $i \in [1 \cdots (k-1)]$, then t_{i+1} is triggerable in h'.*

Based on the fact that the state a'_i reached by t_i in h' equals the restriction on $\mathcal{A}'$ of the state reached by t_i in h, and that subsequent transitions in H_i do not alter elements in a'_i, since t_{i+1} is triggerable in h, the conditions on input event and link saturation hold for t_{i+1} in h' too; in other words, t_{i+1} is triggerable in h', with the state reached being the restriction on $\mathcal{A}'$ of the state reached by t_{i+1} in h. □

In the introduction to this chapter we pointed out that a major reason for pursuing modular diagnosis is to provide candidate diagnoses in real time, thereby avoiding waiting for a sound and complete set of candidates. Although soundness is not guaranteed, the set of real-time diagnoses is expected to be complete. In order to prove this property, we need to introduce the notion of restriction applied to a candidate diagnosis as well as to the solution of a diagnosis problem.

Definition 6.11 (Diagnosis Restriction). Let $\wp(\mathcal{A}) = (a_0, \mathcal{V}, \mathcal{O}, \mathcal{R})$ be a diagnosis problem for an active system $\mathcal{A}$, let $\mathcal{A}' = (C', L', D')$ be a subsystem of $\mathcal{A}$, and let δ be a diagnosis in the solution $\Delta(\wp(\mathcal{A}))$. The *restriction* of δ on $\mathcal{A}'$, denoted $\delta_{[\mathcal{A}']}$, is defined as

$$\delta_{[\mathcal{A}']} = \left\{ f \mid f \in \delta, (t(c), f) \in \mathcal{R}, c \in C' \right\} . \tag{6.14}$$

Example 6.13. Consider the diagnosis problem $\wp(\mathcal{A}_2) = (a_0, \mathcal{V}, \mathcal{O}_2, \mathcal{R})$ defined in Example 4.12, and the subsystem $\mathcal{A}_{pr} \sqsubseteq \mathcal{A}_2$ defined in Example 6.1. According to Example 4.13, $\delta_{lr} = \{\mathbf{ncl}, \mathbf{nor}\}$ is a candidate diagnosis for $\wp(\mathcal{A}_2)$. Hence,

$$\delta_{lr[\mathcal{A}_{pr}]} = \{\mathbf{nor}\} . \tag{6.15}$$

Definition 6.12 (Solution Restriction). Let $\wp(\mathcal{A}) = (a_0, \mathcal{V}, \mathcal{O}, \mathcal{R})$ be a diagnosis problem for an active system $\mathcal{A}$, and let $\mathcal{A}'$ be a subsystem of $\mathcal{A}$. The *restriction* of $\Delta(\wp(\mathcal{A}))$ on $\mathcal{A}'$, denoted $\Delta_{[\mathcal{A}']}(\wp(\mathcal{A}))$, is defined as

$$\Delta_{[\mathcal{A}']}(\wp(\mathcal{A})) = \left\{ \delta' \mid \delta \in \Delta(\wp(\mathcal{A})), \delta' = \delta_{[\mathcal{A}']} \right\} . \tag{6.16}$$

Example 6.14. Consider the diagnosis problem $\wp(\mathcal{A}_2) = (a_0, \mathcal{V}, \mathcal{O}_2, \mathcal{R})$ defined in Example 4.12, and the subsystem $\mathcal{A}_{pr} \sqsubseteq \mathcal{A}_2$ defined in Example 6.1. According to Example 4.13, $\Delta(\wp(\mathcal{A}_2)) = \{\{\mathbf{nor}\}, \{\mathbf{ncl}, \mathbf{nor}\}, \{\mathbf{nor}, \mathbf{fcp}\}\}$. Hence,

$$\Delta_{[\mathcal{A}_{pr}]} = \{\{\mathbf{nor}\}, \{\mathbf{nor}, \mathbf{fcp}\}\} . \tag{6.17}$$

We are now ready to prove the completeness of the set of real-time diagnoses. Given $\mathcal{A}' \sqsubseteq \mathcal{A}$, we prove that the restriction on $\mathcal{A}'$ of a candidate diagnosis in the solution of $\wp(\mathcal{A})$ is a candidate diagnosis in the solution of the restriction of $\wp(\mathcal{A})$ on $\mathcal{A}'$ (Proposition 6.4).

Proposition 6.4. *Let $\wp(\mathcal{A})$ be a diagnosis problem, and $\mathcal{A}'$ a subsystem of $\mathcal{A}$. Then,*

$$\delta \in \Delta(\wp(\mathcal{A})) \implies \delta_{[\mathcal{A}']} \in \Delta(\wp_{[\mathcal{A}']}(\mathcal{A})) . \tag{6.18}$$

Proof. Based on (4.5) in Def. 4.8, we have $\delta = h_{[\mathcal{R}]}$, where h is a trajectory in $Bsp(\mathcal{A})$ such that $h_{[\mathcal{V}]} \in \|\mathcal{O}\|$. Let $h' = h_{[\mathcal{A}']}$. According to Proposition 6.3, $h' \in Bsp(\mathcal{A}')$, where the initial state of $Bsp(\mathcal{A}')$ is the restriction of the initial state of $Bsp(\mathcal{A})$ on $\mathcal{A}'$. Based on Defs. 6.8 and 6.10, $h'_{[\mathcal{R}']} = \delta_{[\mathcal{A}']}$. However, in order for $\delta_{[\mathcal{A}']}$ to belong to the solution $\Delta(\wp_{[\mathcal{A}']}(\mathcal{A}))$, we have to prove the consistency of h' with the restricted observation, namely $h'_{[\mathcal{V}']} \in \|\mathcal{O}_{[\mathcal{A}']}\|$ or, equivalently, by virtue of Proposition 5.1, $h'_{[\mathcal{V}']}$ is a string in the language of $Isp(\mathcal{O}_{[\mathcal{A}']})$ or, equivalently, by virtue of Proposition 6.1, $h'_{[\mathcal{V}']}$ is a string in the language of $Isp_{[\mathcal{A}']}(\mathcal{O})$. Based on

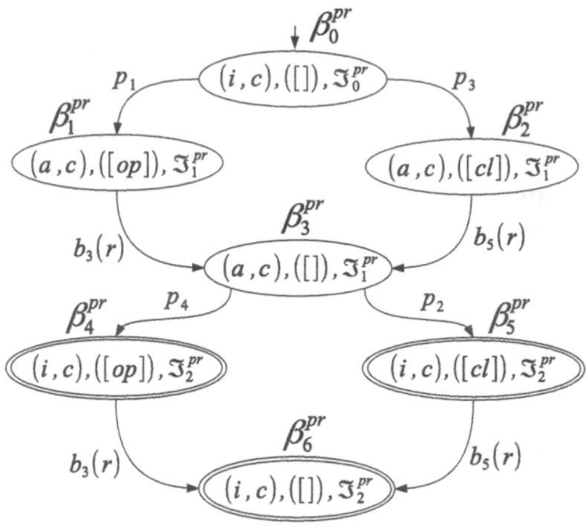

Fig. 6.7 Reconstructed behavior $Bhv(\wp_{[\mathcal{A}_{pr}]}(\mathcal{A}_2))$

Def. 6.7, $Isp(\mathcal{O}_{[\mathcal{A}']})$ is the DFA obtained by determinization of an NFA $\mathcal{N}$. Hence, it suffices to show that $h'_{[\mathcal{V}']}$ is a string in $\mathcal{N}$. Based on the definition of $\mathcal{N}$ in Def. 6.7, for each string in $Isp(\mathcal{O})$, all labels ℓ relevant to the viewer $\mathcal{V}'$ are retained, while all other labels are replaced by ε. Therefore, based on Defs. 6.5 and 6.10, $h'_{[\mathcal{V}']}$ is a string in the language of $\mathcal{N}$ and, as such, it is also a string in the language of $Isp_{[\mathcal{A}']}(\mathcal{O})$. □

Example 6.15. With reference to Example 6.10, consider the restricted diagnosis problem $\wp_{[\mathcal{A}_{pr}]}(\mathcal{A}_2)$, whose reconstructed behavior is displayed in Fig. 6.7. Based on the restricted ruler $\mathcal{R}_{[\mathcal{A}_{pr}]}$ defined in Example 6.9, the final states are decorated as follows: $(\beta_4^{pr}, \{\{\mathbf{nor}, \mathbf{fcp}\}, \{\mathbf{fop}, \mathbf{fcp}\}\})$, $(\beta_5^{pr}, \{\{\mathbf{nor}\}, \{\mathbf{fop}\}\})$, and $(\beta_6^{pr}, \{\{\mathbf{nor}, \mathbf{fcp}\}, \{\mathbf{nor}, \mathbf{fop}\}, \{\mathbf{nor}, \mathbf{fop}, \mathbf{fcp}\}\})$. Therefore, the solution of the restricted diagnosis problem includes six candidate diagnoses, namely

$$\Delta(\wp_{[\mathcal{A}_{pr}]}(\mathcal{A}_2))$$
$$= \{\{\mathbf{nor}\}, \{\mathbf{fop}\}, \{\mathbf{nor}, \mathbf{fop}\}, \{\mathbf{nor}, \mathbf{fcp}\}, \{\mathbf{fop}, \mathbf{fcp}\}, \{\mathbf{nor}, \mathbf{fop}, \mathbf{fcp}\}\}\,.$$

As expected from Proposition 6.4, both diagnoses in the restricted solution $\Delta_{[\mathcal{A}_{pr}]}$ listed in (6.17) are in $\Delta(\wp_{[\mathcal{A}_{pr}]}(\mathcal{A}_2))$, namely $\{\mathbf{nor}\}$ and $\{\mathbf{nor}, \mathbf{fcp}\}$.

A consequence of Proposition 6.4 is that, given a decomposition $\boldsymbol{\mathcal{A}}$ of a system $\mathcal{A}$, any candidate diagnosis δ in the solution of $\wp(\mathcal{A})$ can be assembled by merging a set of candidate diagnoses made up of one specific candidate diagnosis from the solution of each restricted diagnosis problem $\wp_{[\mathcal{A}_i]}(\mathcal{A})$, where $\mathcal{A}_i \in \boldsymbol{\mathcal{A}}$ (Corollary 6.2).

Corollary 6.2. *Let $\wp(\mathcal{A})$ be a diagnostic problem, and $\boldsymbol{\mathcal{A}} = \{\mathcal{A}_1, \ldots, \mathcal{A}_n\}$ a decomposition of $\mathcal{A}$. Then,*

$$\delta \in \Delta(\wp(\mathcal{A})) \implies \forall i \in [1 \cdots n]\left(\delta_i \in \Delta(\wp_{[\mathcal{A}_i]}(\mathcal{A}))\right),\; \delta = \bigcup_{i=1}^{n} \delta_i\,. \tag{6.19}$$

Proof. By virtue of (6.18) in Proposition 6.4, for each subsystem $\mathcal{A}_i \sqsubseteq \mathcal{A}$, $i \in [1 \cdots n]$, there exists a diagnosis $\delta_i \in \Delta(\wp_{[\mathcal{A}_i]}(\mathcal{A}))$ such that $\delta_i = \Delta_{[\mathcal{A}_i]}(\wp(\mathcal{A}))$. Hence, (6.19) follows. □

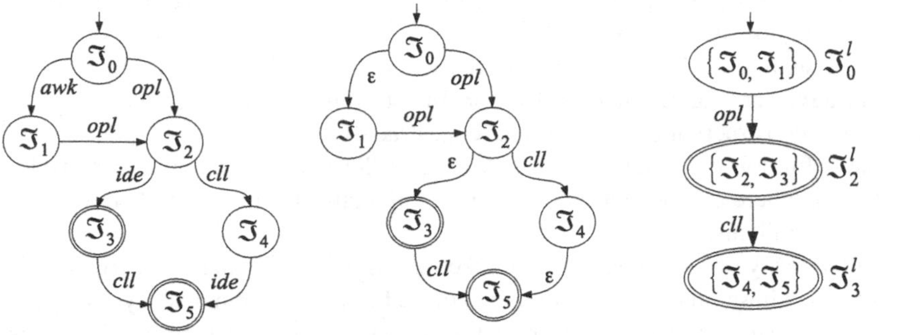

Fig. 6.8 Index space $Isp(\mathcal{O}_2)$ (left), the corresponding nondeterministic restriction on $\mathcal{A}_l$ (center), and index-space restriction $Isp_{[\mathcal{A}_l]}(\mathcal{O}_2)$ (right)

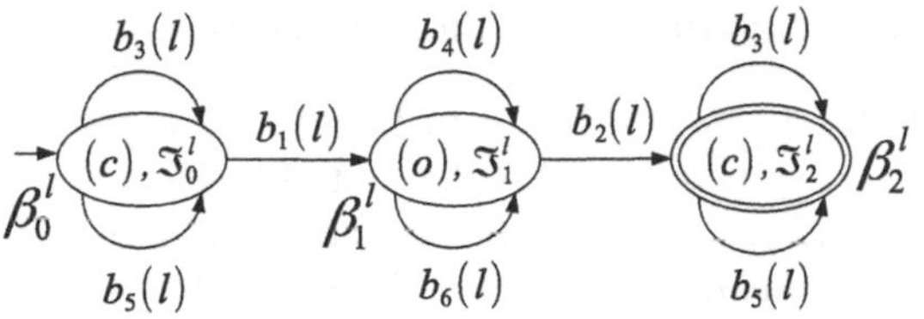

Fig. 6.9 Reconstructed behavior $Bhv(\wp_{[\mathcal{A}_l]}(\mathcal{A}_2))$

Example 6.16. Consider the diagnosis problem $\wp(\mathcal{A}_2) = (a_0, \mathcal{V}, \mathcal{O}_2, \mathcal{R})$ defined in Example 4.12, and the decomposition $\mathcal{A}_2 = \{\mathcal{A}_{pr}, \mathcal{A}_l\}$ defined in Example 6.2. Displayed in Fig. 6.8 is the generation of the restricted index space $Isp_{[\mathcal{A}_l]}(\mathcal{O}_2)$, which, by virtue of Proposition 6.1, equals the index space of the restricted observation, namely $Isp(\mathcal{O}_{2[\mathcal{A}_l]})$. The reconstructed behavior of the restricted diagnosis problem $\wp_{[\mathcal{A}_l]}(\mathcal{A}_2)$ is depicted in Fig. 6.9. Therefore, based on the decoration rules, the solution will be $\Delta(\wp_{[\mathcal{A}_l]}(\mathcal{A}_2)) = \{\emptyset, \{\mathbf{nol}\}, \{\mathbf{ncl}\}, \{\mathbf{nol}, \mathbf{ncl}\}\}$. Based on Corollary 6.2, for each candidate diagnosis $\delta \in \Delta(\wp(\mathcal{A}_2))$, there exist $\delta^{pr} \in \Delta(\wp_{[\mathcal{A}_{pr}]}(\mathcal{A}_2))$ and $\delta^l \in \Delta(\wp_{[\mathcal{A}_l]}(\mathcal{A}_2))$ such that $\delta = \delta^{pr} \cup \delta^l$. In fact, $\{\mathbf{nor}\} = \{\mathbf{nor}\} \cup \emptyset$, $\{\mathbf{ncl}, \mathbf{nor}\} = \{\mathbf{nor}\} \cup \{\mathbf{ncl}\}$, and $\{\mathbf{nor}, \mathbf{fcp}\} = \{\mathbf{nor}, \mathbf{fcp}\} \cup \emptyset$.

6.3 Behavior Merging

Modular reconstruction of the system behavior $Bhv(\wp(\mathcal{A}))$ involves two steps: reconstruction of the subsystem behaviors relevant to a decomposition of $\mathcal{A}$, and merging those behaviors. In Sec. 6.2 we provided a formal framework for coping with the first step, mainly by defining the notion of a restriction of a diagnosis problem. In this section we cope with the second step, namely merging.

To this end, we proceed in a way similar to Chapter 5, where we first defined the behavior of a diagnosis problem in declarative terms and then specified the *Build* algorithm in operational terms.

Likewise, in this section, first we formally define what a merging is and then we provide a relevant algorithm, called *Merge*. In between, we prove the equivalence of modular reconstruction and monolithic reconstruction.

The definition of a merging (Def. 6.13) parallels the definition of a behavior (Def. 5.4), where component models are conceptually substituted by behaviors of restricted problems.

A peculiarity of mergings is that, unlike behaviors, the elicitation of triggerable transitions, whose input/output events are placed in links within a subsystem (and, hence, outside the interface), does not require checking for consistency, as this check is performed during the reconstruction of the subsystem behaviors. This check is instead required for the elicitation of transitions whose input/output events are placed in links within the interface of the decomposition. In fact, although the configurations of links outside subsystems are irrelevant to the reconstruction of subsystem behaviors, they are essential for the triggerability of component transitions in the global behavior of the whole system.

Definition 6.13 (Merging). Let $\wp(\mathcal{A}) = (a_0, \mathcal{V}, \mathcal{O}, \mathcal{R})$ be a diagnosis problem for an active system $\mathcal{A} = (C, L, D)$. Let $\boldsymbol{\mathcal{A}} = \{\mathcal{A}_1, \ldots, \mathcal{A}_n\}$ be a decomposition of $\mathcal{A}$. Let $\boldsymbol{\mathcal{B}} = \{\mathcal{B}_1, \ldots, \mathcal{B}_n\}$ be the set of behaviors such that $\mathcal{B}_i = Bhv(\wp_{[\mathcal{A}_i]}(\mathcal{A}))$, $i \in [1 \cdots n]$.

The *spurious merging* of $\boldsymbol{\mathcal{B}}$, written $Mrg^{\mathrm{s}}(\boldsymbol{\mathcal{B}})$, is a DFA

$$Mrg^{\mathrm{s}}(\boldsymbol{\mathcal{B}}) = (\Sigma, M^{\mathrm{s}}, \tau^{\mathrm{s}}, \mu_0, M_{\mathrm{f}}), \tag{6.20}$$

where:

1. Σ is the alphabet, consisting of the union of the alphabets of $\mathcal{B}_i$, $i \in [1 \cdots n]$,
2. M^{s} is the set of states $(S, Q, \Im)$, with $S = (\beta_1, \ldots, \beta_n)$ being a tuple of states of all behaviors in $\boldsymbol{\mathcal{B}}$, $Q = (q_1, \ldots, q_m)$ a tuple of configurations of all links in $Int(\boldsymbol{\mathcal{A}})$, and $\Im$ a state within $Isp(\mathcal{O})$,
3. $\mu_0 = (S_0, Q_0, \Im_0)$ is the initial state, where S_0 is the tuple of the initial states of all behaviors in $\boldsymbol{\mathcal{B}}$, Q_0 is the tuple of the initial configurations of all links in $Int(\boldsymbol{\mathcal{A}})$, and $\Im_0$ is the initial state of $Isp(\mathcal{O})$,
4. τ^{s} is the transition function $\tau^{\mathrm{s}} : M^{\mathrm{s}} \times \Sigma \mapsto M^{\mathrm{s}}$, where $(S, Q, \Im) \xrightarrow{t(c)} (S', Q', \Im') \in \tau^{\mathrm{s}}$, $S = (\beta_1, \ldots, \beta_n)$, $Q = (q_1, \ldots, q_m)$, $S' = (\beta'_1, \ldots, \beta'_n)$, $Q' = (q'_1, \ldots, q'_m)$, and

$$t(c) = s \xrightarrow{(e,x) \Rightarrow (e_1,y_1),\ldots,(e_p,y_p)} s', \tag{6.21}$$

with l^x denoting either the link in L entering input terminal x of component c, or **nil**, in the latter case when $x \in (D_{\mathrm{on}} \cup \{In\})$, iff:

(4a) $\beta_i \xrightarrow{t(c)} \beta_i'$ is a transition in a $\mathcal{B}_i$, $i \in [1 \cdots n]$,
(4b) Either $l^x \notin Int(\mathbf{A})$ or $l^x \in Int(\mathbf{A})$ and e is ready at terminal x,
(4c) For each $j \in [1 \cdots n]$, we have

$$\beta_j' = \begin{cases} \beta_i' & \text{if } j = i, \\ \beta_j & \text{otherwise}. \end{cases} \tag{6.22}$$

(4d) Based on (3.5), for each $j \in [1 \cdots m]$, we have

$$q_j' = \begin{cases} Tail(q_j) & \text{if } l_j = l^x, \\ Ins(q_j, e_k) & \text{if } l_j \text{ exits terminal } y_k, k \in [1 \cdots m], \\ q_j & \text{otherwise}. \end{cases} \tag{6.23}$$

(4e) Either $t(c)$ is observable in the viewer $\mathcal{V}$ via the label ℓ, $\mathfrak{I} \xrightarrow{\ell} \bar{\mathfrak{I}}$ is a transition in $Isp(\mathcal{O})$, and $\mathfrak{I}' = \bar{\mathfrak{I}}$, or $t(c)$ is unobservable in $\mathcal{V}$ and $\mathfrak{I}' = \mathfrak{I}$.

5. M_{f} is the set of final states $(S_{\text{f}}, Q_{\text{f}}, \mathfrak{I}_{\text{f}})$, where $\mathfrak{I}_{\text{f}}$ is final in $Isp(\mathcal{O})$.

The *merging* of $\mathbf{B}$, written $Mrg(\mathbf{B})$, is the DFA

$$Mrg(\mathbf{B}) = (\Sigma, M, \tau, \mu_0, M_{\text{f}}), \tag{6.24}$$

which results from removing from the spurious merging $Mrg^{\text{s}}(\mathbf{B})$ all states and transitions that are not included in any path from the initial state to a final state.

Example 6.17. Consider the diagnosis problem $\wp(\mathcal{A}_2) = (a_0, \mathcal{V}, \mathcal{O}_2, \mathcal{R})$ defined in Example 4.12, and the decomposition $\mathbf{A}_2 = \{\mathcal{A}_{pr}, \mathcal{A}_l\}$ defined in Example 6.2. Let $\mathbf{B} = \{\mathcal{B}_{pr}, \mathcal{B}_l\}$, where $\mathcal{B}_{pr} = Bhv(\wp_{[\mathcal{A}_{pr}]}(\mathcal{A}_2))$ is displayed in Fig. 6.7, while $\mathcal{B}_l = Bhv(\wp_{[\mathcal{A}_l]}(\mathcal{A}_2))$ is displayed in Fig. 6.9. The spurious merging $Mrg^{\text{s}}(\mathbf{B})$ is depicted in Fig. 6.10, where spurious states and transitions are denoted by gray ovals and dotted arcs, respectively. According to Def. 6.13, each state μ is identified by a triple $(S, Q, \mathfrak{I})$, where $S = (\beta^{pr}, \beta^l)$ is a pair of states in $\mathcal{B}_{pr}$ and $\mathcal{B}_l$, Q is the configuration of link pl (the only link in $Int(\mathbf{A}_2)$, and $\mathfrak{I}$ is a state of the index space $Isp(\mathcal{O}_2)$ (displayed on the left side of Fig. 6.2). For instance, $\mu_0 = ((\beta_0^{pr}, \beta_0^l), ([\,]), \mathfrak{I}_0)$. Note how the Q field changes only upon either insertion or removal of an event into or from the link connecting the protection device p to breaker l. Incidentally, the spurious merging is isomorphic to the spurious behavior $Bhv^{\text{s}}(\wp(\mathcal{A}_2))$ displayed in Fig. 5.13. What is not incidental at all is the equality of the language of $Mrg(\mathbf{B})$ and the language of $Bhv(\wp(\mathcal{A}_2))$, as formally stated in Theorem 6.1.

The equivalence of modular reconstruction and monolithic reconstruction is proven in Theorem 6.1.

Theorem 6.1. *Let $\mathcal{A} = (C, L, D)$ be an active system, $Bhv(\wp(\mathcal{A})) = (\Sigma, B, \tau, \beta_0, B_{\text{f}})$ a behavior, $\mathbf{A} = \{\mathcal{A}_1, \ldots, \mathcal{A}_n\}$ a decomposition of $\mathcal{A}$, $\mathbf{B} = \{\mathcal{B}_1, \ldots, \mathcal{B}_n\}$ the set of behaviors such that $\mathcal{B}_i = Bhv(\wp_{[\mathcal{A}_i]}(\mathcal{A}))$, $i \in [1 \cdots n]$, and $Mrg(\mathbf{B}) = (\Sigma', M, \tau', \mu_0, M_{\text{f}})$. Then, the language of $Bhv(\wp(\mathcal{A}))$ equals the language of $Mrg(\mathbf{B})$.*

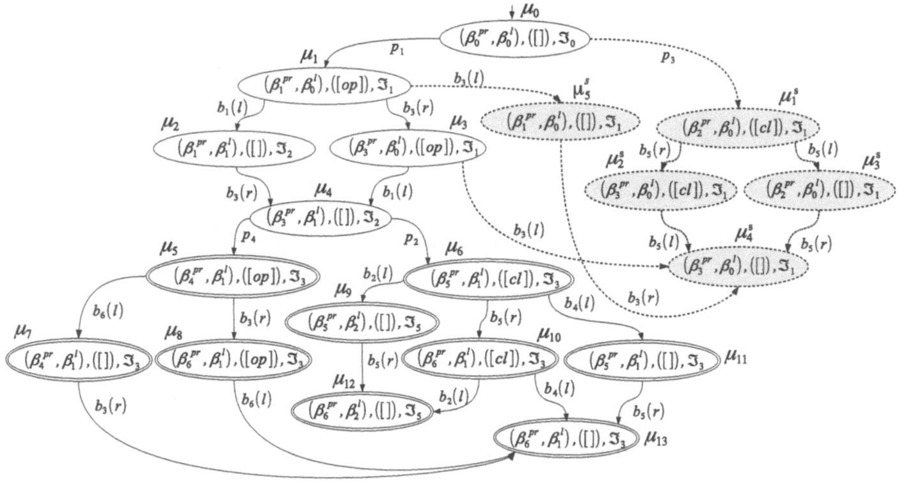

Fig. 6.10 Spurious merging $Mrg^s(\{\mathcal{B}_{pr}, \mathcal{B}_l\})$, where the spurious part is denoted by gray ovals and dotted arcs

Proof. The proof is grounded on Def. 6.14 and a number of lemmas, where $\mathcal{B}$ and $\mathcal{M}$ denote $Bhv(\wp(\mathcal{A}))$ and $Mrg(\mathbf{B})$, respectively.

Definition 6.14 (State Equivalence). Let $\beta = (S, Q, \Im)$ be a state in a behavior $\mathcal{B}$ and $\mu = (S', Q', \Im')$ a state in a merging $\mathcal{M}$, where $S' = (\beta_1, \ldots, \beta_n)$ and $\beta_i = (S_i, Q_i, \Im_i)$, $i \in [1 \cdots n]$. State β is *equivalent* to state μ, written $\beta \simeq \mu$, iff[4]

$$S = \bigcup_{i=1}^{n} S_i, \tag{6.25}$$

$$Q = \left(\bigcup_{i=1}^{n} Q_i \right) \cup Q', \tag{6.26}$$

$$\Im = \Im' . \tag{6.27}$$

Lemma 6.9. $\Sigma = \Sigma'$.

Proof. According to Defs. 5.4 and 6.13, Σ is the set of transitions of components in C, and Σ' is the union of the alphabets of behaviors in $\mathbf{B}$. The equality of Σ and Σ' comes from the fact that the decomposition $\mathbf{\mathcal{A}}$ induces a partition of Σ within $\mathbf{B}$.

Lemma 6.10. $\beta_0 \simeq \mu_0$.

Proof. On the one hand, $\beta_0 = (S_0, Q_0, \Im_0)$, where $a_0 = (S_0, Q_0)$ and $\Im_0$ is the initial state of $Isp(\mathcal{O})$. On the other hand, $\mu_0 = ((\beta_{10}, \ldots, \beta_{n0}), (q_{10}, \ldots, q_{m0}), \Im_0)$, where for each $i \in [1 \cdots n]$ we have $\beta_{i0} = (S_{i0}, Q_{i0}, \Im_{i0})$ and $a_{0[\mathcal{A}_i]} = (S_{i0}, Q_{i0})$, with S_{i0}

[4] With a slight abuse of notation, we treat the tuples S and Q as sets.

being the states in $\mathcal{A}_i$ and Q_{i0} the configurations of links in $\mathcal{A}_i$. Based on Def. 6.8, we have

$$S_0 = \bigcup_{i=1}^{n} S_{i0}, \tag{6.28}$$

$$Q_0 = \left(\bigcup_{i=1}^{n} Q_{i0} \right) \cup \{q_{10}, \ldots, q_{m0}\}, \tag{6.29}$$

which correspond to conditions (6.25) and (6.26), respectively, whereas condition (6.27) holds since $\mathfrak{I} = \mathfrak{I}' = \mathfrak{I}_0$.

Lemma 6.11. *Let* $(S, Q, \mathfrak{I}) \xrightarrow{t(c)} (S', Q', \mathfrak{I}')$ *be a transition in* $\mathcal{M}$*, where* $t(c)$ *is observable via the label* ℓ *by the viewer of* $\wp_{[\mathcal{A}_i]}(\mathcal{A})$*,* $i \in [1 \cdots n]$*.* $Isp(\mathcal{O}_{[\mathcal{A}_i]})$ *includes a transition exiting* $\mathfrak{I}_i$ *and marked by* ℓ*, where* $\beta_i = (S_i, Q_i, \mathfrak{I}_i) \in S$*.*

Proof. The proof is by induction on a path $\mu_0 \rightsquigarrow \mu$ in $\mathcal{M}$.

(*Basis*) Let $\mu = \mu_0 = (S_0, Q_0, \mathfrak{I}_0)$, where $\mathfrak{I}_{i0}$ is the initial state of $Isp(\mathcal{O}_{[\mathcal{A}_i]})$. By construction, $\mathfrak{I}_0 \in \|\mathfrak{I}_{i0}\|$. Besides, since $t(c)$ is observable in $\wp_{[\mathcal{A}_i]}(\mathcal{A})$, a transition marked by ℓ will exit $\mathfrak{I}_{i0}$ in $Isp(\mathcal{O}_{[\mathcal{A}_i]})$.

(*Induction*) If Lemma 6.11 holds for a path $\mu_0 \rightsquigarrow \mu$, then it also holds for the extended path $\mu_0 \rightsquigarrow \mu \xrightarrow{\ell} \mu'$. In fact, based on the induction hypothesis, $Isp(\mathcal{O}_{[\mathcal{A}_i]})$ includes a path from the initial state to $\mathfrak{I}_i$, which generates the subsequence $\mathcal{T}_i$ of observable labels in the viewer of $\wp_{[\mathcal{A}_i]}(\mathcal{A})$ which are relevant to the component transitions marking transitions in $\mu_0 \rightsquigarrow \mu$. Based on the construction of $Isp(\mathcal{O}_{[\mathcal{A}_i]})$ (by determinization of the intermediate NFA derived from $Isp(\mathcal{O})$ in Def. 6.7), we have $\mathfrak{I} \in \|\mathfrak{I}_i\|$. Hence, since ℓ is an observable label in $\wp_{[\mathcal{A}_i]}(\mathcal{A})$, $(\mathcal{T}_i \cup [\ell]) \in Isp(\mathcal{O}_{[\mathcal{A}_i]})$; in other words, $\mathfrak{I}_i$ is exited by a transition marked by ℓ in $Isp(\mathcal{O}_{[\mathcal{A}_i]})$.

Lemma 6.12. *Let* $(S, Q, \mathfrak{I}) \xrightarrow{t(c)} (S', Q', \mathfrak{I}')$ *be a transition in* $\mathcal{M}$*, where* $S = (\beta_1, \ldots, \beta_n)$*,* $S' = (\beta'_1, \ldots, \beta'_n)$*, and* c *is in* $\mathcal{A}_i$*,* $i \in [1 \cdots n]$*.* $\beta_i \xrightarrow{t(c)} \beta'_i$ *is a transition in* $\mathcal{B}_i$*.*

Proof. Considering $\beta_i \xrightarrow{t(c)} \beta'_i$ in the context of Def. 5.4, where $\beta_i = (S_i, Q_i, \mathfrak{I}_i)$ and $\beta'_i = (S'_i, Q'_i, \mathfrak{I}'_i)$, all relevant conditions hold:

(4*a*) If $l^x =$ **nil** in $\mathcal{A}$, then $l^x =$ **nil** in $\mathcal{A}_i$. If $l^x \neq$ **nil** in $\mathcal{A}$ and e is ready at x within l^x, then two cases are possible: either $l^x =$ **nil** in $\mathcal{A}_i$ or $l^x \neq$ **nil** in $\mathcal{A}_i$ and e is ready at x within l^x. In either case, condition (4*a*) is fulfilled.

(4*b*) We define S'_i such that it differs from S_i in the state relevant to component c only, which is set to s'.

(4*c*) We define Q'_i such that it differs from Q_i in the configuration relevant to the links in $\mathcal{A}_i$ only, which are set based on (5.10).

(4*d*) If $t(c)$ is unobservable, then $\mathfrak{I}'_i = \mathfrak{I}_i$. If $t(c)$ is observable in $\mathcal{A}_i$ via the label ℓ, then, according to Lemma 6.11, $Isp(\mathcal{O}_{[\mathcal{A}_i]})$ includes a transition exiting β_i and marked by ℓ. Hence, $\mathfrak{I}'_i$ can be set in accordance with the state entered by such a transition.

Lemma 6.13. *If* $\beta \xrightarrow{t(c)} \beta' \in \tau$ *and* $\beta \simeq \mu$, *then* $\mu \xrightarrow{t(c)} \mu' \in \tau'$ *and* $\beta' \simeq \mu'$.

Proof. According to Def. 5.4, if $(S, Q, \Im) \xrightarrow{t(c)} (S', Q', \Im') \in \tau$, where

$$t(c) = s \xrightarrow{(e,x) \Rightarrow (e_1,y_1),\ldots,(e_p,y_p)} s', \tag{6.30}$$

with l^x denoting either the link in L entering input terminal x of component c or **nil**, in the latter case when $x \in (D_{\text{on}} \cup \{In\})$, then:

$(4a)$ Either $l^x = \textbf{nil}$ or e is ready at terminal x within link l^x,
$(4b)$ S' equals S with the exception of the state relevant to c, which becomes s',
$(4c)$ Q' is obtained by removing x from l^x, if applicable ($l^x \neq \textbf{nil}$), and appending output events $e_1, \ldots, e_p$ to the configurations of links in L (where applicable) exiting output terminals $y_1, \ldots, y_p$, respectively,
$(4d)$ If $t(c)$ is observable via the label ℓ, then $\Im = \bar{\Im}$, where $\Im \xrightarrow{\ell} \bar{\Im}$ is a transition in $Isp(\mathcal{O})$; otherwise, $\Im = \Im'$.

Let $\mathcal{A}_i \in \boldsymbol{\mathcal{A}}$ be the subsystem of $\mathcal{A}$ containing component c. All five conditions listed in Def. 6.13 on transitions in $\mathcal{M}$ occur, specifically:

$(4a)$ According to Lemma 6.12, $\beta_i \xrightarrow{t(c)} \beta_i'$ is a transition in $\mathcal{B}_i$.
$(4b)$ Based on condition $(4a)$ of Def. 5.4, it follows that either $l^x \notin Int(\boldsymbol{\mathcal{A}})$ or e is ready at terminal x within l^x.
$(4c)$ Each β_i, $i \in [1 \cdots n]$, is set based on (6.22).
$(4d)$ Each configuration relevant to a link in $Int(\boldsymbol{\mathcal{A}})$ is set based on (6.23).
$(4e)$ The condition on $\Im'$ is identical to condition $(4d)$ of Def. 5.4.

Hence, $\mu \xrightarrow{t(c)} \mu' \in \tau'$. Now we show that $\beta' \simeq \mu'$, where $\beta' = (S', Q', \Im')$ and $\mu' = ((\beta_1', \ldots, \beta_n'), (q_1', \ldots, q_m'), \Im')$. Notice that β' and μ' share $\Im'$.

Let $S'_{\mathcal{M}}$ be the tuple of component states generated by aggregating the component states in β_i', $i \in [1 \cdots n]$. According to condition $(4c)$ for a merging (see (6.22)), only β_i' differs from β_i, specifically, for the state relevant to component c, which is s', exactly like S'; in other words, $S'_{\mathcal{M}}$ equals S'.

Let $Q'_{\mathcal{M}}$ be the tuple of configurations generated by aggregating the configurations in β_i', $i \in [1 \cdots n]$, along with $(q_1', \ldots, q_m')$. According to condition $(4c)$ of Def. 5.4 and condition $(4d)$ of Def. 6.13, all configurations relevant to links in $\mathcal{A}$ are updated based on $t(c)$ and (5.10); in other words, $Q'_{\mathcal{M}} = Q'$.

Hence, based on Def. 6.14, $\beta' \simeq \mu'$.

Lemma 6.14. *If* $\mu \xrightarrow{t(c)} \mu' \in \tau'$ *and* $\mu \simeq \beta$ *then* $\beta \xrightarrow{t(c)} \beta' \in \tau$ *and* $\mu' \simeq \beta'$.

Proof. Let $\mu = ((\beta_1, \ldots, \beta_n), (q_1, \ldots, q_m), \Im)$ and $\mu' = ((\beta_1', \ldots, \beta_n'), (q_1', \ldots, q_m'), \Im')$. According to condition $(4a)$ of Def. 6.13, $\beta_i \xrightarrow{t(c)} \beta_i'$ is a transition in $\mathcal{B}_i$, $i \in [1 \cdots n]$. As such, according to condition $(4a)$ of Def. 5.4 applied to behavior $\mathcal{B}_i$, and condition $(4b)$ of Def. 6.13, one of the following three cases is possible for event e:

1. Event e is ready at terminal x in a link within subsystem $\mathcal{A}_i$,
2. Event e is ready at terminal x in a link within $Int(\boldsymbol{\mathcal{A}})$,
3. Terminal x belongs to D_{on}; in other words, x is entered by a link outside $\mathcal{A}$.

In all cases, condition $(4a)$ of Def. 5.4, applied to the behavior $\mathcal{B}$, holds. Let $\beta = (S, Q, \mathfrak{I})$ and $\beta' = (S', Q', \mathfrak{I}')$. Condition $(4b)$ of Def. 5.4 applied to $\mathcal{B}$ is met by making S' equal to S with the exception of the state relevant to component c, which is made equal to s'. Condition $(4c)$ of Def. 5.4 applied to $\mathcal{B}$ is met by defining Q' as the result of the application of (5.10) to Q. Finally, condition $(4d)$ on $\mathfrak{I}'$ holds because it is identical to condition $(4e)$ of Def. 6.13, which is assumed to hold. Hence, $\beta \xrightarrow{t(c)} \beta' \in \tau$.

Now we show that $\mu' \simeq \beta'$. Let $S'_{\mathcal{M}}$ be the tuple of component states generated by aggregating the component states in β'_i, $i \in [1 \cdots n]$. According to condition $(4c)$ for a merging (see (6.22)), only β'_i differs from β_i, specifically, for the state relevant to component c, which is s', exactly like S', in other words, $S'_{\mathcal{M}}$ equals S'.

Let $Q'_{\mathcal{M}}$ be the tuple of configurations generated by aggregating the configurations in β'_i, $i \in [1 \cdots n]$, along with $(q'_1, \ldots, q'_m)$. According to condition $(4c)$ of Def. 5.4 and condition $(4d)$ of Def. 6.13, all configurations relevant to links in $\mathcal{A}$ are updated based on $t(c)$ and (5.10); in other words, $Q'_{\mathcal{M}} = Q'$.

Hence, based on Def. 6.14, $\mu' \simeq \beta'$.

Lemma 6.15. *If $\beta \simeq \mu$ then β is final iff μ is final.*

Proof. According to condition (6.27) of Def. 6.14, β and μ share the same final index-space state. Besides, based on Def. 5.4, $\beta = (S, Q, \mathfrak{I})$ is final iff $\mathfrak{I}$ is final in $Isp(\mathcal{O})$. The same applies to μ (Def. 6.13). Hence, β is final iff μ is final.

The proof of Theorem 6.1 is grounded on Lemmas 6.9, 6.10, 6.13, 6.14, and 6.15.

□

Note 6.1. Based on Theorem 6.1, the merging $\mathcal{M} = Mrg(\boldsymbol{\mathcal{B}})$ is unequivocally associated with the behavior $\mathcal{B} = Bhv(\wp(\mathcal{A})$. In fact, based on Def. 6.14, if $\mu \simeq \beta$, then $\mu = (S, Q, \mathfrak{I})$ can be transformed into β by simple restructuring of the fields S and Q. We denote such a behavior as $Bhv(\mathcal{M})$.

A direct consequence of Theorem 6.1 is that, given a diagnosis problem $\wp(\mathcal{A})$ and two different decompositions $\boldsymbol{\mathcal{A}}_1$ and $\boldsymbol{\mathcal{A}}_2$ of system $\mathcal{A}$, the language of the merging based on decomposition $\boldsymbol{\mathcal{A}}_1$ equals the language of the merging based on decomposition $\boldsymbol{\mathcal{A}}_2$ (Corollary 6.3).

Corollary 6.3. *Let $\wp(\mathcal{A})$ be a diagnosis problem, $\boldsymbol{\mathcal{A}}_1$ and $\boldsymbol{\mathcal{A}}_2$ two decompositions of system $\mathcal{A}$, and $\boldsymbol{\mathcal{B}}_1$ and $\boldsymbol{\mathcal{B}}_2$ the set of behaviors relevant to the restricted diagnosis problems on subsystems in $\boldsymbol{\mathcal{A}}_1$ and $\boldsymbol{\mathcal{A}}_2$, respectively. Then, the language of $Mrg(\boldsymbol{\mathcal{B}}_1)$ equals the language of $Mrg(\boldsymbol{\mathcal{B}}_2)$.*

6.3.1 Merging Algorithm

Listed below is the specification of the algorithm *Merge*, which, based on a diagnosis problem $\wp(\mathcal{A})$ and a decomposition $\boldsymbol{\mathcal{A}}$ of system $\mathcal{A}$, merges a set $\boldsymbol{\mathcal{B}}$ of behaviors relevant to the restriction of $\wp(\mathcal{A})$ on subsystems in $\boldsymbol{\mathcal{A}}$, thereby generating $Mrg(\boldsymbol{\mathcal{B}})$, as defined in Def. 6.13.

In a sense, the *Merge* algorithm parallels the *Build* reconstruction algorithm specified in Sec. 5.2.1, where component models are conceptually substituted by subsystem behaviors.

First, the index space of the temporal observation, $Isp(\mathcal{O})$, is generated (line 7). The initial state μ_0 is set to $(S_0, Q_0, \Im_0)$, where S_0 is the tuple of initial states of the behaviors in $\boldsymbol{\mathcal{B}}$, Q_0 is the tuple of initial configurations of the links in $Int(\boldsymbol{\mathcal{A}})$, and $\Im_0$ is the initial state of $Isp(\mathcal{O})$ (line 8). After the initialization of the set M of states and the set τ of transitions, the main loop is performed (lines 10–29).

At each iteration of the loop, an unmarked state $\mu = (S, Q, \Im)$ is selected, where $S = (\beta_1, \ldots, \beta_n)$ (line 11). The goal is to generate all transitions $\mu \xrightarrow{t(c)} \mu'$ exiting μ, where $\mu' = (S', Q', \Im')$.

To this end, based on subsystem behaviors, each triggerable transition $t(c)$ in μ is checked for consistency with $\mathcal{O}$ (lines 14–20).[5] Specifically, if $t(c)$ is not observable, then it is consistent with $\mathcal{O}$ and $\Im'$ equals $\Im$ (lines 14 and 15). If, instead, $t(c)$ is observable via label ℓ and a transition $\Im \xrightarrow{\ell} \bar{\Im}$ is included in the index space, then $\Im'$ is set to $\bar{\Im}$ (lines 16 and 17). Otherwise, $t(c)$ is not consistent with $\mathcal{O}$, and lines 21–27 are skipped (line 19). If $t(c)$ is consistent, then tuples S' and Q' are computed by updating copies of S and Q, respectively (lines 21 and 22).

If the state $\mu' = (S', Q', \Im')$ is new, then it is inserted into M (line 24), while the new transition $\mu \xrightarrow{t(c)} \mu'$ is inserted into τ (line 25).

Once all triggerable transitions have been considered, state μ is marked, as no further exiting transition exists (line 28).

When all states in M are marked, no further state or transition can be generated. Hence, the set M_{f} of final states is determined (line 30).

At this point, M and τ are the states and transitions of the spurious merging $Mrg^{\mathrm{s}}(\boldsymbol{\mathcal{A}})$. To obtain the merging $Mrg(\boldsymbol{\mathcal{A}})$, it suffices to remove the spurious part of the DFA, that is, the states and transitions which are not included in any path connecting the initial state to a final state (line 31).

The specification of the algorithm *Merge* is as follows:

1. **algorithm** *Merge* (**in** $\wp(\mathcal{A})$, **in** $\boldsymbol{\mathcal{A}}$, **in** $\boldsymbol{\mathcal{B}}$, **out** $Mrg(\boldsymbol{\mathcal{B}})$)
2. $\quad\wp(\mathcal{A}) = (a_0, \mathcal{V}, \mathcal{O}, \mathcal{R})$: a diagnosis problem for active system $\mathcal{A}$,
3. $\quad\boldsymbol{\mathcal{A}} = \{\mathcal{A}_1, \ldots, \mathcal{A}_n\}$: a decomposition of $\mathcal{A}$,
4. $\quad\boldsymbol{\mathcal{B}} = \{\mathcal{B}_1, \ldots, \mathcal{B}_n\}$: a set of behaviors where $\mathcal{B}_i = Bhv(\wp_{[\mathcal{A}_i]}(\mathcal{A}))$, $i \in [1 \cdots n]$,
5. $\quad Mrg(\boldsymbol{\mathcal{B}}) = (\Sigma, M, \tau, \nu_0, M_{\mathrm{f}})$: the merging of $\boldsymbol{\mathcal{B}}$;

[5] Unlike the case for monolithic reconstruction, the conditions on configurations refer to links in $Int(\boldsymbol{\mathcal{A}})$ only.

6.   **begin** $\langle Merge \rangle$

7.   Generate the index space $Isp(\mathcal{O}) = (\Sigma, I, \tau', \mathfrak{I}_0, I_f)$;

8. $\mu_0 := (S_0, Q_0, \mathfrak{I}_0)$, where S_0 is the tuple of initial states of the behaviors in $\mathcal{B}$
                  and Q_0 is the tuple of initial configurations of the links in $Int(\mathcal{A})$;

9. $M := \{\mu_0\}$; $\tau := \emptyset$;

10.   **repeat**

11.   Choose an unmarked state $\mu = (S, Q, \mathfrak{I})$ in M, where $S = (\beta_1, \dots, \beta_n)$
                  and $Q = (q_1, \dots, q_m)$;

12.   **foreach** transition $\beta_i \xrightarrow{t(c)} \beta_i'$ in $\mathcal{B}_i$, $i \in [1 \cdots n]$, where x is the input terminal
                  relevant to the input event e triggering transition $t(c)$, **do**

13.   **if** $x = In$ **or** the link entering x is not in $Int(\mathcal{A})$ **or** e is ready at x **then**

14.   **if** $t(c)$ is unobservable in $\mathcal{V}$ **then**

15. $\mathfrak{I}' := \mathfrak{I}$

16.   **elsif** $t(c)$ is observable in $\mathcal{V}$ via label ℓ **and** $\mathfrak{I} \xrightarrow{\ell} \bar{\mathfrak{I}}$ is in $Isp(\mathcal{O})$ **then**

17. $\mathfrak{I}' := \bar{\mathfrak{I}}$

18.   **else**

19.   **continue**

20.   **endif**;

21.   Set S' as a copy of S and replace the i-th element with β_i';

22.   Set Q' as a copy of Q and, based on $t(c)$, update the configurations
                  of links in $Int(\mathcal{A})$ only;

23. $\mu' := (S', Q', \mathfrak{I}')$;

24.   **if** $\mu' \notin M$ **then** insert μ' into M **endif**;

25.   Insert transition $\mu \xrightarrow{t(c)} \mu'$ into τ

26.   **endif**

27.   **endfor**;

28.   Mark μ

29.   **until** all states in M are marked;

30. $M_f := \{ \mu_f \mid \mu_f \in M, \mu_f = (S, Q, \mathfrak{I}_f), \mathfrak{I}_f \in I_f \}$;

31.   Remove from M and τ all states and transitions, respectively,
           which are not included in any path from μ_0 to a final state

32.   **end** $\langle Merge \rangle$.

Example 6.18. An alternative decomposition of system $\mathcal{A}_2$ is shown on the left side of Fig. 6.11, namely $\mathcal{A}_2' = \{\mathcal{A}_{lp}, \mathcal{A}_r\}$, where $\mathcal{A}_{lp}$ includes the protection device p and breaker l, while $\mathcal{A}_r$ includes just breaker r. Reported next to the decomposition is the temporal observation $\mathcal{O}_2$. Note how $Isp_{[\mathcal{A}_{lp}]}(\mathcal{O}_2)$ equals $Isp(\mathcal{O}_2)$ (shown next to $\mathcal{O}_2$), as $\mathcal{O}_2$ does not include any label relevant to breaker r. On the other hand, $Isp_{[\mathcal{A}_r]}(\mathcal{O}_2)$ is composed of just one (both initial and final) state $\mathfrak{I}_0^r$ (right side of Fig. 6.11). The reconstructed behaviors $Bhv(\wp_{[\mathcal{A}_{lp}]}(\mathcal{A}_2))$ and $Bhv(\wp_{[\mathcal{A}_r]}(\mathcal{A}_2))$ are shown in Fig. 6.12 (gray ovals and dotted arcs denote the spurious part). Note how the latter behavior includes just one (both initial and final) state, namely β_0^r. Let $\mathcal{B}' = \{\mathcal{B}_{lp}, \mathcal{B}_r\}$, where $\mathcal{B}_{lp} = Bhv(\wp_{[\mathcal{A}_{lp}]}(\mathcal{A}_2))$ and $\mathcal{B}_r = Bhv(\wp_{[\mathcal{A}_r]}(\mathcal{A}_2))$. The

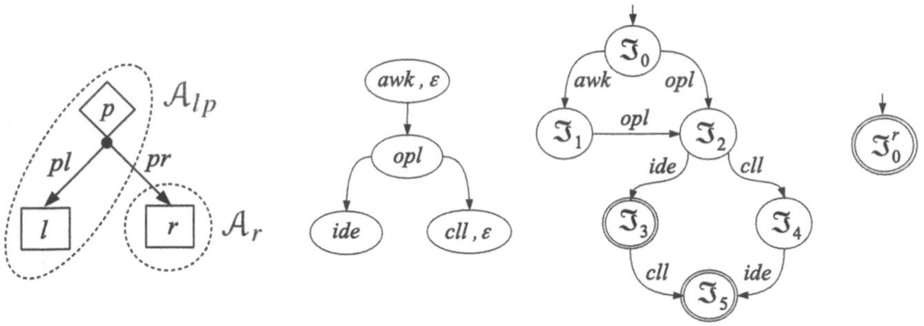

Fig. 6.11 From left to right, $\boldsymbol{\mathcal{A}}_2' = \{\mathcal{A}_{lp}, \mathcal{A}_r\}$, $\mathcal{O}_2$, $Isp(\mathcal{O}_2) = Isp_{[\mathcal{A}_{lp}]}(\mathcal{O}_2)$, and $Isp_{[\mathcal{A}_r]}(\mathcal{O}_2)$

generation of $Mrg(\boldsymbol{\mathcal{B}}')$, based on $\wp(\mathcal{A}_2)$, $\boldsymbol{\mathcal{A}}_2{}'$, and $\boldsymbol{\mathcal{B}}'$, is shown in Fig. 6.13. The resulting graph is isomorphic to $Mrg(\{\mathcal{B}_{pr}, \mathcal{B}_l\})$ shown in Fig. 6.10. Besides, unlike $Mrg(\{\mathcal{B}_{pr}, \mathcal{B}_l\})$, which includes five spurious states and eight spurious transitions, the merging based on $\{\mathcal{B}_{lp}, \mathcal{B}_r\}$ does not include spurious elements. According to Corollary 6.3, $Mrg(\{\mathcal{B}_{pr}, \mathcal{B}_l\})$ and $Mrg(\{\mathcal{B}_{lp}, \mathcal{B}_r\})$ share the same language.

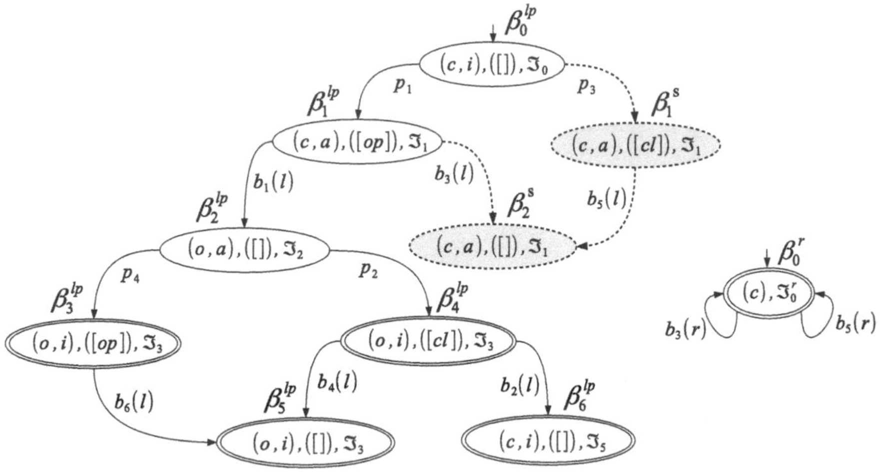

Fig. 6.12 Reconstructed behaviors $Bhv(\wp_{[\mathcal{A}_{lp}]}(\mathcal{A}_2))$ (left) and $Bhv(\wp_{[\mathcal{A}_r]}(\mathcal{A}_2))$ (right)

6.4 Recursive Merging

Up to now, we have considered merging in the context of a simple decomposition $\boldsymbol{\mathcal{A}}$ of an active system $\mathcal{A}$. Based on the decomposition $\boldsymbol{\mathcal{A}}$, the set of subsystem

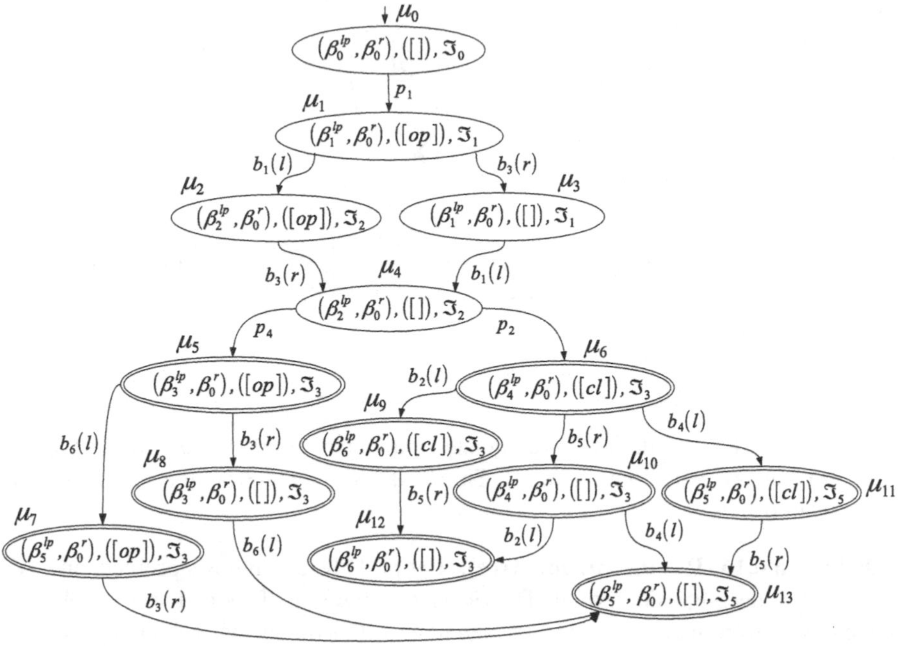

Fig. 6.13 Merging $Mrg(\{\mathcal{B}_{lp}, \mathcal{B}_r\})$ generated by *Merge* algorithm

behaviors is generated (possibly in parallel) and then merged into the global system behavior.

A natural generalization of this approach is to apply merging to other mergings, based on a recursive decomposition of $\mathcal{A}$. A recursive decomposition can be represented by a tree, where each node denotes a (sub)system while arcs denote how a (sub)system is decomposed in turn (leaf nodes represent subsystems that are not decomposed).

Definition 6.15 (Recursive Decomposition). Let $\mathcal{A}$ be an active system. A *recursive decomposition* $\boldsymbol{\mathcal{A}}^*$ of $\mathcal{A}$ is a tree, where the nodes are defined as follows:

1. The root is $\mathcal{A}$,
2. Each child of a node $\mathcal{A}'$ is a subsystem of $\mathcal{A}'$,
3. The set of children of a node $\mathcal{A}'$ is a decomposition of $\mathcal{A}'$.

Example 6.19. Consider the active system $\mathcal{A}_2$ displayed in Fig. 4.2. A relevant recursive decomposition $\boldsymbol{\mathcal{A}}_2^*$ is shown in Fig. 6.14, where system $\mathcal{A}_2$ is decomposed into $\mathcal{A}_{lp}$ and $\mathcal{A}_r$, with subsystem $\mathcal{A}_{lp}$ being further decomposed into $\mathcal{A}_l$ and $\mathcal{A}_p$.

Since a recursive decomposition of a system $\mathcal{A}$ is a tree where each (nonroot) node denotes a subsystem $\mathcal{A}' \sqsubseteq \mathcal{A}$, each node can also be associated with a diagnosis problem relevant to that node. Specifically, root $\mathcal{A}$ is associated with $\wp(\mathcal{A})$, while a

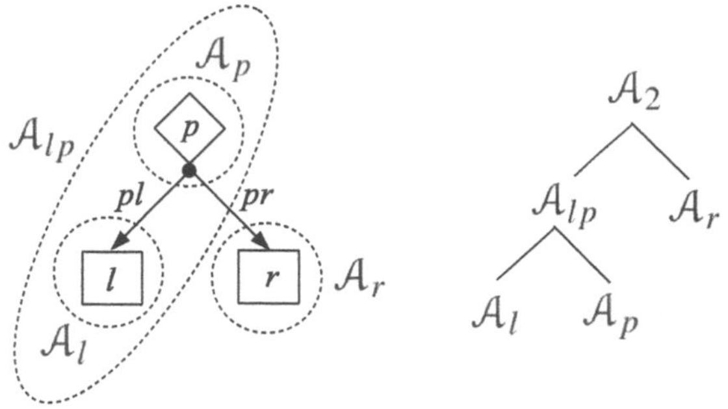

Fig. 6.14 Recursive decomposition $\boldsymbol{\mathcal{A}}_2^*$ of system $\mathcal{A}_2$

node $\mathcal{A}''$ which is a child of $\mathcal{A}'$ is associated with the restriction on $\mathcal{A}''$ of the diagnosis problem associated with node $\mathcal{A}'$. This leads us to the notion of a reconstruction tree (Def. 6.16).

Definition 6.16 (Reconstruction Tree). Let $p(\mathcal{A})$ be a diagnosis problem and $\boldsymbol{\mathcal{A}}^*$ a recursive decomposition of $\mathcal{A}$. The *reconstruction tree* of $\wp(\mathcal{A})$ based on $\boldsymbol{\mathcal{A}}^*$ is a tree obtained from $\boldsymbol{\mathcal{A}}^*$ by replacing each node with a diagnosis problem as follows:

1. The root $\mathcal{A}$ is replaced by $\wp(\mathcal{A})$,
2. If $\mathcal{A}''$ is a child node of $\mathcal{A}'$, then $\mathcal{A}''$ is replaced by $\wp_{[\mathcal{A}'']}(\mathcal{A}')$.

Note 6.2. Based on the transitivity property of problem restriction stated by Proposition 6.2, point 2 of Def. 6.16 can be rephrased as follows: Each nonroot node $\mathcal{A}'$ is replaced by $\wp_{[\mathcal{A}'']}(\mathcal{A})$.

Based on Def. 6.16 and Theorem 6.1, we conclude that, for each internal node $\mathcal{A}'$ of the reconstruction tree, the language of the behavior of the diagnosis problem associated with $\mathcal{A}'$ equals the language of the merging of the behaviors associated with the children of $\mathcal{A}'$ in the tree (Proposition 6.5).

Proposition 6.5. *Let $\mathcal{A}'$ be a node of a reconstruction tree of $\wp(\mathcal{A})$, and $\mathcal{B}'$ the set of behaviors relevant to the diagnosis problems associated in the tree with the children of $\mathcal{A}'$. Then, the language of $Bhv(\wp(\mathcal{A}'))$ equals the language of $Mrg(\mathcal{B}')$.*

A final remark concerns parallelism. As pointed out in the introduction to this chapter, with a simple decomposition $\boldsymbol{\mathcal{A}}$ of $\mathcal{A}$, reconstruction of the behaviors of subsystems in $\boldsymbol{\mathcal{A}}$ is amenable to parallelism, as the behavior of any subsystem in $\boldsymbol{\mathcal{A}}$ does not depend on the behavior of any other subsystem in $\boldsymbol{\mathcal{A}}$.

With a recursive decomposition $\boldsymbol{\mathcal{A}}^*$ of $\mathcal{A}$, parallelism can be intensified. As a general rule, the behavior relevant to a node (be it a merging or a reconstruction) can be generated in parallel with the generation of the behavior of another node provided that one node is not a descendant of the other node. That said, the only constraint on the merging of a set of behaviors is the availability of such behaviors.

In other words, the generation of the system behavior can be accomplished by traversing the reconstruction tree bottom-up, where the processing of each node is either a reconstruction (leaf node) or a merging (internal node), possibly performed in parallel with the processing of other nodes.

6.5 Bibliographic Notes

Modular diagnosis of active systems with (nonrecursive) system decomposition and plain observations was introduced in [11]. Recursive system decomposition was considered in [12, 76, 92].

6.6 Summary of Chapter 6

Subsystem. An active system involving a subset of the components of a given active system and all links between such components.

System Decomposition. A set of subsystems of a given active system, where each component of the system belongs to one and only one subsystem.

Interface. Given a decomposition of a system into subsystems, the set of links in the system that are not included in any subsystem.

State Restriction. Given a state (S, Q) of a system $\mathcal{A}$ and a subsystem $\mathcal{A}' \sqsubseteq \mathcal{A}$, the pair (S', Q'), where S' and Q' are the subsets of states and configurations of components and links, respectively, that are in $\mathcal{A}'$.

Viewer Restriction. Given a viewer $\mathcal{V}$ of a diagnosis problem for a system $\mathcal{A}$ and a subsystem $\mathcal{A}' \sqsubseteq \mathcal{A}$, the subset of pairs in $\mathcal{V}$ that are relevant to components in $\mathcal{A}'$.

Observation Restriction. Given a temporal observation $\mathcal{O}$ of a diagnosis problem for a system $\mathcal{A}$ and a subsystem $\mathcal{A}' \sqsubseteq \mathcal{A}$, the temporal observation of $\mathcal{A}'$ obtained by restricting each node of $\mathcal{O}$ on $\mathcal{A}'$ and by removing the resulting empty nodes (marked by the singleton $\{\emptyset\}$), while retaining the precedence constraints between all other nodes.

Ruler Restriction. Given a ruler $\mathcal{R}$ of a diagnosis problem for a system $\mathcal{A}$ and a subsystem $\mathcal{A}' \sqsubseteq \mathcal{A}$, the subset of pairs in $\mathcal{R}$ that are relevant to components in $\mathcal{A}'$.

Problem Restriction. Given a diagnosis problem $\wp(\mathcal{A}) = (a_0, \mathcal{V}, \mathcal{O}, \mathcal{R})$ and a subsystem $\mathcal{A}' \sqsubseteq \mathcal{A}$, the diagnosis problem $\wp(\mathcal{A}') = (a'_0, \mathcal{V}', \mathcal{O}', \mathcal{R}')$ where each element in $\wp(\mathcal{A}')$ is the restriction on $\mathcal{A}'$ of the corresponding element in $\wp(\mathcal{A})$.

Trajectory Restriction. Given a trajectory h of a system $\mathcal{A}$ and a subsystem $\mathcal{A}' \sqsubseteq \mathcal{A}$, a trajectory of $\mathcal{A}'$ obtained by selecting from h the subsequence of transitions of components in $\mathcal{A}'$.

Diagnosis Restriction. Given a candidate diagnosis in the solution of a diagnosis problem $\wp(\mathcal{A}) = (a_0, \mathcal{V}, \mathcal{O}, \mathcal{R})$ and a subsystem $\mathcal{A}' \sqsubseteq \mathcal{A}$, a candidate diagnosis in the solution of the restriction of the diagnosis problem $\wp(\mathcal{A})$ on $\mathcal{A}'$, obtained by selecting from δ the subset of faults associated in $\mathcal{R}$ with transitions of components in $\mathcal{A}'$.

Solution Restriction. Given the solution $\Delta(\wp(\mathcal{A}))$ and a subsystem $\mathcal{A}' \sqsubseteq \mathcal{A}$, the set of restrictions on $\mathcal{A}'$ of the candidate diagnoses in $\Delta(\wp(\mathcal{A}))$.

Behavior Merging. Given a set $\boldsymbol{\mathcal{B}} = \{\mathcal{B}_1, \ldots, \mathcal{B}_n\}$ of behaviors relevant to a decomposition $\boldsymbol{\mathcal{A}} = \{\mathcal{A}_1, \ldots, \mathcal{A}_n\}$ of a system $\mathcal{A}$, where each $\mathcal{B}_i \in \boldsymbol{\mathcal{B}}$, $i \in [1 \cdots n]$, is the behavior of the restriction of $\wp(\mathcal{A})$ on $\mathcal{A}_i$, a DFA whose language equals the language of $Bhv(\wp(\mathcal{A}))$, obtained by combining the behaviors in $\boldsymbol{\mathcal{B}}$.

Recursive Decomposition. Given a system $\mathcal{A}$, a tree where the root is $\mathcal{A}$ and the set of children of each internal node is a decomposition of $\mathcal{A}$.

Reconstruction Tree. Given a diagnosis problem $\wp(\mathcal{A})$ and a recursive decomposition $\boldsymbol{\mathcal{A}}$ of $\mathcal{A}$, the tree obtained from $\boldsymbol{\mathcal{A}}$ by substituting the root $\mathcal{A}$ by $\wp(\mathcal{A})$ and each child node $\mathcal{A}''$ of $\mathcal{A}'$ by the restriction on $\mathcal{A}''$ of the diagnosis problem replacing $\mathcal{A}'$.

Recursive Merging. Given a reconstruction tree with root $\wp(\mathcal{A})$, based on the tree, the process for generating the DFA equivalent to $Bhv(\wp(\mathcal{A}))$ by traversing the tree bottom-up, with possible parallel computation.

Chapter 7
Reactive Diagnosis

In Chapter 5, we presented a technique for solving a diagnosis problem, called monolithic diagnosis. Given a diagnosis problem $\wp(\mathcal{A}) = (a_0, \mathcal{V}, \mathcal{O}, \mathcal{R})$, this technique generates the solution of $\wp(\mathcal{A})$ in three steps: behavior reconstruction, behavior decoration, and diagnosis distillation. The basic idea is to find all trajectories of $\mathcal{A}$ that conform to the temporal observation $\mathcal{O}$, where these trajectories are accommodated within a DFA called the behavior of $\mathcal{A}$.

An alternative approach to solving the same class of diagnosis problems was introduced in Chapter 6, called modular diagnosis. Unlike monolithic diagnosis, modular diagnosis yields the solution of the diagnosis problem by decomposing $\wp(\mathcal{A})$ into a set of subproblems, each of which is relevant to a subsystem of $\mathcal{A}$. After behavior reconstruction is performed for each subsystem, the resulting behaviors are merged to yield the behavior of $\mathcal{A}$.

In either approach, diagnosis is performed *a posteriori*, based on the whole temporal observation $\mathcal{O}$.

A posteriori diagnosis is adequate for a variety of application domains, including the diagnosis of power network protection systems, as the reaction of the protection apparatus to a short circuit is generally fast, typically generating the temporal observation in a few seconds (or even less). Therefore, it makes sense to wait for the complete reception of a temporal observation before starting a diagnosis task.

However, a posteriori diagnosis is bound to be inadequate in application domains in which continuous supervision is required, such as aircraft and nuclear power plants. These systems continuously generate observable events by means of a network of distributed sensors. Therefore, it does not make sense to wait for the completion of a temporal observation before starting a diagnosis task. Moreover, since such systems operate in a critical environment, each single observed event (or chunk of observation) is expected to be processed in real time in order to carry out relevant recovery actions.

As such, diagnosis is required to be performed at the reception of each chunk of observation. In other words, diagnosis should be *reactive* to each newly received chunk of observation, by updating in real time the set of candidate diagnoses based on the continuous expansion of the temporal observation.

G. Lamperti et al., *Introduction to Diagnosis of Active Systems*,
https://doi.org/10.1007/978-3-319-92733-6_7

A possible (naive) approach to making diagnosis reactive would be to perform a posteriori diagnosis on an instance of a temporal observation resulting from the expansion of the previous instance by a new chunk. This way, a posteriori diagnosis would operate on a temporal observation which became increasingly large.

Clearly, this approach is bound to be impractical, because the larger the observation, the larger the reconstructed behavior and, consequently, the longer the time for completing the diagnosis task, with the result that, sooner or later, the real-time constraints required for the diagnosis response cannot be met.

What makes such a naive approach impractical is that the diagnosis task operates without memory of the previous diagnosis, since it starts from scratch each time a chunk of observation is received, disregarding all the processing performed previously.

To clarify this point, consider the generation of the index space. Once the temporal observation $\mathcal{O}$ has been expanded into $\mathcal{O}'$ by means of a new chunk, both the nondeterministic index space $Nsp(\mathcal{O}')$ and the index space $Isp(\mathcal{O}')$ are generated from scratch, thereby disregarding both the prefix space and the index space of $\mathcal{O}$ generated in the previous diagnosis processing. Since $Isp(\mathcal{O}')$ results from the determinization of $Nsp(\mathcal{O}')$, the larger the temporal observation, the more complex the determinization process.

A solution to this problem is to update $Isp(\mathcal{O})$ based on the expansion of $Nsp(\mathcal{O})$ into $Nsp(\mathcal{O}')$. This incremental determinization of the index space can be performed by the *ISCA* algorithm introduced in Chapter 2.

However, the same problem remains for the reconstruction of the system behavior (as well as for behavior decoration and diagnosis distillation). In fact, for the same reasons as discussed above, it would be impractical to start behavior reconstruction from scratch each time a chunk of observation is received.

So, we need to design a new diagnosis engine that is actually reactive in nature, where model-based reasoning is performed incrementally in order to keep the response time of the diagnosis virtually constant (or, at least, proportional to the extent of the observation chunk).

The notion of a diagnosis problem has to be changed too, as the temporal observation $\mathcal{O}$ is assumed to be received incrementally as a sequence of chunks called *fragments*. This can be formalized by the notion of a temporal-observation fragmentation, denoted by $\hat{\mathcal{O}} = [\mathcal{F}_1, \ldots, \mathcal{F}_n]$, where each $\mathcal{F}_i$, $i \in [1 \cdots n]$, is an observation fragment.

To emphasize the different nature of a diagnosis problem based on a fragmented observation, we introduce the notion of a *monitoring problem*, namely $\mu(\mathcal{A}) = (a_0, \mathcal{V}, \hat{\mathcal{O}}, \mathcal{R})$, which differs from a diagnosis problem $\wp(\mathcal{A})$ in the fragmentation of the observation ($\mathcal{O}$ is replaced by $\hat{\mathcal{O}}$).

Since each fragment $\mathcal{F}_i$ implicitly defines a temporal observation $\mathcal{O}_{[i]}$ made up of the union of all fragments $\mathcal{F}_1, \ldots, \mathcal{F}_i$, the solution of the monitoring problem $\mu(\mathcal{A})$ can be defined as the sequence of the solutions of the diagnosis problems $\wp_{[i]}(\mathcal{A}) = (a_0, \mathcal{V}, \mathcal{O}_{[i]}, \mathcal{R})$, $i \in [1 \cdots n]$.

The challenge is to design a diagnosis engine that provides the solution of $\mu(\mathcal{A})$ without solving each single diagnosis problem $\wp_{[i]}(\mathcal{A})$ by means of the (a posteriori) techniques introduced in Chapters 5 and 6.

7.1 Observation Fragmentation

Reactive diagnosis is expected to output diagnosis results at the reception of each fragment of the temporal observation, where each fragment is a set of nodes and a set of arcs. Therefore, we need to provide a new definition of an observation, called an *observation fragmentation*.

Definition 7.1 (Observation Fragmentation). Let $\mathcal{O} = (L, N, A)$ be a temporal observation. A *fragmentation* $\hat{\mathcal{O}}$ of $\mathcal{O}$ (also called a *fragmented observation*) is a sequence of *fragments*,

$$\hat{\mathcal{O}} = [\mathcal{F}_1, \ldots, \mathcal{F}_k], \tag{7.1}$$

where each fragment $\mathcal{F}_i = (L_i, N_i, A_i)$, $i \in [1 \cdots k]$, is such that $\{N_1, \ldots, N_k\}$ is a partition of N, each L_i, $i \in [1 \cdots k]$, is the set of labels involved in N_i, $\{A_1, \ldots, A_k\}$ is a partition of A, and the following *ordering condition* holds:

$$\forall n \mapsto n' \in A_i \left(n' \in N_i,\ n \in \bigcup_{j=1}^{i} N_j \right). \tag{7.2}$$

$\hat{\mathcal{O}}$ is a *plain fragmentation* of $\mathcal{O}$ when each part N_i is a singleton, $i \in [1 \cdots k]$.

Example 7.1. Consider the temporal observation $\mathcal{O}_2$ introduced in Example 4.8. A fragmentation of $\mathcal{O}_2$ is displayed in Fig. 7.1, namely $\hat{\mathcal{O}}_2 = [\mathcal{F}_1, \mathcal{F}_2]$, where $\mathcal{F}_1 = (L_1, N_1, A_1)$, $\mathcal{F}_2 = (L_2, N_2, A_2)$, $N_1 = \{n_1, n_2\}$, $N_2 = \{n_3, n_4\}$, $A_1 = \{n_1 \mapsto n_2\}$, $A_2 = \{n_2 \mapsto n_3, n_2 \mapsto n_4\}$, $L_1 = \{awk, opl, \varepsilon\}$, and $L_2 = \{ide, cll, \varepsilon\}$.

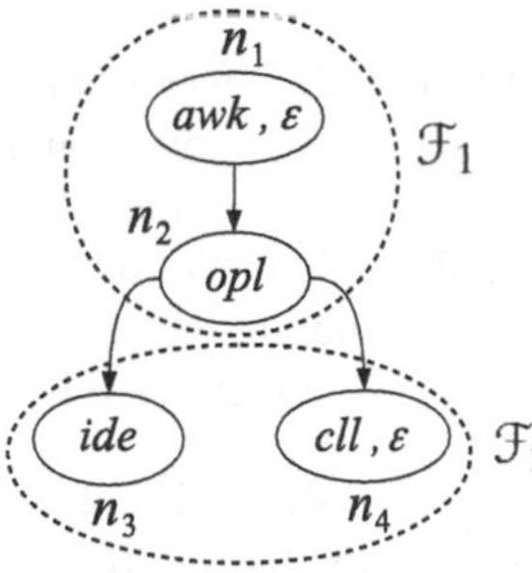

Fig. 7.1 Fragmentation of temporal observation $\mathcal{O}_2$, $\hat{\mathcal{O}}_2 = [\mathcal{F}_1, \mathcal{F}_2]$

Note 7.1. Based on Def. 7.1, the set of nodes of a temporal observation $\mathcal{O}$ is partitioned into parts N_i, $i \in [1 \cdots k]$, with each part being relevant to a different fragment $\mathcal{F}_i$. Hence, given $\mathcal{O}$, each fragment $\mathcal{F}_i$ can be identified by the set of nodes N_i only, as the set of arcs A_i can be inferred to be the arcs entering nodes in N_i.

In a fragmentation of $\mathcal{O}$, each fragment $\mathcal{F}_i$ implicitly identifies a subfragmentation $\hat{\mathcal{O}}_{[i]}$ of $\mathcal{O}$ as the sequence of fragments up to $\mathcal{F}_i$. Also, a subfragmentation $\hat{\mathcal{O}}_{[i]}$ of $\mathcal{O}$ implicitly identifies a temporal observation $\mathcal{O}_{[i]}$ made up of fragments $\mathcal{F}_1, \ldots, \mathcal{F}_i$.

Definition 7.2 (Observation Subfragmentation). Let $\hat{\mathcal{O}} = [\mathcal{F}_1, \ldots, \mathcal{F}_k]$ be a fragmentation of a temporal observation $\mathcal{O}$. A *subfragmentation* of $\mathcal{O}$ is a prefix of $\hat{\mathcal{O}}$ up to the i-th fragment, $i \in [0 \cdots k]$,

$$\hat{\mathcal{O}}_{[i]} = [\mathcal{F}_1, \ldots, \mathcal{F}_i], \tag{7.3}$$

that is, a fragmentation of a temporal observation $\mathcal{O}_{[i]} = (L_{[i]}, N_{[i]}, A_{[i]})$, where

$$L_{[i]} = \bigcup_{j=1}^{i} L_j, \qquad N_{[i]} = \bigcup_{j=1}^{i} N_j, \qquad A_{[i]} = \bigcup_{j=1}^{i} A_j \,. \tag{7.4}$$

If $i = 0$, then $\mathcal{O}_{[i]} = (\emptyset, \emptyset, \emptyset)$ is the *empty* temporal observation.

Based on a subfragmentation $\hat{\mathcal{O}}_{[i]}$ of $\mathcal{O}$ and a diagnosis problem $\wp(\mathcal{A}) = (a_0, \mathcal{V}, \mathcal{O}, \mathcal{R})$, we can define a subproblem of $\wp(\mathcal{A})$ as $\wp_{[i]}(\mathcal{A}) = (a_0, \mathcal{V}, \mathcal{O}_{[i]}, \mathcal{R})$, where $O_{[i]}$ is the temporal observation made up of fragments $\mathcal{F}_1, \ldots, \mathcal{F}_i$.

Definition 7.3 (Diagnosis Subproblem). Let $\wp(\mathcal{A}) = (a_0, \mathcal{V}, \mathcal{O}, \mathcal{R})$ be a diagnosis problem, and $\hat{\mathcal{O}} = [\mathcal{F}_1, \ldots, \mathcal{F}_k]$ a fragmentation of $\mathcal{O}$. A *subproblem* $\wp_{[i]}(\mathcal{A})$ of $\wp(\mathcal{A})$, $i \in [0 \cdots k]$, is a diagnosis problem

$$\wp_{[i]}(\mathcal{A}) = (a_0, \mathcal{V}, \mathcal{O}_{[i]}, \mathcal{R}) \tag{7.5}$$

relevant to temporal observation $\mathcal{O}_{[i]}$.

Example 7.2. Consider the diagnosis problem $\wp(\mathcal{A}) = (a_0, \mathcal{V}, \mathcal{O}_2, \mathcal{R})$ defined in Example 4.12. Displayed on the left side of Fig. 7.2 is a fragmentation of the temporal observation $\mathcal{O}_2$, namely $\hat{\mathcal{O}}_2 = [\mathcal{F}_1, \mathcal{F}_2, \mathcal{F}_3, \mathcal{F}_4]$, where each fragment $\mathcal{F}_i$, $i \in [1 \cdots 4]$, includes node n_i. In other words, each fragment is composed of just one node and one entering arc. Next to $\hat{\mathcal{O}}_2$, from left to right, are the temporal observations $\mathcal{O}_{[i]}$, $i \in [1 \cdots 4]$, which are relevant to the diagnosis subproblems $\wp_{[i]}(\mathcal{A}_2)$, respectively.

The notion of a subproblem is introduced for formal reasons only, in order to precisely define the solution of a monitoring problem (Sec. 7.2). Reactive diagnosis is expected to provide the solution of a monitoring problem as the sequence of solutions of the diagnosis subproblems $\wp_{[i]}(\mathcal{A})$ *without* however performing a posteriori diagnosis.

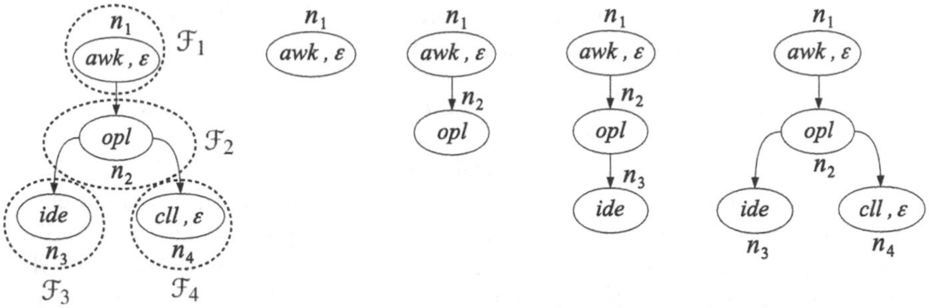

Fig. 7.2 Fragmentation of observation $\mathcal{O}_2$, and relevant observations $\mathcal{O}_{2[1]}$, $\mathcal{O}_{2[2]}$, $\mathcal{O}_{2[3]}$, and $\mathcal{O}_{2[4]}$

7.2 Monitoring Problem

Formally, the notion of a monitoring problem is similar to the notion of a diagnosis problem, with the essential difference that the temporal observation $\mathcal{O}$ (the third element of the quadruple) is substituted by a fragmented observation $\hat{\mathcal{O}}$.

Definition 7.4 (Monitoring Problem). Let $\mathcal{A}$ be an active system, $\mathcal{V}$ a viewer for $\mathcal{A}$, $\hat{\mathcal{O}}$ a fragmentation of a temporal observation $\mathcal{O}$ of $\mathcal{A}$ relevant to a trajectory starting at an initial state a_0, and $\mathcal{R}$ a ruler for $\mathcal{A}$. The quadruple

$$\mu(\mathcal{A}) = (a_0, \mathcal{V}, \hat{\mathcal{O}}, \mathcal{R}) \tag{7.6}$$

is a *monitoring problem* for $\mathcal{A}$. A *plain monitoring problem* is a monitoring problem where $\hat{\mathcal{O}}$ is a fragmentation of a plain temporal observation.

Example 7.3. Consider the diagnosis problem $\wp(\mathcal{A}_2) = (a_0, \mathcal{V}, \mathcal{O}_2, \mathcal{R})$ defined in Example 4.12 and the fragmentation of $\mathcal{O}_2$ defined in Example 7.2 and displayed on the left side of Fig. 7.2. A monitoring problem for system $\mathcal{A}_2$ is $\mu(\mathcal{A}_2) = (a_0, \mathcal{V}, \hat{\mathcal{O}}_2, \mathcal{R})$.

Since reactive diagnosis is expected to output a diagnosis response after the reception of each fragment of observation, the solution of a monitoring problem $\mu(\mathcal{A})$ involving a fragmented observation $\hat{\mathcal{O}} = [\mathcal{F}_1, \ldots, \mathcal{F}_k]$ is the sequence of solutions of the diagnosis subproblems relevant to all intermediate temporal observations made up of all fragments of $\hat{\mathcal{O}}$. In other words, the solution of $\mu(\mathcal{A})$ is defined in terms of the solutions of $\wp_{[i]}(\mathcal{A})$, $i \in [1 \cdots k]$.

Definition 7.5 (Monitoring-Problem Solution). Let $\mu(\mathcal{A}) = (a_0, \mathcal{V}, \hat{\mathcal{O}}, \mathcal{R})$ be a monitoring problem for an active system $\mathcal{A}$, where $\hat{\mathcal{O}} = [\mathcal{F}_1, \ldots, \mathcal{F}_k]$ is a fragmentation of $\mathcal{O}$. The *solution* of $\mu(\mathcal{A})$, written $\Delta(\mu(\mathcal{A}))$, is the sequence of the solutions of the diagnosis subproblems $\wp_{[i]}(\mathcal{A})$ of $\wp(\mathcal{A}) = (a_0, \mathcal{V}, \mathcal{O}, \mathcal{R})$, $i \in [0 \cdots k]$:

$$\Delta(\mu(\mathcal{A})) = \left[\Delta(\wp_{[0]}(\mathcal{A})), \Delta(\wp_{[1]}(\mathcal{A})), \ldots, \Delta(\wp_{[k]}(\mathcal{A}))\right] . \tag{7.7}$$

Example 7.4. Consider the monitoring problem $\mu(\mathcal{A}_2) = (a_0, \mathcal{V}, \hat{\mathcal{O}}_2, \mathcal{R})$ defined in Example 7.3. Based on (7.7), the solution of $\mu(\mathcal{A}_2)$ will be

$$\Delta(\mu(\mathcal{A}_2)) = \left[\Delta(\wp_{[0]}(\mathcal{A}_2)), \Delta(\wp_{[1]}(\mathcal{A}_2)), \Delta(\wp_{[2]}(\mathcal{A}_2)), \Delta(\wp_{[3]}(\mathcal{A}_2)), \Delta(\wp_{[4]}(\mathcal{A}_2))\right],$$

where $\wp(\mathcal{A}_2) = (a_0, \mathcal{V}, \mathcal{O}_2, \mathcal{R})$. Specifically, we need to determine the solutions of the diagnosis problems $\wp_{[i]}(\mathcal{A}_2)$, $i \in [0 \cdots 4]$. Although we might exploit either monolithic or modular diagnosis, we prefer to solve the problem based on (4.5) in Def. 4.8. To this end, for each problem, we consider the behavior space $Bsp(\mathcal{A}_2)$ specified in Table 4.2 and depicted in Fig. 4.4, as well as the viewer $\mathcal{V}$ and ruler $\mathcal{R}$ specified in Table 4.1.

First, we consider $\wp_{[0]}(\mathcal{A}_2) = (a_0, \mathcal{V}, \mathcal{O}_{2[0]}, \mathcal{R})$, where $\mathcal{O}_{2[0]}$ is the empty observation. Since the observation is empty, the only candidate trace is the empty trace (no observable label was generated). Therefore, a trajectory in $Bsp(\mathcal{A}_2)$ is consistent with $\mathcal{O}_{2[0]}$ if and only if it does not involve any observable transition; in our case, this is the empty trajectory $h = [\,]$. Since $h_{[\mathcal{R}]} = \emptyset$, we have

$$\Delta(\wp_{[0]}) = \{\emptyset\}. \tag{7.8}$$

Considering $\wp_{[1]}(\mathcal{A}_2) = (a_0, \mathcal{V}, \mathcal{O}_{2[1]}, \mathcal{R})$, where $\mathcal{O}_{2[1]}$ is composed of just one node marked by the set $\{awk, \varepsilon\}$ of candidate labels, we need to determine the candidate trajectories h that conform to $\mathcal{O}_{2[1]}$, in other words, such that $h_{[\mathcal{V}]}$ is either $[\,]$ or $[awk]$ (these being the two candidate traces in $\|\mathcal{O}_{2[1]}\|$). These candidate trajectories are listed in Table 7.1, along with corresponding traces and diagnoses. Based on these trajectories, we have

$$\Delta(\wp_{[1]}) = \{\emptyset, \{\mathbf{nol}\}, \{\mathbf{nor}\}, \{\mathbf{nol}, \mathbf{nor}\}, \{\mathbf{fop}\}\}. \tag{7.9}$$

Considering $\wp_{[2]}(\mathcal{A}_2) = (a_0, \mathcal{V}, \mathcal{O}_{2[2]}, \mathcal{R})$, where $\mathcal{O}_{2[2]}$ includes nodes n_1 and n_2 (see Fig. 7.2), we need to determine the candidate trajectories h that conform to $\mathcal{O}_{2[2]}$, in other words, such that $h_{[\mathcal{V}]}$ is either $[opl]$ or $[awk, opl]$ (these being the two candidate traces in $\|\mathcal{O}_{2[2]}\|$). These candidate trajectories are listed in Table 7.2, along with corresponding traces and diagnoses.

Based on these trajectories, we have

$$\Delta(\wp_{[2]}) = \{\emptyset, \{\mathbf{nor}\}\}. \tag{7.10}$$

Considering $\wp_{[3]}(\mathcal{A}_2) = (a_0, \mathcal{V}, \mathcal{O}_{2[3]}, \mathcal{R})$, where $\mathcal{O}_{2[3]}$ includes nodes n_1, n_2, and n_3 (see Fig. 7.2), we need to determine the candidate trajectories h that conform to $\mathcal{O}_{2[3]}$, in other words, such that $h_{[\mathcal{V}]}$ is either $[opl, ide]$ or $[awk, opl, ide]$ (these being the two candidate traces in $\|\mathcal{O}_{2[3]}\|$). These candidate trajectories are listed in Table 7.3, along with corresponding traces and diagnoses.

Based on these trajectories, we have

$$\Delta(\wp_{[3]}) = \{\{\mathbf{nor}\}, \{\mathbf{ncl}, \mathbf{nor}\}, \{\mathbf{nor}, \mathbf{fcp}\}\}. \tag{7.11}$$

Table 7.1 Trajectories in $Bsp(\mathcal{A}_2)$ generating traces in $\|\mathcal{O}_{2[1]}\|$

h	*Trajectory*	$h_{[\mathcal{V}]}$	$h_{[\mathcal{R}]}$
h_1	ε	$[\,]$	$\emptyset$
h_2	$0 \xrightarrow{p_1} 1$	$[awk]$	$\emptyset$
h_3	$0 \xrightarrow{p_1} 1 \xrightarrow{b_3(l)} 4$	$[awk]$	$\{\mathbf{nol}\}$
h_4	$0 \xrightarrow{p_1} 1 \xrightarrow{b_3(l)} 4 \xrightarrow{b_3(r)} 9$	$[awk]$	$\{\mathbf{nol}, \mathbf{nor}\}$
h_5	$0 \xrightarrow{p_1} 1 \xrightarrow{b_3(r)} 5$	$[awk]$	$\{\mathbf{nor}\}$
h_6	$0 \xrightarrow{p_1} 1 \xrightarrow{b_3(r)} 5 \xrightarrow{b_3(l)} 9$	$[awk]$	$\{\mathbf{nol}, \mathbf{nor}\}$
h_7	$0 \xrightarrow{p_3} 53$	$[awk]$	$\{\mathbf{fop}\}$
h_8	$0 \xrightarrow{p_3} 53 \xrightarrow{b_5(l)} 44$	$[awk]$	$\{\mathbf{fop}\}$
h_9	$0 \xrightarrow{p_3} 53 \xrightarrow{b_5(l)} 44 \xrightarrow{b_5(r)} 9$	$[awk]$	$\{\mathbf{fop}\}$
h_{10}	$0 \xrightarrow{p_3} 53 \xrightarrow{b_5(r)} 43$	$[awk]$	$\{\mathbf{fop}\}$
h_{11}	$0 \xrightarrow{p_3} 53 \xrightarrow{b_5(r)} 43 \xrightarrow{b_5(l)} 9$	$[awk]$	$\{\mathbf{fop}\}$

Table 7.2 Trajectories in $Bsp(\mathcal{A}_2)$ generating traces in $\|\mathcal{O}_{2[2]}\|$

h	*Trajectory*	$h_{[\mathcal{V}]}$	$h_{[\mathcal{R}]}$
h_1	$0 \xrightarrow{p_1} 1 \xrightarrow{b_1(l)} 2$	$[awk, opl]$	$\emptyset$
h_2	$0 \xrightarrow{p_1} 1 \xrightarrow{b_1(l)} 2 \xrightarrow{b_3(r)} 8$	$[awk, opl]$	$\{\mathbf{nor}\}$
h_3	$0 \xrightarrow{p_1} 1 \xrightarrow{b_3(r)} 5 \xrightarrow{b_1(l)} 8$	$[awk, opl]$	$\{\mathbf{nor}\}$

Finally, considering $\wp_{[4]}(\mathcal{A}_2) = (a_0, \mathcal{V}, \mathcal{O}_{2[4]}, \mathcal{R})$, where $\mathcal{O}_{2[4]} = \mathcal{O}_2$ (see Fig. 7.2), we need to determine the candidate trajectories h that conform to $\mathcal{O}_{2[4]}$, in other words, such that $h_{[\mathcal{V}]}$ is one of the candidate traces in $\|\mathcal{O}_{2[4]}\|$, namely $[awk, opl, ide]$, $[awk, opl, ide, cll]$, $[opl, ide]$, $[opl, ide, cll]$, and $[opl, cll, ide]$. These candidate trajectories are listed in Table 7.4, along with corresponding traces and diagnoses.

Based on these trajectories, we have

$$\Delta(\wp_{[4]}) = \{\{\mathbf{nor}\}, \{\mathbf{ncl}, \mathbf{nor}\}, \{\mathbf{nor}, \mathbf{fcp}\}\}\,. \tag{7.12}$$

Finally, the solution of $\mu(\mathcal{A}_2)$ is

$$\begin{aligned} \Delta(\mu(\mathcal{A}_2)) = [\;\; & \Delta(\wp_{[0]}) = \{\emptyset\}, \\ & \Delta(\wp_{[1]}) = \{\emptyset, \{\mathbf{nol}\}, \{\mathbf{nor}\}, \{\mathbf{nol}, \mathbf{nor}\}, \{\mathbf{fop}\}\}, \\ & \Delta(\wp_{[2]}) = \{\emptyset, \{\mathbf{nor}\}\}, \\ & \Delta(\wp_{[3]}) = \{\{\mathbf{nor}\}, \{\mathbf{ncl}, \mathbf{nor}\}, \{\mathbf{nor}, \mathbf{fcp}\}\}, \\ & \Delta(\wp_{[4]}) = \{\{\mathbf{nor}\}, \{\mathbf{ncl}, \mathbf{nor}\}, \{\mathbf{nor}, \mathbf{fcp}\}\}\;]\,. \end{aligned} \tag{7.13}$$

Table 7.3 Trajectories in $Bsp(\mathcal{A}_2)$ generating traces in $\|\mathcal{O}_{2[3]}\|$

h	*Trajectory*	$h_{[\mathcal{V}]}$	$h_{[\mathcal{R}]}$
h_1	$0 \xrightarrow{p_1} 1 \xrightarrow{b_1(l)} 2 \xrightarrow{b_3(r)} 8 \xrightarrow{p_2} 26$	$[awk, opl, ide]$	$\{\mathbf{nor}\}$
h_2	$0 \xrightarrow{p_1} 1 \xrightarrow{b_1(l)} 2 \xrightarrow{b_3(r)} 8 \xrightarrow{p_2} 26 \xrightarrow{b_4(l)} 48$	$[awk, opl, ide]$	$\{\mathbf{ncl}, \mathbf{nor}\}$
h_3	$0 \xrightarrow{p_1} 1 \xrightarrow{b_1(l)} 2 \xrightarrow{b_3(r)} 8 \xrightarrow{p_2} 26 \xrightarrow{b_4(l)} 48 \xrightarrow{b_5(r)} 16$	$[awk, opl, ide]$	$\{\mathbf{ncl}, \mathbf{nor}\}$
h_4	$0 \xrightarrow{p_1} 1 \xrightarrow{b_1(l)} 2 \xrightarrow{b_3(r)} 8 \xrightarrow{p_2} 26 \xrightarrow{b_5(r)} 12$	$[awk, opl, ide]$	$\{\mathbf{nor}\}$
h_5	$0 \xrightarrow{p_1} 1 \xrightarrow{b_1(l)} 2 \xrightarrow{b_3(r)} 8 \xrightarrow{p_2} 26 \xrightarrow{b_5(r)} 12 \xrightarrow{b_4(l)} 16$	$[awk, opl, ide]$	$\{\mathbf{ncl}, \mathbf{nor}\}$
h_6	$0 \xrightarrow{p_1} 1 \xrightarrow{b_1(l)} 2 \xrightarrow{b_3(r)} 8 \xrightarrow{p_4} 33$	$[awk, opl, ide]$	$\{\mathbf{nor}, \mathbf{fcp}\}$
h_7	$0 \xrightarrow{p_1} 1 \xrightarrow{b_1(l)} 2 \xrightarrow{b_3(r)} 8 \xrightarrow{p_4} 33 \xrightarrow{b_6(l)} 36$	$[awk, opl, ide]$	$\{\mathbf{nor}, \mathbf{fcp}\}$
h_8	$0 \xrightarrow{p_1} 1 \xrightarrow{b_1(l)} 2 \xrightarrow{b_3(r)} 8 \xrightarrow{p_4} 33 \xrightarrow{b_6(l)} 36 \xrightarrow{b_3(r)} 16$	$[awk, opl, ide]$	$\{\mathbf{nor}, \mathbf{fcp}\}$
h_9	$0 \xrightarrow{p_1} 1 \xrightarrow{b_1(l)} 2 \xrightarrow{b_3(r)} 8 \xrightarrow{p_4} 33 \xrightarrow{b_3(r)} 35$	$[awk, opl, ide]$	$\{\mathbf{nor}, \mathbf{fcp}\}$
h_{10}	$0 \xrightarrow{p_1} 1 \xrightarrow{b_1(l)} 2 \xrightarrow{b_3(r)} 8 \xrightarrow{p_4} 33 \xrightarrow{b_3(r)} 35 \xrightarrow{b_6(l)} 16$	$[awk, opl, ide]$	$\{\mathbf{nor}, \mathbf{fcp}\}$
h_{11}	$0 \xrightarrow{p_1} 1 \xrightarrow{b_3(r)} 5 \xrightarrow{b_1(l)} 8 \xrightarrow{p_2} 26$	$[awk, opl, ide]$	$\{\mathbf{nor}\}$
h_{12}	$0 \xrightarrow{p_1} 1 \xrightarrow{b_3(r)} 5 \xrightarrow{b_1(l)} 8 \xrightarrow{p_2} 26 \xrightarrow{b_4(l)} 48$	$[awk, opl, ide]$	$\{\mathbf{ncl}, \mathbf{nor}\}$
h_{13}	$0 \xrightarrow{p_1} 1 \xrightarrow{b_3(r)} 5 \xrightarrow{b_1(l)} 8 \xrightarrow{p_2} 26 \xrightarrow{b_4(l)} 48 \xrightarrow{b_5(r)} 16$	$[awk, opl, ide]$	$\{\mathbf{ncl}, \mathbf{nor}\}$
h_{14}	$0 \xrightarrow{p_1} 1 \xrightarrow{b_3(r)} 5 \xrightarrow{b_1(l)} 8 \xrightarrow{p_2} 26 \xrightarrow{b_5(r)} 12$	$[awk, opl, ide]$	$\{\mathbf{nor}\}$
h_{15}	$0 \xrightarrow{p_1} 1 \xrightarrow{b_3(r)} 5 \xrightarrow{b_1(l)} 8 \xrightarrow{p_2} 26 \xrightarrow{b_5(r)} 12 \xrightarrow{b_4(l)} 16$	$[awk, opl, ide]$	$\{\mathbf{ncl}, \mathbf{nor}\}$
h_{16}	$0 \xrightarrow{p_1} 1 \xrightarrow{b_3(r)} 5 \xrightarrow{b_1(l)} 8 \xrightarrow{p_4} 33$	$[awk, opl, ide]$	$\{\mathbf{nor}, \mathbf{fcp}\}$
h_{17}	$0 \xrightarrow{p_1} 1 \xrightarrow{b_3(r)} 5 \xrightarrow{b_1(l)} 8 \xrightarrow{p_4} 33 \xrightarrow{b_6(l)} 36$	$[awk, opl, ide]$	$\{\mathbf{nor}, \mathbf{fcp}\}$
h_{18}	$0 \xrightarrow{p_1} 1 \xrightarrow{b_3(r)} 5 \xrightarrow{b_1(l)} 8 \xrightarrow{p_4} 33 \xrightarrow{b_6(l)} 36 \xrightarrow{b_3(r)} 16$	$[awk, opl, ide]$	$\{\mathbf{nor}, \mathbf{fcp}\}$
h_{19}	$0 \xrightarrow{p_1} 1 \xrightarrow{b_3(r)} 5 \xrightarrow{b_1(l)} 8 \xrightarrow{p_4} 33 \xrightarrow{b_3(r)} 35$	$[awk, opl, ide]$	$\{\mathbf{nor}, \mathbf{fcp}\}$
h_{20}	$0 \xrightarrow{p_1} 1 \xrightarrow{b_3(r)} 5 \xrightarrow{b_1(l)} 8 \xrightarrow{p_4} 33 \xrightarrow{b_3(r)} 35 \xrightarrow{b_6(l)} 16$	$[awk, opl, ide]$	$\{\mathbf{nor}, \mathbf{fcp}\}$

7.3 Diagnosis Closure

In Sec. 7.2 we defined the notion of a monitoring problem $\mu(\mathcal{A})$, as well as its solution in terms of a sequence of solutions of diagnosis subproblems. However, the real-time constraints imposed on reactive diagnosis prevent us from applying such a definition to actually solve the monitoring problem. Since providing the solution of each subproblem $\wp_{[i]}(\mathcal{A})$ by means of the reconstruction–decoration–distillation technique would be impractical, an alternative diagnosis engine must be devised.

Ideally, the response time of the engine for each observation fragment $\mathcal{F}_i$ should be somehow related to the extent of $\mathcal{F}_i$ rather than to the extent of $\mathcal{O}_{[i]}$ (which is the case in a posteriori diagnosis). Assuming that each fragment has the same extent (size), the response time is expected to be constant. This way, the diagnosis engine can be reactive for a long time (ideally, forever).

A possible approach to reactive diagnosis is to make behavior reconstruction incremental. At the reception of each fragment of observation, instead of generating the behavior of the system from scratch, the behavior relevant to the previous frag-

Table 7.4 Trajectories in $Bsp(\mathcal{A}_2)$ generating traces in $\|\mathcal{O}_{2[4]}\|$

h	*Trajectory*	$h_{[\mathcal{V}]}$	$h_{[\mathcal{R}]}$
h_1	$0 \xrightarrow{p_1} 1 \xrightarrow{b_1(l)} 2 \xrightarrow{b_3(r)} 8 \xrightarrow{p_2} 26$	$[awk, opl, ide]$	$\{\mathbf{nor}\}$
h_2	$0 \xrightarrow{p_1} 1 \xrightarrow{b_1(l)} 2 \xrightarrow{b_3(r)} 8 \xrightarrow{p_2} 26 \xrightarrow{b_4(l)} 48$	$[awk, opl, ide]$	$\{\mathbf{ncl}, \mathbf{nor}\}$
h_3	$0 \xrightarrow{p_1} 1 \xrightarrow{b_1(l)} 2 \xrightarrow{b_3(r)} 8 \xrightarrow{p_2} 26 \xrightarrow{b_4(l)} 48 \xrightarrow{b_5(r)} 16$	$[awk, opl, ide]$	$\{\mathbf{ncl}, \mathbf{nor}\}$
h_4	$0 \xrightarrow{p_1} 1 \xrightarrow{b_1(l)} 2 \xrightarrow{b_3(r)} 8 \xrightarrow{p_2} 26 \xrightarrow{b_5(r)} 12$	$[awk, opl, ide]$	$\{\mathbf{nor}\}$
h_5	$0 \xrightarrow{p_1} 1 \xrightarrow{b_1(l)} 2 \xrightarrow{b_3(r)} 8 \xrightarrow{p_2} 26 \xrightarrow{b_5(r)} 12 \xrightarrow{b_4(l)} 16$	$[awk, opl, ide]$	$\{\mathbf{ncl}, \mathbf{nor}\}$
h_6	$0 \xrightarrow{p_1} 1 \xrightarrow{b_1(l)} 2 \xrightarrow{b_3(r)} 8 \xrightarrow{p_4} 33$	$[awk, opl, ide]$	$\{\mathbf{nor}, \mathbf{fcp}\}$
h_7	$0 \xrightarrow{p_1} 1 \xrightarrow{b_1(l)} 2 \xrightarrow{b_3(r)} 8 \xrightarrow{p_4} 33 \xrightarrow{b_6(l)} 36$	$[awk, opl, ide]$	$\{\mathbf{nor}, \mathbf{fcp}\}$
h_8	$0 \xrightarrow{p_1} 1 \xrightarrow{b_1(l)} 2 \xrightarrow{b_3(r)} 8 \xrightarrow{p_4} 33 \xrightarrow{b_6(l)} 36 \xrightarrow{b_3(r)} 16$	$[awk, opl, ide]$	$\{\mathbf{nor}, \mathbf{fcp}\}$
h_9	$0 \xrightarrow{p_1} 1 \xrightarrow{b_1(l)} 2 \xrightarrow{b_3(r)} 8 \xrightarrow{p_4} 33 \xrightarrow{b_3(r)} 35$	$[awk, opl, ide]$	$\{\mathbf{nor}, \mathbf{fcp}\}$
h_{10}	$0 \xrightarrow{p_1} 1 \xrightarrow{b_1(l)} 2 \xrightarrow{b_3(r)} 8 \xrightarrow{p_4} 33 \xrightarrow{b_3(r)} 35 \xrightarrow{b_6(l)} 16$	$[awk, opl, ide]$	$\{\mathbf{nor}, \mathbf{fcp}\}$
h_{11}	$0 \xrightarrow{p_1} 1 \xrightarrow{b_3(r)} 5 \xrightarrow{b_1(l)} 8 \xrightarrow{p_2} 26$	$[awk, opl, ide]$	$\{\mathbf{nor}\}$
h_{12}	$0 \xrightarrow{p_1} 1 \xrightarrow{b_3(r)} 5 \xrightarrow{b_1(l)} 8 \xrightarrow{p_2} 26 \xrightarrow{b_4(l)} 48$	$[awk, opl, ide]$	$\{\mathbf{ncl}, \mathbf{nor}\}$
h_{13}	$0 \xrightarrow{p_1} 1 \xrightarrow{b_3(r)} 5 \xrightarrow{b_1(l)} 8 \xrightarrow{p_2} 26 \xrightarrow{b_4(l)} 48 \xrightarrow{b_5(r)} 16$	$[awk, opl, ide]$	$\{\mathbf{ncl}, \mathbf{nor}\}$
h_{14}	$0 \xrightarrow{p_1} 1 \xrightarrow{b_3(r)} 5 \xrightarrow{b_1(l)} 8 \xrightarrow{p_2} 26 \xrightarrow{b_5(r)} 12$	$[awk, opl, ide]$	$\{\mathbf{nor}\}$
h_{15}	$0 \xrightarrow{p_1} 1 \xrightarrow{b_3(r)} 5 \xrightarrow{b_1(l)} 8 \xrightarrow{p_2} 26 \xrightarrow{b_5(r)} 12 \xrightarrow{b_4(l)} 16$	$[awk, opl, ide]$	$\{\mathbf{ncl}, \mathbf{nor}\}$
h_{16}	$0 \xrightarrow{p_1} 1 \xrightarrow{b_3(r)} 5 \xrightarrow{b_1(l)} 8 \xrightarrow{p_4} 33$	$[awk, opl, ide]$	$\{\mathbf{nor}, \mathbf{fcp}\}$
h_{17}	$0 \xrightarrow{p_1} 1 \xrightarrow{b_3(r)} 5 \xrightarrow{b_1(l)} 8 \xrightarrow{p_4} 33 \xrightarrow{b_6(l)} 36$	$[awk, opl, ide]$	$\{\mathbf{nor}, \mathbf{fcp}\}$
h_{18}	$0 \xrightarrow{p_1} 1 \xrightarrow{b_3(r)} 5 \xrightarrow{b_1(l)} 8 \xrightarrow{p_4} 33 \xrightarrow{b_6(l)} 36 \xrightarrow{b_3(r)} 16$	$[awk, opl, ide]$	$\{\mathbf{nor}, \mathbf{fcp}\}$
h_{19}	$0 \xrightarrow{p_1} 1 \xrightarrow{b_3(r)} 5 \xrightarrow{b_1(l)} 8 \xrightarrow{p_4} 33 \xrightarrow{b_3(r)} 35$	$[awk, opl, ide]$	$\{\mathbf{nor}, \mathbf{fcp}\}$
h_{20}	$0 \xrightarrow{p_1} 1 \xrightarrow{b_3(r)} 5 \xrightarrow{b_1(l)} 8 \xrightarrow{p_4} 33 \xrightarrow{b_3(r)} 35 \xrightarrow{b_6(l)} 16$	$[awk, opl, ide]$	$\{\mathbf{nor}, \mathbf{fcp}\}$
h_{21}	$0 \xrightarrow{p_1} 1 \xrightarrow{b_1(l)} 2 \xrightarrow{b_3(r)} 8 \xrightarrow{p_2} 26 \xrightarrow{b_2(l)} 28$	$[awk, opl, ide, cll]$	$\{\mathbf{nor}\}$
h_{22}	$0 \xrightarrow{p_1} 1 \xrightarrow{b_1(l)} 2 \xrightarrow{b_3(r)} 8 \xrightarrow{p_2} 26 \xrightarrow{b_2(l)} 28 \xrightarrow{b_5(r)} 0$	$[awk, opl, ide, cll]$	$\{\mathbf{nor}\}$
h_{23}	$0 \xrightarrow{p_1} 1 \xrightarrow{b_1(l)} 2 \xrightarrow{b_3(r)} 8 \xrightarrow{p_2} 26 \xrightarrow{b_5(r)} 12 \xrightarrow{b_2(l)} 0$	$[awk, opl, ide, cll]$	$\{\mathbf{nor}\}$
h_{24}	$0 \xrightarrow{p_1} 1 \xrightarrow{b_3(r)} 5 \xrightarrow{b_1(l)} 8 \xrightarrow{p_2} 26 \xrightarrow{b_2(l)} 28$	$[awk, opl, ide, cll]$	$\{\mathbf{nor}\}$
h_{25}	$0 \xrightarrow{p_1} 1 \xrightarrow{b_3(r)} 5 \xrightarrow{b_1(l)} 8 \xrightarrow{p_2} 26 \xrightarrow{b_2(l)} 28 \xrightarrow{b_5(r)} 0$	$[awk, opl, ide, cll]$	$\{\mathbf{nor}\}$
h_{26}	$0 \xrightarrow{p_1} 1 \xrightarrow{b_3(r)} 5 \xrightarrow{b_1(l)} 8 \xrightarrow{p_2} 26 \xrightarrow{b_5(r)} 12 \xrightarrow{b_2(l)} 0$	$[awk, opl, ide, cll]$	$\{\mathbf{nor}\}$

ment is extended to account for the new fragment. The newly generated part of the behavior can then be decorated and, finally, diagnoses can be distilled.

However, this approach suffers from two shortcomings:

1. Generally speaking, the new index space is not a simple extension of the old index space, as transitions in the old index space may be redirected toward different states in the new index space. Consequently, the new behavior is not a simple extension of the old behavior, thereby making incremental behavior reconstruction problematic (see Example 7.5).

2. The reconstructed behavior may contain parts that differ in the index field only. Consequently, computational resources may be wasted in replicating low-level model-based reasoning in behavior reconstruction (see Example 7.6).

Example 7.5. Displayed in the first line of Fig. 7.3 are, from left to right, a temporal observation $\mathcal{O}$, the corresponding nondeterministic index space $Nsp(\mathcal{O})$, and the index space $Isp(\mathcal{O})$. In the second line, the same graphs are displayed for the observation $\mathcal{O}'$, which is an extension of $\mathcal{O}$ by a fragment involving node n_3. Note how transition $\mathfrak{I}_1 \xrightarrow{opl} \mathfrak{I}_2$ in $Isp(\mathcal{O})$ is redirected toward $\mathfrak{I}_4$ in $Isp(\mathcal{O}')$.

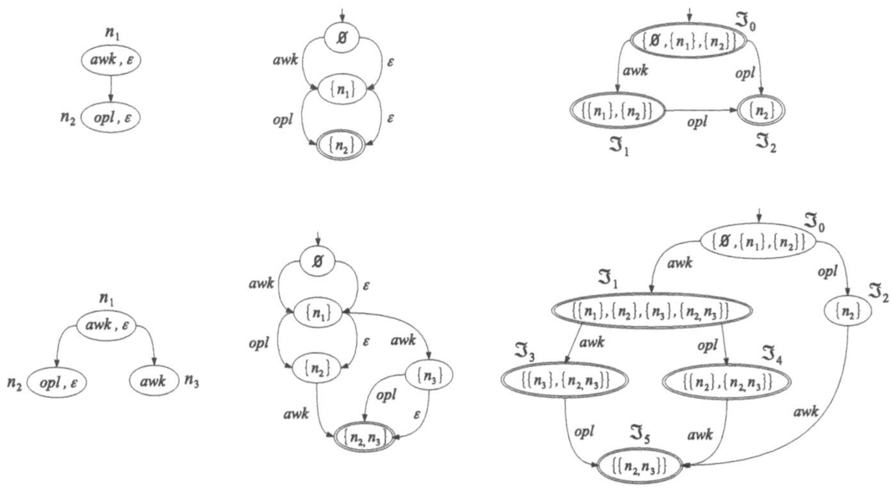

Fig. 7.3 From left to right, temporal observation, nondeterministic index space, and index space

Example 7.6. Consider the diagnosis problem $\wp(\mathcal{A}_2)$ defined in Example 4.12, where the temporal observation $\mathcal{O}_2$ is replaced by the plain temporal observation $[awk, opl, ide, cll, awk, opl]$. Displayed in Fig. 7.4 is the corresponding spurious reconstructed behavior. Each state of the behavior is a pair $(a, \mathfrak{I})$, where a is a state of $Bsp(\mathcal{A}_2)$ (see Table 4.2) and $\mathfrak{I}$ a state of the index space. Note how the two shaded parts of the behavior differ in the values of the index field only, namely $\mathfrak{I}_0$, $\mathfrak{I}_1$, and $\mathfrak{I}_2$ vs. $\mathfrak{I}_4$, $\mathfrak{I}_5$, and $\mathfrak{I}_6$, respectively. This comes from the fact that the two subgraphs are rooted in states $(0, \mathfrak{I}_0)$ and $(0, \mathfrak{I}_4)$, respectively, and from the repetition of the subsequence $[awk, opl]$ in the observation.

A possible solution to the replication of subparts in the reconstructed behavior (Fig. 7.4) is to remove the index field from the states of the behavior, thereby retaining only system states. System states are grouped into clusters involving unobservable transitions only. As such, each cluster is implicitly identified by the entering state, the root of the cluster, as all states and transitions incorporated are the

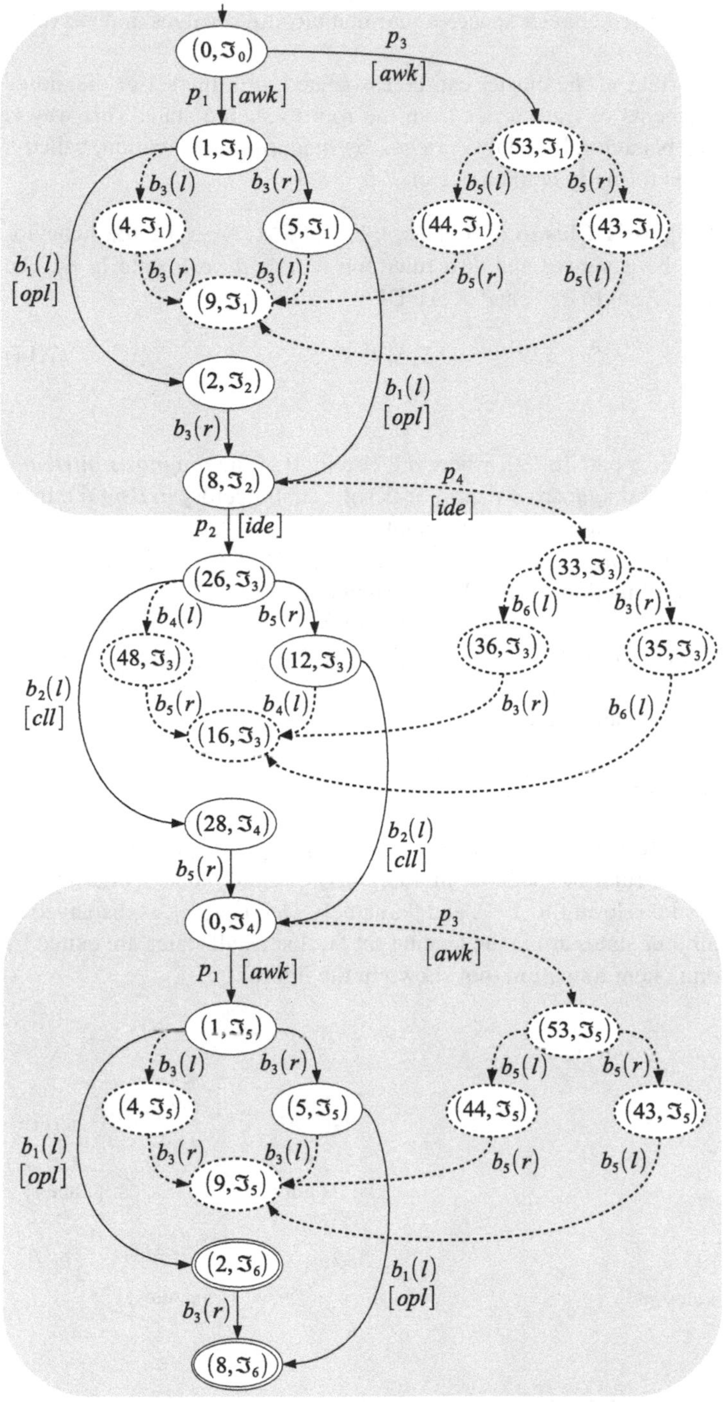

Fig. 7.4 Reconstructed (spurious) behavior of system $\mathcal{A}_2$ based on the plain temporal observation $[awk, opl, ide, cll, awk, opl]$; note how the shaded subgraphs differ in the values of indices only ($\Im_0$, $\Im_1$, and $\Im_2$ vs. $\Im_4$, $\Im_5$, and $\Im_6$, respectively)

closure of the root in the behavior space, where that closure involves unobservable transitions only.

Moreover, each state of the cluster can be associated with the set of diagnoses relevant to the segments of trajectories from the root to such a state. This way, a cluster is a kind of behavioral module enriched by diagnosis information, called a *diagnosis closure*, as formally defined in Def. 7.6.

Definition 7.6 (Diagnosis Closure). Let $Bsp(\mathcal{A}) = (\Sigma, A, \tau, a_0)$ be the behavior space of $\mathcal{A}$. Let $\mathcal{V}$ be a viewer and $\mathcal{R}$ a ruler for $\mathcal{A}$. Let $\bar{a}$ be a state in A. The *diagnosis closure* relevant to $\bar{a}$, $\mathcal{V}$, and $\mathcal{R}$ is a DFA

$$Dcl(\bar{a}, \mathcal{V}, \mathcal{R}) = (\Sigma, S, \tau', s_0, S_l), \tag{7.14}$$

where:

1. Each state $s \in S$ is a pair $(a, \mathcal{D})$, where $a \in A$ and $\mathcal{D}$ is the *diagnosis attribute*, defined as the set of diagnoses δ (based on $\mathcal{R}$) of a subtrajectory in $Bsp(\mathcal{A})$, from $\bar{a}$ to a, where all transitions are unobservable,
2. $s_0 = (\bar{a}, \mathcal{D}_0)$ is the *root*,
3. τ' is the transition function $\tau' : S \times \Sigma \mapsto S$, such that

$$(s, \mathcal{D}) \xrightarrow{t(c)} (s', \mathcal{D}') \in \tau' \iff s \xrightarrow{t(c)} s' \in \tau, t(c) \text{ is unobservable}, \tag{7.15}$$

4. S_l is the *leaving set*, defined as follows:

$$S_l = \{ s \mid s \in S, s = (a, \mathcal{D}), a \xrightarrow{t(c)} a' \in \tau,\ t(c) \text{ is observable} \}. \tag{7.16}$$

Example 7.7. With reference to the behavior space $Bsp(\mathcal{A}_2)$ specified in Table 4.2, and the viewer $\mathcal{V}$ and ruler $\mathcal{R}$ specified in Table 4.1, consider state 1 of $Bsp(\mathcal{A}_2)$. The diagnosis closure relevant to 1, $\mathcal{V}$, and $\mathcal{R}$, namely $Dcl(1, \mathcal{V}, \mathcal{R})$, is displayed in Fig. 7.5, where all four states are in the leaving set S_l; that is, all states are exited by an observable component transition (not shown in the figure).

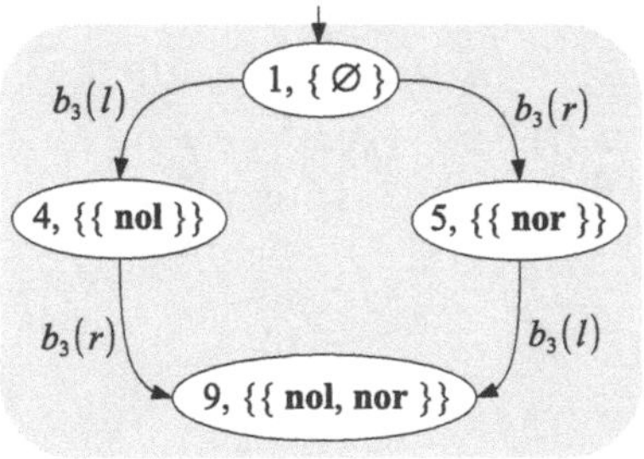

Fig. 7.5 Diagnosis closure $Dcl(1, \mathcal{V}, \mathcal{R})$

A useful operation, called the *diagnosis union*, is introduced for formal reasons.

Definition 7.7 (Diagnosis Union). Let Φ^* be the set of labels in a ruler extended by ε. The *diagnosis union* "$\oplus$" of a set of diagnoses Δ and a label $f \in \Phi^*$ is a set of diagnoses defined as follows:

$$\Delta \oplus f = \begin{cases} \Delta & \text{if } f = \varepsilon, \\ \{\delta' \mid \delta' = \delta \cup \{f\}, \delta \in \Delta\} & \text{otherwise}. \end{cases} \tag{7.17}$$

Example 7.8. Let $\Delta = \{\emptyset, \{\mathbf{nol}\}, \{\mathbf{nor}\}, \{\mathbf{nol}, \mathbf{nor}\}\}$ and $f = \mathbf{nol}$. Based on (7.17), we have (duplicates have been removed)

$$\Delta \oplus f = \{\{\mathbf{nol}\}, \{\mathbf{nol}, \mathbf{nor}\}\}.$$

7.3.1 Diagnosis Closure Algorithm

Listed below is the specification of the algorithm *Dclosure*, which generates the diagnosis closure relevant to a state $\bar{a}$ of a system $\mathcal{A}$, a viewer $\mathcal{V}$ for $\mathcal{A}$, and a ruler $\mathcal{R}$ for $\mathcal{A}$. To this end, it exploits the auxiliary procedure *Propagate*, defined in lines 6–20.

The body of *Dclosure* is listed in lines 21–47. The diagnosis closure is built up incrementally, starting at the initial node $\bar{s} = (\bar{a}, \{\emptyset\})$ (lines 22 and 23). The computation is governed by the loop in lines 24–46. Starting from $\bar{s}$, at each iteration, a new state $s = (a, \mathcal{D})$ of the diagnosis closure is processed (line 25). If appropriate, that state is inserted into the leaving set (lines 26–28).

The inner loop in lines 29–44 considers each unobservable transition triggerable from a, and generates the new diagnosis attribute $\mathcal{D}^*$, associated with the state a' entered by the component transition $t(c)$, as either the diagnosis union between $\mathcal{D}$ and the fault label f associated with $t(c)$ in $\mathcal{R}$, or simply $\mathcal{D}$ (line 31).

At this point, two cases are possible:

1. (Lines 32–38) The subgraph of the diagnosis closure built so far already contains a node $s' = (a', \mathcal{D}')$. First, a new transition $s \xrightarrow{t(c)} s'$ is created (line 33). Then, $\mathcal{D}'$ is possibly extended based on $\mathcal{D}^*$ and $\mathcal{D}'$ (line 36). Finally, the new set of diagnoses inserted into $\mathcal{D}'$ is propagated throughout the current subgraph of the diagnosis closure (line 37).
2. (Lines 39–43) The subgraph of the diagnosis closure built so far does not contain any state involving a'. A new node $s' = (a', \mathcal{D}')$ and a new transition $s \xrightarrow{t(c)} s'$ are generated.

The auxiliary procedure *Propagate* (lines 6–20) takes as input a state s of the current subgraph of the diagnosis closure and a set $\mathcal{D}^*$ of new diagnoses attached to s. It recursively updates the diagnosis attributes associated with the current states of the diagnosis closure reachable from s.

This is performed by the loop in lines 12–19, which considers all transitions leaving s in the current subgraph of the diagnosis closure. For each transition, in line 13, it computes the new set $\mathcal{D}''$ of diagnoses as either the diagnosis union between $\mathcal{D}^*$ and the fault label f associated with $t(c)$ in $\mathcal{R}$ (in this case $t(c)$ is embodied in $\mathcal{R}$) or $\mathcal{D}^*$ (in this case $t(c)$ is not embodied in $\mathcal{R}$). Then, the new set of diagnoses possibly inserted into $\mathcal{D}'$ is recursively propagated (line 17). Note how lines 14–18 of *Propagate* parallel lines 34–38 of *Dclosure*.

The specification of the algorithm *Dclosure* is as follows:

1. **algorithm** *Dclosure* (**in** $\bar{a}$, **in** $\mathcal{V}$, **in** $\mathcal{R}$, **out** $Dcl(\bar{a}, \mathcal{V}, \mathcal{R})$)
2. $\bar{a}$: a state in $Bsp(\mathcal{A})$ with alphabet Σ,
3. $\mathcal{V}$: a viewer for $\mathcal{A}$,
4. $\mathcal{R}$: a ruler for $\mathcal{A}$,
5. $Dcl(\bar{a}, \mathcal{V}, \mathcal{R}) = (\Sigma, S, \tau, s_0, S_l)$: the diagnosis closure relevant to $\bar{a}$, $\mathcal{V}$, and $\mathcal{R}$;

6.   **auxiliary procedure** *Propagate* (**in** s, **in** $\mathcal{D}^*$)
7. s: a state in S,
8. $\mathcal{D}^*$: the additional set of diagnoses attached to s;
9.   **side effects**
10.   Update of the current diagnosis attributes;
11.   **begin** $\langle$*Propagate*$\rangle$
12.   **foreach** transition $s \xrightarrow{t(c)} s' \in \tau$ where $s' = (a', \mathcal{D}')$ **do**
13. $\mathcal{D}''$:= **if** $(t(c), f) \in \mathcal{R}$ **then** $\mathcal{D}^* \oplus f$ **else** $\mathcal{D}^*$ **endif**;
14.   **if** $\mathcal{D}'' \not\subseteq \mathcal{D}'$ **then**
15. $\mathcal{D}^+ := \mathcal{D}'' - \mathcal{D}'$;
16. $\mathcal{D}' := \mathcal{D}' \cup \mathcal{D}^+$;
17.   *Propagate* $(s', \mathcal{D}^+)$
18.   **endif**
19.   **endfor**
20.   **end** $\langle$*Propagate*$\rangle$;

21.   **begin** $\langle$*Dclosure*$\rangle$
22. $\bar{\mathcal{D}} := \{\emptyset\}$; $\bar{s} := (\bar{a}, \bar{\mathcal{D}})$;
23. $S := \{\bar{s}\}$; $\tau := \emptyset$; $S_l; = \emptyset$;
24.   **repeat**
25.   Choose an unmarked state $s = (a, \mathcal{D})$ in S;
26.   **if** there exists an observable transition triggerable from a **then**
27.   Insert s into S_l
28.   **endif**;
29.   **foreach** unobservable transition $t(c)$ triggerable from a **do**
30.   Let a' be the state reached by $t(c)$;
31. $\mathcal{D}^*$:= **if** $(t(c), f) \in \mathcal{R}$ **then** $\mathcal{D} \oplus f$ **else** $\mathcal{D}$ **endif**;
32.   **if** S includes $s' = (a', \mathcal{D}')$ **then**
33.   Insert transition $s \xrightarrow{t(c)} s'$ into τ;
34.   **if** $\mathcal{D}^* \not\subseteq \mathcal{D}'$ **then**
35. $\mathcal{D}^+ := \mathcal{D}^* - \mathcal{D}'$;
36. $\mathcal{D}' := \mathcal{D}' \cup \mathcal{D}^+$;

37. *Propagate* $(s', \mathcal{D}^+)$
38. **endif**
39. **else**
40. $s' := (a', \mathcal{D}^*)$;
41. Insert s' into S;
42. Insert $s \xrightarrow{t(c)} s'$ into τ
43. **endif**
44. **endfor**;
45 Mark s
46. **until** all states in S are marked
47. **end** $\langle$*Dclosure*$\rangle$.

Example 7.9. With reference to the behavior space $Bsp(\mathcal{A}_2)$ defined in Table 4.2, and the viewer $\mathcal{V}$ and ruler $\mathcal{R}$ specified in Table 4.1, consider the diagnosis closure $Dcl(1, \mathcal{V}, \mathcal{R})$ displayed in Fig. 7.5. Traced in Fig. 7.6 is the generation of this diagnosis closure by the algorithm *Dclosure*. Starting from the initial configuration involving only the root $\bar{s} = (1, \{\emptyset\})$, the diagnosis closure is generated by iterating the main loop (lines 24–46) four times, as detailed below:

1. The state $s = (1, \{\emptyset\})$ is chosen and, since it is exited by observable component transitions, it is inserted into S_l. The unobservable component transitions exiting 1 are $b_3(l)$ and $b_3(r)$, reaching system states 4 and 5, respectively. Since both transitions are faulty, involving faults **nol** and **nor**, respectively, and states 4 and 5 are not yet involved in the diagnosis closure, states $(4, \{\{\textbf{nol}\}\})$

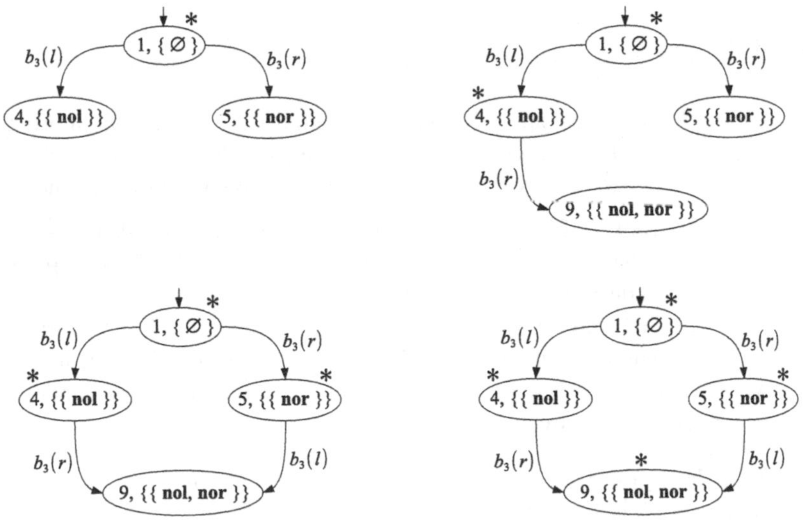

Fig. 7.6 Tracing of algorithm *Dclosure* applied to state 1 of system $\mathcal{A}_2$ and the relevant viewer $\mathcal{V}$ and ruler $\mathcal{R}$ (defined in Table 4.1)

and $(5, \{\{\mathbf{nor}\}\})$ are inserted into S, and transitions $(1, \{\emptyset\}) \xrightarrow{b_3(l)} (4, \{\{\mathbf{nol}\}\})$ and $(1, \{\emptyset\}) \xrightarrow{b_3(r)} (5, \{\{\mathbf{nor}\}\})$ are inserted into τ. Eventually, state $(1, \{\emptyset\})$ is marked (asterisked in Fig. 7.6).
2. The state $s = (4, \{\{\mathbf{nol}\}\})$ is chosen and, since it is exited by an observable component transition, it is inserted into S_l. The only unobservable component transition exiting 4 is $b_3(r)$, reaching system state 9. Since this transition involves the fault **nor**, and state 9 is not yet involved in the diagnosis closure, a state $(9, \{\{\mathbf{nol}, \mathbf{nor}\}\})$ is inserted into S, and a transition $(4, \{\{\mathbf{nol}\}\}) \xrightarrow{b_3(l)} (9, \{\{\mathbf{nol}, \mathbf{nor}\}\})$ is inserted into τ. Eventually, state $(4, \{\{\mathbf{nol}\}\})$ is marked.
3. The state $s = (5, \{\{\mathbf{nor}\}\})$ is chosen and, since it is exited by an observable component transition, it is inserted into S_l. The only unobservable component transition exiting 5 is $b_3(l)$, reaching system state 9. Since this transition involves the fault **nol** and the state $(9, \{\{\mathbf{nol}, \mathbf{nor}\}\})$ is already involved in the diagnosis closure, a transition $(5, \{\{\mathbf{nor}\}\}) \xrightarrow{b_3(l)} (9, \{\{\mathbf{nol}, \mathbf{nor}\}\})$ is inserted into τ. However, since $\mathcal{D}^* = \{\{\mathbf{nol}, \mathbf{nor}\}\} = \mathcal{D}'$ (the diagnosis attribute associated with 9), the latter is not updated and, thus, no propagation occurs. Eventually, state $(5, \{\{\mathbf{nor}\}\})$ is marked.
4. The state $s = (9, \{\{\mathbf{nol}, \mathbf{nor}\}\})$ is chosen and, since it is exited by two observable component transitions, it is inserted into S_l. However, since no unobservable component transition exits 9, no iteration of the loop (lines 29–44) is performed. State $(9, \{\{\mathbf{nol}, \mathbf{nor}\}\})$ is marked.

After the fourth iteration of the main loop (lines 24–46), all states in S are marked, and thereby *Dclosure* terminates.

7.4 Monitor

Starting from the diagnosis closure relevant to the initial state a_0 of a monitoring problem $\mu(\mathcal{A}) = (a_0, \mathcal{V}, \hat{\mathcal{O}}, \mathcal{R})$, it is convenient to define a graph where each node is a diagnosis closure and each arc exits a state $(a, \mathcal{D})$ in the leaving set of a diagnosis closure and enters the root $(\bar{a}, \bar{\mathcal{D}})$ of another diagnosis closure, provided that $a \xrightarrow{t(c)} \bar{a}$ is a transition in the behavior space of $\mathcal{A}$, where $t(c)$ is observable. This graph is called a *monitor*, and is designed for supporting reactive diagnosis efficiently.

However, as with the behavior space, the monitor is defined (Def. 7.8) for formal reasons only, as its complete generation would be impractical in real applications. Therefore, we adopt a lazy approach, where the monitor is generated incrementally, only when new nodes are required.

Definition 7.8 (Monitor). The *monitor* relevant to a system $\mathcal{A}$, an initial state a_0, a viewer $\mathcal{V}$, and a ruler $\mathcal{R}$, is a graph

$$Mtr(\mathcal{A}, a_0, \mathcal{V}, \mathcal{R}) = (\mathbf{N}, \mathbf{L}, \mathbf{A}, N_0), \tag{7.18}$$

where $\mathbf{N}$ is the set of *nodes*, $\mathbf{L}$ is the set of *labels*, $\mathbf{A}$ is the set of *arcs*, and $N_0 \in \mathbf{N}$ is the *initial node*. Each node $N \in \mathbf{N}$ is a diagnosis closure

$$N = Dcl(\bar{a}, \mathcal{V}, \mathcal{R}) = (\Sigma, S, \tau, s_0, S_l), \tag{7.19}$$

where $\bar{a}$ is a state in $Bsp(\mathcal{A})$, with the initial state of the latter being a_0. Let

$$\mathbb{S}_l = \bigcup_{N \in \mathbf{N}} S_l(N), \qquad \mathbb{S}_0 = \bigcup_{N \in \mathbf{N}} \{s_0(N)\}, \tag{7.20}$$

let Ω be the domain of labels in $\mathcal{V}$, and let Φ^* be the domain of labels in $\mathcal{R}$ extended by ε. Each arc $A \in \mathcal{A}$ is marked by a label in $\mathbf{L} \subseteq \mathbb{S}_l \times \Omega \times \Phi^* \times \mathbb{S}_0$. An arc

$$N \xrightarrow{(s,\ell,f,s')} N', \tag{7.21}$$

where $s = (a, \mathcal{D})$ and $s' = (a', \mathcal{D}')$ are internal nodes of N and N', respectively, is such that:

1. s' is the root of N',
2. $a \xrightarrow{t(c)} a'$ is an observable transition in $Bsp(\mathcal{A})$,
3. $(t(c), \ell) \in \mathcal{V}$,
4. Either $(t(c), f) \in \mathcal{R}$ or $f = \varepsilon$.

The root of the diagnosis closure relevant to N_0 is $(a_0, \mathcal{D}_0)$.

Example 7.10. With reference to the behavior space $Bsp(\mathcal{A}_2)$ specified in Table 4.2, and the viewer $\mathcal{V}$ and ruler $\mathcal{R}$ specified in Table 4.1, Fig. 7.7 shows a subpart of the monitor $Mtr(\mathcal{A}_2, 0, \mathcal{V}, \mathcal{R})$, where 0 is the initial state of $Bsp(\mathcal{A}_2)$. The nodes of (this subpart of) the monitor are named $N_0, \ldots, N_7$. Within each node (which, based on Def. 7.8, is a diagnosis closure), states belonging to the leaving set S_l are drawn with thick lines in the figure. Each arc $N \xrightarrow{(s,\ell,f,s')} N'$ (connecting two nodes of the monitor) is represented by a dashed arrow exiting N and entering N', which is marked by the pair (ℓ, f).

Since each node of a monitor relevant to a system $\mathcal{A}$ is a diagnosis closure, internal states carry diagnosis information via diagnosis attributes. Specifically, a state $(a, \mathcal{D})$ indicates that $\mathcal{D}$ is the set of diagnoses relevant to the segments of trajectories of $\mathcal{A}$ starting at the root and ending at a.

In the initial node N_0 of the monitor, these segments are actually trajectories, as they start at the initial state a_0 of $\mathcal{A}$, with a_0 being the root of N_0. Consequently, the union of all the diagnosis attributes in N_0, called the *local candidate set* of $\mathcal{N}_0$, is in fact the solution of the diagnosis problem with an empty observation, namely $\wp_{[0]}(\mathcal{A})$, the first element of the solution of the monitoring problem $\mu(\mathcal{A})$ (Def. 7.5).

Definition 7.9 (Local Candidate Set). The *local candidate set* of a node $N = (\Sigma, S, \tau, s_0, S_l)$ in $Mtr(\mathcal{A}, a_0, \mathcal{V}, \mathcal{R})$ is the union of the diagnosis attributes relevant to the internal states of N:

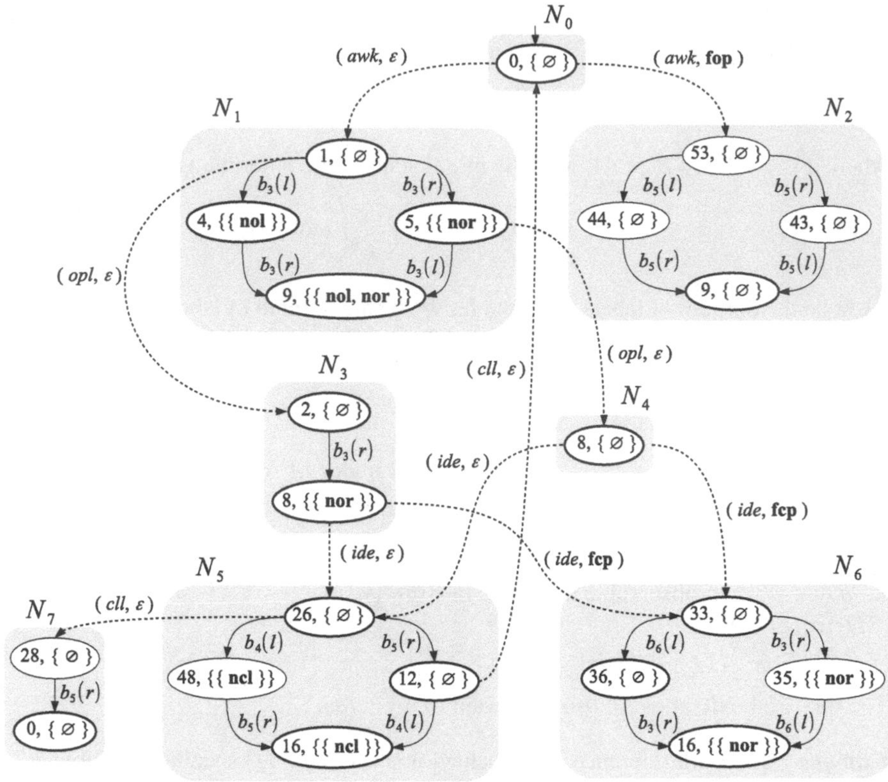

Fig. 7.7 Subpart of the monitor $Mtr(\mathcal{A}_2, a_0, \mathcal{V}, \mathcal{R})$

$$\Delta^{\text{loc}}(N) = \bigcup_{(s,\mathcal{D}) \in N} \mathcal{D} \,. \tag{7.22}$$

Example 7.11. Considering the partial monitor displayed in Fig. 7.7, based on (7.22), the local candidate set of node N_1 is

$$\Delta^{\text{loc}}(N_1) = \left(\bigcup_{(s,\mathcal{D}) \in N_1} \mathcal{D} \right) = \{\emptyset, \{\mathbf{nol}\}, \{\mathbf{nor}\}, \{\mathbf{nol}, \mathbf{nor}\}\} \,.$$

Imagine that we are at the beginning of a reactive diagnosis, when the temporal observation is empty. The monitor is composed of just the initial node N_0, and the first set of candidate diagnoses of the solution of the monitoring problem $\mu(\mathcal{A})$, namely the solution of $\wp_{[0]}(\mathcal{A})$, is the local candidate set of N_0.

Now, assume that we have received the first fragment of observation, which is composed of just one node marked by the label ℓ. The question is: how can we compute the next set of candidate diagnoses, namely the solution of $\wp_{[1]}(\mathcal{A})$?

Based on Def. 4.8, the solution of $\wp_{[1]}(\mathcal{A})$ consists of the diagnoses of the trajectories h such that $h_{[\mathcal{V}]} = [\ell]$. For the sake of simplicity, assume that just one state s in the leaving set of N_0 is exited by an arc marked by the label ℓ, with N_1 being the node of the monitor entered by that arc.

It follows that the trajectories with trace $[\ell]$ are those starting at the root of N_0, passing through state s, and reaching any state in N_1 (as all component transitions in N_1 are unobservable).

Let h be one such trajectory, which ends at state s' in node N_1. Since h passes through state s of the leaving set of N_0, the diagnosis of h includes the union of a diagnosis in the diagnosis attribute of s and a diagnosis in the diagnosis attribute of s'. In addition, if the arc connecting N_0 with N_1 is marked by a fault $f \neq \varepsilon$, the diagnosis of h will include f too.

Since we do not know which is the actual trajectory, in order to generate the solution of $\wp_{[1]}(\mathcal{A})$, we need to combine all candidate diagnoses $\mathcal{D}$ associated with s with the union of all the candidate diagnoses associated with states in N_1, namely $\Delta^{\mathrm{loc}}(N_1)$. The operation combining two sets of diagnoses, called the *diagnosis join*, is formalized in Def. 7.10.

Definition 7.10 (Diagnosis Join). The *diagnosis join* of two sets of diagnoses Δ_1 and Δ_2 is a set of diagnoses defined as follows:

$$\Delta_1 \bowtie \Delta_2 = \{\delta \mid \delta = \delta_1 \cup \delta_2, \delta_1 \in \Delta_1, \delta_2 \in \Delta_2\}\,. \tag{7.23}$$

Example 7.12. Let $\Delta_1 = \{\emptyset, \{\mathbf{nol}\}, \{\mathbf{nor}\}, \{\mathbf{nol}, \mathbf{nor}\}\}$ and $\Delta_2 = \{\{\mathbf{nol}\}, \{\mathbf{nor}, \mathbf{fcp}\}\}$. Based on (7.23), we have

$$\Delta_1 \bowtie \Delta_2 = \{\{\mathbf{nol}\}, \{\mathbf{nor}, \mathbf{fcp}\}, \{\mathbf{nol}, \mathbf{nor}, \mathbf{fcp}\}, \{\mathbf{nol}, \mathbf{nor}\}\}\,.$$

Proposition 7.1. *The following equivalences hold:*

$$\Delta_1 \bowtie \Delta_2 \equiv \Delta_2 \bowtie \Delta_1\,, \tag{7.24}$$

$$\Delta_1 \bowtie (\Delta_2 \cup \Delta_3) \equiv (\Delta_1 \bowtie \Delta_2) \cup (\Delta_1 \bowtie \Delta_3)\,. \tag{7.25}$$

Proof. Equation (7.24) is supported by the commutative property of union:

$$\begin{aligned}\Delta_1 \bowtie \Delta_2 &= \{\delta \mid \delta = \delta_1 \cup \delta_2, \delta_1 \in \Delta_1, \delta_2 \in \Delta_2\}\\ &= \{\delta \mid \delta = \delta_2 \cup \delta_1, \delta_2 \in \Delta_2, \delta_1 \in \Delta_1\} = \Delta_2 \bowtie \Delta_1\,.\end{aligned}$$

Equation (7.25) can be derived as follows:

$$\begin{aligned}\Delta_1 \bowtie (\Delta_2 \cup \Delta_3) =& \{\delta \mid \delta = \delta_1 \cup \delta_{23}, \delta_1 \in \Delta_1, \delta_{23} \in (\Delta_2 \cup \Delta_3)\}\\ =& \{\delta \mid \delta = \delta_1 \cup \delta_2, \delta_1 \in \Delta_1, \delta_2 \in \Delta_2\} \cup\\ & \{\delta \mid \delta = \delta_1 \cup \delta_3, \delta_1 \in \Delta_1, \delta_3 \in \Delta_3\}\\ =& (\Delta_1 \bowtie \Delta_2) \cup (\Delta_1 \bowtie \Delta_3)\,.\end{aligned}$$

□

Imagine that we have a plain monitoring problem where the temporal observation (received in fragments of one node) is $[\ell_1, \ldots, \ell_k]$. Starting from the initial node N_0 of the monitor, the reception of each fragment $\mathcal{F}_i$ (involving a label ℓ_i, $i \in [1 \cdots k]$) moves the state of the monitor from a set of nodes to another set of nodes.

Upon the reception of fragment $\mathcal{F}_i$ carrying label ℓ_i, the new set of nodes of the monitor is defined by the set of nodes reached by the arcs marked by label ℓ_i and exiting nodes of the current set.

However, in order to generate the set of candidate diagnoses on the reception of each fragment, we need to associate with each monitor node N in the current set a set of diagnoses generated by the trajectories up to the root of N. In other words, each node of the monitor in the current set needs to be put in its diagnosis context.

The notions of a *context* and a *monitoring state* are provided in Def. 7.11.

Definition 7.11 (Monitoring State). Let $\mu(\mathcal{A}) = (a_0, \mathcal{V}, \hat{\mathcal{O}}, \mathcal{R})$ be a plain monitoring problem and $Mtr(\mathcal{A}, a_0, \mathcal{V}, \mathcal{R}) = (\mathbf{N}, \mathbf{L}, \mathbf{A}, N_0)$ a relevant monitor. A *context* χ is a pair

$$\chi = (N, \Delta), \tag{7.26}$$

where $N \in \mathbf{N}$ and Δ is a set of diagnoses. A *monitoring state* $\mathcal{M}$ is a set of contexts

$$\mathcal{M} = \{\chi_1, \ldots, \chi_n\}. \tag{7.27}$$

Example 7.13. With reference to the partial monitor displayed in Fig. 7.7, a monitoring state relevant to $Mtr(\mathcal{A}_2, a_0, \mathcal{V}, \mathcal{R})$ is

$$\mathcal{M} = \{(N_5, \{\{\mathbf{nor}\}\}), (N_6, \{\{\mathbf{nor}, \mathbf{fcp}\}\})\}.$$

Consider a plain monitoring problem relevant to a fragmentation of a plain temporal observation $\mathcal{O} = [\ell_1, \ldots, \ell_k]$. In order to perform reactive diagnosis, we need to generate the sequence of monitoring states $[\mathcal{M}_0, \mathcal{M}_1, \ldots, \mathcal{M}_k]$, called the *plain monitoring path*, where $\mathcal{M}_0 = (N_0, \{\emptyset\})$ is the monitoring state before the reception of the first fragment of $\mathcal{O}$.

On the reception of each fragment $\mathcal{F}_i$, $i \in [1 \cdots k]$, the new monitoring state $\mathcal{M}_i$ needs to be determined based on $\mathcal{M}_{i-1}$ and ℓ_i, by combining the set of diagnoses Δ in each context $(N, \Delta) \in \mathcal{M}_{i-1}$ with the diagnosis attributes of the states in N exited by an arc marked by ℓ_i.

The notion of a plain monitoring path is formalized in Def. 7.12.

Definition 7.12 (Plain Monitoring Path). Let $\mu(\mathcal{A})$ be a plain monitoring problem, with $\hat{\mathcal{O}}$ being the fragmentation of a plain observation $\mathcal{O} = [\ell_1, \ldots, \ell_k]$. The *plain monitoring path* of $\mu(\mathcal{A})$ is a sequence of monitoring states

$$Path(\mu(\mathcal{A})) = [\mathcal{M}_0, \mathcal{M}_1, \ldots, \mathcal{M}_k], \tag{7.28}$$

defined as follows:

1. $\mathcal{M}_0 = \{(N_0, \{\emptyset\})\}$,

2. $\forall i \in [1 \cdots k]$, $\mathcal{M}_i$ is the minimal (possibly empty) set of contexts $\chi' = (N', \Delta')$ such that

$$\chi \in \mathcal{M}_{i-1},\ \chi = (N, \Delta), \tag{7.29}$$

$$s = (a, \mathcal{D}) \text{ is a state in the leaving set } S_l \text{ of } N, \tag{7.30}$$

$$N \xrightarrow{(s, \ell_i, f, s_0(N'))} N' \in \mathbf{A}, \tag{7.31}$$

$$\Delta' \supseteq (\mathcal{D} \bowtie \Delta) \oplus f\,. \tag{7.32}$$

Example 7.14. Let $\mu'(\mathcal{A}_2) = (a_0, \mathcal{V}, \hat{\mathcal{O}}', \mathcal{R})$ be a plain monitoring problem for the system $\mathcal{A}_2$ introduced in Example 4.6, where $\mathcal{V}$ and $\mathcal{R}$ are defined in Table 4.1, and $\hat{\mathcal{O}}'$ is the plain fragmentation of the plain temporal observation $\mathcal{O}' = [awk, opl, ide]$. As such, $\hat{\mathcal{O}}'$ is composed of three fragments, one for each node of $\mathcal{O}'$, as displayed in Fig. 7.8. Based on Def. 7.12 and the partial monitor displayed in Fig. 7.7, we have

$$Path(\mu'(\mathcal{A}_2)) = [\,\mathcal{M}_0, \mathcal{M}_1, \mathcal{M}_2, \mathcal{M}_3\,],$$

where $\mathcal{M}_0 = \{(N_0, \{\emptyset\})\}$, $\mathcal{M}_1 = \{(N_1, \{\emptyset\}), (N_2, \{\{\mathbf{fop}\}\})\}$, $\mathcal{M}_2 = \{(N_3, \{\emptyset\}), (N_4, \{\{\mathbf{nor}\}\})\}$, and $\mathcal{M}_3 = \{(N_5, \{\{\mathbf{nor}\}\}), (N_6, \{\{\mathbf{nor}, \mathbf{fcp}\}\})\}$.

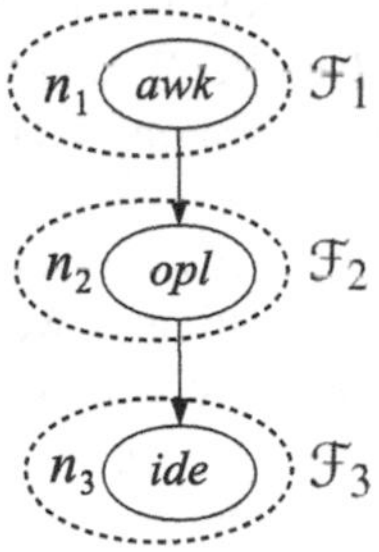

Fig. 7.8 Plain fragmentation of plain temporal observation $\mathcal{O}' = [awk, opl, ide]$ of active system $\mathcal{A}_2$

7.5 Reactive Sequence

A plain monitoring path $[\mathcal{M}_0, \mathcal{M}_1, \ldots, \mathcal{M}_k]$ relevant to a plain monitoring problem $\mu(\mathcal{A})$ is the sequence of monitoring states associated with the sequence $[\mathcal{F}_1, \ldots, \mathcal{F}_k]$ of observation fragments, where each fragment $\mathcal{F}_i$, $i \in [1 \cdots k]$, involves just one observable label ℓ_i.

Within each context (N, Δ) belonging to a monitoring state $\mathcal{M}_i$ of the path, Δ represents the set of diagnoses of the trajectories of $\mathcal{A}$ up to the root of N, with these trajectories being consistent with the trace $[\ell_1, \ldots, \ell_i]$.

Therefore, the i-th element of the solution of $\mu(\mathcal{A})$, namely $\Delta(\wp_{[i]}(\mathcal{A}))$, can be determined by combining the local candidate set of N with Δ, as formalized in Def. 7.13.

Definition 7.13 (Plain Reactive Sequence). Let $\mu(\mathcal{A}) = (a_0, \mathcal{V}, \hat{\mathcal{O}}, \mathcal{R})$ be a plain monitoring problem, with $\hat{\mathcal{O}}$ being the fragmentation of a plain observation $\mathcal{O} = [\ell_1, \ldots, \ell_k]$, $Mtr(\mathcal{A}, a_0, \mathcal{V}, \mathcal{R}) = (\mathbf{N}, \mathbf{L}, \mathbf{A}, N_0)$ being the monitor, and $Path(\mu(\mathcal{A})) = [\mathcal{M}_0, \mathcal{M}_1, \ldots, \mathcal{M}_n]$ being the relevant plain monitoring path. The *plain reactive sequence* of $\mu(\mathcal{A})$ is a sequence of sets of diagnoses

$$React(\mu(\mathcal{A})) = [\Delta_0, \Delta_1, \ldots, \Delta_n], \tag{7.33}$$

where

$$\forall i \in [0 \cdots n] \left(\Delta_i = \bigcup_{(N,\Delta) \in \mathcal{M}_i} \left(\Delta^{\mathrm{loc}}(N) \bowtie \Delta \right) \right). \tag{7.34}$$

Example 7.15. With reference to the plain monitoring problem $\mu'(\mathcal{A}_2) = (a_0, \mathcal{V}, \hat{\mathcal{O}}', \mathcal{R})$ defined in Example 7.14, the relevant plain reactive sequence is

$$React(\mu'(\mathcal{A}_2)) = [\Delta_0, \Delta_1, \Delta_2, \Delta_3],$$

where $\Delta_0 = \{\emptyset\}$, $\Delta_1 = \{\emptyset, \{\mathbf{nol}\}, \{\mathbf{nor}\}, \{\mathbf{nol}, \mathbf{nor}\}, \{\mathbf{fop}\}\}$, $\Delta_2 = \{\emptyset, \{\mathbf{nor}\}\}$, and $\Delta_3 = \{\{\mathbf{nor}\}, \{\mathbf{ncl}, \mathbf{nor}\}, \{\mathbf{nor}, \mathbf{fcp}\}\}$.

The notions of a plain monitoring path (Def. 7.12) and a plain reactive sequence (Def. 7.13) refer to a plain monitoring problem, where a fragmentation of a plain temporal observation is assumed.

Now, we extend both notions to a general monitoring problem, where no restrictions are assumed on the nature of the temporal observation $\mathcal{O}$.

To this end, we associate with each state $\mathfrak{I}$ of the index space of $\mathcal{O}$ the monitoring state involving the whole set of contexts relevant to all the traces up to $\mathfrak{I}$ in the language of $Isp(\mathcal{O})$. In other words, each state of $Isp(\mathcal{O})$ is decorated with a *monitoring attribute*, as formalized in Def. 7.14.

Definition 7.14 (Index Space Decoration). Let $\mu(\mathcal{A}) = (a_0, \mathcal{V}, \hat{\mathcal{O}}, \mathcal{R})$ be a monitoring problem. The *decoration* of the index space $Isp(\mathcal{O})$ based on $\mu(\mathcal{A})$ is a DFA $Isp^*(\mathcal{O})$ isomorphic to $Isp(\mathcal{O})$, where each state $\mathfrak{I}$ is marked by a *monitoring attribute*

$$\mathcal{M} = \bigcup_{\mathcal{O}' \in \|\mathfrak{I}\|} \mathcal{M}_k, \tag{7.35}$$

where $\|\mathfrak{I}\|$ denotes the set of plain observations up to $\mathfrak{I}$ in $Isp(\mathcal{O})$, $\mathcal{O}' = [\ell_1, \ldots, \ell_k]$, $\mu'(\mathcal{A}) = (a_0, \mathcal{V}, \hat{\mathcal{O}}', \mathcal{R})$ is a plain monitoring problem where $\hat{\mathcal{O}}'$ is the plain fragmentation of $\mathcal{O}'$, and $Path(\mu'(\mathcal{A})) = [\mathcal{M}_0, \mathcal{M}_1, \ldots, \mathcal{M}_k]$.

Example 7.16. Consider the monitoring problem $\mu(\mathcal{A}_2) = (a_0, \mathcal{V}, \hat{\mathcal{O}}_2, \mathcal{R})$ defined in Example 7.3, and the partial monitor displayed in Fig. 7.7. Displayed on the left side of Fig. 7.9 is the temporal observation $\mathcal{O}_2$. The decoration of the index space $Isp(\mathcal{O}_2)$ is displayed on the right side of the same figure, where $\mathcal{M}_0 = \{(N_0, \{\emptyset\})\}$, $\mathcal{M}_1 = \{(N_1, \{\emptyset\}), (N_2, \{\{\mathbf{fop}\}\})\}$, $\mathcal{M}_2 = \{(N_3, \{\emptyset\}), (N_4, \{\{\mathbf{nor}\}\})\}$,

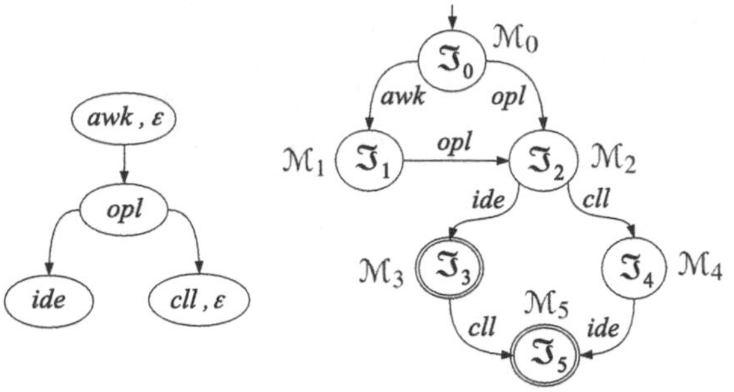

Fig. 7.9 Temporal observation $\mathcal{O}_2$ (left) and relevant index space decoration $Isp^*(\mathcal{O}_2)$ (right)

$\mathcal{M}_3 = \{(N_5, \{\{\mathbf{nor}\}\}), (N_6, \{\{\mathbf{nor}, \mathbf{fcp}\}\})\}, \mathcal{M}_4 = \emptyset$, and $\mathcal{M}_5 = \{(N_7, \{\{\mathbf{nor}\}\}), (N_0, \{\{\mathbf{nor}\}\})\}$.

Based on the decorated index space, we can now generalize the notion of a monitoring path (introduced in Def. 7.12) to any monitoring problem, where the temporal observation $\mathcal{O}$ is received as a sequence $[\mathcal{F}_1, \ldots, \mathcal{F}_k]$ of fragments.

For each fragment $\mathcal{F}_i$, $i \in [1 \cdots k]$, consider the decorated index space $Isp^*(\mathcal{O}_{[i]})$. The i-th monitoring state of the monitoring path, namely $\mathcal{M}_i$, can be determined by aggregating all the monitoring attributes associated with the final states of $Isp^*(\mathcal{O}_{[i]})$, as formalized in Def. 7.15.

Definition 7.15 (Monitoring Path). Let $\mu(\mathcal{A})$ be a monitoring problem involving a fragmentation $\hat{\mathcal{O}} = [\mathcal{F}_1, \ldots, \mathcal{F}_k]$ of an observation $\mathcal{O}$. Let $Isp^*(\mathcal{O}_{[i]})$ be the decorated index space relevant to the subobservation $\mathcal{O}_{[i]}$, $i \in [0 \cdots k]$, with a set of final states $I^*_{\mathrm{f}_i}$. The monitoring *path* of $\mu(\mathcal{A})$ is the sequence of monitoring states

$$Path(\mu(\mathcal{A})) = [\mathcal{M}_0, \mathcal{M}_1, \ldots, \mathcal{M}_k], \tag{7.36}$$

where

$$\forall i \in [0 \cdots k] \left(\mathcal{M}_i = \bigcup_{(\mathfrak{I}, \mathcal{M}) \in I^*_{\mathrm{f}_i}} \mathcal{M} \right). \tag{7.37}$$

Example 7.17. Consider the monitoring problem $\mu(\mathcal{A}_2) = (a_0, \mathcal{V}, \hat{\mathcal{O}}_2, \mathcal{R})$ defined in Example 7.3, and the partial monitor displayed in Fig. 7.7. Displayed in Fig. 7.10 are the decorated index spaces of temporal observations $\mathcal{O}_{2[0]}$, $\mathcal{O}_{2[1]}$, $\mathcal{O}_{2[2]}$, $\mathcal{O}_{2[3]}$, and $\mathcal{O}_{2[4]}$, where $\mathcal{M}_0 = \{(N_0, \{\emptyset\})\}$, $\mathcal{M}_1 = \{(N_1, \{\emptyset\}), (N_2, \{\{\mathbf{fop}\}\})\}$, $\mathcal{M}_2 = \{(N_3, \{\emptyset\}), (N_4, \{\{\mathbf{nor}\}\})\}$, $\mathcal{M}_3 = \{(N_5, \{\{\mathbf{nor}\}\}), (N_6, \{\{\mathbf{nor}, \mathbf{fcp}\}\})\}$, $\mathcal{M}_4 = \emptyset$, and $\mathcal{M}_5 = \{(N_7, \{\{\mathbf{nor}\}\}), (N_0, \{\{\mathbf{nor}\}\})\}$. Based on Def. 7.15, we have:

$$Path(\mu(\mathcal{A}_2)) = [\mathcal{M}_0, (\mathcal{M}_0 \cup \mathcal{M}_1), \mathcal{M}_2, \mathcal{M}_3, (\mathcal{M}_3 \cup \mathcal{M}_5)]\ . \tag{7.38}$$

The notion of a reactive sequence (introduced in Def. 7.13 for plain monitoring problems) can be generalized to any monitoring problem based on the generalized definition of a monitoring path introduced in Def. 7.15.

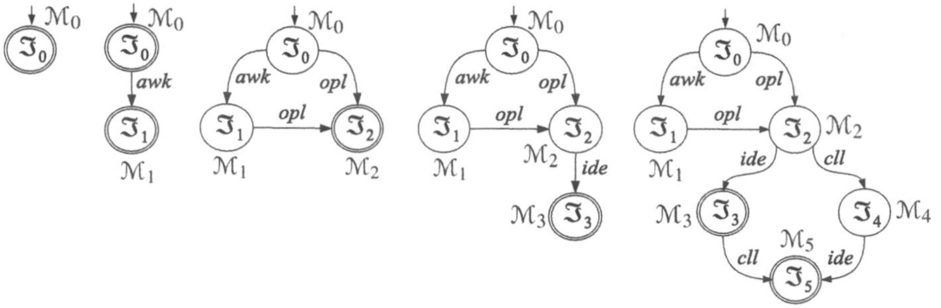

Fig. 7.10 From left to right, $Isp^*(\mathcal{O}_{2[0]})$, $Isp^*(\mathcal{O}_{2[1]})$, $Isp^*(\mathcal{O}_{2[2]})$, $Isp^*(\mathcal{O}_{2[3]})$, and $Isp^*(\mathcal{O}_{2[4]})$

The generalized reactive sequence is formalized in Def. 7.16.[1]

Definition 7.16 (Reactive Sequence). Let $\mu(\mathcal{A}) = (a_0, \mathcal{V}, \hat{\mathcal{O}}, \mathcal{R})$ be a monitoring problem, with $\hat{\mathcal{O}}$ being the fragmentation of an observation $\mathcal{O}$, $Mtr(\mathcal{A}, a_0, \mathcal{V}, \mathcal{R}) = (\mathbf{N}, \mathbf{L}, \mathbf{A}, N_0)$ being the monitor, and $Path(\mu(\mathcal{A})) = [\mathcal{M}_0, \mathcal{M}_1, \ldots, \mathcal{M}_n]$ being the relevant monitoring path. The *reactive sequence* of $\mu(\mathcal{A})$ is a sequence of sets of diagnoses

$$React(\mu(\mathcal{A})) = [\Delta_0, \Delta_1, \ldots, \Delta_n], \tag{7.39}$$

where

$$\forall i \in [0 \cdots n] \left(\Delta_i = \bigcup_{(N,\Delta) \in \mathcal{M}_i} \left(\Delta^{\mathrm{loc}}(N) \bowtie \Delta \right) \right). \tag{7.40}$$

Example 7.18. With reference to the monitoring problem $\mu(\mathcal{A}_2) = (a_0, \mathcal{V}, \hat{\mathcal{O}}_2, \mathcal{R})$ defined in Example 7.3 and the monitoring path in (7.38), based on Def. 7.16, the relevant reactive sequence is

$$React(\mu(\mathcal{A}_2)) = [\Delta_0, \Delta_1, \Delta_2, \Delta_3, \Delta_4], \tag{7.41}$$

where $\Delta_0 = \{\emptyset\}$, $\Delta_1 = \{\emptyset, \{\mathbf{nol}\}, \{\mathbf{nor}\}, \{\mathbf{nol}, \mathbf{nor}\}, \{\mathbf{fop}\}\}$, $\Delta_2 = \{\emptyset, \{\mathbf{nor}\}\}$, $\Delta_3 = \{\{\mathbf{nor}\}, \{\mathbf{ncl}, \mathbf{nor}\}, \{\mathbf{nor}, \mathbf{fcp}\}\}$, and $\Delta_4 = \{\{\mathbf{nor}\}, \{\mathbf{ncl}, \mathbf{nor}\}, \{\mathbf{nor}, \mathbf{fcp}\}\}$.

Notice how the reactive sequence displayed in (7.41) equals the solution of the corresponding monitoring problem $\mu(\mathcal{A}_2)$ determined in Example 7.4 and displayed in (7.13). This is no coincidence, as claimed in Theorem 7.1.

Theorem 7.1. *Let $\mu(\mathcal{A})$ be a monitoring problem and $\Delta(\mu(\mathcal{A}))$ its solution. Then,*

$$React(\mu(\mathcal{A})) = \Delta(\mu(\mathcal{A})). \tag{7.42}$$

Proof. The proof is supported by Lemma 7.1, which is a restriction of Theorem 7.1 to a plain diagnostic problem. Lemma 7.1 is grounded on Lemmas 7.2 and 7.3.

[1] Since the definition of a reactive sequence depends on the monitoring path only, (7.39) in Def. 7.16 equals (7.33) in Def. 7.13.

Lemma 7.1. *Equation (7.42) holds when $\mu(\mathcal{A}) = (a_0, \mathcal{V}, \hat{\mathcal{O}}, \mathcal{R})$ is a plain monitoring problem.*

Proof. Since the monitoring problem is plain, the observation $\mathcal{O}$ is a sequence of labels in $\mathcal{V}$, namely $\mathcal{O} = [\ell_1, \ldots, \ell_k]$. Let *React*$(\mu(\mathcal{A})) = [\Delta_0, \Delta_1, \ldots, \Delta_k]$ and $\Delta(\mu(\mathcal{A})) = [\Delta(\wp_{[0]}(\mathcal{A})), \Delta(\wp_{[1]}(\mathcal{A})), \ldots, \Delta(\wp_{[k]}(\mathcal{A}))]$. We have to show the equality of the two sequences. First we prove

$$\Delta(\mu_{[0]}(\mathcal{A})) = \Delta_0. \tag{7.43}$$

Since $\wp_{[0]}(\mathcal{A}) = (a_0, \mathcal{V}, [\,], \mathcal{R})$ and the language of *Isp*$([\,])$ equals $\{[\,]\}$, based on (7.7) and (7.5),

$$\Delta(\wp_{[0]}(\mathcal{A})) = \{\delta \mid \delta = h_{[\mathcal{R}]}, h \in Bsp(\mathcal{A})|, h_{[\mathcal{V}]} = [\,]\}\,. \tag{7.44}$$

Thus, the trajectories h are unobservable. On the other hand, based on Def. 7.13, the first element of the plain reactive sequence is

$$\Delta_0 = \bigcup_{(N,\Delta) \in \{(N_0, \{\emptyset\})\}} \left(\Delta^{\text{loc}}(N) \bowtie \Delta \right) = \Delta^{\text{loc}}(N_0) \bowtie \{\emptyset\} = \Delta^{\text{loc}}(N_0)\,. \tag{7.45}$$

Let $N_0 = (\Sigma, S, \tau, s_0, S_l)$, where $s_0 = (\bar{a}, \bar{\mathcal{D}})$. From Def. 7.9, we have

$$\Delta^{\text{loc}}(N_0) = \bigcup_{(a,\mathcal{D}) \in S} \mathcal{D}, \tag{7.46}$$

where, based on Def. 7.6, the diagnosis attribute $\mathcal{D}$ is the set of diagnoses δ (based on $\mathcal{R}$) of a subtrajectory in *Bsp*$(\mathcal{A})$, from $\bar{a}$ to a, where all transitions are unobservable. Since such subtrajectories h are rooted in a_0, they are in fact trajectories and the condition may be reformulated as $h \in Bsp(\mathcal{A})$, where h is unobservable. Thus,

$$\Delta_0 = \Delta^{\text{loc}}(N_0) = \{\delta \mid \delta = h_{[\mathcal{R}]}, h \in Bsp(\mathcal{A}), h \text{ is unobservable}\}\,. \tag{7.47}$$

A comparison between (7.44) and (7.47) makes (7.43) proven.

Let *Path*$^*(\mu(\mathcal{A}))$ denote the *connected path* of $\mu(\mathcal{A})$, the graph obtained by connecting the elements of the monitoring states in *Path*$(\mu(\mathcal{A}))$ by means of the relevant arcs of the monitor. Specifically, considering (7.31), an arc from $s \in S_l(N)$ to $s_0(N')$ will be inserted and will be marked by the corresponding observable transition $t(c)$ defined in point 2 of Def. 7.8. This way, *Path*$^*(\mu(\mathcal{A}))$ can be viewed as a directed (and connected) graph rooted in the initial state of node N_0 relevant to $\mathcal{M}_0$.

Lemma 7.2. *Let s_0 be the root of node N_0 in the monitoring state $\mathcal{M}_0$. Let s_i be a state of a node $N_i \in \mathcal{M}_i$. Then, each path $s_0 \rightsquigarrow s_i$ in Path$^*(\mu(\mathcal{A}))$ is a trajectory h such that $h \in Bsp(\mathcal{A})$ and $h_{[\mathcal{V}]} = [\ell_1, \ldots, \ell_i]$, and vice versa.*

Lemma 7.3. *Let (N, Δ) be a context in a monitoring state $\mathcal{M}$ in Path$(\mu(\mathcal{A}))$. Then, Δ is the set of candidate diagnoses relevant to the trajectories in Path$^*(\mu(\mathcal{A}))$ that end at the root s_0 of N.*

Proof. The proof is by induction on the nodes (in the monitoring states) that are involved in the trajectories ending at the root of N. The basis considers (the only) node N_0 in $\mathcal{M}_0$, where $\Delta = \{\emptyset\}$. Owing to the emptiness of the trajectories ending at the root of N_0, the relevant set of candidate diagnoses is in fact empty. In the induction step we show that, if Lemma 7.3 holds for a context $(N, \Delta) \in \mathcal{M}_i$, $i \in [0 \cdots (k-1)]$, then it holds as well for the successive node $(N', \Delta') \in \mathcal{M}_{i+1}$. In fact, based on conditions (7.31) and (7.32), we have

$$\Delta' = \left\{ (\mathcal{D} \bowtie \Delta) \oplus f \mid (N, \Delta) \in \mathcal{M}_i, s = (a, \mathcal{D}), s \in S_l(N), N \xrightarrow{(s, \ell_{i+1}, f, s_0(N'))} N' \in \mathbf{L} \right\}, \tag{7.48}$$

where, based on Def. 7.10 and point 1 in Def. 7.6, and, as Δ is the set of candidate diagnoses relevant to the trajectories in *Path*$^*(\mu(\mathcal{A}))$ that end at the root of N (the induction hypothesis), $\mathcal{D} \bowtie \Delta$ is the set of diagnoses relevant to the trajectories ending in $s \in S_l(N)$. Hence, based on Def. 7.7, $(\mathcal{D} \bowtie \Delta) \oplus f$ is in fact the set of diagnoses relevant to the trajectories ending at the root of N'. This proves the induction step and concludes the proof of Lemma 7.3.

Based on Lemma 7.3, $\forall i \in [1 \cdots k]$, we can now prove

$$\Delta(\mu_{[i]}(\mathcal{A})) = \Delta_i \,. \tag{7.49}$$

Based on (7.34), we have

$$\Delta_i = \bigcup_{(N,\Delta) \in \mathcal{M}_i} \left(\Delta^{\text{loc}}(N) \bowtie \Delta \right), \tag{7.50}$$

where, according to (7.22),

$$\Delta^{\text{loc}}(N) = \bigcup_{(\sigma,\mathcal{D}) \in S} \mathcal{D}, \tag{7.51}$$

with S being the set of internal states of N. Based on the equivalences (7.24) and (7.25) in Proposition 7.1, we have

$$\Delta^{\text{loc}}(N) \bowtie \Delta = \Delta \bowtie \Delta^{\text{loc}}(N) = \Delta \bowtie \bigcup_{(a,\mathcal{D}) \in S} \mathcal{D} = (\Delta \bowtie \mathcal{D}_1) \cup \cdots \cup (\Delta \bowtie \mathcal{D}_k), \tag{7.52}$$

where $\mathcal{D}_1, \ldots, \mathcal{D}_k$ are the diagnosis attributes of states in S. In other words, owing to Lemma 7.3, $\Delta^{\text{loc}}(N) \bowtie \Delta$ is the set of candidate diagnoses relevant to the trajectories ending at any internal state of N. Moving this result to (7.50), we conclude that Δ_i is the set of candidate diagnoses relevant to the trajectories ending at any state of any node N where $(N, \Delta) \in \mathcal{M}_i$. Hence, (7.49) holds. This concludes the proof of Lemma 7.1.

Based on Lemma 7.1, we now prove (7.42) in the general case. First, notice how (7.43) still holds in the general case. Besides, according to (7.37), for each $i \in [1 \cdots k]$

we have

$$\mathcal{M}_i = \bigcup_{(\mathfrak{S},\mathcal{M}) \in I^*_{\mathrm{f}_i}} \mathcal{M}, \tag{7.53}$$

where $\mathcal{M}$ is the monitoring attribute associated with a final state $\mathfrak{S}$ of the decorated index space $Isp^*(\mathcal{O}_{[i]})$. According to (7.35), $\mathcal{M}$ is the union of the last monitoring states $\mathcal{M}^j_i$ of each path $Path(\mu^j_i(\mathcal{A}))$, where $\mu^j_i(\mathcal{A})$ is the plain monitoring problem involving the j-th plain observation $\mathcal{O}^j_i$ in the language of $Isp(\mathcal{O}_{[i]})$. Assuming that m is the cardinality of the language of $Isp(\mathcal{O}_{[i]})$, (7.53) can be rewritten as

$$\mathcal{M}_i = \bigcup_{j=1}^{m} \mathcal{M}^j_i \,. \tag{7.54}$$

Based on (7.34), the i-th element of the reactive sequence is

$$\Delta_i = \bigcup_{(N,\Delta) \in \mathcal{M}_i} \left(\Delta^{\mathrm{loc}}(N) \bowtie \Delta \right), \tag{7.55}$$

which, based on (7.54) and Lemma 7.1, can be rewritten as

$$\Delta_i = \bigcup_{j=1}^{m} \left(\bigcup_{(N,\Delta) \in \mathcal{M}^j_i} \left(\Delta^{\mathrm{loc}}(N) \bowtie \Delta \right) \right) = \bigcup_{j=1}^{m} \Delta(\mu^j_i(\mathcal{A})) \,. \tag{7.56}$$

On the other hand, as by definition the solution $\Delta(\wp_{[i]}(\mathcal{A}))$ is

$$\left\{ \delta \mid \delta = h_{[\mathcal{R}]}, h \in Bsp(\mathcal{A}), h_{[\mathcal{V}]} \text{ is in the language of } Isp(\mathcal{O}_{[i]} \right\}, \tag{7.57}$$

$\Delta(\wp_{[i]}(\mathcal{A}))$ can be rephrased as the union of the solutions relevant to each plain problem,

$$\bigcup_{j=1}^{m} \left\{ \delta \mid \delta = h_{[\mathcal{R}]}, h \in Bsp(\mathcal{A}), h_{[\mathcal{V}]} = \mathcal{O}^j_i \right\} . \tag{7.58}$$

In other words,

$$\Delta(\wp_{[i]}(\mathcal{A})) = \bigcup_{j=1}^{m} \Delta(\wp^j_i(\mathcal{A})) \,. \tag{7.59}$$

Comparison of (7.56) and (7.59) leads to the conclusion $\Delta(\wp_{[i]}(\mathcal{A})) = \Delta_i$, which terminates the proof of Theorem 7.1. □

7.5.1 Reaction Algorithm

Listed below is the specification of the algorithm *React*, which processes the reception of an observation fragment $\mathcal{F}$ for a system $\mathcal{A}$, based on the current temporal observation $\mathcal{O}$ along with the corresponding nondeterministic index space $\mathcal{N}$ and the decorated index space $\mathcal{I}^*$, and the partially generated monitor M of $\mathcal{A}$.

To this end, it exploits the auxiliary procedures *ISCA** (a variant of the *ISCA* incremental determinization algorithm defined in Chapter 2) and *Expmon* (defined in lines 12–34).

The body of *React* is listed in lines 35–62. After the expansion of the temporal observation $\mathcal{O}$ by the fragment $\mathcal{F}$ and the expansion of the corresponding nondeterministic index space $\mathcal{N}$, the decorated index space $\mathcal{I}^*$ is updated by *ISCA**. The latter not only performs the incremental determinization of $\mathcal{N}$ based on $\mathcal{N}$, $\mathcal{I}^*$, and $\Delta\mathcal{N}$, but also marks the states of $\mathcal{I}^*$ whose monitoring attribute requires to be updated.

The need to update the monitoring attribute of a marked state $\mathfrak{I}$ has two possible reasons:

1. State $\mathfrak{I}$ is new (its monitoring attribute is set to $\emptyset$ by *ISCA**),
2. State $\mathfrak{I}$ is not new, but the set of traces up to the existing state $\mathfrak{I}$ is changed because of either the insertion of a new transition entering $\mathfrak{I}$ or the removal of an existing transition entering $\mathfrak{I}$ (or a combination of them).

After the update of the decorated index space $\mathcal{I}^*$ by *ISCA**, marked states in $\mathcal{I}^*$ are processed one by one. At each iteration of the loop (lines 39–61), a marked state $\mathfrak{I}'$, such that all its parent states are unmarked, is chosen and unmarked (line 40). The decoration $\mathcal{M}'$ of $\mathfrak{I}'$ is set to empty after copying it into a temporary variable $\mathcal{M}'_{\text{old}}$ (line 42).

Then, for each parent state $\mathfrak{I}$ of $\mathfrak{I}'$ (lines 43–54), all contexts of the monitoring attribute $\mathcal{M}$ of $\mathfrak{I}$ are considered (lines 45–53). For each context (N, Δ), the monitor is expanded by the auxiliary procedure *Expmon* (line 46), based on node N and the observable label ℓ (which marks the transition exiting $\mathfrak{I}$ and entering $\mathfrak{I}'$). This way, the monitor is expanded from node N by new arcs involving the label ℓ, which are directed toward other (possibly newly created) nodes.

After the expansion of the monitor from node N with arcs involving the label ℓ, each arc $N \xrightarrow{(s,\ell,f,s')} N'$ of the monitor is considered (lines 47–53), with $\mathcal{D}$ being the diagnosis attribute of state s in N. If node N' is already in the current monitoring state $\mathcal{M}'$ of $\mathfrak{I}'$ then the diagnosis attribute Δ' of N' is extended by $(\mathcal{D} \bowtie \Delta) \oplus f$ (lines 48 and 49). Otherwise, the new context $(N', (\mathcal{D} \bowtie \Delta) \oplus f)$ is inserted into $\mathcal{M}'$ (line 51).

Once all transitions entering $\mathfrak{I}'$ have been processed (lines 43–55), the current decoration $\mathcal{M}'$ of $\mathfrak{I}'$ is compared with the old one (line 56): if $\mathcal{M}' \neq \mathcal{M}'_{\text{old}}$, then each successor state $\mathfrak{I}''$ of $\mathfrak{I}'$ is marked in turn (lines 57–59), because it needs to be virtually updated as a consequence of the change in the monitoring attribute of the parent state $\mathfrak{I}'$.

The auxiliary procedure *Expmon* (lines 12–34) takes as input a node N of the partially generated monitor and an observable label ℓ. It expands the monitor by possibly creating new arcs exiting N, marked by the label ℓ, and directed toward possibly newly created nodes.

In so doing, it consider each transition $a \xrightarrow{t(c)} a'$ of system $\mathcal{A}$ such that $(a, \mathcal{D})$ is a state in the leaving set of N and the component transition $t(c)$ is observable via the label ℓ (lines 18–33). In line 19, the label f is set either to the faulty label associated with $t(c)$ in the ruler $\mathcal{R}$ (if $t(c)$ is faulty) or to ε (if $t(c)$ is normal).

If the current subpart of the monitor does not include a node N' with root $(a', \mathcal{D}')$, where a' is the target state of the system transition $a \xrightarrow{t(c)} a'$ considered, then such a node is generated as a diagnosis closure by means of the auxiliary procedure *Dclosure* specified in Sec. 7.3.1 (lines 20–22). Eventually, an arc $N \xrightarrow{(s,f,\ell,s')} N'$ and the relevant label are possibly created (lines 27–32).

The specification of the algorithm *React* is as follows:

1. **algorithm** *React* (**inout** $\mathcal{O}$, **inout** $\mathcal{N}$, **inout** $\mathcal{I}^*$, **in** $\mathcal{F}$, **inout** M)
2. $\mathcal{O}$: a temporal observation,
3. $\mathcal{N}$: the nondeterministic index space of $\mathcal{O}$,
4. $\mathcal{I}^* = (\Sigma, I, \tau, \mathfrak{I}_0, I_f)$: the decorated index space of $\mathcal{O}$,
5. $\mathcal{F}$: the next observation fragment for $\mathcal{O}$,
6. M: the partially generated monitor $Mtr(\mathcal{A}, a_0, \mathcal{V}, \mathcal{R})$;

7. **side effects**
8. Expansion of $\mathcal{O}$,
9. Expansion of $\mathcal{N}$,
10. Update of $\mathcal{I}^*$,
11. Expansion of M;

12. **auxiliary procedure** *Expmon* (**in** N, **in** ℓ)
13. N: a node of monitor $(\mathbf{N}, \mathbf{L}, \mathbf{A}, N_0)$,
14. ℓ: an observable label within viewer $\mathcal{V}$;
15. **side effects**
16. Expansion of the monitor from N;
17. **begin** $\langle$*Expmon*$\rangle$
18. **foreach** transition $a \xrightarrow{t(c)} a'$ of $\mathcal{A}$, where $s \in S_l(N), s = (a, \mathcal{D})$, $t(c)$ is triggerable from a, $(t(c), \ell) \in \mathcal{V}$, **do**
19. $f :=$ **if** $(t(c), \bar{f}) \in \mathcal{R}$ **then** $\bar{f}$ **else** ε **endif**;
20. **if N** does not include a node N' with root $(a', \mathcal{D}')$ **then**
21. *Dclosure* $(a', \mathcal{V}, \mathcal{R}, N')$;
22. Insert N' into **N**
23. **else**
24. $N' :=$ the node in **N** with root $(a', \mathcal{D}')$
25. **endif**;
26. Let s' be the root of N';
27. **if** $(s, \ell, f, s') \notin \mathbf{L}$ **then**
28. Insert (s, ℓ, f, s') into the set **L** of labels of the monitor

29.   **endif**;
30.   **if** $N \xrightarrow{(s,f,\ell,s')} N' \notin \mathbf{A}$ **then**
31.   Insert $N \xrightarrow{(s,f,\ell,s')} N'$ into the set $\mathbf{A}$ of arcs of the monitor
32.   **endif**
33.   **endfor**
34.   **end** $\langle Expmon \rangle$;

35. **begin** $\langle React \rangle$
36.   Expand $\mathcal{O}$ by $\mathcal{F}$;
37.   Let $\Delta \mathcal{N}$ be the corresponding expansion of $\mathcal{N}$;
38. $ISCA^*(\mathcal{N}, \mathcal{I}^*, \Delta \mathcal{N})$;
39.   **while** $\mathcal{I}^*$ includes a marked state **do**
40.   Unmark a state $\mathfrak{I}'$ such that $\forall \mathfrak{I} \xrightarrow{\ell} \mathfrak{I}' \in \tau$, $\mathfrak{I}$ is not marked;
41.   Let $\mathcal{M}'$ be the current decoration of $\mathfrak{I}'$;
42. $\mathcal{M}'_{old} := \mathcal{M}'; \mathcal{M}' := \emptyset$;
43.   **foreach** transition $\mathfrak{I} \xrightarrow{\ell} \mathfrak{I}' \in \tau$ **do**
44.   Let $\mathcal{M}$ be the decoration of $\mathfrak{I}$;
45.   **foreach** context $(N, \Delta) \in \mathcal{M}$ **do**
46. $Expmon(N, \ell)$;
47.   **foreach** $N \xrightarrow{(s,\ell,f,s')} N' \in \mathbf{A}$, where $s = (a, \mathcal{D})$, **do**
48.   **if** $(N', \Delta') \in \mathcal{M}'$ **then**
49.   Extend Δ' by $(\mathcal{D} \bowtie \Delta) \oplus f$
50.   **else**
51.   Insert context $(N', (\mathcal{D} \bowtie \Delta) \oplus f)$ into monitoring state $\mathcal{M}'$
52.   **endif**
53.   **endfor**
54.   **endfor**
55.   **endfor**;
56.   **if** $\mathcal{M}' \neq \mathcal{M}'_{old}$ **then**
57.   **foreach** transition $\mathfrak{I}' \xrightarrow{\ell'} \mathfrak{I}'' \in \tau$ **do**
58.   Mark $\mathfrak{I}''$
59.   **endfor**
60.   **endif**
61.   **endwhile**
62. **end** $\langle React \rangle$.

Example 7.19. Consider the monitoring problem $\mu(\mathcal{A}_2) = (a_0, \mathcal{V}, \hat{\mathcal{O}}_2, \mathcal{R})$ defined in Example 7.3, with the fragmentation $\hat{\mathcal{O}}_2$ displayed on the left side of Fig. 7.2. The effect of the *React* algorithm applied to each fragment of $\hat{\mathcal{O}}_2$ is displayed in Fig. 7.11.

In the center of the first line of the figure is the initial configuration of the relevant monitor (shown partially generated in Fig. 7.7), which is composed of the initial node N_0 only (details of the internal states of the diagnosis closure have been omitted).

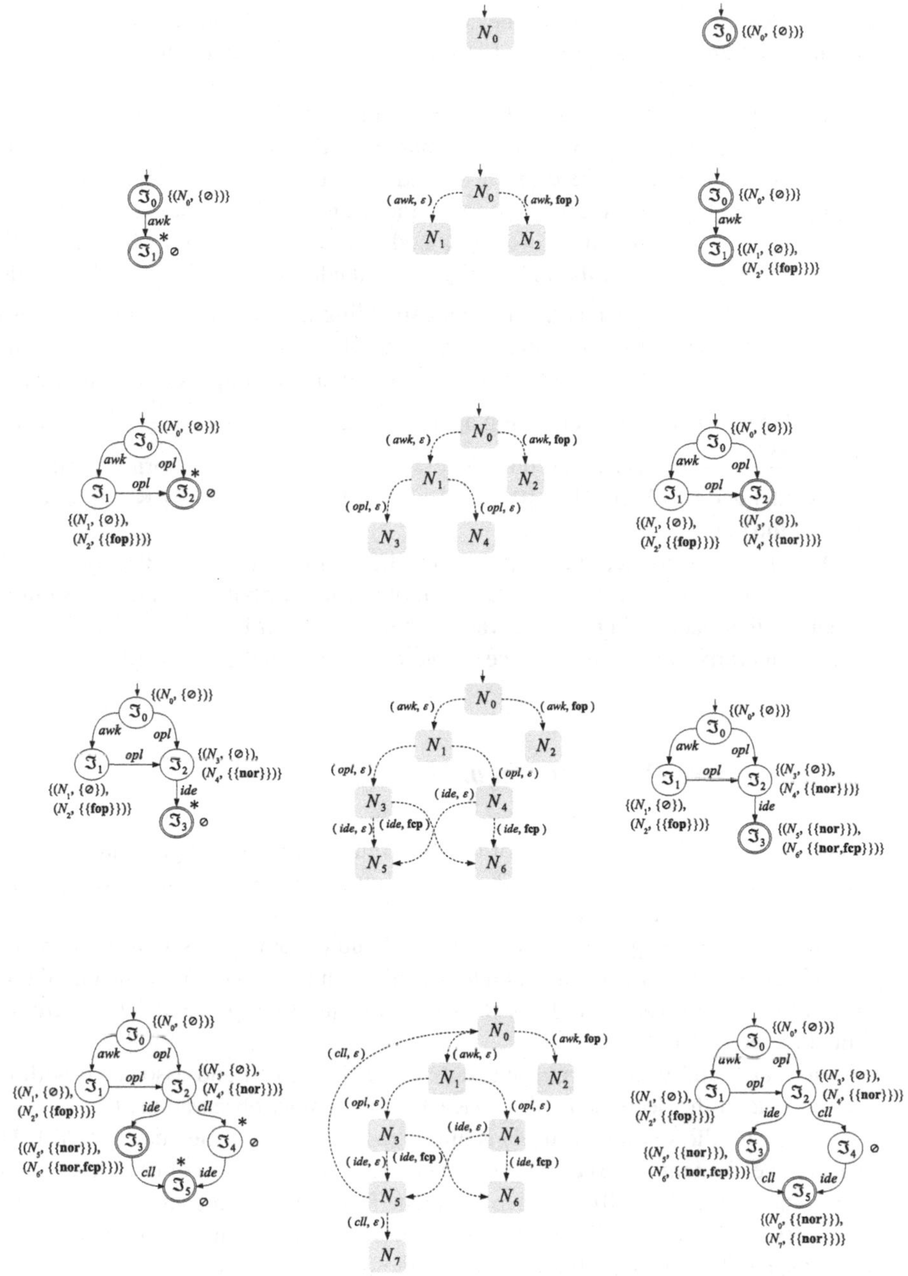

Fig. 7.11 Tracing of *React* algorithm for the fragmented observation $\hat{\mathcal{O}}_2$ displayed in Fig. 7.2

The decorated index space of the empty observation is displayed on the right side of the first line, which is composed of state $\mathfrak{I}_0$ only, associated with the monitoring attribute $\{(N_0, \{\emptyset\})\}$.

Each of the next four lines of the figure displays, from left to right, the expanded index space generated by *ISCA** (where marked nodes are asterisked), the expanded monitor (based on the expansion of the index space), and the complete decorated index space (based on the expansion of the monitor). For instance, after the reception of the first fragment involving node n_1 of the temporal observation $\mathcal{O}_2$, where $\|n_1\| = \{awk, \varepsilon\}$, the index space is extended by a node $\mathfrak{I}_1$ and transition $\mathfrak{I}_0 \xrightarrow{awk} \mathfrak{I}_1$. Since $\mathfrak{I}_1$ is marked, the corresponding monitoring attribute requires to be updated based on the monitoring attribute of $\mathfrak{I}_0$, namely $\{(N_0, \{\emptyset\})\}$. To this end, the monitor is expanded by determining the arcs exiting from N_0 and marked by the label *awk*, thereby generating nodes N_1 and N_2 and arcs $N_0 \xrightarrow{(0,awk,\varepsilon,1)} N_1$ and $N_0 \xrightarrow{(0,awk,\mathbf{fop},53)} N_2$. Based on these two arcs, the (empty) monitoring attribute of $\mathfrak{I}_1$ is extended by $(N_1, (\{\emptyset\} \bowtie \{\emptyset\}) \oplus \varepsilon) = (N_1, \{\emptyset\})$ and $(N_2, (\{\emptyset\} \bowtie \{\emptyset\}) \oplus \mathbf{fop}) = (N_2, \{\{\mathbf{fop}\}\})$.

Finally, the decorated index space after the fourth fragment is displayed on the bottom right of Fig. 7.11. Note that the monitoring attribute of state $\mathfrak{I}_4$ is empty, meaning that each string of observable labels generated by a path from $\mathfrak{I}_0$ to $\mathfrak{I}_4$ (that is, a prefix of a candidate trace) is inconsistent with the monitor.[2]

7.5.2 Reactive Diagnosis Algorithm

Listed below is the specification of the algorithm *Rdiag*, which generates the solution of a monitoring problem $\mu(\mathcal{A}) = (a_0, \mathcal{V}, \hat{\mathcal{O}}, \mathcal{R})$ by reacting to the reception of each observation fragment.

In line 5, a binding between the variable $\mathcal{O}$ and the empty observation is created, with $\mathcal{N}$ being the corresponding nondeterministic index space. Then, the initial node N_0 of the monitor $(\mathcal{A}, a_0, \mathcal{V}, \mathcal{R})$ is created by the auxiliary procedure *Dclosure* (defined in Sec. 7.3.1).

In lines 8 and 9, the monitoring state $\mathcal{M}_0$ is generated and associated with the initial state of the decorated index space $\mathcal{I}^*$ of the empty observation $\mathcal{O}$.

In line 10, the solution of $\mu(\mathcal{A})$ is initialized with the first element $\Delta^{\mathrm{loc}}(N_0)$. The successive elements of $\Delta(\mu(\mathcal{A}))$ are generated at the reception of each fragment of observation (lines 11–20). For each fragment $\mathcal{F}_i$, $i \in [1 \cdots k]$, the auxiliary procedure *React* (specified in Sec. 7.5.1) is called (line 12); consequently, both the decorated index space $\mathcal{I}^*$ and the partially generated monitor M are extended.

Then, the next element Δ_i of the solution of $\mu(\mathcal{A})$ is computed as the union of the diagnosis join $\Delta^{\mathrm{loc}}(N) \bowtie \Delta$, where (N, Δ) belongs to the monitoring attribute of a final state of the decorated index space $\mathcal{I}^*$ (lines 13–20).

[2] If a (prefix of a) candidate trace is inconsistent with the monitor, then it is inconsistent with the behavior space too.

The specification of the algorithm *Rdiag* is as follows:

1. **algorithm** *Rdiag* (**in** $\mu(\mathcal{A})$, **out** $\Delta(\mu(\mathcal{A}))$)
2. $\mu(\mathcal{A}) = (a_0, \mathcal{V}, \hat{\mathcal{O}}, \mathcal{R})$: a monitoring problem, where $\hat{\mathcal{O}} = [\mathcal{F}_1, \ldots, \mathcal{F}_k]$,
3. $\Delta(\mu(\mathcal{A}))$: the solution of $\mu(\mathcal{A})$;

4. **begin** $\langle$*Rdiag*$\rangle$
5. Let $\mathcal{O}$ be the empty observation and $\mathcal{N}$ the nondeterministic index space of $\mathcal{O}$;
6. *Dclosure* $(a_0, \mathcal{V}, \mathcal{R}, N_0)$;
7. Let M be the partial monitor relevant to $\mathcal{A}$, a_0, $\mathcal{V}$, and $\mathcal{R}$, which is composed of the initial state N_0 only;
8. Let $\mathcal{M}_0$ be the monitoring state $\{(N_0, \{\emptyset\})\}$;
9. Let $\mathcal{I}^*$ be the decorated index space of the empty observation $\mathcal{O}$, with the initial state being decorated by $\mathcal{M}_0$;
10. $\Delta(\mu(\mathcal{A})) := \left[\Delta^{\text{loc}}(N_0)\right]$;
11. **foreach** fragment $\mathcal{F}_i \in \hat{\mathcal{O}}$, $i \in [1 \cdots k]$, **do**
12. *React* $(\mathcal{O}, \mathcal{N}, \mathcal{I}^*, \mathcal{F}_i, M)$;
13. $\Delta_i := \emptyset$;
14. **foreach** $(\mathfrak{I}_\text{f}, \mathcal{M})$ in $\mathcal{I}^*$ such that $\mathfrak{I}_\text{f}$ is final **do**
15. **foreach** $(N, \Delta) \in \mathcal{M}$ **do**
16. $\Delta_i := \Delta_i \cup \left(\Delta^{\text{loc}}(N) \bowtie \Delta\right)$
17. **endfor**
18. **endfor**;
19. Append Δ_i to $\Delta(\mu(\mathcal{A}))$
20. **endfor**
21. **end** $\langle$*Rdiag*$\rangle$.

Example 7.20. Consider again the monitoring problem $\mu(\mathcal{A}_2) = (a_0, \mathcal{V}, \hat{\mathcal{O}}_2, \mathcal{R})$ defined in Example 7.3, with the fragmentation $\hat{\mathcal{O}}_2$ displayed on the left side of Fig. 7.2, where the effect of the *React* algorithm applied to each fragment of $\hat{\mathcal{O}}_2$ is displayed in Fig. 7.11. According to the algorithm *Rdiag*, at line 10, $\Delta(\mu(\mathcal{A}_2))$ is initialized with $\left[\Delta^{\text{loc}}(N_0)\right]$, where

$$\Delta^{\text{loc}}(N_0) = \{\emptyset\}\,.$$

Then, within the loop in lines 14–20, each fragment $\mathcal{F}_i$, $i \in [1 \cdots 4]$, is considered and *React* is called. After the call to *React*, the next element of the monitoring-problem solution , namely Δ_i, $i \in [1 \cdots 4]$, is computed by considering the monitoring attribute $\mathcal{M}$ of each final state $\mathfrak{I}_\text{f}$ in the decorated index space. Specifically, for each monitoring state $(N, \Delta) \in \mathcal{M}$, Δ_i is extended by $\Delta^{\text{loc}}(N) \bowtie \Delta$ (line 16).

For fragment $\mathcal{F}_1$, we have

$$\begin{aligned}\Delta_1 &= \left(\Delta^{\text{loc}}(N_1) \bowtie \emptyset\right) \cup \left(\Delta^{\text{loc}}(N_2) \bowtie \{\{\textbf{fop}\}\}\right)\\ &= (\{\emptyset, \{\{\textbf{nol}\}\}, \{\{\textbf{nor}\}\}, \{\{\textbf{nol}, \textbf{nor}\}\} \bowtie \emptyset) \cup (\emptyset \bowtie \{\{\textbf{fop}\}\})\\ &= \{\emptyset, \{\textbf{nol}\}, \{\textbf{nor}\}, \{\textbf{nol}, \textbf{nor}\}, \{\textbf{fop}\}\}\,.\end{aligned}$$

For fragment $\mathcal{F}_2$, we have

$$\begin{aligned}\Delta_2 &= \left(\Delta^{\text{loc}}(N_3) \bowtie \emptyset\right) \cup \left(\Delta^{\text{loc}}(N_4) \bowtie \{\{\textbf{nor}\}\}\right)\\ &= (\{\emptyset, \{\{\textbf{nor}\}\}\} \bowtie \emptyset) \cup (\emptyset \bowtie \{\{\textbf{nor}\}\})\\ &= \{\emptyset, \{\textbf{nor}\}\}\,.\end{aligned}$$

For fragment $\mathcal{F}_3$, we have

$$\begin{aligned}\Delta_3 &= \left(\Delta^{\text{loc}}(N_5) \bowtie \{\{\textbf{nor}\}\}\right) \cup \left(\Delta^{\text{loc}}(N_6) \bowtie \{\{\textbf{nor}, \textbf{fcp}\}\}\right)\\ &= (\{\emptyset, \{\textbf{ncl}\}\} \bowtie \{\{\textbf{nor}\}\}) \cup (\{\emptyset, \{\textbf{nor}\}\} \bowtie \{\{\textbf{nor}, \textbf{fcp}\}\})\\ &= \{\{\textbf{nor}\}, \{\textbf{ncl}, \textbf{nor}\}, \{\textbf{nor}, \textbf{fcp}\}\}\,.\end{aligned}$$

Finally, for fragment $\mathcal{F}_4$, we have

$$\begin{aligned}\Delta_4 = {} & \left(\Delta^{\text{loc}}(N_5) \bowtie \{\{\textbf{nor}\}\}\right) \cup \left(\Delta^{\text{loc}}(N_6) \bowtie \{\{\textbf{nor}, \textbf{fcp}\}\}\right) \cup\\ & \left(\Delta^{\text{loc}}(N_0) \bowtie \{\{\textbf{nor}\}\}\right) \cup \left(\Delta^{\text{loc}}(N_7) \bowtie \{\{\textbf{nor}\}\}\right)\\ = {} & (\{\emptyset, \{\textbf{ncl}\}\} \bowtie \{\{\textbf{nor}\}\}) \cup (\{\emptyset, \{\textbf{nor}\}\} \bowtie \{\{\textbf{nor}, \textbf{fcp}\}\}) \cup\\ & (\{\emptyset\} \bowtie \{\{\textbf{nor}\}\}) \cup (\{\emptyset\} \bowtie \{\{\textbf{nor}\}\})\\ = {} & \{\{\textbf{nor}\}, \{\textbf{ncl}, \textbf{nor}\}, \{\textbf{nor}, \textbf{fcp}\}\}\,.\end{aligned}$$

Hence, the solution generated by *Rdiag* will be

$$\begin{aligned}\Delta(\mu(\mathcal{A}_2)) = [\ & \Delta_0 = \{\emptyset\},\\ & \Delta_1 = \{\emptyset, \{\textbf{nol}\}, \{\textbf{nor}\}, \{\textbf{nol}, \textbf{nor}\}, \{\textbf{fop}\}\},\\ & \Delta_2 = \{\emptyset, \{\textbf{nor}\}\},\\ & \Delta_3 = \{\{\textbf{nor}\}, \{\textbf{ncl}, \textbf{nor}\}, \{\textbf{nor}, \textbf{fcp}\}\},\\ & \Delta_4 = \{\{\textbf{nor}\}, \{\textbf{ncl}, \textbf{nor}\}, \{\textbf{nor}, \textbf{fcp}\}\}\ ]\,.\end{aligned} \tag{7.60}$$

As expected, this coincides with the solution of $\mu(\mathcal{A}_2)$ determined in (7.13) based on Def. 7.5.

7.6 Bibliographic Notes

Reactive diagnosis of active systems with plain observations was introduced in [90], and extended in [93] to polymorphic active systems, that is, DESs incorporating both synchronous and asynchronous behavior. Reactive diagnosis of active systems with (uncertain) temporal observations was considered in [95, 96, 97], while incremental indexing of temporal observations was presented in [98, 100, 101, 105, 113].

7.7 Summary of Chapter 7

Observation Fragmentation. Given a temporal observation $\mathcal{O}$, a sequence $\hat{\mathcal{O}} = [\mathcal{F}_1, \dots, \mathcal{F}_k]$ of observation fragments.

Observation Fragment. Given a fragmentation $\hat{\mathcal{O}} = [\mathcal{F}_1, \dots, \mathcal{F}_k]$ of a temporal observation $\mathcal{O}$, where each $\mathcal{F}_i$, $i \in [1 \cdots k]$, is a fragment of $\mathcal{O}$, a subset of nodes of $\mathcal{O}$ along with entering arcs $n \mapsto n'$, such that n' is in $\mathcal{F}_i$ and n is in a fragment preceding $\mathcal{F}_i$ in $\hat{\mathcal{O}}$.

Plain Observation Fragmentation. An observation fragmentation where each fragment includes just one node of the relevant temporal observation.

Diagnosis Subproblem. Given a diagnosis problem $\wp(\mathcal{A}) = (a_0, \mathcal{V}, \mathcal{O}, \mathcal{R})$ and a fragmentation $\hat{\mathcal{O}} = [\mathcal{F}_1, \dots, \mathcal{F}_k]$ of $\mathcal{O}$, a diagnosis problem $(a_0, \mathcal{V}, \mathcal{O}_{[i]}, \mathcal{R})$, where $\mathcal{O}_{[i]}$ is the temporal observation made up of the fragments $\mathcal{F}_1, \dots, \mathcal{F}_i$.

Monitoring Problem. A quadruple $\mu(\mathcal{A}) = (a_0, \mathcal{V}, \hat{\mathcal{O}}, \mathcal{R})$, where a_0 is the initial state of an active system $\mathcal{A}$, $\mathcal{V}$ a viewer for $\mathcal{A}$, $\hat{\mathcal{O}}$ a fragmentation of a temporal observation of $\mathcal{A}$, and $\mathcal{R}$ a ruler for $\mathcal{A}$.

Monitoring-Problem Solution. Given a monitoring problem $\mu(\mathcal{A}) = (a_0, \mathcal{V}, \hat{\mathcal{O}}, \mathcal{R})$, where $\hat{\mathcal{O}} = [\mathcal{F}_1, \dots, \mathcal{F}_k]$, the sequence $[\Delta(\wp_{[0]}(\mathcal{A})), \dots, \Delta(\wp_{[k]}(\mathcal{A}))]$ of the solutions of the diagnosis subproblems relevant to $\mu(\mathcal{A})$.

Diagnosis Closure. Given a state a of a system $\mathcal{A}$, and a viewer $\mathcal{V}$ and a ruler $\mathcal{R}$ for $\mathcal{A}$, the subgraph of the behavior space of $\mathcal{A}$ rooted in a and encompassing all paths of unobservable transitions (and states involved) rooted in a, where each state is associated with a set of diagnoses, called a diagnosis attribute.

Monitor. Given a state a_0 of a system $\mathcal{A}$, and a viewer $\mathcal{V}$ and a ruler $\mathcal{R}$ for $\mathcal{A}$, a graph where each node is a diagnosis closure and each arc $N \xrightarrow{(s,\ell,f,s')} N'$ is such that s is a state in N, s' is the root state of N', ℓ is an observable label, and f is a (possibly empty) fault label.

Local Candidate Set. Given a node N of a monitor, the union of the diagnosis attributes of all states incorporated into N.

Monitoring State. Given a monitor M, a set of pairs (N, Δ) (called contexts), where N is a node of M and Δ a set of diagnoses.

Index-Space Decoration. Given a monitoring problem $(a_0, \mathcal{V}, \hat{\mathcal{O}}, \mathcal{R})$, a DFA isomorphic to the index space of $\mathcal{O}$ (the temporal observation made up of the fragments in $\hat{\mathcal{O}}$), where each state $\mathfrak{I}$ is marked by a monitoring attribute, that is, the set of contexts consistent with $\mathfrak{I}$.

Monitoring Path. Given a monitoring problem involving a fragmented observation $\hat{\mathcal{O}} = [\mathcal{F}_1, \ldots, \mathcal{F}_k]$, a sequence $[\mathcal{M}_0, \mathcal{M}_1, \ldots, \mathcal{M}_k]$ of monitoring states, where each $\mathcal{M}_i$, $i \in [1 \cdots k]$, is the union of the diagnosis attributes of the final states of the decorated index space of the observation yielded, up to the i-th fragment.

Reactive Sequence. Given a monitoring path $[\mathcal{M}_0, \mathcal{M}_1, \ldots, \mathcal{M}_k]$ relevant to a monitoring problem $\mu(\mathcal{A})$, a sequence $[\Delta_0, \Delta_1, \ldots, \Delta_k]$ of sets of diagnoses, where each Δ_i, $i \in [1 \cdots k]$, is the i-th element of the solution of $\mu(\mathcal{A})$, obtained by combining the local candidate sets of nodes N within pairs (N, Δ) in $\mathcal{M}_i$ with the associated set of diagnoses Δ.

Chapter 8
Monotonic Diagnosis

In Chapter 7 we coped with reactive diagnosis and presented a technique for solving a monitoring problem $\mu(\mathcal{A}) = (a_0, \mathcal{V}, \hat{\mathcal{O}}, \mathcal{R})$, where $\hat{\mathcal{O}} = [\mathcal{F}_1, \ldots, \mathcal{F}_k]$, $i \in [1 \cdots k]$, is a fragmentation of a temporal observation $\mathcal{O}$.

Unlike a posteriori diagnosis, reactive diagnosis requires that candidate diagnoses be generated in real time, soon after the reception of each fragment. Therefore, the solution of $\mu(\mathcal{A})$ is not just a set of candidate diagnoses (as occurs in a posteriori diagnosis). Instead, it is a sequence $[\Delta_0, \Delta_1, \ldots, \Delta_k]$ of sets of candidate diagnoses, where each Δ_i, $i \in [0 \cdots k]$, is the solution of the diagnosis problem $\wp_{[i]}(\mathcal{A}) = (a_0, \mathcal{V}, O_{[i]}, \mathcal{R})$, with $O_{[i]}$ being the temporal observation made up by the sequence of fragments $[\mathcal{F}_1, \ldots, \mathcal{F}_i]$.

Reactive diagnosis is sound and complete: on the reception of each fragment, it produces the whole set of candidate diagnoses that comply with the system model and the temporal observation collected so far. However, from a practical viewpoint, *soundness and completeness are not enough in the monitoring of active systems with fragmented temporal observations*, because the property of soundness and completeness does not in itself guarantee the monotonic growth of the candidate traces, namely $[\, \|\mathcal{O}_{[0]}\|, \|\mathcal{O}_{[1]}\|, \ldots, \|\mathcal{O}_{[k]}\| \,]$.

According to Proposition 4.2, the actual trace $\mathcal{T}$ of the reaction of the system is included in the set of candidate traces of the temporal observation $\mathcal{O}$, that is $\mathcal{T} \in \|\mathcal{O}\|$. This guarantees that the actual diagnosis is in the set of candidate diagnoses.

By contrast, when the temporal observation is received as a sequence of fragments $[\mathcal{F}_1, \ldots, \mathcal{F}_k]$, there is no guarantee that each $\|\mathcal{O}_{[i]}\|$, $i \in [1 \cdots k]$, will include a prefix of $\mathcal{T}$. For instance, assume that $\mathcal{T} = [awk, opl]$, the temporal observation being composed of two disconnected nodes (marked by *awk* and *opl*, respectively), with no temporal precedence defined between them, and $\mathcal{O} = [\mathcal{F}_1, \mathcal{F}_2]$, with $\mathcal{F}_1$ and $\mathcal{F}_2$ being composed of nodes marked by *opl* and *awk*, respectively.

We have $\|\mathcal{O}_{[1]}\| = \{[opl]\}$ and $\|\mathcal{O}_{[2]}\| = \{[opl, awk], [awk, opl]\}$, where $\|\mathcal{O}_{[1]}\|$ does not include any prefix of $\mathcal{T}$. Intuitively, this is due to the fact that the label *opl*, which is generated after *awk*, is received in the fragmented observation before *awk*.

The nonmonotonicity of candidate traces is associated with the nonmonotonicity of candidate diagnoses. Consequently, monitoring results may be disappointing, if

G. Lamperti et al., *Introduction to Diagnosis of Active Systems*,
https://doi.org/10.1007/978-3-319-92733-6_8

not misleading. The challenge is to find some conditions on the mode in which the temporal observation is fragmented such that candidate traces grow monotonically.

8.1 Beyond Soundness and Completeness

As claimed above, soundness and completeness are not enough to make reactive diagnosis meaningful. In what follows, we clarify this point based on various examples.

Example 8.1. With reference to the active system $\mathcal{A}_2$ defined in Example 3.3, consider the monitoring problem $\mu(\mathcal{A}_2) = (a_0, \mathcal{V}, \hat{\mathcal{O}}, \mathcal{R})$, where $\mathcal{V}$ and $\mathcal{R}$ are defined in Table 4.1, and $\hat{\mathcal{O}} = [\mathcal{F}_1, \mathcal{F}_2, \mathcal{F}_3, \mathcal{F}_4]$ is the plain fragmentation of the linear observation $\mathcal{O} = [awk, opl, ide, cll]$. Since $\mathcal{O}$ is linear, each fragment $\mathcal{F}_i$ involves the i-th observable label in $\mathcal{O}$, $i \in [1 \cdots 4]$. Therefore, the solution of $\mu(\mathcal{A}_2)$ will be

$$\Delta(\mu(\mathcal{A}_2)) = [\Delta_0, \Delta_1, \Delta_2, \Delta_3, \Delta_4],$$

where each Δ_i, $i \in [0 \cdots 4]$, is the solution of the diagnosis subproblem $\wp_{[i]}(\mathcal{A}_2)$ relevant to subobservation $\mathcal{O}_{[i]}$. Based on the partial monitor displayed in Fig. 8.1 and the subobservation index spaces displayed on the right side of Fig. 8.2, the solution Δ_i of each subproblem $\wp_{[i]}(\mathcal{A}_2)$ is displayed in Table 8.1.

Since the observation $\mathcal{O}$ is linear, the set of candidate traces for each subobservation $\mathcal{O}_{[i]}$, $i \in [0 \cdots 4]$, includes just one trace, which is the i-th prefix of $\mathcal{O}$. In other words, the actual trace $[awk, opl, ide, cll]$ grows monotonically from one subobservation to the next.

This property translates to a similar property in the corresponding set of candidate diagnoses: the actual diagnosis relevant to $\mathcal{O}$, namely the singleton $\{\{\mathbf{nor}\}\}$, grows monotonically within the set of candidate diagnoses. This means that, starting from the empty diagnosis $\delta_0 \in \Delta_0$, each subsequent set Δ_i, $i \in [1 \cdots 4]$, includes a candidate diagnosis δ_i that is a subset of the actual diagnosis, such that:

$$\delta_0 \subseteq \delta_1 \subseteq \delta_2 \subseteq \delta_3 \subseteq \delta_4,$$

where $\delta_4 = \{\{\mathbf{nor}\}\}$ is the actual diagnosis. In our example, we have

Table 8.1 Solutions of subproblems $\wp_{[i]}(\mathcal{A}_2)$, $i \in [0 \cdots 4]$, in Example 8.1

$\mathcal{O}_{[i]}$	$\lVert \mathcal{O}_{[i]} \rVert$	Δ_i
$\mathcal{O}_{[0]}$	$\{[\,]\}$	$\{\emptyset\}$
$\mathcal{O}_{[1]}$	$\{[awk]\}$	$\{\emptyset, \{\mathbf{nol}\}, \{\mathbf{nor}\}, \{\mathbf{nol}, \mathbf{nor}\}, \{\mathbf{fop}\}\}$
$\mathcal{O}_{[2]}$	$\{[awk, opl]\}$	$\{\emptyset, \{\mathbf{nor}\}\}$
$\mathcal{O}_{[3]}$	$\{[awk, opl, ide]\}$	$\{\{\mathbf{nor}\}, \{\mathbf{ncl}, \mathbf{nor}\}, \{\mathbf{nor}, \mathbf{fcp}\}\}$
$\mathcal{O}_{[4]}$	$\{[awk, opl, ide, cll]\}$	$\{\{\mathbf{nor}\}\}$

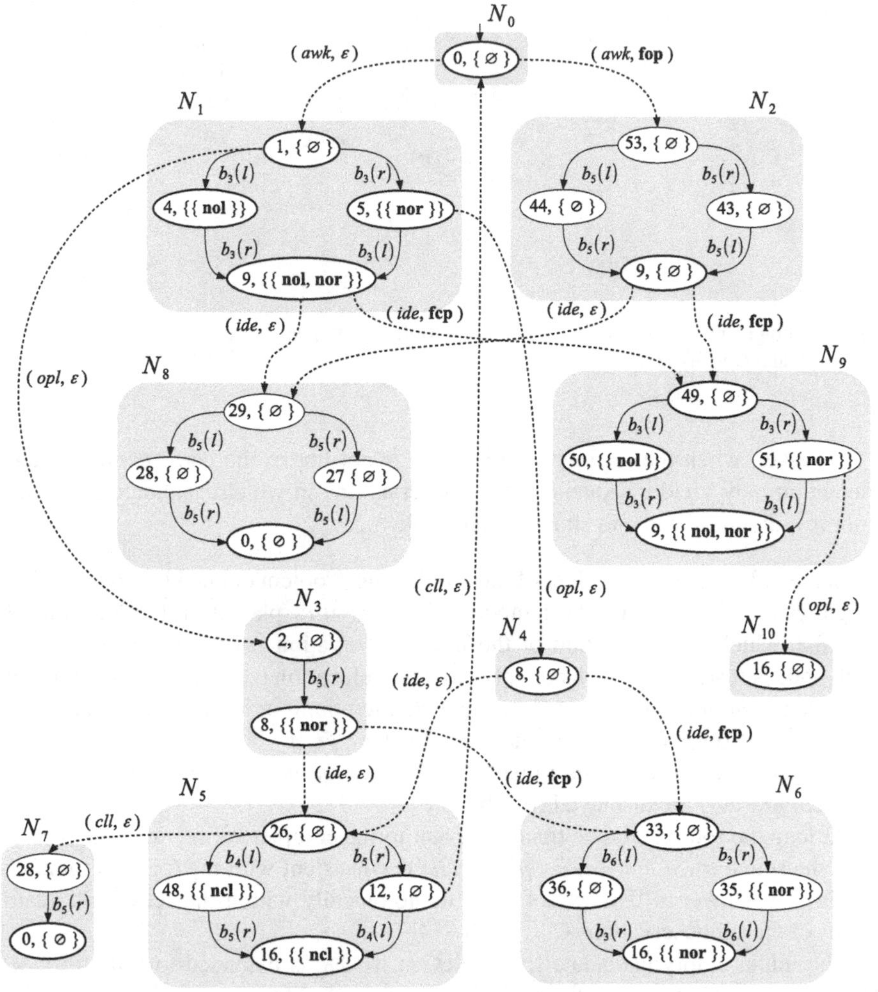

Fig. 8.1 Extended subpart of the monitor $Mtr(\mathcal{A}_2, a_0, \mathcal{V}, \mathcal{R})$

$$\emptyset \subseteq \{\mathbf{nor}\} \subseteq \{\mathbf{nor}\} \subseteq \{\mathbf{nor}\} \subseteq \{\mathbf{nor}\}\,.$$

Note that the set-containment relationship expressing the monotonicity of the diagnosis is not unique. For example, here is another possible instantiation:

$$\emptyset \subseteq \emptyset \subseteq \{\mathbf{nor}\} \subseteq \{\mathbf{nor}\} \subseteq \{\mathbf{nor}\}\,.$$

Whatever the instantiation, such a set-containment relationship certainly exists when the set of candidate traces grows monotonically, which is always true when the temporal observation is linear (as in this example).

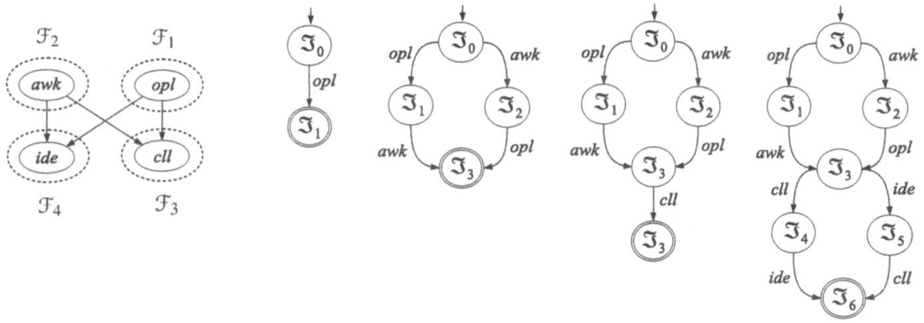

Fig. 8.2 Fragmentation $\hat{\mathcal{O}} = [\mathcal{F}_1, \mathcal{F}_2, \mathcal{F}_3, \mathcal{F}_4]$ (left), and index spaces $Isp(\mathcal{O}_{[1]})$, $Isp(\mathcal{O}_{[2]})$, $Isp(\mathcal{O}_{[3]})$, and $Isp(\mathcal{O}_{[4]})$ (right)

However, when the temporal observation is not linear, the corresponding fragmentation may yield a sequence of subobservations in which candidate traces do not grow monotonically, as shown in Example 8.2.

Example 8.2. Consider a variant of the monitoring problem defined in Example 8.1, where $\mathcal{O}$ is substituted by the temporal observation displayed on the left side of Fig. 8.2, which is a relaxation of the linear observation $[awk, opl, ide, cll]$. Since each node is marked by one label only, no label relaxation occurs. Instead, temporal relaxation is applied, as indicated by the partial temporal ordering of arcs. Based on the figure, the observation is fragmented into $\hat{\mathcal{O}} = [\mathcal{F}_1, \mathcal{F}_2, \mathcal{F}_3, \mathcal{F}_4]$.

Based on the partial monitor displayed in Fig. 8.1, the solutions Δ_i of each subproblem $\wp_{[i]}(\mathcal{A}_2)$ are displayed in Table 8.2.

Among the four candidate traces relevant to the complete observation $\mathcal{O}_{[4]} = \mathcal{O}$, only the actual trace $\mathcal{T} = [awk, opl, ide, cll]$ is consistent with the (partial) monitor in Fig. 8.1. However, $\mathcal{T}$ does not grow monotonically within the set of candidate traces of the subobservations $\mathcal{O}_{[i]}$, $i \in [0 \cdots 4]$.

The initial empty candidate trace relevant to $\mathcal{O}_{[0]}$ is extended to $[opl]$ in $\mathcal{O}_{[1]}$. However, $[opl]$ is not a prefix of the actual trace $\mathcal{T}$. This is possible because the original temporal ordering between *awk* and *opl* in $\mathcal{T}$ (with the former occurring before the latter) is lost owing to temporal relaxation, so that *awk* and *opl* are no longer temporally related.

Table 8.2 Solutions of subproblems $\wp_{[i]}(\mathcal{A}_2)$, $i \in [0 \cdots 4]$, in Example 8.2

$\mathcal{O}_{[i]}$	$\lVert \mathcal{O}_{[i]} \rVert$	Δ_i
$\mathcal{O}_{[0]}$	$\{[\,]\}$	$\{\emptyset\}$
$\mathcal{O}_{[1]}$	$\{[opl]\}$	$\emptyset$
$\mathcal{O}_{[2]}$	$\{[opl, awk], [awk, opl]\}$	$\{\emptyset, \{\mathbf{nor}\}\}$
$\mathcal{O}_{[3]}$	$\{[opl, awk, cll], [awk, opl, cll]\}$	$\emptyset$
$\mathcal{O}_{[4]}$	$\{[opl, awk, cll, ide], [opl, awk, ide, cll], [awk, opl, cll, ide], [awk, opl, ide, cll]\}$	$\{\{\mathbf{nor}\}\}$

The point is that, owing to fragmentation, the observable labels of the actual trace may be received in a different order. Not only can this cause nonmonotonicity of candidate diagnoses, but also a subobservation $\mathcal{O}_i$ may be inconsistent with the system model for all candidate traces in $\|\mathcal{O}_i\|$. Based on Table 8.2, this occurs for both $\mathcal{O}_{[1]}$ and $\mathcal{O}_{[3]}$. Consequently, the corresponding sets of candidate diagnoses are empty (they are shaded in Table 8.2).[1]

Now, imagine that you are the operator of a control room where a diagnosis engine is generating the sequence of candidate sets listed in Table 8.2. After $\Delta_0 = \{\emptyset\}$, you receive $\Delta_1 = \emptyset$, which sounds strange. Then, you receive $\Delta_2 = \{\emptyset, \{\mathbf{nor}\}\}$, which means that either no fault has occurred or just **nor** (breaker *r* did not open). Then, you might possibly expect a refinement of this diagnosis output in the next output. Instead, corresponding to $\mathcal{O}_{[3]}$ you receive an empty candidate set again! Only upon completion of the temporal observation, namely $\mathcal{O}_{[4]}$, will you get $\Delta_4 = \{\{\mathbf{nor}\}\}$, which indicates a single candidate diagnosis involving just the fault **nor**.

What Example 8.2 clearly shows is that, in the worst case, the solution of a monitoring problem relevant to a fragmented observation $\hat{\mathcal{O}}$ may be misleading up to the final set of candidate diagnoses. This is most problematic for those application domains in which observation fragments are continuously received, so that no completion of the temporal observation actually exists. If so, the *raison d'être* of reactive diagnosis would be questionable.

Fortunately, there are other scenarios in which reactive diagnosis makes sense, as shown in Example 8.3.

Example 8.3. Consider another variant of the monitoring problem defined in Example 8.1, where $\mathcal{O}$ is substituted by the temporal observation displayed on the left side of Fig. 8.3, which is a different relaxation of the linear observation

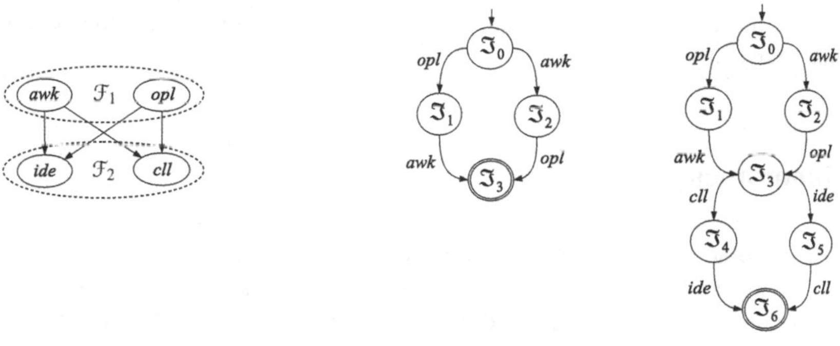

Fig. 8.3 Fragmentation $\hat{\mathcal{O}} = [\mathcal{F}_1, \mathcal{F}_2]$ (left), and index spaces $Isp(\mathcal{O}_{[1]})$, and $Isp(\mathcal{O}_{[2]})$ (right)

[1] An empty set of candidate diagnoses denotes inconsistency of the observation with the system model, as no trajectory of the behavior space is compatible with any of the candidate traces. By contrast, an empty candidate diagnosis indicates that there is at least one trajectory *h* that is consistent with at least one candidate trace, with *h* either empty or involving unobservable transitions only.

Table 8.3 Solutions of subproblems $\wp_{[i]}(\mathcal{A}_2)$, $i \in [0 \cdots 2]$, in Example 8.1

$\mathcal{O}_{[i]}$	$\|\mathcal{O}_{[i]}\|$	Δ_i
$\mathcal{O}_{[0]}$	$\{[\,]\}$	$\{\emptyset\}$
$\mathcal{O}_{[1]}$	$\{[opl, awk], [awk, opl]\}$	$\{\emptyset, \{\mathbf{nor}\}\}$
$\mathcal{O}_{[2]}$	$\{[opl, awk, cll, ide], [opl, awk, ide, cll], [awk, opl, cll, ide], [awk, opl, ide, cll]\}$	$\{\{\mathbf{nor}\}\}$

$[awk, opl, ide, cll]$. Now, the fragmentation is composed of two fragments, $\hat{\mathcal{O}} = [\mathcal{F}_1, \mathcal{F}_2]$, with each fragment including two nodes. Based on the temporal constraints on $\mathcal{O}$, the observable labels *awk* and *opl* are collectively generated before *ide* and *cll*. However, no temporal constraint is defined between the two nodes in each fragment.

Based on the partial monitor displayed in Fig. 8.1 and the subobservation index spaces displayed on the right side of Fig. 8.3, the solution Δ_i of each subproblem $\wp_{[i]}(\mathcal{A}_2)$ is displayed in Table 8.3.

Unlike the fragmentation displayed in Fig. 8.2, the actual trace now grows monotonically, since $[awk, opl] \in \|\mathcal{O}_{[1]}\|$ is a prefix of the actual trace $[awk, opl, ide, cll]$. Consequently, the actual diagnosis $\{\mathbf{nor}\}$ grows monotonically too. As such, the sequence of sets of candidate diagnoses makes sense for the operator in the control room. Specifically, the initial candidate set including just the empty diagnosis is extended in correspondence with the first fragment by the additional candidate diagnosis $\{\mathbf{nor}\}$, which is in fact the only candidate after the reception of the second (final) observation fragment.

Compared with the misleading output in Example 8.2, what makes the reactive diagnosis in Example 8.3 meaningful? In other words, what makes the diagnosis output in Example 8.3 monotonic? As pointed out above, the monotonicity of the diagnosis results is a consequence of the monotonicity of the candidate traces relevant to each subobservation. So, the real question is: *what makes a fragmented observation monotonic?* In order to answer this question, we first need to formalize the notion of monotonicity in general terms.

8.2 Monotonicity

In Sec. 8.1, we adopted the notion of monotonicity for both sequences of candidate traces and sequences of candidate diagnoses. In both cases, we considered sequences of collections. However, candidate traces are sequences, while candidate diagnoses are sets. Therefore, we introduce a notion of monotonicity here that is independent of the kind of collection involved in the sequence.

Definition 8.1 (Monotonic Sequence). Let $Q = [S_0, S_1, \ldots, S_n]$ be a sequence of nonempty sets of collections, with these collections being themselves either sets or sequences. Q is *monotonic* iff $\forall$ sets S_i, $i \in [1 \cdots n]$, $\forall$ collections $C' \in S_i$, $\exists$ a collection $C \in S_{i-1}$ such that $C \subseteq C'$.

Note 8.1. In Def. 8.1, the meaning of the condition $C \subseteq C'$ depends on the nature of the collections involved. If they are sets, then $C \subseteq C'$ is the usual set containment relationship. If they are sequences, then $C \subseteq C'$ means that C is a prefix of C'.

Example 8.4. Let $Q = [\{\{a\},\{b\},\{c\}\},\{\{a,c\},\{b,d\}\},\{\{a,b,c\}\}]$ be a sequence of three sets of sets. According to Def. 8.1, Q is monotonic. In fact, for each collection (set) C' in $\{\{a,c\},\{b,d\}\}$, there exists a collection (set) C in $\{\{a\},\{b\}\}$ such that $C \subseteq C'$. Specifically, $\{a\} \subset \{a,c\}$ and $\{b\} \subset \{b,d\}$. The same applies to the third set, where $\{a,c\} \subset \{a,b,c\}$.

Example 8.5. Let $Q' = [\{[a],[b],[c]\},\{[a,c],[b,d]\},\{[a,b,c]\}]$ be a sequence of three sets of sequences. Note how Q' is obtained from Q in Example 8.4 by transforming inner sets into sequences. Based on Def. 8.1, Q' is *not* monotonic. In fact, the second set, $\{[a,c],[b,d]\}$, does not include any prefix of the sequence $[a,b,c]$ in the third set.

Definition 8.2 (Monotonic Fragmentation). Let $\hat{\mathcal{O}} = [\mathcal{F}_1,\ldots,\mathcal{F}_k]$ be a fragmentation of a temporal observation $\mathcal{O}$. Let $Q = [\,\|\mathcal{O}_{[0]}\|,\|\mathcal{O}_{[1]}\|,\ldots,\|\mathcal{O}_{[k]}\|\,]$ be the sequence of the extensions of temporal observations $\mathcal{O}_{[i]}$ (made up to fragment $\mathcal{F}_i$), $i \in [1\cdots k]$. $\hat{\mathcal{O}}$ is *monotonic* iff Q is monotonic.

Example 8.6. Consider the fragmentation $\hat{\mathcal{O}} = [\mathcal{F}_1,\mathcal{F}_2,\mathcal{F}_3,\mathcal{F}_4]$ displayed in the left side of Fig. 8.4, where each fragment includes just one node. According to Def. 8.2, $\hat{\mathcal{O}}$ is monotonic iff $Q = [\,\|\mathcal{O}_{[0]}\|,\|\mathcal{O}_{[1]}\|,\|\mathcal{O}_{[2]}\|,\|\mathcal{O}_{[3]}\|,\|\mathcal{O}_{[0]}\|\,]$ is monotonic. According to Proposition 5.1, the language of $Isp(\mathcal{O}_{[i]})$ equals $\|\mathcal{O}\|_{[i]}$, $i \in [1\cdots 4]$. We have

$$
\begin{aligned}
\|\mathcal{O}_{[0]}\| &= \{[\,]\},\\
\|\mathcal{O}_{[1]}\| &= \{[awk]\},\\
\|\mathcal{O}_{[2]}\| &= \{[awk, ide]\},\\
\|\mathcal{O}_{[3]}\| &= \{[awk, ide, opl],\ [awk, opl, ide]\ \},\\
\|\mathcal{O}_{[4]}\| &= \{[awk, ide, opl, cll],[awk, opl, ide, cll]\}\ .
\end{aligned}
$$

According to Def. 8.1, Q is *not* monotonic, as for the (shaded) candidate trace $\mathcal{T}' = [awk, opl, ide] \in \|\mathcal{O}_{[3]}\|$ there is no candidate trace $\mathcal{T} \in \|\mathcal{O}_{[2]}\|$ such that $\mathcal{T} \subseteq \mathcal{T}'$.

Proposition 8.1. *Let $\mathcal{O}$ be a temporal observation. There exists a plain fragmentation $\hat{\mathcal{O}}$ of $\mathcal{O}$ that is monotonic.*

Proof. Let $\mathcal{O}$ include k nodes. Let $\mathcal{T}$ be the actual trace relevant to $\mathcal{O}$. According to Proposition 4.2, $\mathcal{T} \in \|\mathcal{O}\|$. In other words, $\mathcal{T} = [\ell_1,\ldots,\ell_k]$, where each $\ell_i \in \|n_i\|$, $i \in [1\cdots k]$. Based on Def. 7.1, a plain fragmentation of $\mathcal{O}$ is a fragmentation $\hat{\mathcal{O}} = [\mathcal{F}_1,\ldots,\mathcal{F}_k]$ where each $\mathcal{F}_i$, $i \in [1\cdots k]$, involves just one node. Let n_i be the node in $\mathcal{F}_i$, $i \in [1\cdots k]$. We prove the thesis by induction on $\hat{\mathcal{O}}$.

(*Basis*) $[\mathcal{F}_1]$ *is monotonic*. That is, $[\,\|\mathcal{O}_{[0]}\|,\|\mathcal{O}_{[1]}\|\,]$ is monotonic. Since $\|\mathcal{O}_{[0]}\| = \{[\,]\}$, the empty sequence $[\,]$ is a prefix of each element in $\|\mathcal{O}_{[1]}\|$.

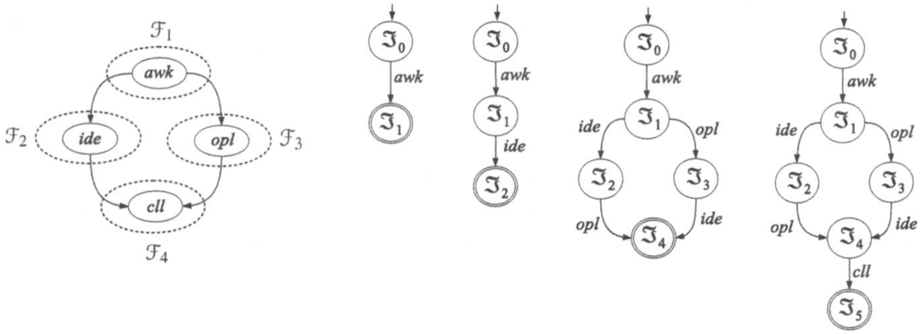

Fig. 8.4 Fragmentation $\hat{\mathcal{O}} = [\mathcal{F}_1, \mathcal{F}_2, \mathcal{F}_3, \mathcal{F}_4]$ (left), and index spaces $Isp(\mathcal{O}_{[1]})$, $Isp(\mathcal{O}_{[2]})$, $Isp(\mathcal{O}_{[3]})$, and $Isp(\mathcal{O}_{[4]})$ (right)

(Induction) If $[\mathcal{F}_1, \ldots, \mathcal{F}_i]$ *is monotonic, then* $[\mathcal{F}_1, \ldots, \mathcal{F}_i, \mathcal{F}_{i+1}]$ *is monotonic.* In fact, $\|\mathcal{O}_{[i]}\|$ includes the prefix $\mathcal{T}_i = [\ell_1, \ldots, \ell_i]$ of $\mathcal{T}$. Likewise, $\|\mathcal{O}_{[i+1]}\|$ includes the prefix $\mathcal{T}_{i+1} = [\ell_1, \ldots, \ell_i, \ell_{i+1}]$ of $\mathcal{T}$. Clearly, $\mathcal{T}_i \subseteq \mathcal{T}_{i+1}$. □

Corollary 8.1. *A fragmentation* $\hat{\mathcal{O}}$ *of a linear temporal observation* $\mathcal{O}$ *is monotonic.*

Definition 8.3 (Monotonic Monitoring Problem). A monitoring problem $\mu(\mathcal{A})$ is *monotonic* iff its solution is monotonic.

Example 8.7. With reference to the active system $\mathcal{A}_2$ defined in Example 3.3, consider the monitoring problem $\mu(\mathcal{A}_2) = (a_0, \mathcal{V}, \hat{\mathcal{O}}, \mathcal{R})$, where $\mathcal{V}$ and $\mathcal{R}$ are defined in Table 4.1, and the fragmented observation $\hat{\mathcal{O}} = [\mathcal{F}_1, \mathcal{F}_2, \mathcal{F}_3, \mathcal{F}_4]$ is displayed on the left side of Fig. 8.4, where each fragment includes just one node. Depicted next to $\hat{\mathcal{O}}$ are the index spaces of the observations relevant to the subfragmentations of $\hat{\mathcal{O}}$, namely $Isp(\mathcal{O}_{[1]})$, $Isp(\mathcal{O}_{[2]})$, $Isp(\mathcal{O}_{[3]})$, and $Isp(\mathcal{O}_{[4]})$. According to Theorem 7.1, $\Delta(\mu(\mathcal{A}))$ equals the reactive sequence

$$React(\mu(\mathcal{A})) = [\Delta_0, \Delta_1, \Delta_2, \Delta_3, \Delta_4].$$

In other words, $\Delta_i = \wp_{[i]}(\mathcal{A}_2)$, $i \in [0 \cdots 4]$. In order to generate the reactive sequence, based on Def. 7.15 and on the partial monitor displayed in Fig. 8.1, we need to compute the monitoring path

$$Path(\mu(\mathcal{A}_2)) = [\mathcal{M}_0, \mathcal{M}_1, \mathcal{M}_2, \mathcal{M}_3, \mathcal{M}_4].$$

We have

$$\begin{aligned}
&\mathcal{M}_0 = \{(N_0, \{\emptyset\})\},\\
&\mathcal{M}_1 = \{(N_1, \{\emptyset\}), (N_2, \{\{\mathbf{fop}\}\})\},\\
&\mathcal{M}_2 = \{(N_8, \{\{\mathbf{nol}, \mathbf{nor}\}, \{\mathbf{fop}\}\}), (N_9, \{\{\mathbf{nol}, \mathbf{nor}, \mathbf{fcp}\}, \{\mathbf{fop}, \mathbf{fcp}\}\})\},\\
&\mathcal{M}_3 = \{(N_5, \{\{\mathbf{nor}\}\}), (N_6, \{\{\mathbf{nor}, \mathbf{fcp}\}\}), (N_{10}, \{\{\mathbf{nol}, \mathbf{nor}, \mathbf{fcp}\}, \{\mathbf{nor}, \mathbf{fop}, \mathbf{fcp}\}\})\},\\
&\mathcal{M}_4 = \{(N_7, \{\{\mathbf{nor}\}\})\}.
\end{aligned}$$

Based on (7.40) in Def. 7.16, we come up with the following sets of candidate diagnoses:

$\Delta_0 = \{\emptyset\}$,
$\Delta_1 = \{\emptyset, \{\mathbf{nol}\}, \{\mathbf{nor}\}, \{\mathbf{nol}, \mathbf{nor}\}, \{\mathbf{fop}\}\}$,
$\Delta_2 = \{\{\mathbf{nol}, \mathbf{nor}, \mathbf{fop}\}, \{\mathbf{nol}, \mathbf{nor}, \mathbf{fcp}\}, \{\mathbf{fop}, \mathbf{fcp}\}, \{\mathbf{nol}, \mathbf{fop}, \mathbf{fcp}\}, \{\mathbf{nor}, \mathbf{fop}, \mathbf{fcp}\}, \{\mathbf{nol}, \mathbf{nor}, \mathbf{fop}, \mathbf{fcp}\}\}$,
$\Delta_3 = \{\{\mathbf{nol}, \mathbf{nor}, \mathbf{fcp}\}, \{\mathbf{nor}, \mathbf{fop}, \mathbf{fcp}\}, \{\mathbf{nor}\}, \{\mathbf{ncl}, \mathbf{nor}\}, \{\mathbf{nor}, \mathbf{fcp}\}\}$,
$\Delta_4 = \{\{\mathbf{nor}\}\}$.

Based on Def. 8.1, $\Delta(\mu(\mathcal{A}))$ is not monotonic, as for each shaded candidate diagnosis $\delta' \in \Delta_3$ there does not exist a candidate diagnosis $\delta \in \Delta_2$ such that $\delta \subseteq \delta'$.

From a practical viewpoint, what is disappointing is that the actual diagnosis $\{\mathbf{nor}\}$ is not monotonically included in each Δ_i, $i \in [0 \cdots 4]$. Specifically, Δ_2 includes neither $\emptyset$ nor $\{\mathbf{nor}\}$. Even though Δ_2 includes four candidate diagnoses involving **nor**, namely $\{\mathbf{nol}, \mathbf{nor}, \mathbf{fop}\}$, $\{\mathbf{nol}, \mathbf{nor}, \mathbf{fcp}\}$, $\{\mathbf{nor}, \mathbf{fop}, \mathbf{fcp}\}$, and $\{\mathbf{nol}, \mathbf{nor}, \mathbf{fop}, \mathbf{fcp}\}$, it nevertheless looks as if the fault **nor** cannot be the only fault in the actual diagnosis, which is not the case, as indicated by Δ_3 and, most evidently, by Δ_4.

Example 8.8. With reference to the active system $\mathcal{A}_2$ defined in Example 3.3, consider the monitoring problem $\mu(\mathcal{A}_2) = (a_0, \mathcal{V}, \hat{\mathcal{O}}, \mathcal{R})$, where $\mathcal{V}$ and $\mathcal{R}$ are defined in Table 4.1, and the fragmented observation $\hat{\mathcal{O}}$ is displayed on the left side of Fig. 8.4. Based on Def. 8.3, $\mu(\mathcal{A}_2)$ is monotonic if and only if $\Delta(\mu(\mathcal{A}_2))$ is monotonic. According to Theorem 7.1, $\Delta(\mu(\mathcal{A}_2))$ equals the reactive sequence

$$React(\mu(\mathcal{A}_2)) = [\Delta_0, \Delta_1, \Delta_2, \Delta_3, \Delta_4] .$$

In other words, $\Delta_i = \wp_{[i]}(\mathcal{A}_2)$, $i \in [0 \cdots 4]$. In order to generate the reactive sequence, based on Def. 7.15 and on the partial monitor displayed in Fig. 8.1, we need to compute the monitoring path

$$Path(\mu(\mathcal{A}_2)) = [\mathcal{M}_0, \mathcal{M}_1, \mathcal{M}_2, \mathcal{M}_3, \mathcal{M}_4] .$$

We have

$\mathcal{M}_0 = \{(N_0, \{\emptyset\})\}$,
$\mathcal{M}_1 = \{(N_1, \{\emptyset\}), (N_2, \{\{\mathbf{fop}\}\})\}$,
$\mathcal{M}_2 = \{(N_3, \{\emptyset\}), (N_4, \{\{\mathbf{nor}\}\})\}$,
$\mathcal{M}_3 = \{(N_5, \{\{\mathbf{nor}\}\}), (N_6, \{\{\mathbf{nor}, \mathbf{fcp}\}\}), (N_{10}, \{\{\mathbf{nol}, \mathbf{nor}, \mathbf{fcp}\}, \{\mathbf{nor}, \mathbf{fop}, \mathbf{fcp}\}\})\}$,
$\mathcal{M}_4 = \{(N_7, \{\{\mathbf{nor}\}\})\}$.

Based on (7.40) in Def. 7.16, we come up with the following sets of candidate diagnoses:

$$
\begin{aligned}
&\Delta_0 = \{\emptyset\}, \\
&\Delta_1 = \{\emptyset, \{\mathbf{nol}\}, \{\mathbf{nor}\}, \{\mathbf{nol}, \mathbf{nor}\}, \{\mathbf{fop}\}\}, \\
&\Delta_2 = \{\emptyset, \{\mathbf{nor}\}\}, \\
&\Delta_3 = \{\{\mathbf{nol}, \mathbf{nor}, \mathbf{fcp}\}, \{\mathbf{nor}, \mathbf{fop}, \mathbf{fcp}\}, \{\mathbf{nor}\}, \{\mathbf{ncl}, \mathbf{nor}\}, \{\mathbf{nor}, \mathbf{fcp}\}\}, \\
&\Delta_4 = \{\{\mathbf{nor}\}\}.
\end{aligned}
$$

Based on Def. 8.1, $\Delta(\mu(\mathcal{A}))$ is monotonic. Hence, $\mu(\mathcal{A})$ is monotonic.

We are now ready to answer the central question asked at the end of Sec. 8.1. What makes a fragmentation monotonic? To this end, we provide a sufficient condition for the monotonicity of a fragmentation, based on the notion of *stratification*.

8.3 Stratification

Consider the fragmentation $\hat{\mathcal{O}}$ displayed on the left side of Fig. 8.3. As shown in Example 8.3, candidate traces in $\hat{\mathcal{O}}$ grow monotonically. A peculiarity of $\hat{\mathcal{O}}$ is that nodes which are temporally unrelated (such that no temporal precedence is defined between them) are included in the same fragment. In fact, the nodes *awk* and *opl* are both in $\mathcal{F}_1$; likewise, the nodes *ide* and *cll* are both in $\mathcal{F}_2$. This way to aggregate nodes within the same fragment leads us to the notion of a stratified fragmentation.

Definition 8.4 (Stratified Fragmentation). Let $\hat{\mathcal{O}} = [\mathcal{F}_1, \ldots, \mathcal{F}_k]$ be a fragmentation of a temporal observation $\mathcal{O}$. Let n be a node of $\mathcal{O}$, and let $Unrl(n)$ denote the set of nodes of $\mathcal{O}$ that are *unrelated* to n, that is, which are neither ancestors nor descendants of n in $\mathcal{O}$. The fragmentation $\hat{\mathcal{O}}$ is *stratified* iff $\forall \mathcal{F}_i = (L_i, N_i, A_i)$, $i \in [1 \cdots k]$, the following condition holds:

$$
\forall n \in N_i \ (Unrl(n) \subseteq N_i) \ . \tag{8.1}
$$

Each fragment in a stratified fragmentation is called a *stratum*.

Example 8.9. Consider the fragmentation $\hat{\mathcal{O}} = [\mathcal{F}_1, \mathcal{F}_2, \mathcal{F}_3]$ shown on the left side of Fig. 8.5. According to Def. 8.4, $\hat{\mathcal{O}}$ is stratified iff condition (8.1) holds for each fragment in $\hat{\mathcal{O}}$. The only unrelated nodes are n_2 and n_3; since they are within the same fragment $\mathcal{F}_2$, condition (8.1) holds for all fragments. Hence, $\hat{\mathcal{O}}$ is stratified.

The proposition below supports the claim that stratification is a sufficient condition for the monotonicity of a fragmented observation.

Proposition 8.2. *A stratified fragmentation is monotonic.*

Proof. Let $\hat{\mathcal{O}} = [\mathcal{F}_1, \ldots, \mathcal{F}_k]$ be a stratified fragmentation. Based on Def. 8.2, $\hat{\mathcal{O}}$ is monotonic iff $\left[\|\mathcal{O}_{[0]}\|, \|\mathcal{O}_{[1]}\|, \ldots, \|\mathcal{O}_{[k]}\|\right]$ is monotonic. That is, $\forall i \in [1 \cdots k]$, $\forall$

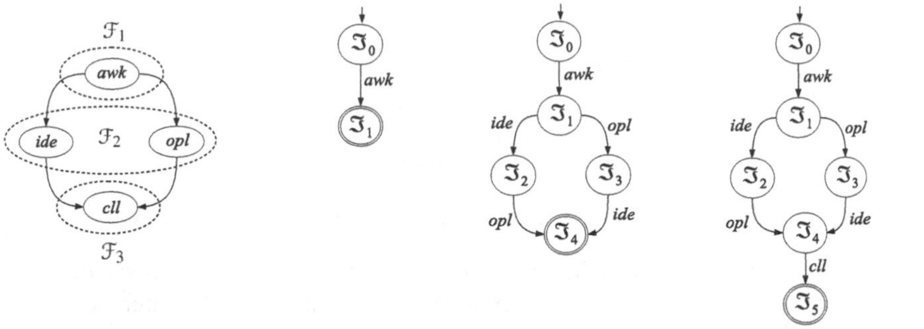

Fig. 8.5 Stratified fragmentation $\hat{\mathcal{O}} = [\mathcal{F}_1, \mathcal{F}_2, \mathcal{F}_3]$ (left), and index spaces $Isp(\mathcal{O}_{[1]})$, $Isp(\mathcal{O}_{[2]})$, and $Isp(\mathcal{O}_{[3]})$ (right)

traces $\mathcal{T}' \in \|\mathcal{O}_{[i]}\|$, there exists $\mathcal{T} \in \|\mathcal{O}_{[i-1]}\|$ such that $\mathcal{T}$ is a prefix of $\mathcal{T}'$. We show this property by induction on $\hat{\mathcal{O}}$.

(*Basis*) $[\mathcal{F}_1]$ *is monotonic.* In fact, for each $\mathcal{T}' \in \|\mathcal{O}_{[i]}\|$, there exists the empty trace $\mathcal{T} = [\,] \in \|\mathcal{O}_{[0]}\|$ such that $\mathcal{T}$ is a prefix of $\mathcal{T}'$.

(*Induction*) *If* $\hat{\mathcal{O}}_{[i]} = [\mathcal{F}_1, \ldots, \mathcal{F}_i]$ *is monotonic, then* $\hat{\mathcal{O}}_{[i+1]}$ *is monotonic.* The proof is grounded on Lemma 8.1.

Lemma 8.1. *Let $\mathcal{F}_i$ and $\mathcal{F}_{i+1}$ be two consecutive fragments in a fragmentation $\hat{\mathcal{O}}$ of $\mathcal{O}$. The following two properties hold for the temporal observation $\mathcal{O}$:*

1. *Each leaf node in $\mathcal{F}_i$ is exited by (at least) one arc entering a node in $\mathcal{F}_{i+1}$,*
2. *Each root node in $\mathcal{F}_{i+1}$ is entered by (at least) one arc exiting a node in $\mathcal{F}_i$.*

Proof. The proof is by contradiction. Assume that the first property is false for a leaf node n in $\mathcal{F}_i$. If so, either no arc exits n or all arcs exiting n do not enter a node in $\mathcal{F}_{i+1}$. In either case, n would be unrelated to the nodes in $\mathcal{F}_{i+1}$, which contradicts condition (8.1). Then, assume that the second property is false for a root node n in $\mathcal{F}_{i+1}$. If so, either no arc enters n or all arcs entering n do not exit any node in $\mathcal{F}_i$. In either case, n would be unrelated to the nodes in $\mathcal{F}_i$, which contradicts condition (8.1).

Based on Lemma 8.1, consider a trace $\mathcal{T}' \in \|\mathcal{O}_{[i+1]}\|$. According to Def. 4.4, $\mathcal{T}'$ is the sequence of labels yielded by choosing one label for each node in $\mathcal{O}_{[i+1]}$ fulfilling the partial temporal order constrained by the arcs. This means that, before choosing a label of a node in $\mathcal{F}_{i+1}$, it is necessary to have chosen a label for each node in all fragments preceding $\mathcal{F}_{i+1}$ in $\hat{\mathcal{O}}$. Hence, $\mathcal{T}'$ is the (possibly empty) extension of a trace $\mathcal{T} \in \mathcal{O}_{[i]}$; in other words, $\mathcal{T}$ is a prefix of $\mathcal{T}'$. □

Example 8.10. Consider the fragmentation $\hat{\mathcal{O}} = [\mathcal{F}_1, \mathcal{F}_2, \mathcal{F}_3]$ displayed on the left side of Fig. 8.5. As shown in Example 8.9, $\hat{\mathcal{O}}$ is stratified. Based on Def. 8.2, $\hat{\mathcal{O}}$ is monotonic iff $Q = [\,\|\mathcal{O}_{[0]}\|, \|\mathcal{O}_{[1]}\|, \|\mathcal{O}_{[2]}\|, \|\mathcal{O}_{[3]}\|\,]$ is monotonic. According to Proposition 5.1, the language of $Isp(\mathcal{O}_{[i]})$ equals $\|\mathcal{O}\|_{[i]}$, $i \in [1 \cdots 3]$. We have

$$\|\mathcal{O}_{[0]}\| = \{[\,]\},$$
$$\|\mathcal{O}_{[1]}\| = \{[awk]\},$$
$$\|\mathcal{O}_{[2]}\| = \{[awk, ide, opl], [awk, opl, ide]\},$$
$$\|\mathcal{O}_{[3]}\| = \{[awk, ide, opl, cll], [awk, opl, ide, cll]\}\ .$$

Based on Def. 8.1, and in accordance with Proposition 8.2, Q is monotonic.

In order to formally show that a monotonic fragmented observation results in a monotonic solution of a monitoring problem involving such a fragmentation, we first take an intermediate step by showing that a monotonic fragmented observation results in a monotonic growth of the traces relevant to the reconstructed behaviors.

Proposition 8.3. *Let* $\mu(\mathcal{A}) = (a_0, \mathcal{V}, \hat{\mathcal{O}}, \mathcal{R})$ *be a monitoring problem, where* $\hat{\mathcal{O}} = [\mathcal{F}_1, \ldots, \mathcal{F}_k]$ *is monotonic. Let* $Q = [H_0, H_1, \ldots, H_k]$ *be the sequence of sets of trajectories of* $\mathcal{A}$ *relevant to* $Bhv(\wp_{[0]}(\mathcal{A})), Bhv(\wp_{[1]}(\mathcal{A})), \ldots, Bhv(\wp_{[k]}(\mathcal{A}))$*, respectively.* Q *is monotonic.*

Proof. The proof is by induction on $\hat{\mathcal{O}}$.

(*Basis*) $[H_0, H_1]$ *is monotonic.* In fact, since $[\,] \in H_0$, the empty trajectory in H_0 is a prefix of each trajectory in H_1.

(*Induction*) *If* $[H_0, H_1, \ldots, H_i]$ *is monotonic, then* $[H_0, H_1, \ldots, H_i, H_{i+1}]$ *is monotonic.* Consider a trajectory $h' \in H_{i+1}$. Based on Def. 5.4, we have $h' \in Bsp(\mathcal{A})$ and $h'_{[\mathcal{V}]} \in \|O_{[i+1]}\|$. According to Def. 8.2, $\exists\, \mathcal{T} \in \|\mathcal{O}_{[i]}\|$ such that $\mathcal{T}$ is a prefix of $\mathcal{T}'$. Let h be the shortest prefix of h' such that $h_{[\mathcal{V}]} = \mathcal{T}$. As such, $h \in Bsp(\mathcal{A})$ and $h_{[\mathcal{V}]} \in \|\mathcal{O}_{[i]}\|$. Hence, $h \in H_i$. In other words, for each $h' \in H_{i+1}$ there exists $h \in H_i$ which is a prefix of h'. □

Example 8.11. With reference to the active system $\mathcal{A}_2$ defined in Example 3.3, consider the monitoring problem $\mu(\mathcal{A}_2) = (a_0, \mathcal{V}, \hat{\mathcal{O}}, \mathcal{R})$, where $\mathcal{V}$ and $\mathcal{R}$ are defined in Table 4.1, and the fragmented observation $\hat{\mathcal{O}}$ is displayed on the left side of Fig. 8.5. As shown in Example 8.9, $\hat{\mathcal{O}}$ is stratified; hence, according to Proposition 8.2, $\hat{\mathcal{O}}$ is monotonic. Let $Q = [H_0, H_1, H_2, H_3]$ be the sequence of sets of trajectories of $\mathcal{A}_2$ relevant to $Bhv(\wp_{[0]}(\mathcal{A}_2)), Bhv(\wp_{[1]}(\mathcal{A}_2)), Bhv(\wp_{[2]}(\mathcal{A}_2)), Bhv(\wp_{[3]}(\mathcal{A}_2))$, respectively. We expect that Q will be monotonic.

In order to concisely represent the sets H_i of trajectories, $i \in [1 \cdots 3]$, rather than considering the behavior space of $\mathcal{A}_2$, we look at the partial monitor displayed in Fig. 8.1. After all, a monitor is an enriched representation of a behavior space, where segments of unobservable trajectories are clustered in diagnosis closures. Specifically, each path from the root of N_0 to a state s within a node N_i (the diagnosis closure) identifies a trajectory.[2]

Besides, we may represent a set of trajectories concisely by the extension of a path $N_0 \rightsquigarrow N_i$ within the monitor, denoted $\|N_0 \rightsquigarrow N_i\|$. In other words, $\|N_0 \rightsquigarrow N_i\|$ implicitly identifies the set of trajectories within $N_0 \rightsquigarrow N_i$. For instance, $\|N_0 \xrightarrow{awk} N_1\|$

[2] To this end, considering Fig. 8.1, we have to assume that dashed arcs connecting nodes are implicitly marked by the relevant component transition.

identifies trajectories $0 \xrightarrow{p_1} 1, 0 \xrightarrow{p_1} 1 \xrightarrow{b_3(l)} 4, 0 \xrightarrow{p_1} 1 \xrightarrow{b_3(l)} 4 \xrightarrow{b_3(r)} 9, 0 \xrightarrow{p_1} 1 \xrightarrow{b_3(r)} 5$, and $0 \xrightarrow{p_1} 1 \xrightarrow{b_3(r)} 5 \xrightarrow{b_3(l)} 9$. In contrast, $\|N_0 \xrightarrow{awk} N_1 \xrightarrow{opl} N_4\|$ identifies the single trajectory $0 \xrightarrow{p_1} 1 \xrightarrow{b_3(r)} 5 \xrightarrow{b_1(l)} 8$.

This representation also allows for a simple sufficient condition for monotonicity of trajectories. Specifically, if $\|N \rightsquigarrow N'\|$ is included in H_i and $\|N \rightsquigarrow N' \rightsquigarrow N''\|$ is included in H_{i+1}, then for each trajectory h' in $\|N \rightsquigarrow N' \rightsquigarrow N''\|$ there exists a trajectory h in $\|N \rightsquigarrow N'\|$ such that $h \subseteq h'$.

Now, based on the candidate traces $\|\mathcal{O}_{[i]}\|$, $i \in [1 \cdots 3]$, determined in Example 8.11, we have

$$
\begin{aligned}
H_0 &= \{[\,]\},\\
H_1 &= \|N_0 \xrightarrow{awk} N_1\| \cup \|N_0 \xrightarrow{awk} N_2\|,\\
H_2 &= \|N_0 \xrightarrow{awk} N_1 \xrightarrow{ide} N_9 \xrightarrow{opl} N_{10}\| \cup \|N_0 \xrightarrow{awk} N_2 \xrightarrow{ide} N_9 \xrightarrow{opl} N_{10}\| \ \cup\\
&\quad \|N_0 \xrightarrow{awk} N_1 \xrightarrow{opl} N_3 \xrightarrow{ide} N_5\| \cup \|N_0 \xrightarrow{awk} N_1 \xrightarrow{opl} N_3 \xrightarrow{ide} N_6\| \ \cup\\
&\quad \|N_0 \xrightarrow{awk} N_1 \xrightarrow{opl} N_4 \xrightarrow{ide} N_5\| \cup \|N_0 \xrightarrow{awk} N_1 \xrightarrow{opl} N_4 \xrightarrow{ide} N_6\|,\\
H_3 &= \|N_0 \xrightarrow{awk} N_1 \xrightarrow{opl} N_3 \xrightarrow{ide} N_5 \xrightarrow{cll} N_7\| \cup \|N_0 \xrightarrow{awk} N_1 \xrightarrow{opl} N_4 \xrightarrow{ide} N_5 \xrightarrow{cll} N_7\| .
\end{aligned}
$$

According to the sufficient condition for monotonicity of trajectories discussed above, $Q = [H_0, H_1, H_2, H_3]$ is monotonic.

We are now ready to prove that a monotonic fragmentation is a sufficient condition for the monotonicity of the solution of the monitoring problem.

Proposition 8.4. *A monitoring problem with a monotonic fragmentation is monotonic.*

Proof. Let $\mu(A) = (a_0, \mathcal{V}, \hat{\mathcal{O}}, \mathcal{R})$ be a monitoring problem, where $\hat{\mathcal{O}} = [\mathcal{F}_1, \ldots, \mathcal{F}_k]$. Based on Def. 8.3, $\mu(A)$ is monotonic iff $\Delta(\mu(\mathcal{A}))$ is monotonic. According to Proposition 8.3, if $\hat{\mathcal{O}}$ is monotonic then the sequence $[H_0, H_1, \ldots, H_k]$ of sets of trajectories of $\mathcal{A}$ relevant to $Bhv(\wp_{[0]}(\mathcal{A})), Bhv(\wp_{[1]}(\mathcal{A})), \ldots, Bhv(\wp_{[k]}(\mathcal{A}))$ is monotonic. We show that $\Delta(\mu(\mathcal{A})) = [\Delta(\wp_{[0]}(\mathcal{A})), \Delta(\wp_{[1]}(\mathcal{A})), \ldots, \Delta(\wp_{[k]}(\mathcal{A}))] = [\Delta_0, \Delta_1, \ldots, \Delta_k]$ is monotonic by induction on $\hat{\mathcal{O}}$.

(Basis) $[\Delta_0, \Delta_1]$ *is monotonic.* In fact, since the empty trajectory belongs to H_0, the empty diagnosis $\emptyset$ belongs to Δ_0. Therefore, each diagnosis $\delta' \in \Delta_1$ is an extension of $\emptyset$, that is, $\emptyset \subseteq \delta'$.

(Induction) If $[\Delta_0, \ldots, \Delta_i]$ *is monotonic, then* $[\Delta_0, \ldots, \Delta_i, \Delta_{i+1}]$ *is monotonic.* Let $\delta' \in \Delta_{i+1}$, and $h' \in H_{i+1}$ such that $h'_{[\mathcal{R}]} = \delta'$. Based on Def. 8.1, $\forall h' \in H_{i+1}$, $\exists h \in H_i$ such that $h \subseteq h'$. Let $h_{[\mathcal{R}]} = \delta$. Since h is a prefix of h', we have $\delta \subseteq \delta'$. In other words, $\forall \delta' \in \Delta_{i+1}$, $\exists \delta \in \Delta_i$ such that $\delta \subseteq \delta'$. □

Corollary 8.2. *A monitoring problem $\mu(\mathcal{A})$ with a stratified fragmentation $\hat{\mathcal{O}}$ is monotonic.*

Proof. Since $\hat{\mathcal{O}}$ is stratified, according to Proposition 8.2, $\hat{\mathcal{O}}$ is monotonic. Hence, according to Proposition 8.4, $\mu(\mathcal{A})$ is monotonic. □

Example 8.12. With reference to the active system $\mathcal{A}_2$ defined in Example 3.3, consider the monitoring problem $\mu(\mathcal{A}_2) = (a_0, \mathcal{V}, \hat{\mathcal{O}}, \mathcal{R})$, where $\mathcal{V}$ and $\mathcal{R}$ are defined in Table 4.1, and the fragmented observation $\hat{\mathcal{O}}$ is displayed on the left side of Fig. 8.5. Based on Def. 8.3, $\mu(\mathcal{A}_2)$ is monotonic if and only if $\Delta(\mu(\mathcal{A}_2))$ is monotonic. According to Theorem 7.1, $\Delta(\mu(\mathcal{A}_2))$ equals the reactive sequence

$$React(\mu(\mathcal{A}_2)) = [\Delta_0, \Delta_1, \Delta_2, \Delta_3],$$

where $\Delta_i = \wp_{[i]}(\mathcal{A}_2)$, $i \in [0 \cdots 3]$. In order to generate the reactive sequence, based on Def. 7.15 and on the partial monitor displayed in Fig. 8.1, we need to compute the monitoring path

$$Path(\mu(\mathcal{A}_2)) = [\mathcal{M}_0, \mathcal{M}_1, \mathcal{M}_2, \mathcal{M}_3].$$

We have

$$\mathcal{M}_0 = \{(N_0, \{\emptyset\})\},$$
$$\mathcal{M}_1 = \{(N_1, \{\emptyset\}), (N_2, \{\{\mathbf{fop}\}\})\},$$
$$\mathcal{M}_2 = \{(N_5, \{\{\mathbf{nor}\}\}), (N_6, \{\{\mathbf{nor}, \mathbf{fcp}\}\}), (N_{10}, \{\{\mathbf{nol}, \mathbf{nor}, \mathbf{fcp}\}, \{\mathbf{nor}, \mathbf{fop}, \mathbf{fcp}\}\})\},$$
$$\mathcal{M}_3 = \{(N_7, \{\{\mathbf{nor}\}\})\}.$$

Based on (7.40) in Def. 7.16, we come up with the following sets of candidate diagnoses:

$$\Delta_0 = \{\emptyset\},$$
$$\Delta_1 = \{\emptyset, \{\mathbf{nol}\}, \{\mathbf{nor}\}, \{\mathbf{nol}, \mathbf{nor}\}, \{\mathbf{fop}\}\},$$
$$\Delta_2 = \{\{\mathbf{nol}, \mathbf{nor}, \mathbf{fcp}\}, \{\mathbf{nor}, \mathbf{fop}, \mathbf{fcp}\}, \{\mathbf{nor}\}, \{\mathbf{ncl}, \mathbf{nor}\}, \{\mathbf{nor}, \mathbf{fcp}\}\},$$
$$\Delta_3 = \{\{\mathbf{nor}\}\}.$$

Based on Def. 8.1, $\Delta(\mu(\mathcal{A}))$ is monotonic. Hence, $\mu(\mathcal{A})$ is monotonic.

Corollary 8.3. *Let $\mathcal{O}$ be a temporal observation. There exists a plain fragmentation $\hat{\mathcal{O}}$ of $\mathcal{O}$ such that each monitoring problem involving $\hat{\mathcal{O}}$ is monotonic.*

Proof. According to Proposition 8.1, there exists a plain fragmentation $\hat{\mathcal{O}}$ of $\mathcal{O}$ that is monotonic. Let $\mu(\mathcal{A})$ be a monitoring problem involving $\hat{\mathcal{O}}$. According to Proposition 8.4, $\mu(\mathcal{A})$ is monotonic. □

Example 8.13. Consider a variation of Example 8.7, where the monitoring problem $\mu(\mathcal{A}_2) = (a_0, \mathcal{V}, \hat{\mathcal{O}}, \mathcal{R})$ involves the plain fragmentation $\hat{\mathcal{O}} = [\mathcal{F}_1, \mathcal{F}_2, \mathcal{F}_3, \mathcal{F}_4]$ displayed on the left side of Fig. 8.6. Depicted next to $\hat{\mathcal{O}}$ are the index spaces of the observations relevant to the subfragmentations of $\hat{\mathcal{O}}$, namely $Isp(\mathcal{O}_{[1]})$, $Isp(\mathcal{O}_{[2]})$, $Isp(\mathcal{O}_{[3]})$, and $Isp(\mathcal{O}_{[4]})$. According to Theorem 7.1, we have

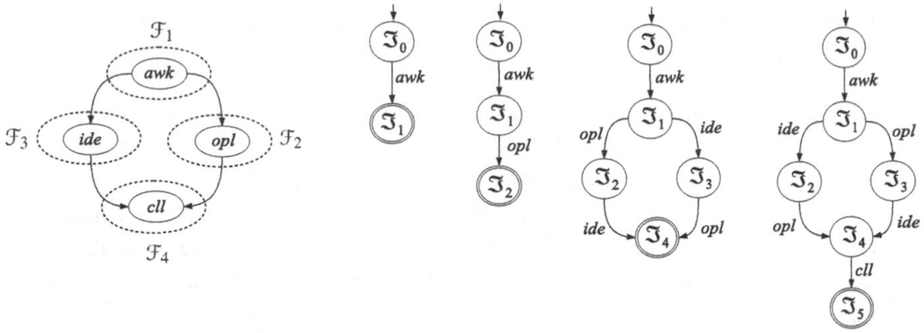

Fig. 8.6 Fragmentation $\hat{\mathcal{O}} = [\mathcal{F}_1, \mathcal{F}_2, \mathcal{F}_3, \mathcal{F}_4]$ (left), and index spaces $Isp(\mathcal{O}_{[1]})$, $Isp(\mathcal{O}_{[2]})$, $Isp(\mathcal{O}_{[3]})$, and $Isp(\mathcal{O}_{[4]})$ (right)

$$\Delta(\mu(\mathcal{A})) = React(\mu(\mathcal{A})) = [\Delta_0, \Delta_1, \Delta_2, \Delta_3, \Delta_4],$$

where $\Delta_i = \wp_{[i]}(\mathcal{A}_2)$, $i \in [0 \cdots 4]$. In order to generate the reactive sequence, based on Def. 7.15 and on the partial monitor displayed in Fig. 8.1, we need to compute the monitoring path

$$Path(\mu(\mathcal{A}_2)) = [\mathcal{M}_0, \mathcal{M}_1, \mathcal{M}_2, \mathcal{M}_3, \mathcal{M}_4].$$

We have

$$\mathcal{M}_0 = \{(N_0, \{\emptyset\})\},$$
$$\mathcal{M}_1 = \{(N_1, \{\emptyset\}), (N_2, \{\{\mathbf{fop}\}\})\},$$
$$\mathcal{M}_2 = \{(N_3, \{\emptyset\}), (N_4, \{\{\mathbf{nor}\}\})\},$$
$$\mathcal{M}_3 = \{(N_5, \{\{\mathbf{nor}\}\}), (N_6, \{\{\mathbf{nor}, \mathbf{fcp}\}\}), (N_{10}, \{\{\mathbf{nol}, \mathbf{nor}, \mathbf{fcp}\}, \{\mathbf{nor}, \mathbf{fop}, \mathbf{fcp}\}\})\},$$
$$\mathcal{M}_4 = \{(N_7, \{\{\mathbf{nor}\}\})\}.$$

Based on (7.40) in Def. 7.16, we have

$$\Delta_0 = \{\emptyset\},$$
$$\Delta_1 = \{\emptyset, \{\mathbf{nol}\}, \{\mathbf{nor}\}, \{\mathbf{nol}, \mathbf{nor}\}, \{\mathbf{fop}\}\},$$
$$\Delta_2 = \{\{\emptyset\}, \{\mathbf{nor}\}\},$$
$$\Delta_3 = \{\{\mathbf{nol}, \mathbf{nor}, \mathbf{fcp}\}, \{\mathbf{nor}, \mathbf{fop}, \mathbf{fcp}\}, \{\mathbf{nor}\}, \{\mathbf{ncl}, \mathbf{nor}\}, \{\mathbf{nor}, \mathbf{fcp}\}\},$$
$$\Delta_4 = \{\{\mathbf{nor}\}\}.$$

Based on Def. 8.1, $\Delta(\mu(\mathcal{A}))$ is monotonic. Hence, $\mu(\mathcal{A})$ is monotonic.

Given a temporal observation $\mathcal{O}$, there always exists a trivial stratification of $\mathcal{O}$ which is composed of just one fragment made up of the whole of $\mathcal{O}$. However, we are interested in nontrivial stratifications only. The proposition below provides a necessary condition for the existence of a nontrivial stratification.

Proposition 8.5. *Let $\mathcal{O}$ be a temporal observation embodying at least two nodes. If a nontrivial stratification exists for $\mathcal{O}$, then $\mathcal{O}$ is connected.*

Proof. By contradiction, assume that $\mathcal{O}$ is disconnected and that a nontrivial stratification $\hat{\mathcal{O}} = [\mathcal{F}_1, \ldots, \mathcal{F}_k]$, where $k > 1$, exists for $\mathcal{O}$. Let $\mathcal{O}$ be composed of d disconnected subgraphs $\mathcal{O}_1, \ldots, \mathcal{O}_d$. Let n be a node in subgraph $\mathcal{O}_j$, $j \in [1 \cdots d]$, and $\mathcal{F}_l$, $l = [1 \cdots k]$, the stratum of $\hat{\mathcal{O}}$ including n. According to Def. 8.4, all nodes in the other subgraphs $\mathcal{O}_i$, $i \neq j$, belong to $\mathcal{F}_l$, as all these nodes are unrelated to n. Moreover, since all nodes in $\mathcal{O}_j$ are unrelated to all nodes in subgraphs $\mathcal{O}_i$, $i \neq j$, all such nodes are included in $\mathcal{F}_l$ too. Consequently, $\mathcal{F}_l$ includes all nodes of $\mathcal{O}$. Hence, the stratification is composed of stratum $\mathcal{F}_l$ only. In other words, the stratification is trivial, a contradiction. □

Note 8.2. Proposition 8.5 asserts that, in order for an observation $\mathcal{O}$ to be nontrivially stratified, $\mathcal{O}$ must be connected. In other words, the connection of $\mathcal{O}$ is a necessary condition for its nontrivial stratification. One may ask whether connection is a sufficient condition too for a nontrivial stratification. The answer is given in Example 8.14.

Example 8.14. Displayed in Fig. 8.7 is a connected observation $\mathcal{O}$ composed of a set N of four nodes. It is easy to check that, for each nontrivial partition $\mathcal{N}$ of N, there exists a part in $\mathcal{N}$ for which condition (8.1) is not fulfilled. In other words, the only stratification of $\mathcal{O}$ is trivial, even if $\mathcal{O}$ is connected. Hence, connection of the observation is not a sufficient condition for the existence of a nontrivial stratification.

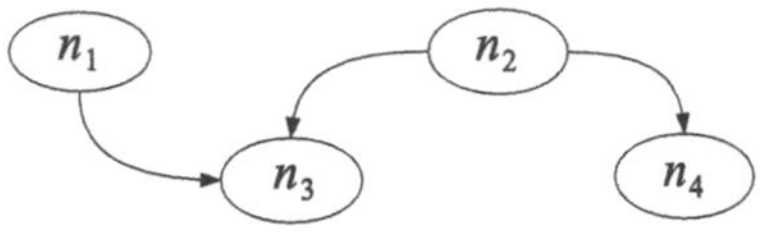

Fig. 8.7 Connected temporal observation without nontrivial stratification

8.4 Minimal Stratification

Given a nontrivial stratified observation $\hat{\mathcal{O}}$, it always possible to obtain another stratification by merging two contiguous strata, that is, two strata $\mathcal{F}$ and $\mathcal{F}'$ such that each arc exiting $\mathcal{F}$ enters a node in $\mathcal{F}'$.

Example 8.15. Consider the stratified fragmentation of $\mathcal{O}$ displayed on the left side of Fig. 8.5, namely $\hat{\mathcal{O}} = [\mathcal{F}_1, \mathcal{F}_2, \mathcal{F}_3]$. A new stratified observation $\hat{\mathcal{O}}'$ can be obtained by merging fragments $\mathcal{F}_1$ and $\mathcal{F}_2$, that is, $\hat{\mathcal{O}}' = [\mathcal{F}_1 \cup \mathcal{F}_2, \mathcal{F}_3]$.

However, if we split a fragment into two fragments, the resulting fragmentation is not guaranteed to be stratified.

Example 8.16. Consider again the stratified fragmentation of $\mathcal{O}$ displayed on the left side of Fig. 8.5, namely $\hat{\mathcal{O}} = [\mathcal{F}_1, \mathcal{F}_2, \mathcal{F}_3]$. If we split $\mathcal{F}_2$ into $\mathcal{F}_2'$ (marked by *ide*) and $\mathcal{F}_2''$ (marked by *opl*), the resulting fragmented observation $\hat{\mathcal{O}}'' = [\mathcal{F}_1, \mathcal{F}_2', \mathcal{F}_2'', \mathcal{F}_3]$ is not stratified, as condition (8.1) is violated as the two nodes in $\mathcal{F}_2'$ and $\mathcal{F}_2''$ are unrelated to one another.

Definition 8.5 (Minimal Stratification). A stratified fragmentation is *minimal* iff it is impossible to split a stratum without losing stratification.

Example 8.17. With reference to Fig. 8.5, consider the stratified fragmentation of $\mathcal{O}$ displayed on the left side, namely $\hat{\mathcal{O}} = [\mathcal{F}_1, \mathcal{F}_2, \mathcal{F}_3]$. Based on Def. 8.5, $\hat{\mathcal{O}}$ is minimal. In fact, the only possibly splittable stratum is $\mathcal{F}_2$, which includes two nodes. However, as pointed out in Example 8.16, the resulting fragmentation is no longer stratified. Hence, $\hat{\mathcal{O}}$ is minimal.

Minimal stratification is convenient not only in reactive diagnosis but also in a posteriori diagnosis. This may sound strange, as fragmented observations have been introduced for monitoring problems only.

Consider a posteriori diagnosis, where the temporal observation $\mathcal{O}$ is received as a whole by the diagnosis engine before the reconstruction of the system behavior is started. If $\mathcal{O}$ is large, the chances are that the reconstruction of the system behavior may require a considerable time before the generation of the solution of the diagnosis problem, namely the (sound and complete) set of candidate diagnoses.

From a practical viewpoint, a more convenient approach is to generate the set of candidate diagnoses incrementally, based on a fragmentation of $\mathcal{O}$. Note how such a fragmentation is not relevant to reactive diagnosis: it is only an artifice aimed at providing diagnosis information in several steps rather than in one shot (at the end of the reconstruction phase).

Since we are seeking monotonic diagnosis, the fragmentation of $\mathcal{O}$ is expected to be a stratification. Moreover, in order to minimize the delay between consecutive diagnosis outputs, such a stratification should be minimal. This way, the diagnosis problem relevant to the observation $\mathcal{O}$ is translated into a monitoring problem relevant to a minimal stratification $\hat{\mathcal{O}}$. Consequently, the solution of the diagnosis problem is translated into the solution of the monitoring problem, with each set of candidate diagnoses being relevant to each subobservation of $\mathcal{O}$.

In the next section we provide an algorithm for the minimal stratification of a temporal observation.

8.4.1 Minimal Stratification Algorithm

Written below is the specification of the algorithm *Minstra*, which takes as input a temporal observation $\mathcal{O}$, and generates a partition $\mathcal{N}$ of the nodes of $\mathcal{O}$ identifying the minimal stratification of $\mathcal{O}$.

At line 5, $\mathcal{N}$ is initialized as the set of singletons involving all nodes $n_1, \ldots, n_l$ in $\mathcal{O}$, where a sequential order is defined for nodes.[3]

Then, each node n_i, $i \in [1 \cdots (l-1)]$, is considered in the main loop (lines 6–17) along with the current part N' that includes n_i (line 7).

For each node n_i, each successive node n_j, $j > i$, is considered in the nested loop (lines 8–16), along with the current part N'' that includes n_j (line 7). Specifically, if n_j is unrelated to n_i, that is, if n_j is neither an ancestor nor a descendant of n_i, then N' is extended by N'' (provided that $N' \neq N''$), with N'' being removed from $\mathcal{N}$ (lines 12 and 13).

Note 8.3. Within the nested loop (lines 8–16), only successive nodes of n_i are considered, because the unrelatedness relationship is symmetric: if n is unrelated to n', then n' is unrelated to n. As a matter of fact, the nodes preceding n_i have been tested already for unrelatedness in previous iterations of the main loop.

The specification of the algorithm *Minstra* is as follows:

```
1.  algorithm Minstra (in 𝒪, out 𝒩)
2.    𝒪 = (L,N,A): a connected temporal observation, where N = {n1,...,nl},
3.    𝒩: a partition of N identifying the minimal stratification of 𝒪;

4.    begin ⟨Minstra⟩
5.      𝒩 := {{n1},...,{nl}};
6.      for i ∈ [1···(l−1)] do
7.        Let N′ be the part in 𝒩 including ni;
8.        for j ∈ [(i+1)···l] do
9.          if nj is unrelated to ni then
10.           Let N″ be the part in 𝒩 including nj;
11.           if N′ ≠ N″ then
12.             N′ := N′ ∪ N″;
13.             Remove N″ from 𝒩
14.           endif
15.         endif
16.       endfor
17.     endfor
18.   end ⟨Minstra⟩
```

Example 8.18. Consider the temporal observation $\mathcal{O}$ displayed in Fig. 8.8, composed of nodes $n_1, \ldots, n_{16}$ (observable labels are omitted). The application of the algorithm *Minstra* to $\mathcal{O}$ is traced in Table 8.4, where the first line indicates the initial configuration of the partition $\mathcal{N}$. The effect of each iteration of the main loop is detailed below:

1. Since n_1 has no unrelated nodes, $\mathcal{N}$ is unchanged.

---

[3] The actual minimal stratification of $\mathcal{O}$ cannot be more refined than this set of singletons.

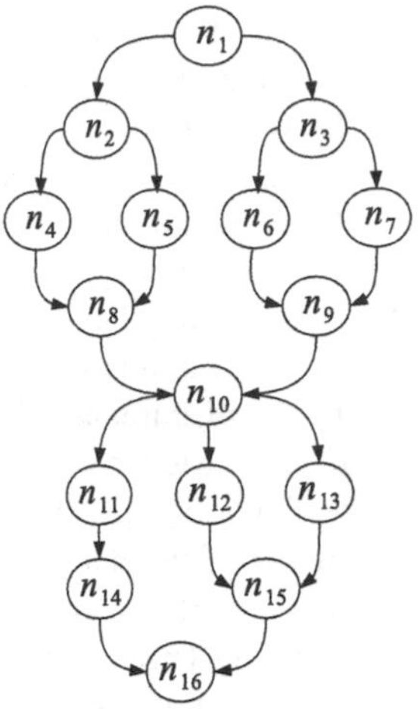

Fig. 8.8 Temporal observation to be stratified by *Minstra* algorithm (observable labels are omitted)

2. The set of unrelated nodes n_j of n_2, $j > 2$, is $\{n_3, n_6, n_7, n_9\}$. Hence, the part $\{n_2\}$ is extended by $\{n_3, n_6, n_7, n_9\}$, thereby becoming $\{n_2, n_3, n_6, n_7, n_9\}$, with the singletons $\{n_3\}$, $\{n_6\}$, $\{n_7\}$, and $\{n_9\}$ being removed.
3. The set of unrelated nodes n_j of n_3, $j > 3$, is $\{n_4, n_5, n_8\}$. Hence, $\{n_2, n_3, n_6, n_7, n_9\}$ is extended by $\{n_4, n_5, n_8\}$, thereby becoming $\{n_2, n_3, n_4, n_5, n_6, n_7, n_8, n_9\}$, with the singletons $\{n_4\}$, $\{n_5\}$, and $\{n_8\}$ being removed.
4. The set of unrelated nodes n_j of n_4, $j > 4$, is $\{n_5, n_6, n_7, n_9\}$. However, all these nodes are in the same part of n_4 already. Hence, $\mathcal{N}$ is unchanged.
5. The set of unrelated nodes n_j of n_5, $j > 5$, is $\{n_6, n_7, n_9\}$. However, all these nodes are in the same part of n_5 already. Hence, $\mathcal{N}$ is unchanged.

Table 8.4 Trace of *Minstra* algorithm applied to the temporal observation displayed in Fig. 8.8

i	$\mathcal{N}$
	$\{\{n_1\},\{n_2\},\{n_3\},\{n_4\},\{n_5\},\{n_6\},\{n_7\},\{n_8\},\{n_9\},\{n_{10}\},\{n_{11}\},\{n_{12}\},\{n_{13}\},\{n_{14}\},\{n_{15}\},\{n_{16}\}\}$
1	$\{\{n_1\},\{n_2\},\{n_3\},\{n_4\},\{n_5\},\{n_6\},\{n_7\},\{n_8\},\{n_9\},\{n_{10}\},\{n_{11}\},\{n_{12}\},\{n_{13}\},\{n_{14}\},\{n_{15}\},\{n_{16}\}\}$
2	$\{\{n_1\},\{n_2,n_3,n_6,n_7,n_9\},\{n_4\},\{n_5\},\{n_8\},\{n_{10}\},\{n_{11}\},\{n_{12}\},\{n_{13}\},\{n_{14}\},\{n_{15}\},\{n_{16}\}\}$
3	$\{\{n_1\},\{n_2,n_3,n_4,n_5,n_6,n_7,n_8,n_9\},\{n_{10}\},\{n_{11}\},\{n_{12}\},\{n_{13}\},\{n_{14}\},\{n_{15}\},\{n_{16}\}\}$
4	$\{\{n_1\},\{n_2,n_3,n_4,n_5,n_6,n_7,n_8,n_9\},\{n_{10}\},\{n_{11}\},\{n_{12}\},\{n_{13}\},\{n_{14}\},\{n_{15}\},\{n_{16}\}\}$
5	$\{\{n_1\},\{n_2,n_3,n_4,n_5,n_6,n_7,n_8,n_9\},\{n_{10}\},\{n_{11}\},\{n_{12}\},\{n_{13}\},\{n_{14}\},\{n_{15}\},\{n_{16}\}\}$
6	$\{\{n_1\},\{n_2,n_3,n_4,n_5,n_6,n_7,n_8,n_9\},\{n_{10}\},\{n_{11}\},\{n_{12}\},\{n_{13}\},\{n_{14}\},\{n_{15}\},\{n_{16}\}\}$
7	$\{\{n_1\},\{n_2,n_3,n_4,n_5,n_6,n_7,n_8,n_9\},\{n_{10}\},\{n_{11}\},\{n_{12}\},\{n_{13}\},\{n_{14}\},\{n_{15}\},\{n_{16}\}\}$
8	$\{\{n_1\},\{n_2,n_3,n_4,n_5,n_6,n_7,n_8,n_9\},\{n_{10}\},\{n_{11}\},\{n_{12}\},\{n_{13}\},\{n_{14}\},\{n_{15}\},\{n_{16}\}\}$
9	$\{\{n_1\},\{n_2,n_3,n_4,n_5,n_6,n_7,n_8,n_9\},\{n_{10}\},\{n_{11}\},\{n_{12}\},\{n_{13}\},\{n_{14}\},\{n_{15}\},\{n_{16}\}\}$
10	$\{\{n_1\},\{n_2,n_3,n_4,n_5,n_6,n_7,n_8,n_9\},\{n_{10}\},\{n_{11}\},\{n_{12}\},\{n_{13}\},\{n_{14}\},\{n_{15}\},\{n_{16}\}\}$
11	$\{\{n_1\},\{n_2,n_3,n_4,n_5,n_6,n_7,n_8,n_9\},\{n_{10}\},\{n_{11},n_{12},n_{13},n_{15}\},\{n_{14}\},\{n_{16}\}\}$
12	$\{\{n_1\},\{n_2,n_3,n_4,n_5,n_6,n_7,n_8,n_9\},\{n_{10}\},\{n_{11},n_{12},n_{13},n_{14},n_{15}\},\{n_{16}\}\}$
13	$\{\{n_1\},\{n_2,n_3,n_4,n_5,n_6,n_7,n_8,n_9\},\{n_{10}\},\{n_{11},n_{12},n_{13},n_{14},n_{15}\},\{n_{16}\}\}$
14	$\{\{n_1\},\{n_2,n_3,n_4,n_5,n_6,n_7,n_8,n_9\},\{n_{10}\},\{n_{11},n_{12},n_{13},n_{14},n_{15}\},\{n_{16}\}\}$
15	$\{\{n_1\},\{n_2,n_3,n_4,n_5,n_6,n_7,n_8,n_9\},\{n_{10}\},\{n_{11},n_{12},n_{13},n_{14},n_{15}\},\{n_{16}\}\}$

6. The set of unrelated nodes n_j of n_6, $j > 6$, is $\{n_7, n_8\}$. However, these two nodes are in the same part of n_6 already. Hence, $\mathcal{N}$ is unchanged.
7. The set of unrelated nodes n_j of n_7, $j > 7$, is $\{n_8\}$. However, n_8 is in the same part of n_7 already. Hence, $\mathcal{N}$ is unchanged.
8. The set of unrelated nodes n_j of n_8, $j > 8$, is $\{n_9\}$. However, n_9 is in the same part of n_6 already. Hence, $\mathcal{N}$ is unchanged.
9. The set of unrelated nodes n_j of n_9, $j > 9$, is empty. Hence, $\mathcal{N}$ is unchanged.
10. The set of unrelated nodes n_j of n_{10}, $j > 10$, is empty. Hence, $\mathcal{N}$ is unchanged.
11. The set of unrelated nodes n_j of n_{11}, $j > 11$, is $\{n_{12}, n_{13}, n_{15}\}$. Hence, the part $\{n_{11}\}$ is extended by $\{n_{12}, n_{13}, n_{15}\}$, thereby becoming $\{n_{11}, n_{12}, n_{13}, n_{15}\}$, with the singletons $\{n_{12}\}$, $\{n_{13}\}$, and $\{n_{15}\}$ being removed.
12. The set of unrelated nodes n_j of n_{12}, $j > 12$, is $\{n_{13}, n_{14}\}$, where n_{13} is in the same part of n_{12} already. Hence, the part $\{n_{12}\}$ is extended by $\{n_{14}\}$ only, thereby becoming $\{n_{11}, n_{12}, n_{13}, n_{14}, n_{15}\}$, with the singleton $\{n_{14}\}$ being removed.
13. The set of unrelated nodes n_j of n_{13}, $j > 13$, is $\{n_{14}\}$. However, n_{14} is in the same part of n_{13} already. Hence, $\mathcal{N}$ is unchanged.
14. The set of unrelated nodes n_j of n_{14}, $j > 14$, is $\{n_{15}\}$. However, n_{15} is in the same part of n_{14} already. Hence, $\mathcal{N}$ is unchanged.
15. The set of unrelated nodes n_j of n_{15}, $j > 15$, is empty. Hence, $\mathcal{N}$ is unchanged.

Eventually, the minimal stratification of $\mathcal{O}$ is identified by a partition (shaded in Table 8.4) composed of five parts, namely

$$\mathcal{N} = \{\{n_1\}, \{n_2, n_3, n_4, n_5, n_6, n_7, n_8, n_9\}, \{n_{10}\}, \{n_{11}, n_{12}, n_{13}, n_{14}, n_{15}\}, \{n_{16}\}\}\ .$$

Example 8.19. Consider the connected observation $\mathcal{O}$ displayed in Fig. 8.7. As claimed in Example 8.14, $\mathcal{O}$ cannot be nontrivially stratified. In other words, the only possible stratification for $\mathcal{O}$ is composed of a single stratum. This can be checked by applying the algorithm *Minstra* to $\mathcal{O}$ as detailed below, where $\mathcal{N}$ is initialized to $\{\{n_1\}, \{n_2\}, \{n_3\}, \{n_4\}\}$.

1. The set of unrelated nodes n_j of n_1, $j > 1$, is $\{n_2, n_3\}$. Hence, the part $\{n_1\}$ is extended into $\{n_1, n_2, n_3\}$, with the singletons $\{n_2\}$ and $\{n_3\}$ being removed.
2. The set of unrelated nodes n_j of n_2, $j > 2$, is empty. Hence, $\mathcal{N}$ is unchanged.
3. The set of unrelated nodes n_j of n_3, $j > 3$, is $\{n_4\}$. Hence, the part $\{n_1, n_2, n_3\}$ is extended into $\{n_1, n_2, n_3, n_4\}$, with the singleton $\{n_4\}$ being removed.

As expected, the minimal stratification of $\mathcal{O}$ is $\mathcal{N} = \{n_1, n_2, n_3, n_4\}$, that is, the trivial stratification.

8.5 Bibliographic Notes

Monotonic monitoring of DESs with uncertain temporal observations was proposed and analyzed in [102, 103, 106, 110].

8.6 Summary of Chapter 8

Monotonic Sequence. The property of a sequence Q of sets of collections (either sets or sequences) such that, for each set S' in Q and for each collection C' in S', there exists a collection C in the set preceding S' in Q such that $C \subseteq C'$.

Subobservation. Given a fragmented observation $\hat{\mathcal{O}} = [\mathcal{F}_1, \ldots, \mathcal{F}_k]$, the temporal observation $\mathcal{O}_{[i]}$, $i \in [0 \cdots k]$, up to the i-th fragment.

Monotonic Fragmentation. A fragmentation $\hat{\mathcal{O}} = [\mathcal{F}_1, \ldots, \mathcal{F}_k]$ where the sequence $[\,\|\mathcal{O}_{[0]}\|, \|\mathcal{O}_{[1]}\|, \ldots, \|\mathcal{O}_{[k]}\|\,]$ is monotonic.

Monotonic Monitoring Problem. A monitoring problem such that its solution is monotonic.

Unrelated Nodes. Given a temporal observation, the nodes between which no temporal precedence is defined.

Stratified Fragmentation. A fragmentation where unrelated nodes are included in the same fragment (called a stratum).

Minimal Stratification. A stratification in which no fragment can be split without losing stratification.

Chapter 9
Reusable Diagnosis

This chapter was contributed by Federica Vivenzi[1]

In Chapter 5 we introduced a technique for a posteriori diagnosis, where the temporal observation is available as a whole to the diagnosis engine. A variant of such a technique was presented in Chapter 6, where the reconstruction of the system behavior is accomplished in a modular way, rather than monolithically. Both approaches consist of three phases: behavior reconstruction, behavior decoration, and diagnosis distillation. What distinguishes modular diagnosis from monolithic diagnosis is the mode in which behavior reconstruction is accomplished, either in one shot or in several steps (based on merging techniques), respectively.

A diagnosis problem for a system $\mathcal{A}$ is a quadruple $\wp(\mathcal{A}) = (a_0, \mathcal{V}, \mathcal{O}, \mathcal{R})$, where a_0 is the initial state of $\mathcal{A}$, $\mathcal{V}$ a ruler for $\mathcal{A}$, $\mathcal{O}$ a temporal observation of $\mathcal{A}$, and $\mathcal{R}$ a ruler for $\mathcal{A}$.

In behavior reconstruction, the behavior of $\wp(\mathcal{A})$ is generated by reasoning on the model of $\mathcal{A}$, based on the observation $\mathcal{O}$ and viewer $\mathcal{V}$. Behavior decoration and diagnosis distillation are carried out on the reconstructed behavior based on the ruler $\mathcal{R}$.

In all cases, given a new diagnosis problem, all three phases need to be accomplished from scratch in order to produce the diagnosis solution, this being the set of candidate diagnoses.

The question we address in this chapter is: *Is it always necessary for the diagnosis engine to perform all three phases (reconstruction, decoration, and distillation) to solve the diagnosis problem?*

Apparently, the answer is *yes*. This is the correct answer as long as the diagnosis engine works without memory of the past, that is, without considering the nature of previously solved diagnosis problems. By contrast, if we assume the diagnosis engine to have memory of previous diagnosis sessions, the answer is *no*.

To understand why, consider the trivial case in which a new diagnosis problem $\wp'(\mathcal{A})$ equals a diagnosis problem $\wp(\mathcal{A})$ already solved in the past. In other words, assume that, given $\wp'(\mathcal{A}) = (a'_0, \mathcal{V}', \mathcal{O}', \mathcal{R}')$ to be solved, $\wp(\mathcal{A}) = (a_0, \mathcal{V}, \mathcal{O}, \mathcal{R})$ has

[1] Federica Vivenzi carried out research with the authors on this topic at the University of Brescia.

G. Lamperti et al., *Introduction to Diagnosis of Active Systems*,
https://doi.org/10.1007/978-3-319-92733-6_9

already been solved, such that $a_0' = a_0$, $\mathcal{V} = \mathcal{V}'$, $\mathcal{O} = \mathcal{O}'$, and $\mathcal{R} = \mathcal{R}'$. Clearly, the solution of $\wp'(\mathcal{A})$ equals the solution of $\wp(\mathcal{A})$.

Provided that we keep the solutions of previous diagnosis problems, solving $\wp'(\mathcal{A})$ amounts to merely *reusing* the solution of $\wp(\mathcal{A})$ as is. That is, the diagnosis engine is able to solve $\wp'(\mathcal{A})$ by skipping all the three phases it normally performs.

One may argue that $\wp(\mathcal{A}) = \wp'(\mathcal{A})$ is too unrealistic an assumption in real applications. As such, this reuse-based technique is unlikely to be carried out in practice.

We agree with this objection. However, even if it is not identical, the chances are that the new diagnosis problem $\wp'(\mathcal{A})$ is *similar* to a previously solved problem $\wp(\mathcal{A})$. Similarity refers to any of the four elements of the problem, namely the initial state, the viewer, the temporal observation, and the ruler.

Assuming the same initial state, namely $a_0 = a_0'$, is realistic, as this normally refers to the state of the system when it starts operating. On the other hand, generally speaking, if $a_0 \neq a_0'$ then the behavior of $\wp(\mathcal{A})$ will have very little (if anything) to do with the behavior of $\wp'(\mathcal{A})$. Consequently, the solution of $\wp(\mathcal{A})$ is bound to be unrelated to the solution of $\wp'(\mathcal{A})$. This is why we consider $a_0 = a_0'$ a prerequisite for the reuse of $\wp(\mathcal{A})$ in order to solve $\wp'(\mathcal{A})$.

Hence, similarity actually refers to the other three parameters, namely $\mathcal{V}$, $\mathcal{O}$, and $\mathcal{R}$. The initial question can now be reformulated as follows: *What are the conditions under which $\wp(\mathcal{A}) = (a_0, \mathcal{V}, \mathcal{O}, \mathcal{R})$ can be reused to solve $\wp'(\mathcal{A}) = (a_0, \mathcal{V}', \mathcal{O}', \mathcal{R}')$?*

First, we consider the scenario in which, besides the initial state, $\wp(\mathcal{A})$ and $\wp'(\mathcal{A})$ share the same viewer and observation, while $\mathcal{R} \neq \mathcal{R}'$. Since $\mathcal{V}$ and $\mathcal{O}$ are the same in the two problems, the reconstructed behavior will also be the same, as this is generated independently from the ruler. Therefore, in order to solve $\wp'(\mathcal{A})$, the diagnosis engine can skip behavior reconstruction and jump directly to behavior decoration, where the behavior being decorated is (a copy of) $Bhv(\wp(\mathcal{A}))$.

This represents a conspicuous saving of computation, as behavior reconstruction is the most complex phase in the diagnosis process. However, a natural question that arises at this point is whether there exists a particular relationship between $\mathcal{R}$ and $\mathcal{R}'$ such that behavior decoration too can be skipped, so that the solution of $\wp'(\mathcal{A})$ amounts to mapping the solution of $\wp(\mathcal{A})$ to $\mathcal{R}'$.

In the simplest case, in which the set of faulty transitions involved in $\mathcal{R}$ equals the set of faulty transitions involved in $\mathcal{R}'$, this mapping can be straightforwardly accomplished by renaming each fault label in the solution of $\wp(A)$ as the corresponding fault label in $\mathcal{R}'$.

In the most general case (when the set of faulty transitions in $\mathcal{R}$ differs from the set of faulty transitions in $\mathcal{R}'$), two conditions are expected to be fulfilled:

1. The set of transitions involved in $\mathcal{R}$ contains the set of transitions involved in $\mathcal{R}'$.
2. For each association $(t(c), f) \in \mathcal{R}$ such that $(t(c), f') \in \mathcal{R}$, the set of transitions associated with f' in $\mathcal{R}'$ contains the set of transitions associated with f in $\mathcal{R}$.

If the first condition is violated, there is no guarantee that a faulty transition in $\mathcal{R}'$ which is not involved in $\mathcal{R}$ will be represented by the associated fault label in $\mathcal{R}'$.

The reason for the second condition can be explained by an abstract example. Let $T' = \{t_1(c_1), t_2(c_2)\}$ be the set of faulty transitions associated with f' in $\mathcal{R}'$. Let

$T = \{t_1(c_1), t_2(c_2), t_3(c_3)\}$ be the set of faulty transitions associated with f in $\mathcal{R}$. Clearly, the second condition is violated, the reason being that $T' \not\supseteq T$, precisely because $t_3(c_3) \notin T'$. Consider a diagnosis $\delta \in \Delta(\wp(\mathcal{A}))$ such that $f \in \delta$, $\delta = h_{[\mathcal{R}]}$, $t_3(c_3) \in h$, $t_1(c_1) \notin h$, $t_2(c_2) \notin h$. As such, δ is the candidate diagnosis generated by a trajectory h which contains $t_3(c)$, without containing $t_1(c_1)$ or $t_2(c_2)$. Hence, the label $f \in \delta$ is due only to the transition $t_3(c_3) \in h$.

To generate the solution $\Delta(\wp'(\mathcal{A}))$ we need to map each candidate diagnosis in $\Delta(\wp(\mathcal{A}))$ to the corresponding label in $\mathcal{R}'$. The question is: considering δ, which is the label in $\mathcal{R}'$ corresponding to f? In theory, this is expected to be the label associated with $t_3(c_3)$ in $\mathcal{R}'$. Two cases are possible:

1. $t_3(c)$ is not involved in $\mathcal{R}'$; that is, $t_3(c)$ is not cataloged as faulty in $\wp'(\mathcal{A})$,
2. $t_3(c)$ is associated in $\mathcal{R}'$ with a fault label f''.

In the first case, we could map f to either nothing (correct) or f' (incorrect). The point is that we are mapping δ, not h. Consequently, we do not know the actual transition associated with f in δ. In the second case, we could map f to either f'' (correct) or f' (incorrect). Even in this case, we do not know the actual transition associated with f in δ. Hence, in either case, it is impossible to know the correct mapping of f to $\mathcal{R}'$.

By contrast, assume that the second condition on the relationship between $\mathcal{R}$ and $\mathcal{R}'$ is true. For example, let $T' = \{t_1(c_1), t_2(c_2), t_3(c_3)\}$ be the set of faulty transitions associated with f' in $\mathcal{R}'$, and $T = \{t_1(c_1), t_2(c_2)\}$ the set of faulty transitions associated with f in $\mathcal{R}$. Hence, $T' \supseteq T$. Assume that $\delta \in \Delta(\wp(\mathcal{A}))$, $f \in \delta$. In this case, the mapping of f to $\mathcal{R}'$ is f' and only f'. In fact, since $T' \supseteq T$, whichever transition $t(c) \in T$ generates f, the fault label associated with $t(c)$ in $\mathcal{R}'$ is invariably f'.

When the relationship between $\mathcal{R}$ and $\mathcal{R}'$ fulfills the two conditions listed above, we say that $\mathcal{R}$ *subsumes* $\mathcal{R}'$, written $\mathcal{R} \Supset \mathcal{R}'$.

In summary, given two similar diagnosis problems $\wp(\mathcal{A}) = (a_0, \mathcal{V}, \mathcal{O}, \mathcal{R})$ and $\wp'(\mathcal{A}) = (a_0, \mathcal{V}, \mathcal{O}, \mathcal{R}')$, the solution of $\wp'(\mathcal{A})$ can be generated by mapping each fault label in each candidate diagnosis of $\wp(A)$ to the corresponding (possibly empty) fault label in $\mathcal{R}'$.

Interestingly, diagnosis reuse is not limited to ruler subsumption. The subsumption relationship can be defined for temporal observations too.

Assume that $\wp(\mathcal{A}) = (a_0, \mathcal{V}, \mathcal{O}, \mathcal{R})$ and $\wp'(\mathcal{A}) = (a_0, \mathcal{V}, \mathcal{O}', \mathcal{R})$, with the behavior of the previously-solved $\wp(\mathcal{A})$ being available to the diagnosis engine. As such, $\wp'(A)$ differs from $\wp(\mathcal{A})$ in the temporal observation $\mathcal{O}'$ only.

As we know, the extension of the behavior of $\wp(\mathcal{A})$ includes the trajectories h of $\mathcal{A}$ that conform to $\mathcal{O}$, that is, such that $h_{[\mathcal{V}]}$ is a candidate trace in $\|\mathcal{O}\|$. The same applies to the behavior of $\wp'(\mathcal{A})$, which contains the trajectories h' such that $h'_{[\mathcal{V}]}$ is a candidate trace in $\|\mathcal{O}'\|$.

Now, suppose $\|\mathcal{O}\| \supseteq \|\mathcal{O}'\|$, that is, each candidate trace of $\mathcal{O}'$ is also a candidate trace of $\mathcal{O}'$. If so, each trajectory in the behavior of $\wp'(\mathcal{A})$ is also a trajectory in the behavior of $\wp(\mathcal{A})$. In other words, $\|Bhv(\wp(\mathcal{A}))\| \supseteq \|Bhv(\wp'(\mathcal{A}))\|$. We say that $Bhv(\wp(\mathcal{A}))$ *subsumes* $Bhv(\wp'(\mathcal{A}))$, written $Bhv(\wp(\mathcal{A})) \Supset Bhv(\wp'(\mathcal{A}))$.

Behavior subsumption is an interesting property, as the behavior of $\wp'(\mathcal{A})$ can be generated by selecting in the behavior of $\wp(\mathcal{A})$ the trajectories conforming to $\mathcal{O}'$. $Bhv(\wp(\mathcal{A}))$ is the portion of the behavior space of $\mathcal{A}$ that contains all the behavior of $\wp'(\mathcal{A})$. As such, $Bhv(\wp(\mathcal{A}))$ can be matched with $\mathcal{O}'$ in order to generate $Bhv(\wp'(\mathcal{A}))$.

Instead of reconstructing $Bhv(\wp'(\mathcal{A}))$ by reasoning on the model of $\mathcal{A}$ in terms of system topology and component models, $Bhv(\wp'(\mathcal{A}))$ can be generated directly as a matching between $Bhv(\wp(\mathcal{A}))$ and $\mathcal{O}'$. The advantage comes from the fact that such a matching is computationally less expensive than the low-level model-based reasoning performed by the reconstruction algorithm.

When $\|\mathcal{O}\| \supseteq \|\mathcal{O}'\|$, we say that $\mathcal{O}$ *subsumes* $\mathcal{O}'$, written $\mathcal{O} \Supset \mathcal{O}'$. Therefore, given $\wp(\mathcal{A}) = (a_0, \mathcal{V}, \mathcal{O}, \mathcal{R})$ and $\wp'(\mathcal{A}) = (a_0, \mathcal{V}, \mathcal{O}', \mathcal{R})$, where $\mathcal{O} \Supset \mathcal{O}'$, $Bhv(\wp'(\mathcal{A}))$ is the DFA resulting from the matching of $Bhv(\wp(\mathcal{A}))$ with $\mathcal{O}'$.

One may ask whether diagnosis reuse can be still achieved when the viewers $\mathcal{V}$ and $\mathcal{V}'$ are different. Assume that $\wp(\mathcal{A}) = (a_0, \mathcal{V}, \mathcal{O}, \mathcal{R})$ and $\wp'(\mathcal{A}) = (a_0, \mathcal{V}', \mathcal{O}', \mathcal{R})$. Since $\mathcal{V} \neq \mathcal{V}'$, candidate traces in $\mathcal{O}$ are incomparable with candidate traces in $\mathcal{O}'$. Hence, generally speaking, no subsumption relationship exists between $\mathcal{O}$ and $\mathcal{O}'$.

However, consider the following scenario: for each candidate trace $\mathcal{T}' \in \|\mathcal{O}'\|$ there exists a candidate trace $\mathcal{T} \in \|\mathcal{O}\|$ such that the set of trajectories of $\mathcal{A}$ consistent with $\mathcal{T}$ contains the set of trajectories of $\mathcal{A}$ consistent with $\mathcal{T}'$. If so, behavior subsumption occurs, namely $Bhv(\wp(\mathcal{A})) \Supset Bhv(\wp'(\mathcal{A}))$. Again, $Bhv(\wp'(\mathcal{A}))$ can be generated by matching $Bhv(\wp(\mathcal{A}))$ with $\mathcal{O}'$.

The question is: *What are the relationships between $\mathcal{V}$, $\mathcal{V}'$, and $\mathcal{O}$, $\mathcal{O}'$, such that $Bhv(\wp(\mathcal{A})) \Supset Bhv(\wp'(\mathcal{A}))$?*

Based on the scenario discussed above, such relationships are expected to guarantee that for each trace $\mathcal{T}'$ in $\mathcal{O}'$ there exists a trace $\mathcal{T}$ in $\mathcal{O}$ such that the set of trajectories generating $\mathcal{T}$ is a superset of the set of trajectories generating $\mathcal{T}'$.

Notice that the definition of a viewer is structurally identical to the definition of a ruler, namely a set of associations between a component transition and a label (in the case of a viewer, the latter is an observable label). Therefore, the notion of viewer subsumption can be defined exactly like the notion of ruler subsumption. Specifically, $\mathcal{V}'$ *subsumes* $\mathcal{V}$, written $\mathcal{V}' \Supset \mathcal{V}$, iff the following conditions hold:

1. The set of transitions involved in $\mathcal{V}'$ contains the set of transitions involved in $\mathcal{V}$.
2. For each association $(t(c), \ell') \in \mathcal{V}'$ such that $(t(c), \ell) \in \mathcal{V}$, the set of transitions associated with ℓ in $\mathcal{V}$ contains the set of transitions associated with ℓ' in $\mathcal{V}'$.

Assume that $\mathcal{V}' \Supset \mathcal{V}$, and consider a trace $\mathcal{T}' \in \|\mathcal{O}'\|$. Can we expect that there exists a trace $\mathcal{T} \in \|\mathcal{O}\|$ such that the set of trajectories generating $\mathcal{T}$ is a superset of the set of trajectories generating $\mathcal{T}'$?

Intuitively, $\mathcal{V}' \Supset \mathcal{V}$ means that the label ℓ in $\mathcal{V}$ corresponding to a label ℓ' in $\mathcal{V}'$ is less constraining than ℓ', in other words, the set of component transitions consistent with ℓ contains the set of component transitions consistent with ℓ'.

For example, let $T = \{t_1(c_1), t_2(c_2), t_3(c)\}$ be the set of observable transitions associated with ℓ in $\mathcal{V}$, and $T' = \{t_1(c_1), t_2(c_2)\}$ the set of observable transitions associated with ℓ' in $\mathcal{V}'$. Since $T \supseteq T'$, ℓ is less constraining than ℓ'. In fact, besides $t_1(c_1)$ and $t_2(c_2)$, ℓ is consistent with $t_3(c_3)$ too.

Given a trace $\mathcal{T}' \in \|\mathcal{O}'\|$, consider the mapping of $\mathcal{T}'$ to $\mathcal{V}'$, resulting in a trace $\mathcal{T}$, where each label $\ell' \in \mathcal{T}'$ is replaced by either ε (when no transition associated with ℓ' in $\mathcal{V}'$ is involved in $\mathcal{V}$) or ℓ (the label associated in $\mathcal{V}$ with all the transitions associated with ℓ' in $\mathcal{V}'$).

Without the removal of labels ε, $\mathcal{T}$ is isomorphic to $\mathcal{T}'$. Since each label in $\mathcal{T}$ is less constraining than the corresponding label in $\mathcal{T}'$, it follows that $\mathcal{T}$ is less constraining than $\mathcal{T}'$. Hence, the set of trajectories consistent with $\mathcal{T}$ is a superset of the trajectories consistent with $\mathcal{T}'$. In other words, $Bhv(\wp(\mathcal{A})) \Supset Bhv(\wp'(\mathcal{A}))$.

However, a question is still dangling: *What is the condition such that, for each $\mathcal{T}' \in \|\mathcal{O}'\|$, there exists $\mathcal{T} \in \|\mathcal{O}\|$, with $\mathcal{T}$ being less constraining than $\mathcal{T}'$?*

Assuming $\mathcal{V}' \Supset \mathcal{V}$, it suffices that the mapping of each candidate trace in $\mathcal{O}'$ to $\mathcal{V}$ is also a candidate trace in $\mathcal{O}$. This way, we are guaranteed that each trajectory in $Bhv(\wp'(\mathcal{A}))$ is also a trajectory in $Bhv(\wp(\mathcal{A}))$.

Formally, this condition is expressed by $\mathcal{O} \Supset \mathcal{O}'_{[\mathcal{V}]}$, where $\mathcal{O}'_{[\mathcal{V}]}$ is the *projection* of the observation $\mathcal{O}'$ on the viewer $\mathcal{V}$, that is, the observation obtained from $\mathcal{O}'$ by mapping each label in $\mathcal{O}'$ to either the corresponding label in $\mathcal{V}$ or ε.

So, if $\wp(\mathcal{A}) = (a_0, \mathcal{V}, \mathcal{O}, \mathcal{R})$ and $\wp'(\mathcal{A}) = (a_0, \mathcal{V}', \mathcal{O}', \mathcal{R})$,[2] where $\mathcal{V}' \Supset \mathcal{V}$ and $\mathcal{O} \Supset \mathcal{O}'_{[\mathcal{V}]}$, then $Bhv(\wp(\mathcal{A})) \Supset Bhv(\wp'(\mathcal{A}))$. As such, $Bhv(\wp'(\mathcal{A}))$ can be generated by matching $Bhv(\wp(\mathcal{A}))$ with $\mathcal{O}'$. We reuse $Bhv(\wp(\mathcal{A}))$ to generate $Bhv(\wp'(\mathcal{A}))$.

9.1 Subsumption

As discussed above, in order to solve a diagnosis problem $\wp(\mathcal{A}) = (a_0, \mathcal{V}, \mathcal{O}, \mathcal{R})$ by reusing a previously solved diagnosis problem $\wp'(\mathcal{A}) = (a_0, \mathcal{V}', \mathcal{O}', \mathcal{R}')$, a subsumption relationship is required between corresponding elements of the diagnosis problems, specifically between viewers, temporal observations, and rulers (in addition, the initial states are required to be equal).

Definition 9.1 agrees with the fact that diagnosis reuse requires $\mathcal{V}'$ to subsume $\mathcal{V}$, and not vice versa.

Definition 9.1 (Viewer Subsumption). Let $\mathcal{V}$ and $\mathcal{V}'$ be two viewers for an active system. $\mathcal{V}'$ *subsumes* $\mathcal{V}$, written $\mathcal{V}' \Supset \mathcal{V}$, iff the following conditions hold:

1. The set of transitions involved in $\mathcal{V}'$ contains all the transitions involved in $\mathcal{V}$,
2. For each association $(t, \ell') \in \mathcal{V}'$ such that $(t, \ell) \in \mathcal{V}$, the set of transitions associated with ℓ in $\mathcal{V}$ contains all the transitions associated with ℓ' in $\mathcal{V}'$.

The rationale for the two conditions in Def. 9.1 can be grasped by contradiction. If the first condition is false, then there exists a transition $t(c)$ observable in $\mathcal{V}$ which is not observable in $\mathcal{V}'$. Consequently, generally speaking, the behavior of $\wp(\mathcal{A})$

[2] Since the ruler is irrelevant to the behavior, the ruler of $\wp'(\mathcal{A})$ may differ from the ruler of $\wp(\mathcal{A})$.

Table 9.1 Viewers $\mathcal{V}$ and $\mathcal{V}'$ and rulers $\mathcal{R}$ and $\mathcal{R}'$ for system $\mathcal{A}_2$

$t(c)$	$\mathcal{V}$	$\mathcal{V}'$	$\mathcal{R}$	$\mathcal{R}'$
$b_1(l)$	*opl*	*lopen*		
$b_2(l)$	*cll*	*lclose*		
$b_3(l)$			**nol**	**bfail**
$b_4(l)$			**ncl**	**bfail**
$b_5(l)$		*lnop*		
$b_6(l)$		*lnop*		
p_1	*awk*	*awake*		
p_2	*ide*	*idle*		
p_3	*awk*	*abawk*	**fop**	**pfail**
p_4	*ide*	*abide*	**fcp**	**pfail**
$b_1(r)$	*opr*	*ropen*		
$b_2(r)$	*clr*	*rclose*		
$b_3(r)$			**nor**	**bfail**
$b_4(r)$			**ncr**	**bfail**
$b_5(r)$		*rnop*		
$b_6(r)$		*rnop*		

will be constrained by the observable label associated with t in $\mathcal{V}$. By contrast, the behavior of $\wp'(\mathcal{A})$ is not constrained by any label associated with t. Therefore, the behavior of $\wp'(\mathcal{A})$ is bound to include some trajectories which are not in the behavior of $\wp(\mathcal{A})$. Hence, based on what we have pointed out above, $Bhv(\wp(\mathcal{A}))$ cannot be reused to generate $Bhv(\wp'(\mathcal{A}))$.

If the second condition in Def. 9.1 is false, then, for an association $(t, \ell') \in \mathcal{V}'$ such that $(t, \ell) \in \mathcal{V}$, there exists an association $(t', \ell') \in \mathcal{V}', t \neq t'$, such that $(t', \bar{\ell}) \in \mathcal{V}$, $\bar{\ell} \neq \ell'$.[3]

Assume for simplicity that $\wp(\mathcal{A})$ and $\wp'(A)$ share the same actual trajectory, and that $\mathcal{O}$ and $\mathcal{O}'$ equal the actual respective traces. If the actual trajectory includes the component transition t', then $\mathcal{O}'$ will include ℓ' while $\mathcal{O}$ will include $\bar{\ell}$.

Therefore, based on ℓ' (generated by transition t'), the behavior of $\wp'(\mathcal{A})$ is bound to include at least two candidate trajectories, one involving t and another involving t' (the actual transition).

By contrast, for the corresponding label $\bar{\ell}$ in $\mathcal{O}$, the behavior of $\wp(\mathcal{A})$ is bound to include the trajectory involving t' (the actual transition), but not the trajectory involving t, because t is associated with ℓ rather than $\bar{\ell}$. Consequently, $Bhv(\wp(\mathcal{A}))$ is bound to miss some trajectories included in $Bhv(\wp'(\mathcal{A}))$. As such, $Bhv(\wp(\mathcal{A}))$ cannot be reused to generate $Bhv(\wp'(\mathcal{A}))$.

Example 9.1. With reference to the active system $\mathcal{A}_2$ introduced in Example 4.6, displayed on the left side of Table 9.1 is the specification of the viewers $\mathcal{V}$ and $\mathcal{V}'$, where $\mathcal{V}$ is inherited from Table 4.1, while $\mathcal{V}'$ is a different viewer. According to Def. 9.1, $\mathcal{V}' \Supset \mathcal{V}$. In fact, (1) the set of transitions involved in $\mathcal{V}'$ contains all the

[3] The reason being that the first condition in Def. 9.1 requires that all observable transitions in $\mathcal{V}$ are observable in $\mathcal{O}'$ too.

transitions involved in $\mathcal{V}$, and (2) for each association $(t,\ell') \in \mathcal{V}'$ such that $(t,\ell) \in \mathcal{V}$, the set of transitions associated with ℓ in $\mathcal{V}$ contains all the transitions associated with ℓ' in $\mathcal{V}'$. For instance, considering point (2), for the association $(p_1, awake) \in \mathcal{V}'$, the set of transitions associated with *awk* in $\mathcal{V}$ are $\{p_1, p_3\}$, which is a superset of the set of transitions associated with *awake* in $\mathcal{V}'$, namely the singleton $\{p_1\}$.

Proposition 9.1. *Viewer subsumption is reflexive.*

Proof. Assuming $\mathcal{V} = \mathcal{V}'$, both conditions hold in Def. 9.1, where the containment relationship becomes an equality. □

In order to define the notion of a subsumption relationship between observations, we need to introduce the notion of observation projection (Def. 9.3), which is based on the notion of observable-label renaming (Def. 9.2).

Definition 9.2 (Observable-Label Renaming). Let $\mathcal{V}$ and $\mathcal{V}'$ be two viewers for an active system, with Ω and Ω' being the corresponding domains of observable labels, where $\mathcal{V}' \Supset \mathcal{V}$. The *renaming* of an observable label $\ell' \in \Omega'$ based on $\mathcal{V}$ is an observable label in $\Omega \cup \{\varepsilon\}$ defined as follows:

$$Ren(\ell', \mathcal{V}) = \begin{cases} \ell & \text{if } (t,\ell') \in \mathcal{V}', (t,\ell) \in \mathcal{V}, \\ \varepsilon & \text{otherwise .} \end{cases} \tag{9.1}$$

Example 9.2. With reference to the viewers $\mathcal{V}$ and $\mathcal{V}'$ defined in Table 9.1 (Example 9.1), where $\mathcal{V}' \Supset \mathcal{V}$, we have $Ren(abawk, \mathcal{V}) = awk$, and $Ren(lnop, \mathcal{V}) = \varepsilon$.

As pointed out in the introduction to this chapter, given two diagnosis problems $\wp(\mathcal{A}) = (a_0, \mathcal{V}, \mathcal{O}, \mathcal{R})$ and $\wp'(\mathcal{A}) = (a_0, \mathcal{V}', \mathcal{O}', \mathcal{R}')$, the behavior of $\wp'(\mathcal{A})$ can be generated by reusing the behavior of $\wp(\mathcal{A})$ if $\mathcal{V}'$ subsumes $\mathcal{V}$ and $\mathcal{O}$ subsumes the projection of $\mathcal{O}'$ on $\mathcal{V}$. The notion of observation projection is formalized in Def. 9.3.

Definition 9.3 (Observation Projection). Let $\mathcal{O}'$ be a temporal observation for a system $\mathcal{A}$, based on a viewer $\mathcal{V}'$, and N' the set of nodes of $\mathcal{O}'$. Let $\mathcal{V}$ be another viewer for $\mathcal{A}$ such that $\mathcal{V}' \Supset \mathcal{V}$. The *projection* of a node n' in $\mathcal{O}'$ on $\mathcal{V}$, denoted $n'_{[\mathcal{V}]}$, is the node n that includes a label ℓ for each label ℓ' in n', where[4]

$$\ell = \begin{cases} \varepsilon & \text{if } \ell' = \varepsilon, \\ Ren(\ell', \mathcal{V}) & \text{otherwise .} \end{cases} \tag{9.2}$$

The *projection* of $\mathcal{O}'$ on $\mathcal{V}$, denoted $\mathcal{O}'_{[\mathcal{V}]}$, is a temporal observation $\mathcal{O}$ based on $\mathcal{V}$ with the set N of nodes being defined as follows:

$$N = \{n \mid n' \in N', n = n'_{[\mathcal{V}]}, \|n\| \neq \{\varepsilon\}\}, \tag{9.3}$$

while the set of arcs A is defined as follows: $n_1 \to n_2 \in A$, where $n_1 = n'_{1[\mathcal{V}]}$ and $n_2 = n'_{2[\mathcal{V}]}$, iff:

[4] The number of labels in n may decrease because of possible duplicate removals.

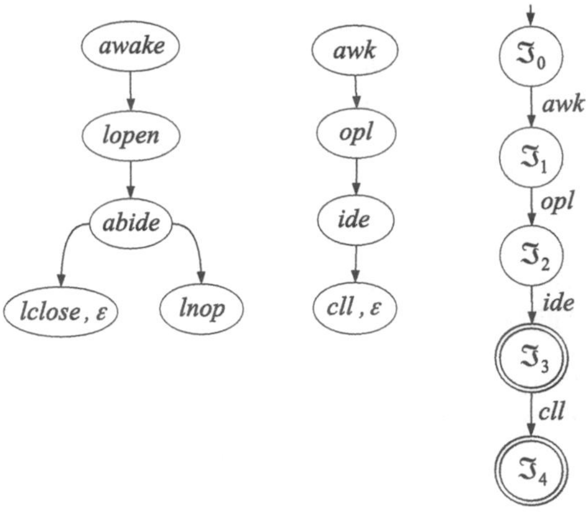

Fig. 9.1 Temporal observation $\mathcal{O}'$ (left) based on viewer $\mathcal{V}'$ (Table 9.1), the relevant projection $\mathcal{O}'_{[\mathcal{V}]}$ on viewer $\mathcal{V}$ (center), and corresponding index space $Isp(\mathcal{O}'_{[\mathcal{V}]})$ (right)

1. (*Precedence*) $n'_1 \prec n'_2$ in $\mathcal{O}'$,
2. (*Canonicity*) $\nexists n_3 \in N, n_3 = n'_{3[\mathcal{V}]}$, such that $n'_1 \prec n'_3 \prec n'_2$ in $\acute{\mathcal{O}}'$.

Intuitively, the projection of the observation $\mathcal{O}'$ on the viewer $\mathcal{V}$ is obtained by renaming all labels (within nodes of $\mathcal{O}'$) based on the viewer $\mathcal{V}$, where the resulting nodes with an extension equal to $\{\varepsilon\}$ are removed, but preserving, however, the temporal precedences between the other nodes, as well as the canonical form of the resulting observation.

Example 9.3. With reference to the viewers $\mathcal{V}$ and $\mathcal{V}'$ defined in Table 9.1, where $\mathcal{V}' \Supset \mathcal{V}$, the left side of Fig. 9.1 shows a temporal observation $\mathcal{O}'$ based on $\mathcal{V}'$. The projection of $\mathcal{O}'$ on the viewer $\mathcal{V}$ according to Def. 9.3, namely $\mathcal{O}'_{[\mathcal{V}]}$, is shown on the right side of Fig. 9.1. Notably, since the renaming of *lnop* on $\mathcal{V}$ is ε, the corresponding projected node (marked by ε only) is not included in $\mathcal{O}'_{[\mathcal{V}]}$.

As asserted by Proposition 5.1, the extension of a temporal observation $\mathcal{O}$ equals the language of the index space of $\mathcal{O}$, namely $\|\mathcal{O}\| = \|Isp(\mathcal{O})\|$. This means that each candidate trace in $\mathcal{O}$ is a string in the language of $Isp(\mathcal{O})$, and vice versa. So, we can extend the notion of projection to the index space, as formalized in Def. 9.4.

Definition 9.4 (Index-Space Projection). Let $\mathcal{O}'$ be a temporal observation of a system $\mathcal{A}$, based on a viewer $\mathcal{V}'$. Let $\mathcal{V}$ be another viewer of $\mathcal{A}$ such that $\mathcal{V}' \Supset \mathcal{V}$. The *projection* of the index space $Isp(\mathcal{O}')$ on $\mathcal{V}$, denoted $Isp_{[\mathcal{V}]}(\mathcal{O}')$, is the DFA obtained from $Isp(\mathcal{O}')$ as follows:

1. Replace each label in $Isp(\mathcal{O}')$ with the new label $\ell = Ren(\ell', \mathcal{V})$, thereby obtaining an NFA $Isp^{\mathrm{n}}(\mathcal{O}')$,
2. Determinize $Isp^{\mathrm{n}}(\mathcal{O}')$ into $Isp_{[\mathcal{V}]}(\mathcal{O}')$.

Example 9.4. With reference to the viewers $\mathcal{V}$ and $\mathcal{V}'$ defined in Table 9.1, where $\mathcal{V}' \Supset \mathcal{V}$, consider the temporal observation $\mathcal{O}'$ displayed on the left side of Fig. 9.1.

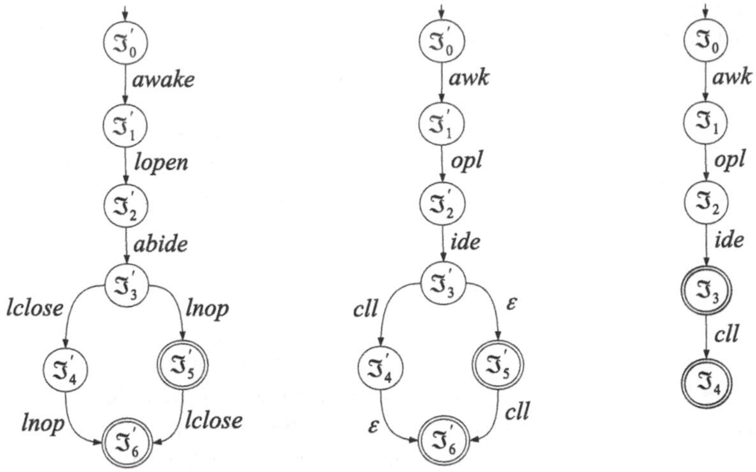

Fig. 9.2 With reference to the temporal observation $\mathcal{O}'$ displayed on the left side of Fig. 9.1, and viewers $\mathcal{V}$ and $\mathcal{V}'$ defined in Table 9.1, from left to right, $Isp(\mathcal{O}')$, $Isp^{\mathrm{n}}(\mathcal{O}')$, and $Isp_{[\mathcal{V}]}(\mathcal{O}')$

In accordance with Def. 9.4, displayed from left to right in Fig. 9.2 are $Isp(\mathcal{O}')$, $Isp^{\mathrm{n}}(\mathcal{O}')$, and $Isp_{[\mathcal{V}]}(\mathcal{O}')$.

A natural question arising at this point is whether the projection of an index space of a temporal observation equals the index space of the projection of the temporal observation. The answer is given by Proposition 9.2.

Proposition 9.2. *Let $\mathcal{O}'$ be a temporal observation of a system $\mathcal{A}$, based on a viewer $\mathcal{V}'$. Let $\mathcal{V}$ be another viewer of $\mathcal{A}$ such that $\mathcal{V}' \Supset \mathcal{V}$. The language of the index space of the projection of $\mathcal{O}'$ on $\mathcal{V}$ equals the language of the projection on $\mathcal{V}$ of the index space of $\mathcal{O}'$:*

$$\left\| Isp(\mathcal{O}'_{[\mathcal{V}]}) \right\| = \left\| Isp_{[\mathcal{V}]}(\mathcal{O}') \right\| . \tag{9.4}$$

Proof. Equation (9.4) is grounded on the equivalence of $Isp_{[\mathcal{V}]}(\mathcal{O}')$ and $Isp^{\mathrm{n}}(\mathcal{O}')$. When the renaming introduces neither ε nor duplicates within nodes of $\mathcal{O}'$, (9.4) is supported by the fact that $Isp^{\mathrm{n}}(\mathcal{O}')$ still represents all possible candidate traces of $\mathcal{O}'_{[\mathcal{V}]}$. If a label is mapped to ε, the same mapping will hold in $Isp^{\mathrm{n}}(\mathcal{O}')$, with no effect on the candidate trace. Finally, if two labels of the same node are mapped to a single label, the replication of the new label on two different transitions of $Isp^{\mathrm{n}}(\mathcal{O}')$ will result in a single transition after the determinization of $Isp^{\mathrm{n}}(\mathcal{O}')$ into $Isp_{[\mathcal{V}]}(\mathcal{O}')$. □

Example 9.5. With reference to the viewers $\mathcal{V}$ and $\mathcal{V}'$ defined in Table 9.1, where $\mathcal{V}' \Supset \mathcal{V}$, consider the temporal observation $\mathcal{O}'$ displayed on the left side of Fig. 9.1. According to $Isp(\mathcal{O}'_{[\mathcal{V}]})$ (displayed on the right side of Fig. 9.1) and $Isp_{[\mathcal{V}]}(\mathcal{O}')$ (displayed on the right side of Fig. 9.2), in conformity with Proposition 9.2, $Isp(\mathcal{O}'_{[\mathcal{V}]}) = Isp_{[\mathcal{V}]}(\mathcal{O}')$.

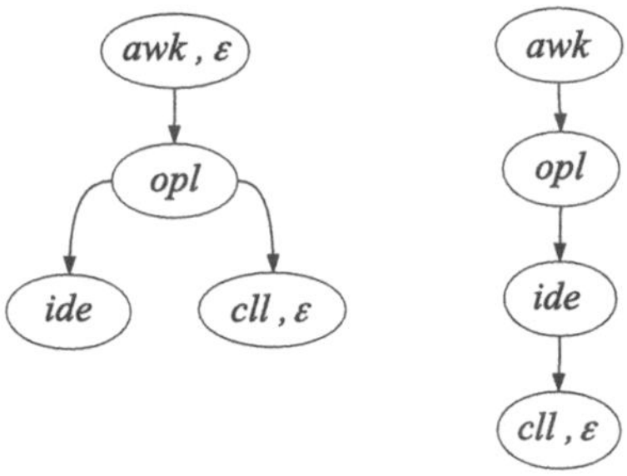

Fig. 9.3 Temporal observations $\mathcal{O}_2$ (left) and $\mathcal{O}'_2$ (right), where $\mathcal{O}_2 \Supset \mathcal{O}'_2$

Let $\wp(\mathcal{A}) = (a_0, \mathcal{V}, \mathcal{O}, \mathcal{R})$ and $\wp'(\mathcal{A}) = (a_0, \mathcal{V}, \mathcal{O}', \mathcal{R}')$ be two diagnosis problems sharing the same viewer $\mathcal{V}$. The behavior of $\wp(\mathcal{A})$ includes all the trajectories of $\mathcal{A}$ whose trace is a candidate trace of $\mathcal{O}$. Likewise, the behavior of $\wp'(\mathcal{A})$ includes all the trajectories of $\mathcal{A}$ whose trace is a candidate trace of $\mathcal{O}'$. Hence, in order for $Bhv(\wp(\mathcal{A}))$ to include all the trajectories of $Bhv(\wp'(\mathcal{A}))$, it suffices that each candidate trace of $\mathcal{O}'$ is also a candidate trace of $\mathcal{O}$. In other words, it suffices that $\|\mathcal{O}\| \supseteq \|\mathcal{O}'\|$. As formalized in Def. 9.5, this set-containment relationship is in fact the condition defining observation subsumption.

Definition 9.5 (Observation Subsumption). Let $\mathcal{O}$ and $\mathcal{O}'$ be two temporal observations. $\mathcal{O}$ *subsumes* $\mathcal{O}'$, written $\mathcal{O} \Supset \mathcal{O}'$, iff the extension of $\mathcal{O}$ includes the extension of $\mathcal{O}'$:

$$\mathcal{O} \Supset \mathcal{O}' \iff \|\mathcal{O}\| \supseteq \|\mathcal{O}'\| \,. \tag{9.5}$$

Example 9.6. Displayed in Fig. 9.3 are two temporal observations of the system $\mathcal{A}_2$ (defined in Example 3.3), namely $\mathcal{O}_2$ (left) and $\mathcal{O}'_2$ (right). It is easy to check that each candidate trace in $\|\mathcal{O}'_2\|$ is also a candidate trace in $\|\mathcal{O}_2\|$. Hence, based on Def. 9.5, $\mathcal{O}_2$ subsumes $\mathcal{O}'_2$, namely $\mathcal{O}_2 \Supset \mathcal{O}'_2$.

Proposition 9.3. *Observation subsumption is reflexive.*

Proof. Assuming $\mathcal{O} = \mathcal{O}'$, (9.5) holds, where the containment relationship of observation extensions becomes an equality. □

Proposition 9.4. $\mathcal{O} \Supset \mathcal{O}' \iff \|Isp(\mathcal{O})\| \supseteq \|Isp(\mathcal{O}')\|$.

Proof. The proof is grounded on Proposition 5.1, which asserts that $\|Isp(\mathcal{O})\| = \|\mathcal{O}\|$. □

In order to reuse the behavior of a diagnosis problem $\wp(\mathcal{A})$ to generate the behavior of a diagnosis problem $\wp'(\mathcal{A})$, we must be sure that each trajectory in $Bhv(\wp(\mathcal{A}'))$ is also a trajectory in $Bhv(\wp(\mathcal{A}))$. This set-containment relationship is the condition for behavior subsumption formalized in Def. 9.6.

Definition 9.6 (Behavior Subsumption). Let $\mathcal{B}$ and $\mathcal{B}'$ be two behaviors of an active system. $\mathcal{B}$ *subsumes* $\mathcal{B}'$, written $\mathcal{B} \Supset \mathcal{B}'$, iff $\mathcal{B}$ contains the trajectories of $\mathcal{B}'$.

The precise relationships between viewers and temporal observations that constitute a sufficient condition for behavior subsumption are provided in Proposition 9.5.

Proposition 9.5. *Let $\wp(\mathcal{A}) = (a_0, \mathcal{V}, \mathcal{O}, \mathcal{R})$ and $\wp'(\mathcal{A}) = (a_0, \mathcal{V}', \mathcal{O}', \mathcal{R}')$ be two diagnosis problems for $\mathcal{A}$. If $\mathcal{V}' \Supset \mathcal{V}$ and $\mathcal{O} \Supset \mathcal{O}'_{[\mathcal{V}]}$, then $Bhv(\wp(\mathcal{A})) \Supset Bhv(\wp'(\mathcal{A}))$.*

Proof. We first introduce Lemma 9.1.

Lemma 9.1. *If $\ell = Ren(\ell', \mathcal{V})$, $\ell \neq \varepsilon$, $T_\ell = \{t \mid (t, \ell) \in \mathcal{V}\}$, $T_{\ell'} = \{t \mid (t, \ell') \in \mathcal{V}'\}$, then $T_\ell \supseteq T_{\ell'}$.*

Proof. This is grounded on Def. 9.1, specifically condition 2, which establishes that, for each $(t, \ell) \in \mathcal{V}'$ such that $(t, \ell) \in \mathcal{V}$, the set of transitions associated with ℓ in $\mathcal{V}$ contains the transitions associated with ℓ' in $\mathcal{V}'$.

To prove Proposition 9.5, we show that each trajectory $h = [t_1, \ldots, t_k]$ in $Bhv(\wp'(\mathcal{A}))$ is a trajectory in $Bhv(\wp(\mathcal{A}))$ too. Let $\mathcal{T}' = [\ell'_1, \ldots, \ell'_k]$ be the sequence of labels isomorphic to h, where each label ℓ'_i, $i \in [1 \cdots k]$, is defined as follows:

$$\ell'_i = \begin{cases} \ell & \text{if } (t, \ell) \in \mathcal{V}', \\ \varepsilon & \text{otherwise} . \end{cases} \tag{9.6}$$

Let $\mathcal{T} = [\ell_1, \ldots, \ell_k]$ be the sequence of labels isomorphic to $\mathcal{T}'$, where, $\forall i \in [1 \cdots k]$, $\ell_i = Ren(\ell'_i, \mathcal{V})$. Since $\mathcal{O} \Supset \mathcal{O}'_{[\mathcal{V}]}$, based on Defs. 9.3 and 9.5, the string of observable labels obtained from $\mathcal{T}$ by removing labels ε is a candidate trace in $\mathcal{O}$. Since $\wp(\mathcal{A})$ and $\wp'(\mathcal{A})$ share the same initial state a_0, to prove that h is a trajectory in $Bhv(\wp(\mathcal{A}))$ it suffices to show that $h_{[\mathcal{V}]} \in \|\mathcal{O}\|$, in other words, that each component transition in h is consistent with each corresponding (possibly empty) label in $\mathcal{T}$. The proof is by induction on h, where $h(i)$, $i \in [0 \cdots k]$, denotes the (possibly empty) prefix of h up to the i-th component transition.

(*Basis*) *$h(0)$ is a trajectory in $Bhv(\wp(\mathcal{A}))$*. This comes trivially from the assumption that the initial state a_0 is shared with $\wp'(\mathcal{A})$.

(*Induction*) *If $h(i)$ is a trajectory in $Bhv(\wp(\mathcal{A}))$, $i \in [0 \cdots (k-1)]$, then $h(i+1)$ is a trajectory in $Bhv(\wp(\mathcal{A}))$*. Since $\mathcal{V}' \Supset \mathcal{V}$, three cases are possible for the component transition t_{i+1}:

1. t_{i+1} is unobservable in $\mathcal{V}'$. Since $\mathcal{V}' \Supset \mathcal{V}$, based on the first condition of Def. 9.1, t_{i+1} is unobservable in $\mathcal{V}$ too. Hence, based on (9.6), t_{i+1} is consistent with the corresponding label $\ell_{i+1} = \varepsilon$ in $\mathcal{T}$. In other words, $h(i+1)$ is a trajectory in $Bhv(\wp(\mathcal{A}))$.
2. t_{i+1} is observable in $\mathcal{V}'$ via the label ℓ'_{i+1} and unobservable in $\mathcal{V}$. Based on Def. 9.2, $\ell_{i+1} = Ren(\ell'_{i+1}, \mathcal{V}) = \varepsilon$. Hence, t_{i+1} is consistent with the corresponding empty label ℓ_{i+1} in $\mathcal{T}$. In other words, $h(i+1)$ is a trajectory in $Bhv(\wp(\mathcal{A}))$.
3. t_{i+1} is observable in both $\mathcal{V}'$ and $\mathcal{V}$. In this case, the corresponding observable label in $\mathcal{V}'$ is ℓ'_{i+1}, while, based on Def. 9.1 and the way in which $\mathcal{T}$ is defined, the corresponding observable label in $\mathcal{V}$ is ℓ_{i+1}. Hence, t_{i+1} is consistent with the corresponding label ℓ_{i+1} in $\mathcal{T}$. In other words, $h(i+1)$ is a trajectory in $Bhv(\wp(\mathcal{A}))$.

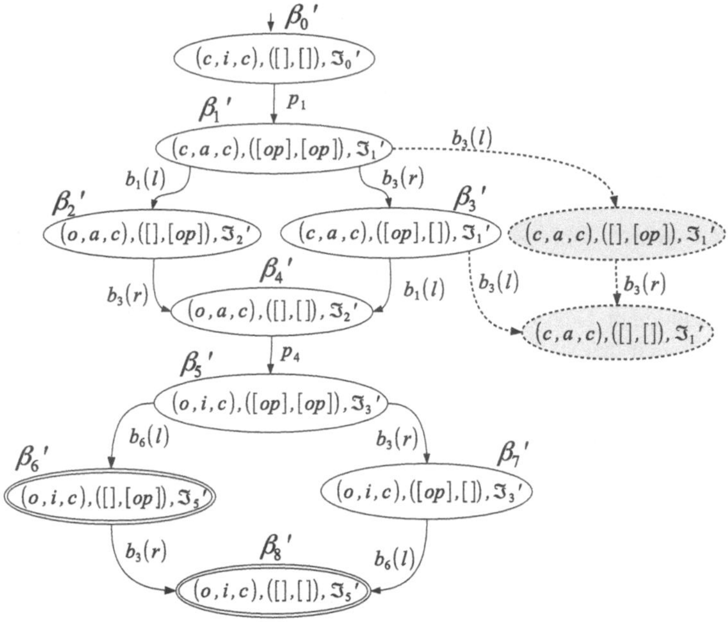

Fig. 9.4 Reconstructed behavior $Bhv(\wp'(\mathcal{A}_2)$ based on temporal observation $\mathcal{O}'$ displayed on the left side of Fig. 9.1, with $Isp(\mathcal{O}')$ displayed on the left side of Fig. 9.2

□

Example 9.7. Consider the diagnosis problem $\wp(\mathcal{A}_2) = (a_0, \mathcal{V}, \mathcal{O}_2, \mathcal{R})$ defined in Example 4.12. Let $\wp'(\mathcal{A}_2) = (a_0, \mathcal{V}', \mathcal{O}', \mathcal{R})$ be another diagnosis problem for the system $\mathcal{A}_2$, where $\mathcal{V}'$ is defined in Table 9.1, while the observation $\mathcal{O}'$ is displayed on the left side of Fig. 9.1, with the index space $Isp(\mathcal{O}')$ displayed on the left side of Fig. 9.2. The projection of $\mathcal{O}'$ on the viewer $\mathcal{V}$, namely $O'_{[\mathcal{V}]}$, is displayed in the center of Fig. 9.1. Based on Example 9.6 and Fig. 9.3, $\mathcal{O}_2$ subsumes $\mathcal{O}'_2$, with the latter being equal to $\mathcal{O}'_{[\mathcal{V}]}$. In summary, $\mathcal{V}' \Supset \mathcal{V}$ and $\mathcal{O}_2 \Supset \mathcal{O}'_{[\mathcal{V}]}$. Hence, according to Proposition 9.5, we expect $Bhv(\wp(\mathcal{A}_2)) \Supset Bhv(\wp'(\mathcal{A}_2))$, where the reconstruction of $Bhv(\wp(\mathcal{A}_2))$ by the *Build* algorithm is displayed in Fig. 5.13. To generate $Bhv(\wp'(\mathcal{A}_2))$, we can make use of the same algorithm, based on the index space $Isp(\mathcal{O}')$ displayed on the left side of Fig. 9.2. The result is displayed in Fig. 9.4, where the spurious part is denoted by dashed lines. Comparing Figs 5.13 and 9.4, it is easy to check that the DFA in Fig. 9.4 is a subgraph of the DFA in Fig. 5.13. In other words, each trajectory in $Bhv(\wp'(\mathcal{A}_2))$ is also a trajectory in $Bhv(\wp(\mathcal{A}_2))$, with the latter including several trajectories that are not included in the former. Hence, based on Def. 9.6 and as expected from Proposition 9.5, $Bhv(\wp(\mathcal{A}_2)) \Supset Bhv(\wp'(\mathcal{A}_2))$.

As pointed out in the introduction to this chapter, if two diagnosis problems $\wp(\mathcal{A})$ and $\wp'(\mathcal{A})$ share the same initial state, viewer, and observation, but have different rulers, namely $\mathcal{R} \neq \mathcal{R}'$, then the solution of $\wp'(\mathcal{A})$ can be determined by projecting the solution of $\wp(\mathcal{A})$ on $\mathcal{R}'$ provided that $\mathcal{R}$ subsumes $\mathcal{R}'$. Since the notion of a ruler is conceptually isomorphic to the notion of a viewer (both of them are defined as a set of associations between component transitions and labels), it should come as no surprise that the notion of ruler subsumption is conceptually isomorphic to the notion of viewer subsumption. However, unlike viewer subsumption, the condition for reuse depends on $\mathcal{R}$ subsuming $\mathcal{R}'$ (and not vice versa, as is the case for viewer subsumption, with $\mathcal{V}'$ subsuming $\mathcal{V}$). Ruler subsumption is formalized in Def. 9.7.

Definition 9.7 (Ruler Subsumption). Let $\mathcal{R}$ and $\mathcal{R}'$ be two rulers for an active system. *R subsumes* $\mathcal{R}'$, written $\mathcal{R} \Supset \mathcal{R}'$, iff the following conditions hold:

1. The set of transitions involved in $\mathcal{R}$ contains all the transitions involved in $\mathcal{R}'$,
2. For each association $(t, f) \in \mathcal{R}$ such that $(t, f') \in \mathcal{R}'$, the set of transitions associated with f' in $\mathcal{R}'$ contains all the transitions associated with f in $\mathcal{R}$.

Example 9.8. With reference to the active system $\mathcal{A}_2$ introduced in Example 4.6, displayed on the left side of Table 9.1 is the specification of the rulers $\mathcal{R}$ and $\mathcal{R}'$, where $\mathcal{R}$ is inherited from Table 4.1, while $\mathcal{R}'$ is a different ruler. According to Def. 9.7, $\mathcal{R} \Supset \mathcal{R}'$. In fact, (1) the set of transitions involved in $\mathcal{R}$ contains all the transitions involved in $\mathcal{R}'$ (the two sets are equal), and (2) for each association $(t, f) \in \mathcal{R}$ such that $(t, f') \in \mathcal{R}'$, the set of transitions associated with f' in $\mathcal{R}'$ contains all the transitions associated with f in *R*. For instance, considering point (2), for the association $(p_3, \mathbf{fop}) \in \mathcal{R}$, the set of transitions associated with **pfail** in $\mathcal{R}'$ is $\{p_3, p_4\}$, which is a superset of the set of transitions associated with **fop** in $\mathcal{R}$ (the singleton $\{p_3\}$).

Proposition 9.6. *Ruler subsumption is reflexive.*

Proof. Assuming $\mathcal{R} = \mathcal{R}'$, both conditions in Def. 9.7 hold, where the containment relationship becomes an equality. □

As with observable labels, we need to define the notion of fault-label renaming, which is formalized in Def. 9.8. Conceptually, this definition parallels the notion of observable-label renaming provided in Def. 9.2.

Definition 9.8 (Fault-Label Renaming). Let $\mathcal{R}$ and $\mathcal{R}'$ be two rulers for an active system, with Φ and Φ' being the corresponding domains of fault labels, where $\mathcal{R} \Supset \mathcal{R}'$. The *renaming* of a fault label $f \in \Phi$ based on $\mathcal{R}'$ is a fault label in $\mathcal{R}' \cup \{\varepsilon\}$ defined as follows:

$$Ren(f, \mathcal{R}') = \begin{cases} f' & \text{if } (t, f) \in \mathcal{R}, (t, f') \in \mathcal{R}', \\ \varepsilon & \text{otherwise} . \end{cases} \tag{9.7}$$

Example 9.9. With reference to the rulers $\mathcal{R}$ and $\mathcal{R}'$ defined in Table 9.1 (Example 9.8), where $\mathcal{R} \Supset \mathcal{R}'$, we have $Ren(\mathbf{nol}, \mathcal{R}') = \mathbf{bfail}$ and $Ren(\mathbf{fop}, \mathcal{V}) = \mathbf{pfail}$.

Having introduced the different notions of subsumption and projection, we are now ready to cope with diagnosis reuse, as detailed in Sec. 9.2.

9.2 Diagnosis with Reuse

Let $\wp(\mathcal{A})$ and $\wp'(\mathcal{A})$ be two diagnosis problems, with rulers $\mathcal{R}$ and $\mathcal{R}'$, respectively. If we know that the behavior of $\wp'(\mathcal{A})$ to be solved will equal the behavior of $\wp(\mathcal{A})$, which has already been solved, then the solution of $\wp'(\mathcal{A})$, namely $\Delta(\wp'(\mathcal{A}))$, can be generated by projecting on $\mathcal{R}'$ the candidate diagnoses of $\Delta(\wp(\mathcal{A}))$, provided that $\mathcal{R}$ subsumes $\mathcal{R}'$. The operation of diagnosis projection is formalized in Def. 9.9.

Definition 9.9 (Diagnosis Projection). Let δ be a diagnosis of a system $\mathcal{A}$ based on a ruler $\mathcal{R}$, and $\mathcal{R}'$ another ruler for $\mathcal{A}$ such that $\mathcal{R} \Supset \mathcal{R}'$. The *projection* of δ on $\mathcal{R}'$ is defined as follows:

$$\delta_{[\mathcal{R}']} = \{f' \mid f \in \delta, f' = Ren(f, \mathcal{R}'), f' \neq \varepsilon\}\,. \tag{9.8}$$

Let Δ be a set of diagnoses of $\mathcal{A}$ based on $\mathcal{R}$. The *projection* of Δ on $\mathcal{R}'$ is defined as follows:

$$\Delta_{[\mathcal{R}']} = \{\delta' \mid \delta \in \Delta, \delta' = \delta_{[\mathcal{R}']}\}\,. \tag{9.9}$$

Example 9.10. Consider the rulers $\mathcal{R}$ and $\mathcal{R}'$ defined in Table 9.1, where, as shown in Example 9.8, $\mathcal{R} \Supset \mathcal{R}'$. Let $\Delta = \{\{\mathbf{nor}\}, \{\mathbf{ncl}, \mathbf{nor}\}, \{\mathbf{nor}, \mathbf{fcp}\}\}$ be a set of diagnoses based on $\mathcal{R}$. According to Def. 9.9, $\Delta_{[\mathcal{R}']} = \{\{\mathbf{bfail}\}, \{\mathbf{bfail}, \mathbf{pfail}\}\}$ (duplicates are removed). Compared with Δ, $\Delta_{[\mathcal{R}']}$ is less informative, as a consequence of the conditions on subsumption between rulers. Specifically, both of the fault labels **ncl** and **nor** in $\mathcal{R}$ are mapped to a single label in $\mathcal{R}'$, namely **bfail**, meaning a failing breaker.[5]

As pointed out above, if two diagnosis problems $\wp(\mathcal{A})$ and $\wp'(\mathcal{A})$, with rulers $\mathcal{R}$ and $\mathcal{R}'$, respectively, share the same behavior, then the solution of $\wp'(\mathcal{A})$ can be generated by projecting the solution of $\wp(\mathcal{A})$ on the ruler $\mathcal{R}'$, provided that $\mathcal{R}$ subsumes $\mathcal{R}'$. Intuitively, the need for $\mathcal{R} \Supset \mathcal{R}'$ can be grasped by contradiction.

Assume that $\mathcal{R} \not\Supset \mathcal{R}'$. This is possible if either of the two conditions in Def. 9.7 is violated. The violation of the first condition means that $\mathcal{R}'$ includes an association (t, f'), where transition t is not included in $\mathcal{R}$. If so, a candidate diagnosis $\delta \in \Delta(\wp(\mathcal{A}))$ generated by a trajectory involving t will not include a corresponding fault label, as t is not a faulty transition in $\mathcal{R}$. Consequently, the projection of δ on $\mathcal{R}'$ is bound not to include f', even though t is faulty in $\mathcal{R}'$.

The violation of the second condition means that, for an association $(t, f) \in \mathcal{R}$ such that $(t, f') \in \mathcal{R}'$, there exists an association $(\bar{t}, f) \in \mathcal{R}, t \neq \bar{t}$, such that $(\bar{t}, \bar{f}) \in \mathcal{R}'$, $\bar{f} \neq f$.[6] A candidate diagnosis $\delta \in \Delta(\wp(\mathcal{A}))$ generated by a trajectory involving t will include a fault label f (associated with t in $\mathcal{R}$). However, the projection of δ on $\mathcal{R}'$ is ambiguous, since f is associated in $\mathcal{R}$ with two transitions, namely t and $\bar{t}$, which are associated in $\mathcal{R}'$ with fault labels f' and $\bar{f}$, respectively.

---

[5] Because of duplicate removals, the fault label **bfail** may in fact denote the failure of both breakers l and r, possibly in several ways. In our example, the first occurrence of **bfail** is the projection of both **ncl** (breaker l did not close) and **nor** (breaker r did not open).

[6] In fact, the first condition in Def. 9.7 requires that all faulty transitions in $\mathcal{R}'$ are faulty in $\mathcal{R}$ too.

Projecting δ on two different diagnoses, one including f' and the other including $\bar{f}$, allows a complete yet unsound solution of the diagnosis problem $\wp'(\mathcal{A})$. In other words, the set of candidate diagnoses for $\wp'(\mathcal{A})$ generated by projection of $\Delta(\wp(\mathcal{A}))$ is bound to differ from the (sound and complete) solution of $\wp'(\mathcal{A})$.

Proposition 9.7. *If $\wp(\mathcal{A}) = (a_0, \mathcal{V}, \mathcal{O}, \mathcal{R})$ and $\wp'(\mathcal{A}) = (a_0, \mathcal{V}, \mathcal{O}, \mathcal{R}')$ are two diagnosis problems such that $\mathcal{R} \Supset \mathcal{R}'$, then*

$$\Delta(\wp'(\mathcal{A})) = (\Delta(\wp(\mathcal{A})))_{[\mathcal{R}']} \,. \tag{9.10}$$

Proof. Based on (4.5), a candidate diagnosis $\delta \in \Delta(\wp(\mathcal{A}))$ is defined as $\delta = h_{[\mathcal{R}]}$, $h \in Bsp(\mathcal{A})$, $h_{[\mathcal{V}]} \in \|\mathcal{O}\|$. Likewise, a candidate diagnosis $\delta' \in \Delta(\wp'(\mathcal{A}))$ is defined as $\delta' = h_{[\mathcal{R}']}$, $h \in Bsp(\mathcal{A})$, $h_{[\mathcal{V}]} \in \|\mathcal{O}\|$. As such, δ and δ' differ only in the term $h_{[\mathcal{R}]}$ vs. $h_{[\mathcal{R}']}$, as the two conditions on the trajectory h are identical. Based on Def. 4.6, $h_{[\mathcal{R}]} = \{f \mid t \in h, (t, f) \in \mathcal{R}\}$. Hence, $\delta_{[\mathcal{R}']} = \{f' \mid f \in \delta, f' = Ren(f, \mathcal{R}'), f' \neq \varepsilon\} = \{f' \mid t \in h, (t, f') \in \mathcal{R}'\} = \delta'$. □

Example 9.11. Let $\wp(\mathcal{A}_2) = (a_0, \mathcal{V}, \mathcal{O}_2, \mathcal{R})$ be the diagnosis problem defined in Example 4.12, and $\wp'(\mathcal{A}_2) = (a_0, \mathcal{V}, \mathcal{O}_2, \mathcal{R}')$ be another diagnosis problem for the system $\mathcal{A}_2$, where $\mathcal{R}'$ is defined in Table 9.1. Based on Example 9.8, $\mathcal{R} \Supset \mathcal{R}'$. Hence, according to Proposition 9.7, we expect $\Delta(\wp'(\mathcal{A}_2)) = (\Delta(\wp(\mathcal{A}_2)))_{[\mathcal{R}']}$. Based on Example 4.13, $\Delta(\wp(\mathcal{A}_2)) = \{\{\mathbf{nor}\}, \{\mathbf{ncl}, \mathbf{nor}\}, \{\mathbf{nor}, \mathbf{fcp}\}\}$, which is exactly the set Δ of diagnoses introduced in Example 9.10. Therefore, according to Example 9.10, we expect $\Delta(\wp'(\mathcal{A}_2)) = \{\{\mathbf{bfail}\}, \{\mathbf{bfail}, \mathbf{pfail}\}\}$. On the other hand, $Bhv(\wp'(\mathcal{A}_2))$ equals $Bhv(\wp(\mathcal{A}_2))$, as $\wp'(\mathcal{A}_2)$ differs from $\wp(\mathcal{A}_2)$ in the ruler $\mathcal{R}'$ only, which is not relevant to the reconstruction of the behavior. Hence, to generate $\Delta(\wp'(\mathcal{A}_2))$ it suffices to decorate the reconstructed behavior $Bhv(\wp(\mathcal{A}_2))$, as shown in Sec. 5.3, based on $\mathcal{R}'$. Displayed in Fig. 5.14 is the consistent part of the reconstructed behavior $Bhv(\wp(\mathcal{A}_2))$, with faulty transitions marked by corresponding fault labels based on $\mathcal{R}$. To obtain the same behavior with faulty transitions marked by fault labels based on $\mathcal{R}'$, it suffices to project each fault label to $\mathcal{R}'$, as displayed in Fig. 9.5. According to the decoration algorithm introduced in Sec. 5.3.1, we come up with the final nodes that are marked by diagnosis sets in three different ways:

1. $\beta_5, \beta_7, \beta_8$: $\{\{\mathbf{bfail}, \mathbf{pfail}\}\}$,
2. $\beta_6, \beta_9, \beta_{10}, \beta_{11}, \beta_{12}$: $\{\{\mathbf{bfail}\}\}$,
3. β_{13}: $\{\{\mathbf{bfail}\}, \{\mathbf{bfail}, \mathbf{pfail}\}\}$.

Therefore, according to Theorem 5.1, the solution of $\wp'(\mathcal{A}_2)$ is the union of the diagnosis sets associated with the final nodes, namely $\Delta(\wp'(\mathcal{A}_2)) = \{\{\mathbf{bfail}\}, \{\mathbf{bfail}, \mathbf{pfail}\}\}$. As expected, $\Delta(\wp'(\mathcal{A}_2))$ equals $(\Delta(\wp(\mathcal{A}_2)))_{[\mathcal{R}']}$.

Given two diagnosis problems $\wp(\mathcal{A}) = (a_0, \mathcal{V}, \mathcal{O}, \mathcal{R})$ and $\wp'(\mathcal{A}) = (a_0, \mathcal{V}', \mathcal{O}', \mathcal{R}')$, we expect, generally speaking, that the corresponding behaviors will be different, namely $Bhv(\wp(\mathcal{A})) \neq Bhv(\wp'(\mathcal{A}))$. Therefore, even if $\mathcal{R} \Supset \mathcal{R}'$, we are not allowed to generate the solution of $\wp'(\mathcal{A})$ by projecting the solution of $\wp(\mathcal{A})$ on $\mathcal{R}'$. On the other hand, if we know that $Bhv(\wp(\mathcal{A}))$ subsumes $Bhv(\wp'(\mathcal{A}))$, then $Bhv(\wp'(\mathcal{A}))$

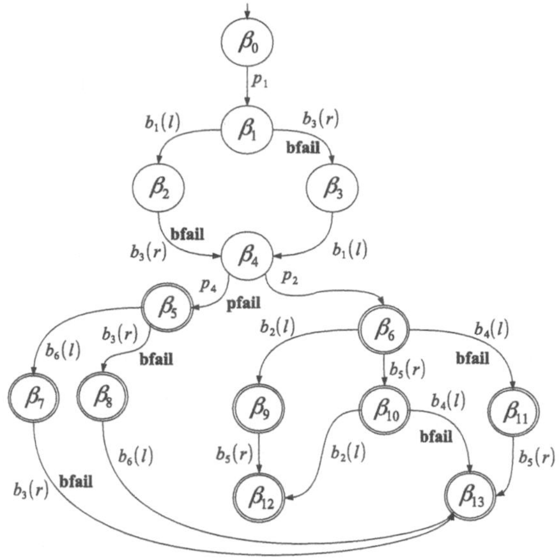

Fig. 9.5 Reconstructed behavior $Bhv(\wp(\mathcal{A}_2))$ displayed in Fig. 5.14, with faulty transitions marked by corresponding fault labels based on ruler $\mathcal{R}'$ (Table 9.1)

can be generated by matching $Bhv(\wp(\mathcal{A}))$ with $\mathcal{O}'$ (based on $\mathcal{V}'$). Such a matching operation is formalized in Def. 9.10.

Definition 9.10 (Behavior Matching). Let $\mathcal{B}$ be a behavior of a system $\mathcal{A}$, and $\mathcal{O}$ a temporal observation of $\mathcal{A}$ based on a viewer $\mathcal{V}$. The *matching* of $\mathcal{B}$ with $\mathcal{O}$ based on $\mathcal{V}$, denoted $\mathcal{B} \asymp (\mathcal{O}, \mathcal{V})$, is a behavior of $\mathcal{A}$ composed of the trajectories h in $\mathcal{B}$ such that $h_{[\mathcal{V}]} \in \|\mathcal{O}\|$.

According to the discussion in the introduction to this chapter, given two diagnosis problems $\wp(\mathcal{A}) = (a_0, \mathcal{V}, \mathcal{O}, \mathcal{R})$ and $\wp'(\mathcal{A}) = (a_0, \mathcal{V}', \mathcal{O}', \mathcal{R})$, where $\mathcal{V}' \Supset \mathcal{V}$ and $\mathcal{O} \Supset \mathcal{O}'_{[\mathcal{V}]}$, we are guaranteed that the behavior of $\wp(\mathcal{A}))$ will subsume the behavior of $\wp'(\mathcal{A}))$. In other words, the set of trajectories in $Bhv(\wp(\mathcal{A}))$ will include all the trajectories in $Bhv(\wp'(\mathcal{A}))$. Hence, $Bhv(\wp'(\mathcal{A}))$ can be generated by matching $Bhv(\wp(\mathcal{A}))$ with $\mathcal{O}'$, as specified in Def. 9.10. This is formalized in Proposition 9.8.

Proposition 9.8. *If $\wp(\mathcal{A}) = (a_0, \mathcal{V}, \mathcal{O}, \mathcal{R})$ and $\wp'(\mathcal{A}) = (a_0, \mathcal{V}', \mathcal{O}', \mathcal{R})$ are two diagnosis problems such that $\mathcal{V}' \Supset \mathcal{V}$ and $\mathcal{O} \Supset \mathcal{O}'_{[\mathcal{V}]}$, then*

$$Bhv(\wp'(\mathcal{A})) = Bhv(\wp(\mathcal{A})) \asymp (\mathcal{O}', \mathcal{V}') . \tag{9.11}$$

Proof. As $\mathcal{V} \Supset \mathcal{V}$ and $\mathcal{O} \Supset \mathcal{O}'_{[\mathcal{V}]}$, based on Proposition 9.6, $Bhv(\wp(\mathcal{A})) \Supset Bhv(\wp'(\mathcal{A}))$. Hence, based on Def. 9.10, $\|Bhv(\wp(\mathcal{A})) \asymp (\mathcal{O}', \mathcal{V}')\| = \{h \mid h \in Bhv(\wp(\mathcal{A})), h_{[\mathcal{V}']} \in \|\mathcal{O}'\|\} = \{h \mid h \in Bsp(\mathcal{A}), h_{[\mathcal{V}']} \in \|\mathcal{O}'\|\} = Bhv(\wp'(\mathcal{A}))$. □

Example 9.12. Consider the diagnosis problem $\wp(\mathcal{A}_2) = (a_0, \mathcal{V}, \mathcal{O}_2, \mathcal{R})$ defined in Example 4.12. Let $\wp'(\mathcal{A}_2) = (a_0, \mathcal{V}', \mathcal{O}', \mathcal{R})$ be another diagnosis problem for the system $\mathcal{A}_2$, where $\mathcal{V}'$ is defined in Table 9.1 and $\mathcal{O}'$ is displayed on the left side of

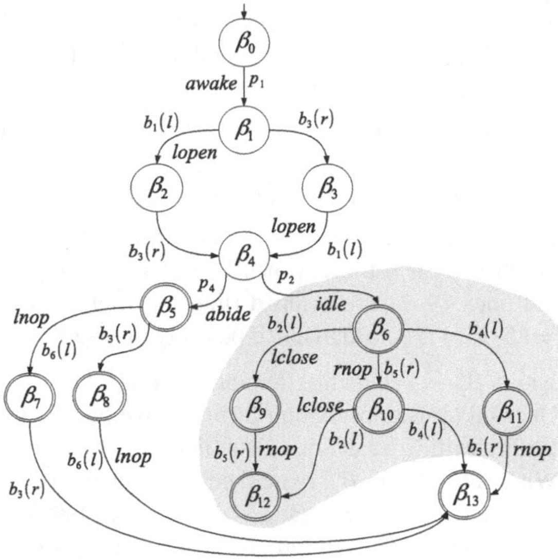

Fig. 9.6 Matching of the reconstructed behavior $Bhv(\wp(\mathcal{A}_2))$ with the temporal observation $\mathcal{O}'$ (displayed on the left side of Fig. 9.1), where observable transitions are marked by corresponding labels in viewer $\mathcal{V}'$ (Table 9.1); the unmatched part of the behavior is depicted in gray

Fig. 9.1. As shown in Example 9.7, $\mathcal{V}' \Supset \mathcal{V}$ and $\mathcal{O}_2 \Supset \mathcal{O}'_{[\mathcal{V}]}$. Hence, according to Proposition 9.8, we expect the behavior of $\wp'(\mathcal{A}_2)$ to be the matching of the behavior of $\wp(\mathcal{A}_2)$ with $\mathcal{O}'$ based on $\mathcal{V}'$, namely $Bhv(\wp'(\mathcal{A}_2)) = Bhv(\wp(\mathcal{A}_2)) \asymp (\mathcal{O}', \mathcal{V}')$. Reported in Fig. 9.6 is the reconstructed behavior of $\wp(\mathcal{A}_2)$, where observable transitions in $\mathcal{V}'$ are marked by corresponding labels. Based on the index space $Isp(\mathcal{O}')$ displayed on the left of Fig. 9.2, it is easy to check that such a matching corresponds to the plain part of the DFA, with the gray part denoting the unmatched behavior. As expected, apart from state identification, the matching result in Fig. 9.6 equals the reconstructed behavior $Bhv(\wp'(\mathcal{A}_2))$ displayed in Fig. 9.4.

In Sec. 9.2.1, we provide a pseudocoded implementation of the matching operation specified in Def. 9.10, namely the *Match* algorithm.

9.2.1 Matching Algorithm

Listed below is the specification of the algorithm *Match*, which implements the matching operation introduced in Def. 9.10, by generating the behavior $\mathcal{B}'$ of a diagnosis problem $\wp'(\mathcal{A}) = (a_0, \mathcal{V}', \mathcal{O}', \mathcal{R})$ based on the behavior $\mathcal{B}$ of a diagnosis problem $\wp(\mathcal{A}) = (a_0, \mathcal{V}, \mathcal{O}, \mathcal{R})$, where $\mathcal{V}' \Supset \mathcal{V}$ and $\mathcal{O} \Supset \mathcal{O}'_{[\mathcal{V}]}$.

To this end, the index space of the temporal observation $\mathcal{O}'$, $Isp(\mathcal{O}')$, is generated (line 7). Each state of $\mathcal{B}'$ is a pair $(\beta, \Im)$, where β is a state in $\mathcal{B}$ and $\Im$ a state in $Isp(\mathcal{O}')$. The initial state β'_0 is set to $(\beta_0, \Im_0)$, where β_0 is the initial state of $\mathcal{B}$ and $\Im_0$ is the initial state of $Isp(\mathcal{O}')$ (line 6). In the same line, the set B' of states is initialized as the singleton $\{\beta'_0\}$, while the set τ' of transitions is set to empty.

Within the main loop of the algorithm (lines 9–24), at each iteration, an unmarked state $\beta' = (\beta_1, \Im_1)$ is selected (line 10). The goal is to generate all transitions $\beta' \xrightarrow{t(c)} \beta''$, with $\beta'' = (\beta_2, \Im_2)$.

Thus, each component transition $t(c)$ marking a transition exiting β_1 in $\mathcal{B}$ is checked for consistency with $\mathcal{O}'$ (lines 11–18). Specifically, if $t(c)$ is not observable then it is consistent with $\mathcal{O}'$ and $\Im_2$ equals $\Im_1$ (lines 12 and 13). If, instead, $t(c)$ is observable via a label ℓ in $\mathcal{V}'$ and a transition $\Im_1 \xrightarrow{\ell} \bar{\Im}$ is included in the index space, then $\Im_2$ is set to $\bar{\Im}$ (lines 14 and 15). Otherwise, $t(c)$ is not consistent with $\mathcal{O}$ and lines 19–21 are skipped (line 17). If $t(c)$ is consistent, then the new state $\beta'' = (\beta_2, \Im_2)$ is considered and possibly inserted into B' (lines 19 and 20). The new transition $\beta' \xrightarrow{t(c)} \beta''$ is inserted into τ' (line 21).

Once all transitions exiting β_1 in $\mathcal{B}$ have been considered, state β' is marked, as no further exiting transition exists (line 23).

When all states in B' are marked, no further state or transition can be generated. Hence, the set B'_{f} of final states is determined (line 25).

At this point, B' and τ' are the states and transitions of the spurious behavior of $\wp'(\mathcal{A})$. To obtain the actual behavior of $\wp'(\mathcal{A})$, it suffices to remove the spurious part of the DFA, that is, the states and transitions which are not included in any path connecting the initial state to a final state (line 26).

The specification of the algorithm *Match* is as follows:

1. **algorithm** *Match* (**in** $\mathcal{B}$, **in** $\mathcal{V}'$, **in** $\mathcal{O}'$, **out** $\mathcal{B}'$)
2. $\quad$ $\mathcal{B} = (\Sigma, B, \tau, \beta_0, B_{\mathrm{f}})$: the behavior of a diagnosis problem $\wp(\mathcal{A}) = (a_0, \mathcal{V}, \mathcal{O}, \mathcal{R})$,
3. $\quad$ $\mathcal{V}'$: a viewer for $\mathcal{A}$ such that $\mathcal{V}' \Supset \mathcal{V}$,
4. $\quad$ $\mathcal{O}'$: a temporal observation for $\mathcal{A}$ such that $\mathcal{O} \Supset \mathcal{O}'$,
5. $\quad$ $\mathcal{B}' = (\Sigma, B', \tau', \beta_0, B'_{\mathrm{f}})$: the behavior of $\wp'(\mathcal{A}) = (a_0, \mathcal{V}', \mathcal{O}', \mathcal{R})$;

6. $\quad$ **begin** $\langle$*Match*$\rangle$
7. $\quad\quad$ Generate the index space $Isp(\mathcal{O}') = (\Sigma, I, \tau'', \Im_0, I_{\mathrm{f}})$;
8. $\quad\quad$ $\beta'_0 := (\beta_0, \Im_0)$; $B' := \{\beta'_0\}$; $\tau' := \emptyset$;
9. $\quad\quad$ **repeat**
10. $\quad\quad\quad$ Choose an unmarked state $\beta' = (\beta_1, \Im_1) \in B'$;
11. $\quad\quad\quad$ **foreach** transition $\beta_1 \xrightarrow{t(c)} \beta_2 \in \tau$ **do**
12. $\quad\quad\quad\quad$ **if** $t(c)$ is unobservable in $\mathcal{V}'$ **then**
13. $\quad\quad\quad\quad\quad$ $\Im_2 := \Im_1$
14. $\quad\quad\quad\quad$ **elsif** $(t(c), \ell) \in \mathcal{V}'$ **and** $\Im_1 \xrightarrow{\ell} \bar{\Im}$ is in $Isp(\mathcal{O}')$ **then**
15. $\quad\quad\quad\quad\quad$ $\Im_2 := \bar{\Im}$
16. $\quad\quad\quad\quad$ **else**
17. $\quad\quad\quad\quad\quad$ **continue**
18. $\quad\quad\quad\quad$ **endif**;
19. $\quad\quad\quad\quad$ $\beta'' := (\beta_2, \Im_2)$;
20. $\quad\quad\quad\quad$ **if** $\beta'' \notin B'$ **then** insert β'' into B' **endif**;

Insert transition $\beta' \xrightarrow{t(c)} \beta''$ into τ'
endfor;
Mark β'
until all states in B' are marked;
$B'_{\mathrm{f}} := \{\, \beta'_{\mathrm{f}} \mid \beta'_{\mathrm{f}} \in B', \beta'_{\mathrm{f}} = (\beta, \Im_{\mathrm{f}}), \Im_{\mathrm{f}} \in I_{\mathrm{f}} \,\}$;
Remove from B' and τ' all states and transitions, respectively, which are not included in any path from β'_0 to a final state
end $\langle$*Match*$\rangle$.

Example 9.13. Let $\wp(\mathcal{A}_2) = (a_0, \mathcal{V}, \mathcal{O}, \mathcal{R})$ and $\wp'(\mathcal{A}_2) = (a_0, \mathcal{V}', \mathcal{O}', \mathcal{R})$ be the diagnosis problems considered in Example 9.12. As shown, by virtue of Proposition 9.7, the behavior of $\wp'(\mathcal{A}_2)$ can be generated by matching the behavior of $\wp(\mathcal{A}_2)$ with $\mathcal{O}'$ and $\mathcal{V}'$, with $Bhv(\wp(\mathcal{A}_2))$ being displayed on the left side of Fig. 9.7. The index space $Isp(\mathcal{O}')$ is displayed on the left of Fig. 9.2.

We make use of the *Match* algorithm, whose result is displayed on the right side of Fig. 9.7. At the beginning, the initial state $\beta'_0 = (\beta_0, \Im'_0)$ is created. Then, the main loop is iterated by selecting, at each iteration, a state β' not yet processed.

Considering the initial state $(\beta_0, \Im'_0)$, the only transition exiting β_0 in $Bhv(\wp(\mathcal{A}_2))$ is p_1, which is consistent with $Isp(\mathcal{O}')$. Therefore, the new state $(\beta_1, \Im'_1)$ is created in $Bsp(\wp'(\mathcal{A}_2))$, along with the new transition $(\beta_0, \Im'_0) \xrightarrow{p_1} (\beta_1, \Im'_1)$.

Notice that, when the state $(\beta_4, \Im'_2)$ is being processed, only one of the two transitions exiting β_4 in $Bhv(\wp(\mathcal{A}_2))$ is consistent with $Isp(\mathcal{O}')$, namely the transition marked by p_4. The reason lies in the next observable label in $Isp(\mathcal{O}')$, namely *abide* (rather than *awk*), which is associated with p_4 only (as p_2 is associated in $\mathcal{V}'$ with the observable label *idle*, which is not the next label in $Isp(\mathcal{O}')$).

Eventually, two states are qualified as final in $Bsp(\wp'(\mathcal{A}_2))$, namely $(\beta_7, \Im'_5)$ and $(\beta_{13}, \Im'_5)$, with $\Im'_5$ being final in $Isp(\mathcal{O}')$.

As expected, apart from state identification, the DFA generated by *Match* (displayed on the right side of Fig. 9.7) equals the behavior $Bsp(\wp'(\mathcal{A}_2))$ displayed in Fig. 9.4 (where the spurious part is irrelevant).

9.3 Checking Observation Subsumption

Reuse-based diagnosis is grounded on subsumption relationships between viewers, temporal observations, and rulers. Whilst viewer and ruler subsumption are not computationally problematic, checking observation subsumption based on Def. 9.5 is bound to be computationally prohibitive because of the large number of candidate traces involved in the observation.

Consider the scenario of reuse-based diagnosis at work. We have to solve a new diagnosis problem $\wp'(\mathcal{A})$ and, instead of applying the three canonical steps of behavior reconstruction, behavior decoration, and diagnosis distillation, we search for possible reusable diagnosis problems in a knowledge base. In doing so, we are

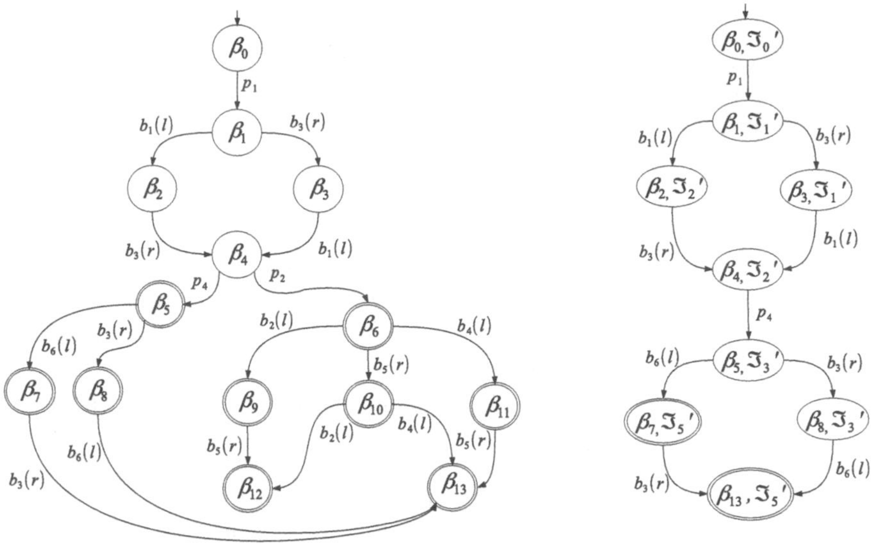

Fig. 9.7 Based on the behavior $Bhv(\wp(\mathcal{A}_2))$ (left), reconstruction of the behavior $Bhv(\wp'(\mathcal{A}_2))$ by the *Match* algorithm (right)

bound to perform subsumption checking between temporal observations. In order for the whole approach to work in practice, checking observation subsumption (on possibly several instances) is required to be no more complex than performing behavior reconstruction from scratch. Therefore, it is convenient to envisage alternative techniques for checking observation subsumption, which do not rely on Def. 9.5 and which will provide acceptable performance.

Checking observation subsumption via set containment of candidate traces, as expressed in Def. 9.5, disregards any reasoning on the given observations. Consider the problem of checking whether $\mathcal{O} \Supset \mathcal{O}'$, called the *observation-subsumption problem*. The idea is to look for some conditions that either imply or are implied by such a relationship. If these conditions can be checked using a reasonable amount of computational (space and time) resources, then the chances are that we can give an answer to the subsumption problem efficiently.

Specifically, if a necessary condition N_c is violated, then the answer to the subsumption problem will be *no*. Dually, if a sufficient condition S_c holds, then the answer will be *yes*. However, if N_c holds and S_c is violated, then the subsumption problem remains unanswered.

Let $\mathcal{O} = (L,N,A)$ and $\mathcal{O}' = (L',N',A')$ be two temporal observations such that $\mathcal{O} \Supset \mathcal{O}'$. Let k and k' be the numbers of nodes in N and N', respectively. Let k_ε and k'_ε be the numbers of ε-nodes (nodes including a label ε) in N and N', respectively. Let σ (the number of *spare nodes*) and σ_ε be defined as

$$\sigma = k - k', \tag{9.12}$$

$$\sigma_\varepsilon = k_\varepsilon - k'_\varepsilon. \tag{9.13}$$

Let ℓ denote a label in $(L \cup L')$. Let k_ℓ and k'_ℓ be the numbers of nodes that include the label ℓ in N and N', respectively. Let k_ℓ^ε be the number of nodes in N that include both of the labels ℓ and ε. Let r_ℓ be the *residue* of ℓ, defined as the number of nodes in $\mathcal{O}$ containing the label ε without however containing the label ℓ, namely,

$$r_\ell = k_\varepsilon - k_\ell^\varepsilon. \tag{9.14}$$

Let β_ℓ be the *balance* of ℓ, defined as

$$\beta_\ell = \begin{cases} k_\ell & \text{if } \sigma \leq r_\ell, \\ k_\ell - (\sigma - r_\ell) & \text{otherwise}. \end{cases} \tag{9.15}$$

Then, four necessary conditions for the subsumption $\mathcal{O} \Supset \mathcal{O}'$ can be stated:

1. The number of nodes in N is at least equal to the number of nodes in N', namely $k \geq k'$. Otherwise, there would exist candidate traces in $\mathcal{O}'$ longer than (and, thereby, different from) any candidate trace in $\mathcal{O}$.
2. The number of ε-nodes in N is at least σ more than the number of ε-nodes in N', that is, $\sigma_\varepsilon \geq \sigma$. If not, all candidate traces in $\mathcal{O}$ would be longer than any candidate trace in $\mathcal{O}'$.
3. The set of labels L contains the set of labels L'. Otherwise, there would exist traces in $\mathcal{O}'$ that are not in $\mathcal{O}$.
4. For each label ℓ in L', we have $\beta_\ell \geq k'_\ell$.

The last condition requires some clarification. One may argue, *why not simply* $k_\ell \geq k'_\ell$? After all, this is in fact a necessary condition for subsumption, allowing candidate traces in $\mathcal{O}$ to have (at least) k'_ℓ occurrences of the label ℓ (otherwise, there would exist candidate traces in $\mathcal{O}'$ which are not in $\mathcal{O}$). So, in the condition, why do we replace k_ℓ by β_ℓ, with the latter being defined in (9.15)?

The point is, the condition $k_\ell \geq k'_\ell$ can be conveniently refined (restricted) by considering the fact that not all k_ℓ labels ℓ can actually be chosen so as to yield a candidate trace of $\mathcal{O}$ that has the same length as candidate traces in $\mathcal{O}'$. Based on (9.15), this is true only when $\sigma \leq r_\ell$ (as, in such a case, $\beta_\ell = k_\ell$).

The explanation lies in the definition of the residue given in (9.14). If the number σ of spare nodes is not greater than r_ℓ, then the label ε can be chosen in σ nodes without losing any label ℓ (this choice allows candidate traces in $\mathcal{O}$ to have the same length as candidate traces in $\mathcal{O}'$).

In contrast, if $\sigma > r_\ell$, in order to obtain the same length, the construction of the candidate traces in $\mathcal{O}$ requires choosing the label ℓ from nodes in $\mathcal{O}$ that contain the label ε too. Consequently, the number of labels ℓ actually selectable in $\mathcal{O}$ decreases exactly by $(\sigma - r_\ell)$ occurrences. This is why, according to (9.15), $\beta_\ell = k_\ell - (\sigma - r_\ell)$ when $\sigma > r_\ell$.

Necessary conditions for the subsumption problem are formalized in Proposition 9.9.

Proposition 9.9. *Based on the notation given above, if $\mathcal{O} \Supset \mathcal{O}'$, then*

$$\sigma \geq 0, \tag{9.16}$$
$$\sigma_\varepsilon \geq \sigma, \tag{9.17}$$
$$L \supseteq L', \tag{9.18}$$
$$\forall \ell \in L' \; (\beta_\ell \geq n'_\ell) \,. \tag{9.19}$$

Example 9.14. Displayed in Fig. 9.8 are two temporal observations of the system $\mathcal{A}_2$ (defined in Example 3.3), namely $\mathcal{O}_1$ (left) and $\mathcal{O}_2$ (right). It is easy to check that each candidate trace in $\|\mathcal{O}_2\|$ is also a candidate trace in $\|\mathcal{O}_1\|$. Hence, based on Def. 9.5, $\mathcal{O}_1$ subsumes $\mathcal{O}_2$. So, the conditions relevant to Proposition 9.9 are expected to hold for $\mathcal{O}_1$ and $\mathcal{O}_2$. We have $k = 5$, $k' = 4$, $n_\varepsilon = 3$, $k'_\varepsilon = 2$. Indeed, both of the conditions (9.16) and (9.17) hold. Moreover, since $L = \{awk, opl, cll, ide, opr, \varepsilon\}$ and $L' = \{awk, opl, cll, ide, \varepsilon\}$, the condition (9.18) holds. Finally, the condition (9.19) can be checked based on Table 9.2, where, for each label in L', the relevant parameters are listed. Since, for each row, $\beta_\ell \geq k'_\ell$, the condition (9.19) holds.

Table 9.2 Parameters for checking condition (9.19)

ℓ	k_ℓ	k_ℓ^ε	r_ℓ	β_ℓ	k'_ℓ
awk	3	2	1	3	2
opl	4	2	1	4	1
cll	1	0	3	1	1
ide	1	0	3	1	1
ε	3	3	0	2	2

The necessary conditions for subsumption stated in Proposition 9.9 can be easily checked. Thus, they correspond to the first actions of the checking algorithm. If any of them is violated, the algorithm terminates immediately with a negative answer. Otherwise, the algorithm continues by checking a sufficient condition for subsumption based on the notion of coverage defined in Sec. 9.4. The notion of coverage, along with some relevant properties, is formalized in the next section.

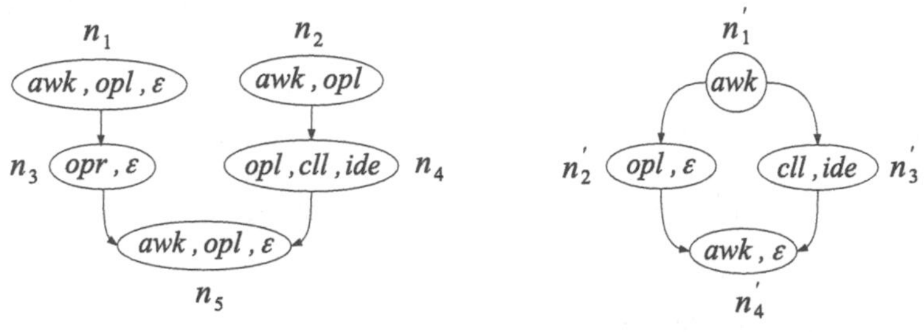

Fig. 9.8 Temporal observations $\mathcal{O}_1$ (left) and $\mathcal{O}_2$ (right)

9.4 Observation Coverage

Intuitively, an observation $\mathcal{O}$ covers an observation $\mathcal{O}'$ when the topology of $\mathcal{O}$ somehow contains the topology of $\mathcal{O}'$. Specifically, it is possible to define an isomorphism between a subset of the nodes of $\mathcal{O}$ and the set of nodes of $\mathcal{O}'$ such that relevant topology properties hold. Theorem 9.1 below shows that coverage is a sufficient condition for subsumption.

Definition 9.11 (Observation Coverage). Let $\mathcal{O} = (L,N,A)$ and $\mathcal{O}' = (L',N',A')$ be two temporal observations, where $N = \{n_1,\dots,n_k\}$ and $N' = \{n'_1,\dots,n'_{k'}\}$. We say that $\mathcal{O}$ *covers* $\mathcal{O}'$, written

$$\mathcal{O} \trianglerighteq \mathcal{O}', \tag{9.20}$$

iff there exists a subset $\bar{N}$ of N, with $\bar{N} = \{\bar{n}_1,\dots,\bar{n}_{k'}\}$ having the same cardinality as N', such that, writing $N^{\varepsilon} = (N - \bar{N})$, we have:

1. (ε-coverage) $\forall n \in N^{\varepsilon}$ $(\varepsilon \in \|n\|)$,
2. (Label coverage) $\forall i \in [1 \cdots k']$ $(\|\bar{n}_i\| \supseteq \|n'_i\|)$,
3. (Temporal coverage) $\forall$ pairs of nodes $(\bar{n}_i, \bar{n}_j)$ in $\bar{N}$ such that $\bar{n}_i \prec \bar{n}_j$ in $\mathcal{O}$, we have $n'_i \prec n'_j$ in $\mathcal{O}'$.

Example 9.15. With reference to the observations in Fig. 9.8, it is easy to show that $\mathcal{O}_1 \trianglerighteq \mathcal{O}_2$. Assume the subset of N_1 to be $\bar{N}_1 = \{n_2,n_1,n_4,n_5\}$, thereby corresponding to $N' = \{n'_1,n'_2,n'_3,n'_4\}$. Hence, $N_1^{\varepsilon} = \{n_3\}$. Clearly, ε-coverage holds, as $\varepsilon \in \|n_3\|$. Label coverage holds, as $\|n_2\| \supseteq \|n'_1\|$, $\|n_1\| \supseteq \|n'_2\|$, $\|n_4\| \supseteq \|n'_3\|$, and $\|n_5\| \supseteq \|n'_4\|$. It is easy to check that temporal coverage occurs. For instance, for $n_1 \prec n_5$ in $\mathcal{O}_1$ we have $n'_2 \prec n'_4$ in $\mathcal{O}_2$.

Theorem 9.1. *Coverage entails subsumption:*

$$\mathcal{O} \trianglerighteq \mathcal{O}' \implies \mathcal{O} \Supset \mathcal{O}' . \tag{9.21}$$

Proof. In addition to Def. 9.11, the proof is based on Def. 9.12.

Definition 9.12 (Sterile Sequence). Let $\tilde{N} = [\tilde{n}_1,\dots,\tilde{n}_{k'}]$ be a strict total ordering of nodes in $\bar{N}$. The *sterile sequence* of $\tilde{N}$,

$$\left[N_0^{\varepsilon}, N_1^{\varepsilon}, \dots, N_{n'}^{\varepsilon}\right], \tag{9.22}$$

is a sequence of subsets of N^{ε}, called *sterile sets*, inductively defined as follows.

(*Basis*) N_0^{ε} is defined by the following two rules:

1. If $n \in N^{\varepsilon}$, n is a root of $\mathcal{O}$, and then $n \in N_0^{\varepsilon}$,
2. If $n \in N^{\varepsilon}$, all parents of n are in N_0^{ε}, and then $n \in N_0^{\varepsilon}$.

(*Induction*) Given N_i^{ε}, $i \in [0 \cdots (k'-1)]$, N_{i+1}^{ε} is defined by the following two rules:

1. If $n \in N^{\varepsilon}$, all parents of n are in $(N_i^{*} \cup \{\tilde{n}_{i+1}\})$, and then $n \in N_{i+1}^{\varepsilon}$,

2. If $n \in N^{\varepsilon}$, all parents of n are in $\left(N_i^* \cup \{\tilde{n}_{i+1}\} \cup N_{i+1}^{\varepsilon}\right)$, and then $n \in N_{i+1}^{\varepsilon}$.

Here, N_i^*, $i \in [0 \cdots k']$, is recursively defined as follows:

$$N_i^* = \begin{cases} N_0^{\varepsilon} & \text{if } i = 0, \\ N_{i-1}^* \cup \{\tilde{n}_i\} \cup N_i^{\varepsilon} & \text{otherwise}. \end{cases} \tag{9.23}$$

To prove the theorem, it suffices to show that each candidate trace $\mathcal{T}$ in the index space of $\mathcal{O}'$ is also a candidate trace in the index space of $\mathcal{O}$, namely,

$$\forall \mathcal{T} \in \|Isp(\mathcal{O}')\| \; (\mathcal{T} \in \|Isp(\mathcal{O})\|). \tag{9.24}$$

According to Def. 4.4, a candidate trace $\mathcal{T}$ is the sequence of labels obtained by selecting, in compliance with the partial ordering of nodes in A', one label from each node in N', and by removing all the null labels ε. Let

$$\mathbb{N}' = [\tilde{N}_1', \ldots, \tilde{N}_{n'}'] \tag{9.25}$$

be the strict total ordering of N' relevant to the choices of such labels. Accordingly, the sequence $\mathbb{L}'$ of the chosen labels can be written as

$$\mathbb{L}' = [\ell \mid \ell \in \|\tilde{n}_i'\|, i \in [1 \cdots k']], \tag{9.26}$$

while the candidate trace $\mathcal{T}$ is in fact

$$\mathcal{T} = [\ell \mid \ell \in \mathbb{L}', \ell \neq \varepsilon]. \tag{9.27}$$

We need to show that there exists a strict total ordering $\mathbb{N}$ of N fulfilling the partial ordering imposed by A, from which it is possible to select a sequence $\mathbb{L}$ of labels,

$$\mathbb{L} = [\ell_1, \ell_2, \ldots, \ell_k], \tag{9.28}$$

such that the subsequence of nonnull labels in $\mathbb{L}$ equals $\mathcal{T}$:

$$[\ell \mid \ell \in \mathbb{L}, \ell \neq \varepsilon] = \mathcal{T}. \tag{9.29}$$

Note how $\mathbb{N}$ (as well as any other strict total ordering of N) can be represented as a sequence of nodes in $\bar{N}$, with each node being interspersed with (possibly empty) subsequences $\mathbb{N}_i^{\varepsilon}$ of nodes in N^{ε}, specifically

$$\mathbb{N} = \mathbb{N}_0^{\varepsilon} \cup [\tilde{n}_1] \cup \mathbb{N}_1^{\varepsilon} \cup [\tilde{n}_2] \cup \mathbb{N}_2^{\varepsilon} \ldots [\tilde{n}_{k'}] \cup \mathbb{N}_{k'}^{\varepsilon}, \tag{9.30}$$

where

$$\bigcup_{i=1}^{k'} \{\tilde{n}_i\} = \bar{N}, \quad \bigcup_{i=0}^{k'} \mathbb{N}_i^{\varepsilon} = N^{\varepsilon}, \quad \bigcap_{i=0}^{k'} \mathbb{N}_i^{\varepsilon} = \emptyset. \tag{9.31}$$

The proof is by induction on $\mathbb{L}'$. Let $\mathbb{L}_i'$ denote the subsequence of $\mathbb{L}'$ up to the i-th label, $i \in [1 \cdots k']$. Let $\mathbb{L}_i$ denote the subsequence of $\mathbb{L}$ relevant to the choices of labels made in correspondence with the labels in $\mathbb{L}'_i$. Let $\mathcal{T}_i$ and $\mathcal{T}'_i$ denote the candidate traces corresponding to $\mathbb{L}_i$ and $\mathbb{L}'_i$, respectively.[7]

(*Basis*) No label is chosen in $\mathcal{O}'$, that is, $\mathbb{L}'_0 = []$. We choose a sequence of empty labels for all the nodes in N_0^ε, which is clearly possible according to the property that N_0^ε is a sterile set composed of nodes having ancestors in N^ε only. In other words, $\mathbb{N}_0^\varepsilon$ is a strict total ordering of N_0^ε, while $\mathbb{L}_0 = [\varepsilon, \varepsilon, \ldots, \varepsilon]$; hence, $\mathcal{T}_0 = \mathcal{T}'_0 = []$.

(*Induction*) We assume that $\mathbb{L}_i$ and $\mathbb{L}'_i$, $i \in [0 \cdots (k'-1)]$, are such that $\mathcal{T}_i = \mathcal{T}'_i$. We also assume that, given the sequence $[\tilde{n}'_1, \ldots, \tilde{n}'_i]$ of chosen nodes in $\mathcal{O}'$, the corresponding sequence of chosen nodes in $\mathcal{O}$ is $\mathbb{N}_0^\varepsilon \cup [\tilde{n}_1] \cup \mathbb{N}_1^\varepsilon \cup [\tilde{n}_2] \cup \mathbb{N}_2^\varepsilon \ldots [\tilde{n}_i] \cup \mathbb{N}_i^\varepsilon$, where, $\forall j \in [1 \cdots i]$, if $\tilde{n}'_j$ is the node n'_h in N', then $\tilde{n}_j$ is the node $\bar{n}_h$ in $\bar{N}$, and each $\mathbb{N}_j^\varepsilon$ is a strict total ordering of N_j^ε. We have to show that, once the next label $\ell \in \|\tilde{n}'_{i+1}\|$ has been chosen, thereby determining $\mathbb{L}'_{i+1}$ and $\mathcal{T}'_{i+1}$, it is possible to choose a node $\tilde{n}_{i+1} \in \bar{N}$ that includes ℓ, and to choose $\mathbb{N}_{i+1}^\varepsilon$ as a strict total ordering of N_{i+1}^ε from which ε is chosen, thereby determining $\mathbb{L}_{i+1}$ such that $\mathcal{T}_{i+1} = \mathcal{T}'_{i+1}$.

Let N'_m be the node in $N' = \{n'_1, \ldots, n'_{k'}\}$ corresponding to $\tilde{n}'_{i+1}$. According to the condition for label coverage in Def. 9.11, there exists a node $\bar{n}_m$ in $\bar{N} = \{\bar{n}_1, \ldots, \bar{n}_{k'}\}$ such that $\|\bar{n}_m\| \supseteq \|\bar{n}'_m\|$, in other words, $\bar{n}_m$ includes ℓ. We choose $\tilde{n}_{i+1} = n_m$. In order for $\bar{n}_m$ to be actually chosen, we have to show that each parent node n of $\bar{n}_m$ in $\mathcal{O}$ has already been considered, that is, n belongs to the prefix of $\mathbb{N}$ relevant to $\mathbb{L}_i$. Two cases are possible for n:

(*a*) n is a node $\bar{n}_j \in \bar{N}$. On the one hand, because of temporal coverage, $\bar{n}_j \rightarrow \bar{n_m}$ in $\mathcal{O}$ entails $n'_j \prec n'_m$ in $\mathcal{O}'$. On the other hand, since n'_m has been chosen in $\mathcal{O}'$, all its parent nodes must have been considered already, that is, $n'_j \in [\tilde{n}'_1, \ldots, \tilde{n}'_i]$. Since, based on the induction assumption, we always choose for each node in $n'_p \in N'$ the corresponding node $\bar{n}_p \in \bar{N}$, it is possible to claim that $\bar{n}_j$ has already been considered in $\mathcal{O}$, that is, $\bar{n}_j \in [\tilde{n}_1, \ldots, \tilde{n}_i]$.

(*b*) $n \in N_\varepsilon$. We consider each path $n_{\mathrm{a}} \rightsquigarrow n$ in $\mathcal{O}$ such that n_{a} is the first ancestor of n (possibly n itself), where either n_{a} is a root of $\mathcal{O}$ or $n_{\mathrm{a}} \in \bar{N}$. Let N_{a} be the set of such ancestors. We show that each node $n_{\mathrm{a}} \in N_{\mathrm{a}}$ has been considered already. Two cases are possible: either $n_{\mathrm{a}} \in N_\varepsilon$ or $n_{\mathrm{a}} \in \bar{N}$. In the first case, n_{a} is a node in the sterile set N_0^ε and, hence, it has been considered in $\mathbb{N}_0^\varepsilon$ already (see the basis). In the second case ($n_{\mathrm{a}} \in \bar{N}$), let $\bar{n}_h$ be the node in $\bar{N}$ corresponding to n_{a}. We consider a path $\bar{n}_h \rightsquigarrow n \rightarrow \bar{n}_m$. Since $\bar{n}_h \prec \bar{n}_m$ in $\mathcal{O}$, temporal coverage implies that $n'_h \prec n'_m$ in $\mathcal{O}'$, where n'_h is the node in N' corresponding to $\bar{n}_h$. Thus, n'_h has already been considered in $\mathcal{O}'$. As, based on the induction assumption, we always choose in $\mathcal{O}$ the node corresponding to that chosen in $\mathcal{O}'$, this implies that $\bar{n}_h$ has already been considered in $\mathcal{O}$. We conclude that all nodes in N_{a} have been considered. Now, it is clear that either n is in N_{a} or n is a node belonging to the sterile set of some node in N_{a}. In either case, because of the induction assumption, n must have been considered already. In other words, all parents of $\bar{n}_m$ have been chosen already, thereby allowing $\bar{n}_m$ itself, alias $\tilde{n}_{i+1}$, to be chosen. Furthermore, based on the definition of a sterile sequence, we may also

[7] Note how $\mathbb{L}'_i$ includes exactly i labels, while, because of the ε selected for the nodes in N_ε, the number of labels in $\mathbb{L}_i$ may possibly be greater than i.

Fig. 9.9 Observations $\mathcal{O}$ (left) and $\mathcal{O}'$ (right)

consider a strict total ordering $\mathbb{N}^{\varepsilon}_{i+1}$ of N^{ε}_{i+1} and choose a label ε for each of those nodes, thereby leading to the conclusion that $\mathcal{T}_{i+1} = \mathcal{T}'_{i+1}$. □

Theorem 9.2. *Subsumption does not entail coverage:*

$$\mathcal{O} \Supset \mathcal{O} \;\not\Rightarrow\; \mathcal{O} \trianglerighteq \mathcal{O}' . \tag{9.32}$$

Proof. It suffices to show a counterinstance in which subsumption holds while coverage does not. Consider the two observations $\mathcal{O}$ and $\mathcal{O}'$ displayed in Fig. 9.9. Notice how, unlike $\mathcal{O}$, $\mathcal{O}'$ does not force any temporal constraint between the two nodes. Incidentally, both observations involve just one candidate trace, namely $\mathcal{T} = [awk, awk]$. Thus, since $\|\mathcal{O}\| = \|\mathcal{O}'\| = \{[awk, awk]\}$, both observations subsume each other; in particular, $\mathcal{O} \Supset \mathcal{O}'$. However, it is clear that $\mathcal{O}$ does *not* cover $\mathcal{O}'$; namely, $\mathcal{O} \not\trianglerighteq \mathcal{O}'$. □

Theorems 9.1 and 9.2 prove that coverage is only a sufficient condition for subsumption, not a necessary one. Since coverage entails subsumption, conditions (9.16) to (9.19) in Proposition 9.9 are necessary for coverage too. However, there exists a condition, formalized in Proposition 9.10, that is necessary for temporal coverage while it is not necessary for subsumption (as this condition constrains entities that are defined for coverage only).

Proposition 9.10. *If $\mathcal{O} \trianglerighteq \mathcal{O}'$, then $\forall n' \in N'$, where n' is a root node of $\mathcal{O}'$, the corresponding node $n \in \bar{N}$ is such that, $\forall n_i \in N$ with $n_i \prec n$, we have $n_i \in N^{\varepsilon}$.*

Informally, this condition requires that each root node of $\mathcal{O}'$ matches either a root node of $\mathcal{O}$ or a node whose ancestors are all spare nodes, thus being reachable through a chain of nodes in N^{ε} starting from a root node of $\mathcal{O}$ that belongs to N^{ε}.

Proof. The proof is by contradiction. Let $\bar{N}$ be the subset of N, having the same cardinality as N', containing all and only the nodes of $\mathcal{O}$ corresponding to the nodes of $\mathcal{O}'$ in the coverage relation $\mathcal{O} \trianglerighteq \mathcal{O}'$. Suppose there exists a root node n' of $\mathcal{O}'$, corresponding to a node $n \in \bar{N}$ in the coverage relation, that does not fulfill the condition stated in Proposition 9.10. Then there exists a node $\bar{n}_i \in \bar{N}$, which corresponds to a node $\bar{n'}_i \in N'$, where $\bar{n}_i$ is distinct from n in the same way as $\bar{n'}_i$ is distinct from n', such that $\bar{n}_i \prec n$. However, $\bar{n'}_i \nprec n'$, since either $\bar{n'}_i$ is a root node of $\mathcal{O}'$, in which case it does not precede n', since n' is a root node of $\mathcal{O}'$ and all root nodes are temporally unrelated, or it is not a root node of $\mathcal{O}'$, in which case it cannot precede n', the latter being a root node. Since $\bar{n'}_i \nprec n'$ while $\bar{n}_i \prec n$, we can conclude that temporal coverage does not hold, a contradiction of the hypothesis $\mathcal{O} \trianglerighteq \mathcal{O}'$. □

algorithm *SBS*(**in** $\mathcal{O}$, **in** $\mathcal{O}'$)
begin
Compute the parameters of $\mathcal{O}$ and $\mathcal{O}'$ relevant to Proposition 9.9;
If there exists a condition of Proposition 9.9 that is violated then return **no**;
Set $\mathcal{R}$ to the empty stack and choose a root node n' of $\mathcal{O}'$;
Return *Recurse*(n');
end.

auxiliary function *Recurse*(**in** n')
begin
If the number of pairs in $\mathcal{R}$ equals the number of nodes in $\mathcal{O}'$ then return **yes**;
For each node n in $\mathcal{O}$ whose parents have been consumed do
If $\|n\| \supseteq \|n'\|$ and
necessary conditions of Proposition 9.9 still hold after consumption of n and
for each (n_a, n'_a) in $\mathcal{R}$, with n_a ancestor of n, we have $n'_a \prec n'$ in $\mathcal{O}'$ then
Push (n, n') into $\mathcal{R}$;
Choose a new node n'' in $\mathcal{O}'$ whose parents are in $\mathcal{R}$;[a]
If *Recurse*(n'') = **yes** then return **yes**
end
end;
Pop the pair on top of $\mathcal{R}$;
For each ε-node n_ε in $\mathcal{O}$ whose parents have been consumed do
If assuming n_ε in N^ε all conditions of Proposition 9.9 still hold then
If *Recurse*(n') = **yes** then return **yes**
end;
Return **unknown**
end.

[a] Formally, if there is no n'' to be chosen, then n'' = **nil**.

Fig. 9.10 Abstract specification of *SBS* algorithm

9.5 Checking Observation Subsumption via Coverage

We first give an abstract specification (Fig. 9.10) of an algorithm for subsumption checking via coverage. The algorithm, called *SBS*, tests both the conditions of Proposition 9.9 and the coverage relationship. Coverage checking aims to uncover a set $\bar{N} \subseteq N$, with the same cardinality as N', such that there exists a correspondence between nodes of $\bar{N}$ and N' for which ε-coverage, label coverage, and temporal coverage hold.

The abstract specification of the *SBS* algorithm makes use of an auxiliary function *Recurse*. The associations (n, n') between nodes in $\mathcal{O}$ and $\mathcal{O}'$ (establishing, according to Def. 9.11, the correspondence between $\bar{N}$ and N') are generated on the stack (relation) $\mathcal{R}$. Recursion is performed on nodes of $\mathcal{O}'$: in compliance with Proposition 9.10, a new root node n' of the remaining portion of $\mathcal{O}'$ is chosen and a corresponding root node n is chosen in the remaining portion of $\mathcal{O}$. Moreover, the association (n, n') has to be such that the conditions of Proposition 9.9 hold assuming the consumption of n. As such, these conditions are not checked once and for all. Rather, they are checked after each association (n, n') is generated, based on the recursive nature of the algorithm.[8] If no candidate node in $\mathcal{O}$ makes the algorithm succeed at a given computation point, assumptions about spare nodes are tried (ε-nodes can be assumed to be in N^ε). If this fails too, **unknown** is returned.

Example 9.16. Displayed in Fig. 9.11 are observations $\mathcal{O}_1$ and $\mathcal{O}_2$ (introduced in Example 9.14) on which $\mathcal{O}_1 \Supset \mathcal{O}_2$ is checked by means of *SBS*. Pairs of nodes involved in $\mathcal{R}$ are linked by dotted lines. The pair (n_1, n'_1) is first tried and subsequently discarded. Moreover, node n_3 is assumed to be in N^ε (as it cannot cover with labels any node in $\mathcal{O}_2$). The answer from *SBS* is therefore **yes** (which is consistent with the result obtained in Example 9.15 based on Def. 9.11).

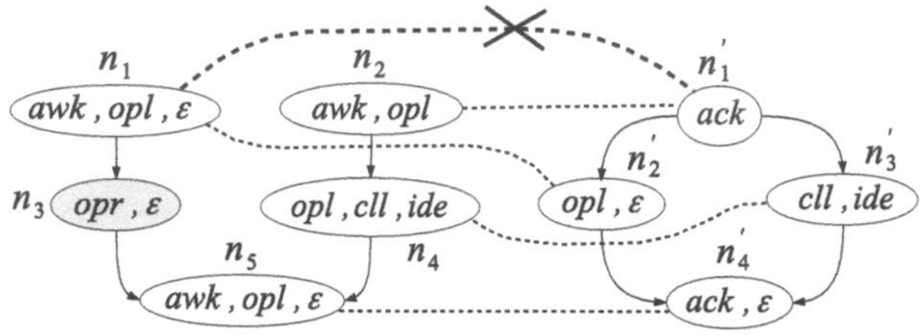

Fig. 9.11 Observations $\mathcal{O}_1$ (left) and $\mathcal{O}_2$ (right), with node associations (dotted lines) given by *SBS*

9.5.1 Detailed Specification of SBS *Algorithm*

The detailed pseudocode of the function *SBS* is provided below (page 220). It takes as input two observations, $\mathcal{O}$ and $\mathcal{O}'$, and outputs a pseudo-Boolean value (either **yes**, **no**, or **unknown**) indicating whether or not $\mathcal{O}$ subsumes $\mathcal{O}'$. In particular, since **unknown** means a lack of coverage, it also indicates uncertainty as far as subsumption is concerned.[9]

The body of *SBS* is listed in lines 86–101 (after the definition of the three auxiliary functions *InitSymbTab*, *UpdateSymbTab*, and *Recurse*). In lines 87 and 88, the observation parameters for $\mathcal{O}$ and $\mathcal{O}'$ are set. Then, at line 89, the first three necessary conditions for subsumption (given in Proposition 9.9, namely (9.16), (9.17), and (9.18)), are checked. At line 92, each ε-node in N is marked as a *candidate spare node*.[10] At line 93, all labels in L that are not in L' are removed both from

[8] After all, Def. 9.11 (coverage) may be defined equivalently in recursive terms.

[9] Since coverage is only a sufficient condition for subsumption, nothing about subsumption can be derived from a lack of coverage, hence **unknown**.

[10] Such nodes may possibly be considered as belonging to N^ε during the computation.

L and from nodes in N.[11] At line 94, the symbol table $\mathcal{S}$ is initialized by the *InitSymbTab* auxiliary function.

As detailed below, $\mathcal{S}$ is a dynamic catalog that keeps information on the consumption of labels during the computation. Both the initialization of $\mathcal{S}$ and each subsequent update check the consistency of the consumption of labels, specifically, condition (9.19) in Proposition 9.9.[12] If consistency is preserved, the new configuration of the symbol table is returned, otherwise the value returned is **nil**. If all conditions of Proposition 9.9 hold, a root node n' of $\mathcal{O}'$ is chosen at line 97.[13] Then, the algorithm yields $\bar{N}$, the subset of N that is associated with N' in Def. 9.11, by building the sequence $\mathcal{R}$ of associations through a call to the auxiliary function *Recurse* at line 98.

The specification of *Recurse* is given in lines 51–85. The input parameters are:

1. $\mathcal{S}$: the symbol table.
2. $\mathcal{O}$: the first observation.
3. $\mathcal{O}'$: the second observation.
4. $\mathcal{C}$: the set of nodes consumed (up to now) for $\mathcal{O}$.
5. $\mathcal{C}'$: the set of nodes consumed (up to now) for $\mathcal{O}'$.
6. n: a root node of the portion of $\mathcal{O}'$ not yet considered.
7. σ: the number of nodes in N^{ε} not yet considered (spare nodes).
8. $\mathcal{W}$: the *worthless set*, a subset of nodes in N that, because of previous failures, cannot be associated with n' in this state of computation.
9. $\mathcal{R}$: the ordered set of associations between nodes in $\bar{N}$ and N' built so far.

Recurse returns either **yes** or **unknown**, depending on whether coverage holds or it does not, respectively. The term **unknown** is used in reference to subsumption (as coverage is only a sufficient condition for subsumption).

The body of *Recurse* starts at line 61, where the cardinality of $\mathcal{R}$ is tested: if $\mathcal{R}$ contains k' pairs, this means that all nodes in N' have been considered and $\bar{N}$ has been completed; thereby, coverage holds (and, hence, subsumption too). Otherwise, at line 64, the set $\mathcal{F}$ of (frontier) nodes in $\mathcal{O}$ (they are compliant with Proposition 9.10) is created, which includes all nodes N of $\mathcal{O}$ such that:

1. n has not yet been consumed, namely $n \notin \mathcal{C}$.
2. n does not belong to the worthless set, namely $n \notin \mathcal{W}$.
3. All parents of n have been consumed already, that is, they are in $\mathcal{C}$.

A loop for each node n in $\mathcal{F}$ is iterated in lines 65–75. First, label coverage is tested (line 66), along with the update of the symbol table $\mathcal{S}$ by means of the *UpdateSymbTab* auxiliary function, which also checks the consistency of the consumption of such labels, based on condition (9.19) in Proposition 9.9. Then, the set N_{a} of the nearest ancestors of n which have already been involved in the associations of $\mathcal{R}$ is instantiated (line 67). More precisely, the set N_{a} is determined as follows. N_{a} is

[11] This is possible because label coverage does not depend on them.

[12] The condition (9.19) is not checked once and for all: it is checked at any update of the symbol table by the *UpdateSymbTab* auxiliary function instead.

[13] Formally, in the special case in which $\mathcal{O}'$ is empty ($N' = \emptyset$), the chosen node n' is **nil**.

first initialized to $\emptyset$. Then, for each node n_a in $\mathcal{R}(N)$ which is an ancestor of n, two update rules for N_a are considered:

1. If n_a is unrelated to any node in N_a, then n_a is inserted into N_a;
2. If n_a is the descendant of a node $n_a^* \in N_a$, then n_a^* is replaced by n_a.

This allows temporal-coverage checking (line 68). If the latter succeeds, at line 69, a new node n'' that is a root node of the remaining portion of $\mathcal{O}'$ is chosen from among the nodes in N' this way:[14]

1. n'' has not yet been consumed, namely $n'' \notin \mathcal{C}'$.
2. All parents of N'' are in $\mathcal{C}' \cup \{n'\}$.

Then, *Recurse* is recursively called at line 70, with the new actual parameters being the new symbol table $\mathcal{S}'$, the sets $\mathcal{C}$ and $\mathcal{C}'$ of the nodes considered extended with n and n', respectively, the new node n'', σ,[15] the empty worthless set $\mathcal{W}$, and the new instance of $\mathcal{R}$, which is extended by the pair (n, n'). If such a call succeeds, the current activation of *Recurse* succeeds too (line 71).

If none of the nodes in $\mathcal{F}$ makes *Recurse* succeed, in compliance with Proposition 9.10, chances still remain if each node $n \in \mathcal{F}$ is assumed to be a spare node, namely $n \in N^{\varepsilon}$: this is viable only on condition that n is marked as a candidate spare node, $\sigma > 0$ (at least one spare node is available), and the symbol table, once updated in consequence of the consumption of labels of n, remains consistent (line 77).[16] If so, a further recursive call to *Recurse* is performed at line 78 in order to match node n', with the changed parameters being the new symbol table $\mathcal{S}'$, the (extended) set $\mathcal{C}$ of consumed nodes in $\mathcal{O}$, the decremented value of σ,[17] and the worthless set extended with $\mathcal{F} - \{n\}$.

If such a call succeeds, the current activation of *Recurse* will succeed too. Otherwise, the loop is iterated and a new node in $\mathcal{F}$ is tried. If the computation exits the loop in a natural way, this means that no node can be associated with n' in this computational context, thereby causing the current activation of *Recurse* to fail (line 83).[18]

The other two auxiliary functions, *InitSymbTab* and *UpdateSymbTab*, operate on the symbol table. Each row of the symbol table maintains information on a label in L' through a quadruple $(\ell, d_\ell, r_\ell, h_\ell)$, where ℓ is the specific label concerned. The second field, d_ℓ, is the difference between the number k_ℓ of nodes of N containing ℓ and the number k'_ℓ of nodes of N' containing the same label ℓ, that is,

$$d_\ell = k_\ell - k'_\ell. \tag{9.33}$$

The third field, r_ℓ, is the residue of ℓ, namely, according to (9.14),

[14] Formally, if all nodes of N' have been considered already (namely $\mathcal{C}' = N'$), then $n'' =$ **nil**.

[15] The value of σ does not change, because no spare node is consumed in this case.

[16] In the actual parameters of the call to *UpdateSymbTab*, σ is decremented, as a new spare node is assumed, while $n' =$ **nil**, as no node is consumed in N'.

[17] Here, σ is decremented as, in the current call to *Recurse*, n is assumed to be a spare node, $n \in N^{\varepsilon}$.

[18] As pointed out above, the returned value is **unknown**, denoting uncertainty about subsumption.

$$r_\ell = k_\varepsilon - k_\ell^\varepsilon. \tag{9.34}$$

The last field, h_ℓ, is defined as

$$h_\ell = \begin{cases} 0 & \text{if } \sigma \le r_\ell, \\ \sigma - r_\ell & \text{otherwise}. \end{cases} \tag{9.35}$$

Comparing (9.35) with (9.15), it follows that

$$h_\ell = k_\ell - \beta_\ell. \tag{9.36}$$

The auxiliary function *InitSymbTab* generates either the initial configuration of the symbol table or **nil**, depending on whether the number of labels in N and N' is consistent with coverage or not, respectively. In addition to N and N', it takes as input k and k', the numbers of nodes in N and N', respectively, and k_ε, the number of nodes in N that include ε. The symbol table is first set to the empty set (line 9). The instantiation of $\mathcal{S}$ is performed by the loop in lines 11–22. Specifically, the parameters k_ℓ, k'_ℓ, k_ℓ^ε, d_ℓ, r_ℓ, and h_ℓ are computed (lines 12–17). Then, consistency for label ℓ is checked at line 18, based on condition (9.19) in Proposition 9.9, namely

$$\beta_\ell \ge k'_\ell, \tag{9.37}$$

which, in terms of the parameters in the symbol table, is equivalent to

$$d_\ell \ge h_\ell\,. \tag{9.38}$$

In fact, based on (9.36), the body of the quantified expression of condition (9.19) becomes $k_\ell - h_\ell \ge k'_\ell$, in other words, $k_\ell - k'_\ell \ge h_\ell$, that is, $d_\ell \ge h_\ell$, namely the *balancing condition*. Thus, if and only if the balancing condition holds ($d_\ell \ge h_\ell$) is the symbol table extended with the row $(\ell, d_\ell, r_\ell, h_\ell)$; otherwise, a **nil** value is returned. In the end, after the natural termination of the loop, the constructed symbol table $\mathcal{S}$ is returned at line 23.

Once initialized, the symbol table is required to be updated whenever a node is consumed by *Recurse* (line 66 or 77). The update is performed by the auxiliary function *UpdateSymbTab* (lines 25–50). This function takes as input the current symbol table $\mathcal{S}$, σ, and the two nodes n and n' that are involved in the consumption, with n' possibly being **nil** (the call to *UpdateSymbTab* at line 77). In lines 30–32, the d_ℓ field of each row relevant to labels that are in $\|n\|$ but not in $\|n'\|$ is decremented.[19]

Then, in lines 33–37, if n is a candidate spare node, the values of r_ℓ and h_ℓ are updated for all labels $\ell \in (L' - \|n\|)$.[20] Specifically, r_ℓ is decremented, while h_ℓ is set according to (9.35).

Furthermore, if $n' = \textbf{nil}$ (line 38), this means that n has been assumed to be a spare node ($n \in N^\varepsilon$), thereby causing σ to be decremented before being passed to

[19] In fact, if a label belongs to both $\|n\|$ and $\|n'\|$, d_ℓ does not change.

[20] In fact, based on (9.34), for labels in $\|n\|$, the value of r_ℓ does not change, as both k_ε and k_ℓ^ε decrement by one.

UpdateSymbTab (line 77). Based on (9.35), decrementing σ affects the field h_ℓ. However, since h_ℓ has just been updated at line 38 for all labels in $L' - \|n\|$, the only relevant labels are now those in $\|n\|$, that is, those not involved in the previous update of h_ℓ.

Finally, in lines 44–48, the balancing condition (9.38) can be checked for each label relevant to a row that has been updated so far (that is, in lines 30–43): if it is violated for a row, a **nil** value is returned. Otherwise, the updated symbol table $\mathcal{S}$ is returned at line 49.

The pseudocode of the function *SBS* is as follows:

1. **algorithm** *SBS* (**in** $\mathcal{O}$, **in** $\mathcal{O}'$, **out** *answer*)
2. $\mathcal{O} = (N, L, A)$, $\mathcal{O}' = (N', L', A')$: two temporal observations,
3. *answer*: either **yes**, **no**, or **unknown**;

4. **auxiliary function** *InitSymbTab* $(N, N', k, k', k_\varepsilon)$: either the initialized symbol table or **nil**
5. N, N': the sets of nodes for observations $\mathcal{O}$ and $\mathcal{O}'$, respectively,
6. k, k': the numbers of nodes in N and N', respectively,
7. k_ε: the number of candidate spare nodes in N;
8. **begin** $\langle$*InitSymbTab*$\rangle$
9. $\mathcal{S} := \emptyset$;
10. $\sigma := (k - k')$;
11. **foreach** $\ell \in L'$ **do**
12. k_ℓ := the number of nodes $n \in N$ such that $\ell \in \|n\|$;
13. k'_ℓ := the number of nodes $n' \in N'$ such that $\ell \in \|n'\|$;
14. k^ε_ℓ := the number of candidate spare nodes $n_\varepsilon \in N$ such that $\ell \in \|n_\varepsilon\|$;
15. $d_\ell := (k_\ell - k'_\ell)$;
16. $r_\ell := (k_\varepsilon - k^\varepsilon_\ell)$;
17. $h_\ell :=$ **if** $\sigma \leq r_\ell$ **then** 0 **else** $\sigma - r_\ell$ **endif**;
18. **if** $d_\ell \geq h_\ell$ **then**
19. Insert row $(\ell, d_\ell, r_\ell, h_\ell)$ into $\mathcal{S}$
20. **else**
21. **return nil**
22. **endfor**;
23. **return** $\mathcal{S}$
24. **end** $\langle$*InitSymbTab*$\rangle$;

25. **auxiliary function** *UpdateSymbTab* $(\mathcal{S}, \sigma, n, n')$: either **nil** or the updated symbol table
26. $\mathcal{S}$: the symbol table,
27. σ: the number of spare nodes (still consumable),
28. n, n': two node identifiers, where $n \in N$ and $n' \in (N' \cup \{\mathbf{nil}\})$;
29. **begin** $\langle$*UpdateSymbTab*$\rangle$
30. **foreach** $\ell \in (\|n\| - \|n'\|)$, $(\ell, d_\ell, r_\ell, h_\ell) \in \mathcal{S}$ **do**
31. $d_\ell := d_\ell - 1$
32. **endfor**;

```
    if n is a candidate spare node then
      foreach ℓ ∈ (L′ − ‖n‖), (ℓ, d_ℓ, r_ℓ, h_ℓ) ∈ 𝒮 do
        r_ℓ := r_ℓ − 1;
        h_ℓ := if σ ≤ r_ℓ then 0 else σ − r_ℓ endif
      endfor;
      if n′ = nil then
        foreach ℓ ∈ ‖n‖, (ℓ, d_ℓ, r_ℓ, h_ℓ) ∈ 𝒮 do
          h_ℓ := if σ ≤ r_ℓ then 0 else σ − r_ℓ endif
        endfor
      endif
    endif;
    foreach row (ℓ, r_ℓ, d_ℓ, h_ℓ) ∈ 𝒮 that was updated so far do
      if d_ℓ < h_ℓ then
        return nil
      endif
    endfor;
    return 𝒮
  end ⟨UpdateSymbTab⟩;

  auxiliary function Recurse (𝒮, 𝒪, 𝒪′, 𝒞, 𝒞′, n′, σ, 𝒲, ℛ): either yes or
                                                                  unknown
    𝒮: the symbol table,
    𝒪 = (L, N, A): a temporal observation (with possibly removed labels),
    𝒪′ = (L′, N′, A′): a temporal observation,
    𝒞, 𝒞′: the sets of consumed nodes for 𝒪 and 𝒪′, respectively,
    n′: a node in (N′ − 𝒞′),
    σ: the number of candidate spare nodes still consumable in N_ε,
    𝒲 ⊂ N: the set of worthless nodes,
    ℛ ⊆ (N × N′): a relation on N and N′;
  begin ⟨Recurse⟩
    if |ℛ| = k′ then
      return yes
    else
      ℱ := the set of n ∈ (N − (𝒞 ∪ 𝒲)) where all parents of n are in 𝒞;
      foreach n ∈ ℱ do
        if ‖n‖ ⊇ ‖n′‖ and (𝒮′ := UpdateSymbTab(𝒮, σ, n, n′)) ≠ nil then
          N_a := the set of nearest ancestors of n which are in ℛ(N);
          if ∀n_a ∈ N_a, (n_a, n′_a) ∈ ℛ (n′_a ≺ n′) then
            Choose a node n″ ∈ (N′ − (𝒞′ ∪ {n′})),
            where all parents of n″ are in 𝒞′ ∪ {n′}, if any, or n″ := nil;
            if Recurse (𝒮′, 𝒪, 𝒪′, 𝒞 ∪ {n}, 𝒞′ ∪ {n′}, n″, σ, ∅, ℛ ∪ {(n, n′)}) = yes then
              return yes
            endif
          endif
        endif
```

75. **endfor**;
76. **foreach** $n \in \mathcal{F} \cup \mathcal{W}$ marked as a candidate spare node **do**
77. **if** $\sigma > 0$ **and** $(\mathcal{S}' := \mathit{UpdateSymbTab}(\mathcal{S}, \sigma - 1, n, \textbf{nil})) \neq$ **nil then**
78. **if** $\mathit{Recurse}(\mathcal{S}', \mathcal{O}, \mathcal{O}', \mathcal{C} \cup \{n\}, \mathcal{C}', n', \sigma - 1, (\mathcal{W} \cup \mathcal{F}) - \{n\}, \mathcal{R}) =$ **yes then**
79. **return yes**
80. **endif**
81. **endif**
82. **endfor**;
83. **return unknown**
84. **endif**
85. **end** $\langle \mathit{Recurse} \rangle$;

86. **begin** $\langle \mathit{SBS} \rangle$
87. $k := |N|$; $k_\varepsilon := |\{n \mid n \in N, \varepsilon \in \|n\|\}|$;
88. $k' := |N'|$; $k'_\varepsilon := |\{n' \mid n' \in N', \varepsilon \in \|n'\|\}|$;
89. **if** $k < k'$ **or** $k_\varepsilon - k'_\varepsilon < k - k'$ **or** $L \not\supseteq L'$ **then**
90. **return no**
91. **else**
92. Mark as a *candidate spare node* each node in N that includes ε;
93. Remove from L, and from all nodes in N, all the labels (ε aside) that are not in L';
94. **if** $(\mathcal{S} := \mathit{InitSymbTab}(N, N', k, k', k_\varepsilon)) =$ **nil then**
95. **return no**
96. **else**
97. Choose a root node n' of $\mathcal{O}'$;
98. **return** $\mathit{Recurse}(\mathcal{S}, \mathcal{O}, \mathcal{O}', \emptyset, \emptyset, n', k - k', \emptyset, \emptyset)$
99. **endif**
100. **endif**
101. **end** $\langle \mathit{SBS} \rangle$.

Example 9.17. With reference to the observations displayed on the left of Fig. 9.12, consider a possible run of $\mathit{SBS}(\mathcal{O}_1, \mathcal{O}_2)$. According to Example 9.14, all the necessary conditions of Proposition 9.9 hold. Therefore, computation continues at line 92, where N_1, N_3, and N_5 are marked as candidate spare nodes. Then, at line 93, the label *wait* is removed from N_3 (as well as from $\mathcal{L}$). The symbol table $\mathcal{S}$ generated at line 94 is shown on the right of Fig. 9.12 (compare with Table 9.2).

After choosing the (unique) root node n'_1 of $\mathcal{O}'$ at line 97, the first call to *Recurse* is performed at line 98, where we assume[21] $n' = n'_1$, and thereby $\mathcal{F} = \{n'_1, n_2\}$. In the loop at line 65, choosing $n = n_1$ causes a call to *UpdateSymbTab* (line 66). The set of labels to be considered is $\{opl, \varepsilon\}$ (those not included in $\|n'_1\|$). In the loop at line 30, considering $\ell = opl$, d_{opl} is decremented, and thereby $d_{opl} = 2$. Then, considering $\ell = \varepsilon$, d_ε is decremented, and thereby $d_\varepsilon = 0$. Since n_1 is a candidate spare node, both r_{cll} and r_{ide} are decremented (line 35), and thereby $r_{cll} = 2$ and

[21] The result of the algorithm does not depend on the order in which we choose a node within the set of root nodes.

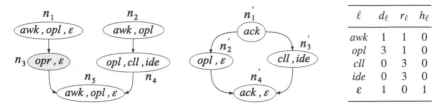

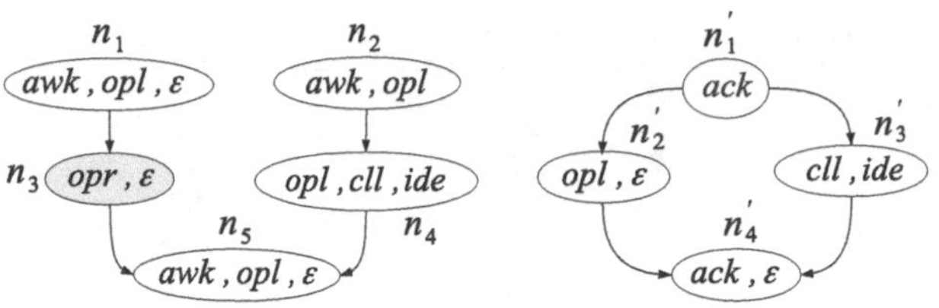

ℓ	d_ℓ	r_ℓ	h_ℓ
awk	1	1	0
opl	3	1	0
cll	0	3	0
ide	0	3	0
ε	1	0	1

Fig. 9.12 Observations $\mathcal{O}_1$, $\mathcal{O}_2$ (left), and initial configuration of the symbol table $\mathcal{S}$ (right)

$r_{ide} = 2$. Furthermore, both h_{cll} and h_{ide} are updated (line 35), and thereby $h_{cll} = 0$ and $h_{ide} = 0$. Eventually, the balancing condition ($d_\ell \geq h_\ell$) is checked at line 45 for *opl*, *cll*, *ide*, and ε. This is violated for $\ell = \varepsilon$ only, which suffices to return **nil** (line 46).

To understand why the association between n_1 and n'_1 fails, consider Fig. 9.8, where the consumption of nodes n_1 and n'_1 leaves the remainder of the two observations unbalanced from the viewpoint of label coverage. In fact, the numbers of remaining nodes in $\mathcal{O}_1$ and $\mathcal{O}_2$ are 4 and 3, respectively. Consequently, one of the two spare nodes (necessarily n_3) will not be associated with any node in $\mathcal{O}_2$. The point is, choosing ε in both n'_2 and n'_4 in $\mathcal{O}_2$ requires choosing both n_3 and n_5 in $\mathcal{O}_1$ (in order to fulfill the label coverage relevant to the label ε), thereby causing $\mathcal{O}_1$ to remain with two (nonspare) nodes (n_2 and n_4) and $\mathcal{O}_2$ with just one node (n'_3). Thus, based on Def. 9.11, a subset $\bar{N}$ of nodes in $\mathcal{O}_1$ isomorphic to nodes in $\mathcal{O}_2$ cannot exist (as the remaining nodes in $\mathcal{O}_1$ do not include ε).

Continuing the loop in *Recurse* (line 65), considering $n = n_2$ causes a new call to *UpdateSymbtab* (line 66). Now, the only label to be considered is *open*. At line 31, d_{opl} is decremented, and thereby $d_{opl} = 2$. As n_2 is not a candidate spare node, neither r_{opl} nor h_{opl} is updated. Since the balancing condition holds, $d_{opl} \geq h_{opl}$, the update of the symbol table now succeeds (line 49).

Then, the computation of *Recurse* proceeds from line 67, where $N_a = \emptyset$. Since N_a is empty, the condition at line 68 is true, and a new node $n'' = n'_2$ is picked up from among the nodes not yet considered in N' (line 69). A second, recursive call to *Recurse* is performed at line 70, whose changed parameters are the updated symbol table $\mathcal{S}'$, the sets of consumed nodes $\mathcal{C}$ and $\mathcal{C}'$ extended by $n = n_2$ and $n' = n'_1$, respectively, the new node $n'' = n'_2$ in N', and the singleton $\mathcal{R} = \{(n_2, n'_1)\}$, establishing the first correspondence pair.

Shown in Table 9.3 are the details relevant to the different calls to *Recurse* leading to the positive final result (where observation nodes are identified by the relevant indices). Specifically, for each call, a triple (d_ℓ, r_ℓ, h_ℓ) is indicated for each label ℓ (the row of the symbol table relevant to label ℓ), along with the other actual parameters, namely σ, $\mathcal{C}$, $\mathcal{C}'$, n', and $\mathcal{R}$ ($\mathcal{W} = \emptyset$ in every call).

According to Table 9.3, the last call to *Recurse* succeeds in (positively) solving the subsumption problem (line 62), where, according to the final instance of $\mathcal{R}$, the subset of nodes in N corresponding to N' in Def. 9.11 is $\bar{N} = \{n_2, n_1, n_4, n_5\}$ (the first nodes of pairs in $\mathcal{R}$).

Table 9.3 Tracing of calls to the *Recurse* auxiliary function in Example 9.17

Call	*awk*	*opl*	*cll*	*ide*	ε	σ	$\mathcal{C}$	$\mathcal{C}'$	n'	$\mathcal{R}$
1	110	310	030	030	101	1	$\emptyset$	$\emptyset$	1	$\emptyset$
2	110	210	030	030	101	1	2	1	2	$(2,1)$
3	010	210	020	020	101	1	12	12	3	$(2,1)(1,2)$
4	010	110	020	020	101	1	124	123	4	$(2,1)(1,2)(4,3)$
5	000	100	010	010	000	0	1234	123	4	$(2,1)(1,2)(4,3)$
6	000	000	000	000	000	0	12345	1234	**nil**	$(2,1)(1,2)(4,3)(5,4)$

9.6 Bibliographic Notes

The notion of *flexible diagnosis* was introduced in [99], based on a list of *flexibility requirements*, among which is *diagnosis reusability*. Coverage techniques for checking temporal-observation subsumption were presented in [41]. An implemented diagnostic environment for active systems was illustrated in [27]. Relaxation of temporal observations was considered in [82]. Techniques for checking temporal-observation subsumption were provided in [81, 83].

9.7 Summary of Chapter 9

Subsumption. A relationship defined between viewers, temporal observations, rulers, and behaviors, which supports diagnosis reuse.

Viewer Subsumption. A relationship between two viewers $\mathcal{V}'$ and $\mathcal{V}$ for the same active system. Specifically, $\mathcal{V}' \Supset \mathcal{V}$ iff (1) the set of observable transitions in $\mathcal{V}'$ includes all the observable transitions in $\mathcal{V}$, and (2) the projection of each observable label in $\mathcal{V}'$ to $\mathcal{V}$ is not ambiguous.

Observation Projection. Given a temporal observation $\mathcal{O}'$ based on a viewer $\mathcal{V}'$, and a viewer $\mathcal{V}$, the temporal observation $\mathcal{O}'_{[\mathcal{V}]}$ based on $\mathcal{V}$, obtained by replacing each label in each node of $\mathcal{O}'$ with either the corresponding observable label in $\mathcal{V}$ or ε.

Index-Space Projection. Given an index space $Isp(\mathcal{O}')$ and a viewer $\mathcal{V}$, the index space $Isp_{[\mathcal{V}]}(\mathcal{O}')$ generated by determinizing the NFA obtained by replacing each observable label marking each arc of $Isp(\mathcal{O}')$ with either the corresponding observable label in $\mathcal{V}$ or ε.

Observation Subsumption. A relationship between two temporal observations $\mathcal{O}$ and $\mathcal{O}'$ for the same active system. Specifically, $\mathcal{O} \Supset \mathcal{O}'$ iff $\|\mathcal{O}\| \supseteq \|\mathcal{O}'\|$.

Behavior Subsumption. A relationship between two behaviors $Bhv(\wp(\mathcal{A}))$ and $Bhv(\wp'(\mathcal{A}))$ of two different diagnosis problems for the same active system. Specifically, $Bhv(\wp(\mathcal{A})) \Supset Bhv(\wp'(\mathcal{A}))$ iff $\|Bhv(\wp(\mathcal{A}))\| \supseteq \|Bhv(\wp'(\mathcal{A}))\|$.

Ruler Subsumption. A relationship between two rulers $\mathcal{R}$ and $\mathcal{R}'$ for the same active system. Specifically, $\mathcal{R} \Supset \mathcal{R}'$ iff (1) the set of faulty transitions in $\mathcal{R}$ includes all the faulty transitions in $\mathcal{R}'$, and (2) the projection of each observable label in $\mathcal{R}$ to $\mathcal{R}'$ is not ambiguous.

Diagnosis Projection. Given a diagnosis $\delta \in \Delta(\wp(\mathcal{A}))$ based on a ruler $\mathcal{R}$, and another ruler $\mathcal{R}'$ for $\mathcal{A}$, the diagnosis $\delta_{[\mathcal{R}']}$ based on the ruler $\mathcal{R}'$, obtained by replacing each fault label in δ with the corresponding fault label in $\mathcal{R}'$.

Behavior Matching. Given a behavior $Bhv(\wp(\mathcal{A}))$ and a temporal observation $\mathcal{O}$ based on a viewer $\mathcal{V}$, the subpart of $Bhv(\wp(\mathcal{A}))$ consistent with $\mathcal{O}$. This matching operation is expressed as $Bhv(\wp(\mathcal{A})) \asymp (\mathcal{O}, \mathcal{V})$.

Observation-Subsumption Problem. The practical problem of checking the subsumption relationship between two temporal observations, namely $\mathcal{O} \Supset \mathcal{O}'$.

Observation Coverage. A relationship between two temporal observations, namely $\mathcal{O} \trianglerighteq \mathcal{O}'$, which implies $\mathcal{O} \Supset \mathcal{O}'$.

Chapter 10
Lazy Diagnosis

In human society, laziness is normally considered a negative feature, if not a capital fault. Not so in computer science. On the contrary, and significantly, lazy computation may save computational resources, in both space and time. For example, when we have to evaluate the result of an expression involving a Boolean operator, such as the logical disjunction in the expression E **or** E', where E and E' are (possibly complex) Boolean expressions themselves, two approaches are possible:

1. *Greedy evaluation.* Both expressions E and E' are evaluated, giving rise to corresponding Boolean values $\bar{E}$ and $\bar{E}'$, respectively. Then, the **or** operator is applied to them, namely $\bar{E}$ **or** $\bar{E}'$, to compute the final result.
2. *Lazy evaluation.* The first expression E is evaluated, yielding the Boolean value $\bar{E}$. If $\bar{E}$ equals **true**, then the final result will be **true**; otherwise, E' is evaluated and the final result will be $\bar{E}'$.

The advantage of the lazy approach (also called *short-circuit* evaluation in the context of Boolean expressions) is clear: the second expression E' is evaluated only if the value of the first expression E is **false**. More generally, since a Boolean expression may involve several logical operators (possibly of different nature) at different levels of the expression tree, lazy evaluation is applied at each such level.

Lazy evaluation is also adopted in modern functional languages, including Haskell [157]. In this realm, the notion of laziness goes beyond the short-circuit evaluation of operators. It is a more general principle which states that every evaluation should be performed only when necessary. This allows powerful (and strange) computations, such as the manipulation of lists of infinite length. For example, in Haskell it is possible to define the list of the squared integer numbers as

```
integers = [1,2 ..]
squares = [ n*n | n <- integers ]
```

where `integers` is the infinite sequence of integer numbers, and `squares` is the infinite sequence of (ordered) squared integer numbers, isomorphic to `integers`.[1]

[1] The symbol '<-' is the textual representation of the set-theoretic membership symbol ∈.

G. Lamperti et al., *Introduction to Diagnosis of Active Systems*,
https://doi.org/10.1007/978-3-319-92733-6_10

```
2 3 4 5 6 7 8 9 10 11 12 13 14 15 16 17 18 19 20 ...
  3   5   7   9    11    13    15    17    19    ...
      5   7        11    13          17    19    ...
          7        11    13                      ...
                   11    13                      ...
                         13                      ...
```

Fig. 10.1 Tracing of the first steps of the sieve of Eratosthenes' algorithm, which computes the (infinite) list of prime numbers (numbers in gray are those deleted in step 3 of the algorithm)

Although it may sound odd, there is nothing magic in this code: the materialization of such lists only occurs when necessary, and only to the necessary extent. For example, we may define a function which tests whether a number n is a square of an integer. A possible implementation is to test the membership of n in `squares`. Since n is finite and `squares` is ordered, the search for n in `squares` requires a limited number of steps. This way, `squares` is (automatically) materialized only up to the element n, if included, or the successive one, if not included, which, in all cases, causes the search to stop.

Another interesting example of the use of lazy evaluation in Haskell is the implementation of the algorithm of Eratosthenes for the generation of the (infinite) list of prime numbers, called the *sieve of Eratosthenes*. This algorithm can be described by the following steps:

1. Write the infinite sequence of integers $2, 3, 4, 5, 6, \ldots$,
2. Mark the first number p of the list as prime,
3. Delete all multiples of p from the list,
4. Go to step 2.

The effect of the sieve of Eratosthenes is outlined in Fig. 10.1, where the lines correspond to the successive iterations of steps 2 and 3.

In Haskell, we can define the infinite list of prime numbers as follows:

```
primes :: [Int]
primes = sieve [2..]
```

where `primes` is first declared as a list of integers, and then assigned with the result of the `sieve` function applied to the infinite list of integers $n \geq 2$.

The implementation of the `sieve` function is astonishingly simple, namely

```
sieve :: [Int] -> [Int]
sieve (p:rest) = p: sieve [ x | x <- rest, x `mod` p /= 0 ]
```

The function `sieve` is first declared as a mapping from a list of integers (the natural numbers $n \geq 2$) to a list of integers (the prime numbers sifted from the input list). Then, in the actual body of `sieve`, the input list is identified by the pattern `(p:rest)`, with `p` and `rest` denoting the head and the tail, respectively, of the list.

The body of `sieve` is defined recursively as a list with head `p` and a tail resulting from the application of `sieve` to the infinite list defined by the generator

```
[ x | x <- rest, x `mod` p /= 0 ]
```

corresponding to the numbers `x` in `rest` which are not multiples of `p` (as required by step 2 of the algorithm).

By virtue of the use of lazy evaluation, all infinite lists in the input to the recursive calls to `sieve` are materialized up to the necessary point, thereby allowing an effective evaluation of the function. For example, if we submit to the Haskell interpreter the following expression:

```
take 15 primes
```

where the library function `take` generates the prefix of the input list (`primes`) up to the given number (15) of elements, we expect as output the list of the first 15 prime numbers, namely `[2,3,5,7,11,13,17,19,23,29,31,37,41,43,47]`.

Significantly, the idea of lazy computation may be injected in some way into the diagnosis of active systems. Until two decades ago, diagnosis methods for discrete-event systems required the (offline) explicit generation of the system model in order to perform (online) diagnosis. Unfortunately, as pointed out in previous chapters, this systematic approach is impractical when the system is large and distributed.

The reconstruction of the system behavior without the need for a global system model can be seen as a sort of lazy computation, in the sense that only a portion of the behavior space is in fact materialized during reconstruction.

However, a similar problem still exists when diagnosis involves a temporal observation represented by a DAG. As highlighted in Chapter 5, in order to reconstruct the system behavior based on the temporal observation, an index space must be generated as the determinization of the nondeterministic index space. The point is that the index space may suffer from the same computational limitations as the system model, with the aggravating factor that it must be generated at problem-solving time (online).

Generally speaking, as not all strings (trajectories) in the behavior space of the system are consistent with the temporal observation, so not all strings of observable labels in the language of the index space are consistent with the behavior space of the system.

The goal is therefore to confine the explosion of the index space based on the constraints offered by the reconstructed behavior. However, a circularity arises: on the one hand, we need the index space to reconstruct the system behavior; on the other hand, we need the reconstructed system behavior to confine the index space.

To overcome this circularity, we present in this chapter a pruning technique which allows the online lazy generation of both the system behavior and the index space.

10.1 Laziness Requirements

As pointed out above, when the temporal observation is large and temporally underconstrained, the greedy approach to diagnosis of active systems may become inappropriate, owing to the explosion of the nondeterministic index space and, consequently, of the index space.[2]

The reason for the huge number of states can be understood by analyzing how the index space is generated. As shown in Sec. 5.1, given a temporal observation $\mathcal{O}$, we have to build the nondeterministic index space of $\mathcal{O}$, namely $Nsp(\mathcal{O})$, by considering all possible ways in which nodes of $\mathcal{O}$ can be selected, based on the precedence constraints imposed by the arcs of $\mathcal{O}$.

At each choice, we create new transitions in the nondeterministic index space, marked by the labels within the extension of the selected node, and connect them to a new state of $Nsp(\mathcal{O})$. This state is marked by a set of nodes (a prefix) of $\mathcal{O}$ that identifies the whole set of nodes already chosen in $\mathcal{O}$. Roughly speaking, the less temporally constrained $\mathcal{O}$ is, the larger the set of possible sequences of choices. The exact number of states in $Nsp(\mathcal{O})$ is given by Proposition 10.1.

Proposition 10.1. *Let $\mathcal{O}$ be a temporal observation. The cardinality of the set of states in $Nsp(\mathcal{O})$ equals the cardinality of the whole set of prefixes of $\mathcal{O}$.*

Proof. The proof is based on the fact that each possible way to select labels in nodes of $\mathcal{O}$ is represented by a path in $Nsp(\mathcal{O})$, from the initial to the final state. By contradiction, assume that the number n_{p} of nodes in $Nsp(\mathcal{O})$ does not equal the number n_{x} of possible prefixes of $\mathcal{O}$. Then, two cases are possible: either $n_{\mathrm{p}} > n_{\mathrm{x}}$ or $n_{\mathrm{p}} < n_{\mathrm{x}}$. If $n_{\mathrm{p}} > n_{\mathrm{x}}$, then $Nsp(\mathcal{O})$ will include either duplicate nodes or nodes that are not prefixes, which is impossible. If $n_{\mathrm{p}} < n_{\mathrm{x}}$, then there is (at least) one prefix $\mathcal{P}$ which is *not* involved in $Nsp(\mathcal{O})$. However, based on the definition in Def. 5.1 of a prefix $\mathcal{P}$, since a prefix identifies the set of consumed nodes involving all the ancestors of the nodes in $\mathcal{P}$, it follows that there is a sequence of choices in the consumed nodes which is embodied in $Nsp(\mathcal{O})$; in other words, $\mathcal{P}$ is included in $Nsp(\mathcal{O})$, a contradiction. □

Corollary 10.1. *Let $\mathcal{O}$ be a temporal observation involving n nodes. If $\mathcal{O}$ is linear (nodes totally ordered), then the number of states in $Nsp(\mathcal{O})$ equals $n+1$. If $\mathcal{O}$ is totally disconnected (nodes temporally unconstrained), then the number of states in $Nsp(\mathcal{O})$ equals 2^n.*

Based on Corollary 10.1, when $\mathcal{O}$ is totally disconnected (no temporal constraints between nodes), all combinations of nodes are possible, giving rise to 2^n prefixes. Hence, in the worst case, the number of states in $Nsp(\mathcal{O})$ grows exponentially with the number of nodes in $\mathcal{O}$. In practice, even for disconnected observations of the moderate size of 40 nodes, the nondeterministic index space contains 2^{40} states,

[2] We faced this problem when experimenting with algorithms for subsumption checking of temporal observations [104] (see Chapter 9). To test the proposed technique, we had to generate a large set of observations and construct the relevant index spaces.

corresponding to more than 10^{12} states. With such numbers, if the generation of the nondeterministic index space is impractical, the transformation of it into the index space is simply out of the question. So, what to do? After all, we need some sort of observation-indexing for reconstructing the system behavior.

Generally speaking, not all candidate traces included in $Isp(\mathcal{O})$ are consistent with the behavior space of the system, just as not all the trajectories included in the behavior space are consistent with $Isp(\mathcal{O})$. In fact, in the reconstruction phase, we implicitly filter the trajectories in the behavior space $Bsp(\mathcal{A})$ based on the constraints imposed by $Isp(\mathcal{O})$, thereby yielding the behavior $Bhv(\wp(\mathcal{A}))$.

Now, we might try to perform some sort of pruning of $Isp(\mathcal{O})$ based on the constraints imposed by the behavior space $Bsp(\mathcal{A})$. However, this would work only assuming the availability of the latter, which is not the case.

A better idea is to filter the index space based on the reconstructed behavior $Bhv(\wp(\mathcal{A}))$. This allows us to avoid the generation of $Bsp(\mathcal{A})$. Instead, however, the problem is now that $Bhv(\wp(\mathcal{A}))$ is itself generated based on $Isp(\mathcal{O})$, giving rise to a circularity: we need $Isp(\mathcal{O})$ to generate $Bhv(\wp(\mathcal{A}))$ and we need $Bhv(\wp(\mathcal{A}))$ to generate $Isp(\mathcal{O})$.

We can cope with this circularity by building the (pruned) index space and the reconstructed behavior adopting a lazy approach, where the constructions of the two automata are intertwined. So, the reciprocal constraints can be checked at each step of the building process. This is summarized in Requirement 10.1.

Requirement 10.1 (Index-Space Pruning). The index space should be pruned while the system behavior is being restructured.

A second shortcoming of the greedy approach to diagnosis-problem solving concerns the structure of the reconstructed behavior. As described in Sec. 5.2, the behavior $Bhv(\wp(\mathcal{A}))$ consists of a DFA, with each state being a triple $(S,Q,\mathfrak{I})$, where (S,Q) is a system state and $\mathfrak{I}$ a state of the index space. Within a transition $(S,Q,\mathfrak{I}) \xrightarrow{t(c)} (S',Q',\mathfrak{I}')$ in $Bhv(\wp(\mathcal{A}))$, $\mathfrak{I}'$ differs from $\mathfrak{I}$ when the component transition $t(c)$ is observable. Conversely, when $t(c)$ is unobservable, we have $\mathfrak{I}' = \mathfrak{I}$. This property indicates the following two topological peculiarities of $Bhv(\wp(\mathcal{A}))$:

1. Let $\beta = (S,Q,\mathfrak{I})$ be either the initial state or a state reached by an observable transition in $Bhv(\wp(\mathcal{A}))$. Let $Unobs(\beta)$ be the subgraph of $Bhv(\wp(\mathcal{A}))$ rooted in β and embodying all paths of unobservable transitions rooted in β. Then, all states in $Unobs(\beta)$ will share the same index $\mathfrak{I}$.
2. Let $\beta_1 = (S,Q,\mathfrak{I}_1)$ and $\beta_2 = (S,Q,\mathfrak{I}_2)$ be two states in $Bhv(\wp(\mathcal{A}))$ sharing the same system state (S,Q). Then, the projections of $Unobs(\beta_1)$ and $Unobs(\beta_2)$ on $Bsp(\mathcal{A})$ are identical. In other words, if we remove the indices $\mathfrak{I}_1$ and $\mathfrak{I}_2$ from $Unobs(\beta_1)$ and $Unobs(\beta_2)$, respectively, we obtain the same fragment of behavior space.

These peculiarities suggest that there is some redundancy in the reconstruction of the behavior. On the one hand, states in $Bhv(\wp(\mathcal{A}))$ marked by the same index $\mathfrak{I}$ can be grouped to form a fragment of $Bsp(\mathcal{A})$ involving unobservable transitions only. This way, the index $\Im$ can be associated with the whole fragment rather than with each state within the fragment.

On the other hand, and more importantly, since each fragment depends functionally on its root β (either the initial state of $Bhv(\wp(\mathcal{A}))$ or a state reached by an observable transition), a previous generation of the fragment can be reused without any need for model-based reasoning when β is generated as the next state in $Bhv(\wp(\mathcal{A}))$.[3] This way, we avoid regenerating the duplicate fragment of behavior. This is summarized in Requirement 10.2.

Requirement 10.2 (Behavior Factorization). The reconstructed behavior should be factorized based on observation indices.

The above requirements, namely pruning and factorization, apply to observation indexing and behavior reconstruction, respectively. We now give a third requirement that is relevant to decoration (and distillation) of candidate diagnoses.

In greedy problem-solving, the decoration of the reconstructed behavior is performed top-down in a systematic way. This raises a problem similar to that of the redundant reconstruction of the behavior: two identical fragments of $Bhv(\wp(\mathcal{A}))$ involving different observation indices are decorated independently of one another.

In lazy diagnosis, we require that, based on a factorized representation of the reconstructed behavior, each fragment should be decorated in a two-phase fashion: first, a decoration *relative* to the fragment is performed, and then an additional *absolute* decoration completes the task by extending the relative decoration with the faults entailed by the prefixes of the trajectories entering the fragment.[4] This supports reuse of local decoration for topologically identical fragments of behavior. This is summarized in Requirement 10.3.

Requirement 10.3 (Two-Phase Decoration). The decoration of the reconstructed behavior should be performed in two phases: relative decoration and absolute decoration.

We now consider each of the three *laziness requirements* introduced above and provide a formal framework for the lazy diagnosis of active systems.

10.2 Behavior Factorization

We start by considering Requirement 10.2, namely behavior factorization. We have to formalize the subgraph of the behavior space that is rooted in the initial state or a state reached by an observable transition and comprises all states reached by unobservable transitions only.

---

[3] Unlike the case for the behavior space, the same system state can appear in the reconstructed behavior several times, although associated with different observation indices.

[4] The prefix of a trajectory should not be confused with the prefix of an observation. The former is simply a contiguous subsequence of the trajectory rooted in the first transition.

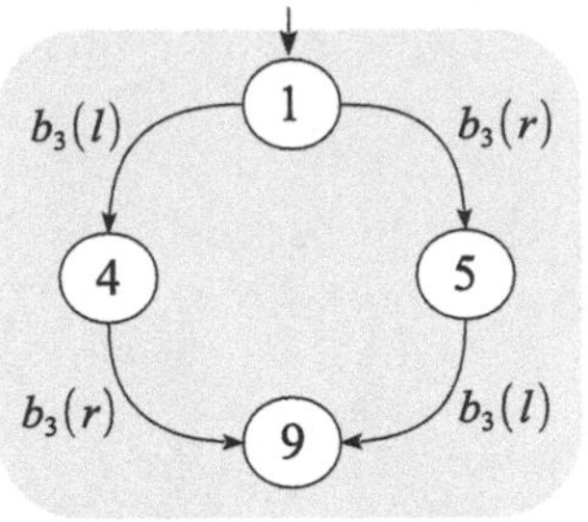

Fig. 10.2 Condensation $Cond(1,\mathcal{V})$ relevant to $Bsp(\mathcal{A}_2)$ specified in Table 4.2 and viewer $\mathcal{V}$ specified in Table 4.1

Definition 10.1 (Condensation). Let $Bsp(\mathcal{A}) = (\Sigma, A, \tau, a_0)$ be the behavior space of $\mathcal{A}$, $\mathcal{V}$ a viewer for $\mathcal{A}$, and $\bar{a}$ a state in A. The *condensation* relevant to $\bar{a}$ and $\mathcal{V}$ is a DFA

$$Cond(\bar{a}, \mathcal{V}) = (\Sigma, \bar{A}, \bar{\tau}, \bar{a}, A_l), \tag{10.1}$$

where $\bar{A} \subseteq A$ is the set of states; $\bar{a}$ is the *root*; $\bar{\tau}$ is the transition function $\bar{\tau} : \bar{A} \times \Sigma \mapsto \bar{A}$, such that $a \xrightarrow{t(c)} a' \in \bar{\tau}$ iff $a \xrightarrow{t(c)} a' \in \tau$, where $t(c)$ is unobservable; and A_l is the *leaving set*, defined as follows:

$$A_l = \{a \mid a \in \bar{A}, a \xrightarrow{t(c)} a' \in \tau,\ t(c) \text{ is observable}\}\ . \tag{10.2}$$

A state in the leaving set A_l is a *leaving state*.

Example 10.1. With reference to the behavior space $Bsp(\mathcal{A}_2)$ specified in Table 4.2 and the viewer $\mathcal{V}$ specified in Table 4.1, consider state 1 of $Bsp(\mathcal{A}_2)$. The condensation relevant to 1 and $\mathcal{V}$, namely $Cond(1,\mathcal{V})$, is displayed in Fig. 10.2, where all four states are in the leaving set, that is, all states are exited by an observable transition.

The notion of a condensation allows us to define a condensed behavior space, where each state is a condensation and transitions between condensations are marked by observable transitions in the behavior space.

Definition 10.2 (Condensed Behavior Space). Let $\mathcal{V}$ be a viewer relevant to a system $\mathcal{A}$. A *condensed behavior space* relevant to $\mathcal{A}$ and $\mathcal{V}$ is an automaton

$$\mathbf{Bsp}(\mathcal{A}, \mathcal{V}) = (\Sigma, \mathbf{C}, \tau, \mathcal{C}_0), \tag{10.3}$$

where Σ is the set of observable component transitions in the alphabet of $Bsp(\mathcal{A})$; $\mathbf{C}$ is the set of states, with each state being a condensation;

$$\mathcal{C}_0 = Cond(a_0, \mathcal{V}) \tag{10.4}$$

is the initial state, where a_0 is the initial state of $Bsp(\mathcal{A})$; and τ is the transition function $\tau : \mathbf{C} \times A \times \Sigma \mapsto \mathbf{C}$, where A is the set of states in $Bsp(\mathcal{A})$, such that $(\mathcal{C}, a) \xrightarrow{t(c)} \mathcal{C}' \in \tau$ iff $a \xrightarrow{t(c)} a'$ is a transition in $Bsp(\mathcal{A})$, where $t(c)$ is observable, a is a leaving state of $\mathcal{C}$, and a' is the initial state of $\mathcal{C}'$.

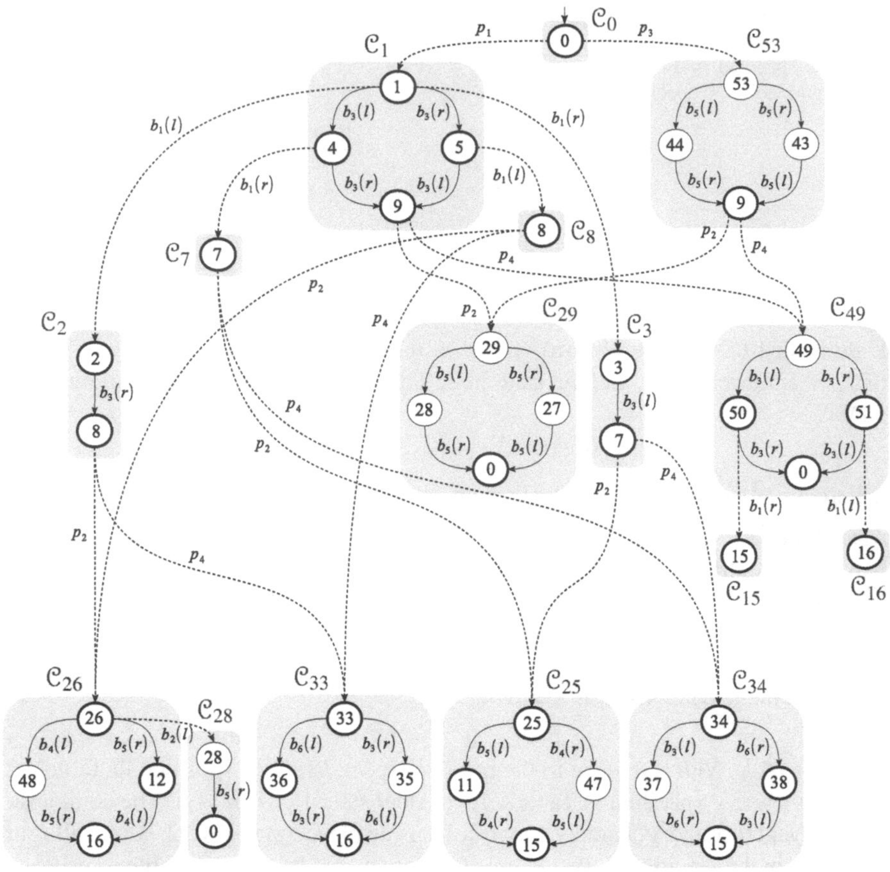

Fig. 10.3 Condensed behavior space $\mathbf{Bsp}(\mathcal{A}_2, \mathcal{V})$ relevant to system $\mathcal{A}_2$ displayed in Fig. 4.2 and viewer $\mathcal{V}$ specified in Table 4.1 (cf. behavior space $Bsp(\mathcal{A}_2)$ defined in Table 4.2)

Example 10.2. With reference to the behavior space $Bsp(\mathcal{A}_2)$ specified in Table 4.2 and the viewer $\mathcal{V}$ in Table 4.1, Fig. 10.3 shows a portion of the condensed behavior space $\mathbf{Bsp}(\mathcal{A}_2, \mathcal{V})$, where 0 is the initial state of $Bsp(\mathcal{A}_2)$. Within each state (condensation), states in the leaving set are shown in bold. Each transition between condensations is represented by a dashed arrow.

The condensed behavior space has been introduced for formal reasons only as, similarly to the behavior space, generally speaking, its generation is prohibitive in practice. Based on it, we can define a condensed behavior, where each state is a state of the condensed behavior space associated with a state of the index space.

Definition 10.3 (Condensed Behavior). Let $\wp(\mathcal{A}) = (a_0, \mathcal{V}, \mathcal{O}, \mathcal{R})$ be a diagnosis problem. The relevant *condensed behavior* is a DFA

$$\mathbf{Bhv}(\wp(\mathcal{A})) = (\Sigma, \mathbf{B}, \tau, \beta_0, \mathbf{B}_f), \tag{10.5}$$

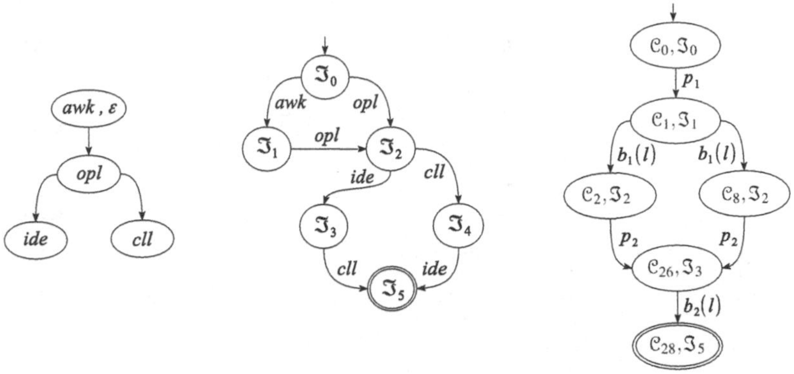

Fig. 10.4 From left to right: temporal observation $\mathcal{O}'_2$ for system $\mathcal{A}_2$ (the latter being displayed in Fig. 4.2), index space $Isp(\mathcal{O}'_2)$, and condensed behavior $\mathbf{Bhv}(\wp'(\mathcal{A}_2))$, where the diagnosis problem $\wp'(\mathcal{A}_2) = (a_0, \mathcal{V}, \mathcal{O}'_2, \mathcal{R})$ is a variant of $\wp(\mathcal{A}_2)$ defined in Example 4.12, with $\mathcal{O}_2$ being replaced by $\mathcal{O}'_2$ (cf. condensed behavior $Bsp(\mathcal{A}_2, \mathcal{V})$ displayed in Fig. 10.3)

where Σ is a subset of the alphabet of $\mathbf{Bsp}(\mathcal{A}, \mathcal{V})$; $\mathbf{B}$ is the set the states, with each state $\beta = (\mathcal{C}, \mathfrak{I})$ being an association between a state $\mathcal{C}$ in $\mathbf{Bsp}(\mathcal{A}, \mathcal{V})$ and a state $\mathfrak{I}$ in $Isp(\mathcal{O})$;

$$\beta_0 = (\mathcal{C}_0, \mathfrak{I}_0) \tag{10.6}$$

is the initial state, with $\mathcal{C}_0$ and $\mathfrak{I}_0$ being the initial states of $\mathbf{Bsp}(\mathcal{A}, \mathcal{V})$ and $Isp(\mathcal{O})$, respectively; τ is the transition function $\tau : \mathbf{B} \times A \times \Sigma \mapsto \mathbf{B}$, where A is the set of states in $Bsp(\mathcal{A})$, such that $(\mathcal{C}, \mathfrak{I}, a) \xrightarrow{t(c)} (\mathcal{C}', \mathfrak{I}') \in \tau$ iff $(\mathcal{C}, a) \xrightarrow{t(c)} \mathcal{C}'$ is a transition in $\mathbf{Bsp}(\mathcal{A}, \mathcal{V})$, ℓ is the observable label relevant to $t(c)$, and $\mathfrak{I} \xrightarrow{\ell} \mathfrak{I}'$ is a transition in $Isp(\mathcal{O})$; and $\mathbf{B}_\mathrm{f}$ is the set of final states, defined as follows:

$$\mathbf{B}_\mathrm{f} = \{\beta \mid \beta \in \mathbf{B}, \beta = (\mathcal{C}, \mathfrak{I}), \mathfrak{I} \text{ is final in } Isp(\mathcal{O})\}\,. \tag{10.7}$$

Example 10.3. Displayed on the left-hand side of Fig. 10.4 is the temporal observation $\mathcal{O}'_2$ for the system $\mathcal{A}_2$ (displayed in Fig. 4.2). The corresponding index space $Isp(\mathcal{O}'_2)$ is displayed in the center of the figure. Shown on the right-hand side is the condensed behavior $\mathbf{Bhv}(\wp'(\mathcal{A}_2))$, where $\wp'(\mathcal{A}_2) = (a_0, \mathcal{V}, \mathcal{O}'_2, \mathcal{R})$ is a variant of the diagnosis problem $\wp(\mathcal{A}_2)$ defined in Example 4.12, where $\mathcal{O}_2$ is replaced by $\mathcal{O}'_2$.[5]

10.3 Lazy Index Space Generation

A basic property of $Nsp(\mathcal{O})$ and $Isp(\mathcal{O})$ is *reachability*: each state belongs to (at least) one path from the initial state to a final state. Besides, both $Nsp(\mathcal{O})$ and $Isp(\mathcal{O})$

[5] The only difference between $\mathcal{O}_2$ and $\mathcal{O}'_2$ is that, in the latter, the label *cll* is alone within its node (ε is no longer a candidate label). This causes $\mathfrak{I}_3$ to be no longer a final state in the index space.

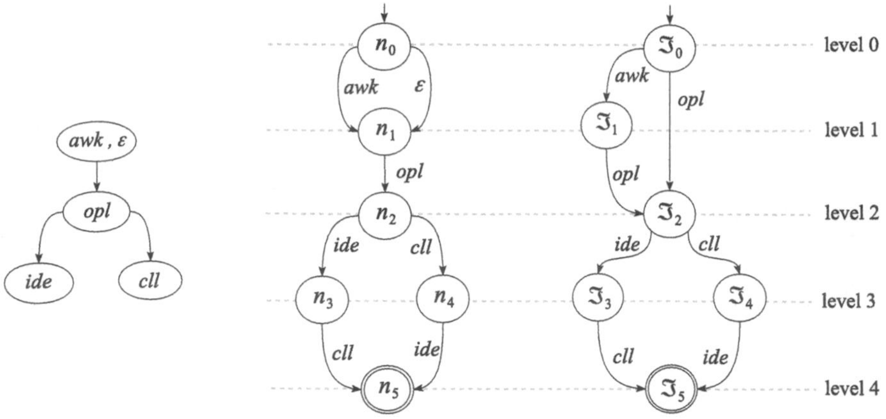

Fig. 10.5 From left to right: temporal observation $\mathcal{O}'_2$, nondeterministic index space $Nsp(\mathcal{O}'_2)$, and index space $Isp(\mathcal{O}'_2)$, with state levels relevant to the hierarchical partition

are *acyclic* automata. The acyclicity of $Isp(\mathcal{O})$ comes from the acyclicity of $Nsp(\mathcal{O})$. The acyclicity of $Nsp(\mathcal{O})$ comes from the acyclicity of $\mathcal{O}$.

For such acyclic automata, we can define the notion of a *level* of a state s, denoted $Level(s)$, as the maximum length of the paths $s_0 \rightsquigarrow s$ connecting the initial state s_0 to s. In particular, $Level(s_0) = 0$.

Accordingly, we can define a partition of the set of states where each part is composed of states at the same level. This leads to the property of *hierarchical partition*: the set of states is partitioned into a hierarchy of parts, from level 0 to level n, with the latter being the deepest level.

Example 10.4. Reported in Fig. 10.5 is the temporal observation $\mathcal{O}'_2$ (left), along with the nondeterministic index space $Nsp(\mathcal{O}'_2)$ (center) and the index space $Isp(\mathcal{O}'_2)$ (right). In both $Nsp(\mathcal{O}'_2)$ and $Isp(\mathcal{O}'_2)$, states at the same level $i \in [0 \cdots 4]$ in the hierarchical partition are aligned on the same dashed line.

10.3.1 Lazy ISCA

Listed below is the specification of the algorithm *LISCA*, which takes as input a portion of the nondeterministic index space $Nsp(\mathcal{O})$ up to level $k \geq 0$, the corresponding index space $Isp(\mathcal{O})$, and an expansion $\Delta\mathcal{N}$ of $Nsp(\mathcal{O})$ to level $k+1$. As a result, it updates $Isp(\mathcal{O})$ to make it equivalent to the expanded $Nsp(\mathcal{O})$, and returns the sequence $\mathcal{U}$ of relevant updates performed on $Isp(\mathcal{O})$.

Roughly speaking, *LISCA* operates like the algorithm *ISCA* (listed in Sec. 2.7.1), including the rules $\mathcal{R}_1, \ldots, \mathcal{R}_6$. However, since *LISCA* is specialized for lazy determinization of nondeterministic index spaces, it differs from *ISCA* in two main respects.

First, the expansion $\Delta\mathcal{N}$ is precisely the next layer of $Nsp(\mathcal{O})$, not a generic expansion.[6] Second, and more important, the relevant updates performed on $Isp(\mathcal{O})$ are collected into a sequence $\mathcal{U}$, which is eventually returned (line 61). Five types of actions are considered, namely:

1. $Ext(\Im)$: the extension of index $\Im$ is enlarged (line 24),[7]
2. $Mrg(\Im, \Im')$: indices $\Im$ and $\Im'$ are merged into a single index (line 22),
3. $New(\Im \xrightarrow{\ell} \Im')$: transition $\Im \xrightarrow{\ell} \Im'$ is created, possibly after creating $\Im'$ (line 44),
4. $Red(\Im \xrightarrow{\ell} \Im', \Im'')$: transition $\Im \xrightarrow{\ell} \Im'$ is redirected toward $\Im''$ (line 51),
5. $Dup(\Im \xrightarrow{\ell} \Im', \Im'')$: state $\Im''$ is created as an extended duplication of $\Im'$, with transition $\Im \xrightarrow{\ell} \Im'$ being redirected toward $\Im''$ (line 55).

Other details of *LISCA* can be grasped by analyzing the algorithm *ISCA* presented in Sec. 2.7.1.

The specification of the *LISCA* algorithm is as follows:

1. **algorithm** *LISCA*(**inout** $\mathcal{N}$, **inout** $\mathcal{I}$, **in** $\Delta\mathcal{N}$, **out** $\mathcal{U}$)
2. $\mathcal{N} = (\Sigma, N, \tau_n, n_0, N_f)$: a portion of a (possibly pruned) nondeterministic index space up to level k,
3. $\mathcal{I} = (\Sigma, I, \tau_d, \Im_0, I_f)$: the (possibly pruned) portion of index space equivalent to $\mathcal{N}$,
4. $\Delta\mathcal{N} = (\Delta N, \Delta\tau_n)$: an expansion of $\mathcal{N}$ to level $k+1$,
5. $\mathcal{U}$: the sequence of relevant updates performed in $\mathcal{I}$;

6. **side effects**
7. $\mathcal{N}$ is extended by $\Delta\mathcal{N}$,
8. $\mathcal{I}$ becomes the portion of the index space equivalent to $\mathcal{N} \cup \Delta\mathcal{N}$;

9. **auxiliary procedure** *Expand*(**inout** $\Im$, **in** $\mathbb{N}$)
10. $\Im$: a state in I,
11. $\mathbb{N}$: a subset of N;
12. **begin** $\langle$*Expand*$\rangle$
13. **if** $\mathbb{N} \not\subseteq \|\Im\|$ **then**
14. Enlarge $\|\Im\|$ by $\mathbb{N}$;
15. **if** $\Im \notin I_f, \mathbb{N} \cap N_f \neq \emptyset$ **then** Insert $\Im$ into I_f **endif**;
16. **if** I includes a state $\Im'$ such that $\|\Im'\| = \|\Im\|$ **then**
17. Redirect to $\Im'$ all transitions entering $\Im$;
18. Redirect from $\Im'$ all transitions exiting $\Im$;
19. **if** $\Im \in I_f$ **then** Remove $\Im$ from I_f **endif**;
20. Remove $\Im$ from I;
21. Convert to $\Im'$ the buds in $\mathcal{B}$ relevant to $\Im$;

[6] Unlike the case for *ISCA*, the expansion $\Delta\mathcal{N}$ does not involve information about final states, namely ΔF_n^+ and ΔF_n^-, as $Nsp(\mathcal{O})$ contains just one final state, positioned at the last level (e.g., state n_5 in Fig. 10.5)

[7] Only if $\Im$ is not a newly created state.

22.   Append $Mrg(\mathfrak{I},\mathfrak{I}')$ to $\mathcal{U}$
23.   **else**
24.   **if** $\|\mathfrak{I}\|$ in input was not empty **then** append $Ext(\mathfrak{I})$ to $\mathcal{U}$ **endif**
25.   **endif**
26.   **endif**
27.   **end** $\langle Expand \rangle$;

28.   **begin** $\langle LISCA \rangle$
29. $\mathcal{U} := [\,]$;
30. $\bar{\mathbb{N}} :=$ the set of states in $\mathcal{N}$ exited by transitions in $\Delta\tau_n$;
31.   Extend $\mathcal{N}$ based on $\Delta\mathcal{N}$;
32. $\mathcal{B} := [\,(\mathfrak{I},\ell,\mathbb{N}) \mid \mathfrak{I} \in I, n \in \|\mathfrak{I}\| \cap \bar{\mathbb{N}}, n \xrightarrow{\ell} n' \in \Delta\tau_n, \ell \notin Inc(\mathfrak{I}),$
                                    $\mathbb{N} = \ell\text{-closure}(\|\mathfrak{I}\| \cap \bar{\mathbb{N}})\,]$;
33.   **repeat**
34.   Pop bud $(\mathfrak{I},\ell,\mathbb{N})$ from $\mathcal{B}$;
35. $(\mathcal{R}_1)$ **if** $\ell = \varepsilon$ **then**
36. $Expand(\mathfrak{I},\mathbb{N})$
37.   **elsif** no ℓ-transition exits $\mathfrak{I}$ **then**
38. $(\mathcal{R}_2)$ **if** I includes a state $\mathfrak{I}'$ such that $\|\mathfrak{I}'\| = \mathbb{N}$ **then**
39.   Insert a new transition $\mathfrak{I} \xrightarrow{\ell} \mathfrak{I}'$ into I
40. $(\mathcal{R}_3)$ **else**
41.   Create a new state $\mathfrak{I}'$ and insert $\mathfrak{I} \xrightarrow{\ell} \mathfrak{I}'$ into I;
42. $Expand(\mathfrak{I}',\mathbb{N})$
43.   **endif**;
44.   Append $New(\mathfrak{I} \xrightarrow{\ell} \mathfrak{I}')$ to $\mathcal{U}$
45.   **else**
46.   **foreach** transition $t = \mathfrak{I} \xrightarrow{\ell} \mathfrak{I}'$ where $\mathbb{N} \not\subseteq \|\mathfrak{I}'\|$ **do**
47. $(\mathcal{R}_4)$ **if** no other transition enters $\mathfrak{I}'$ **then**
48. $Expand(\mathfrak{I}',\mathbb{N})$
49. $(\mathcal{R}_5)$ **elsif** there is $\mathfrak{I}'' \in I$ such that $\|\mathfrak{I}''\| = \|\mathfrak{I}'\| \cup \mathbb{N}$ **then**
50.   Redirect t toward $\mathfrak{I}''$;
51.   Append $Red(\mathfrak{I} \xrightarrow{\ell} \mathfrak{I}',\mathfrak{I}'')$ to $\mathcal{U}$;
52. $(\mathcal{R}_6)$ **else**
53.   Create a copy $\mathfrak{I}''$ of $\mathfrak{I}'$, along with related buds;
54.   Redirect t toward $\mathfrak{I}''$;
55.   Append $Dup(\mathfrak{I} \xrightarrow{\ell} \mathfrak{I}',\mathfrak{I}'')$ to $\mathcal{U}$;
56. $Expand(\mathfrak{I}'',\mathbb{N})$
57.   **endif**
58.   **endfor**
59.   **endif**
60.   **until** $\mathcal{B}$ is empty;

```
61.        return 𝔘
62.    end ⟨LISCA⟩.
```

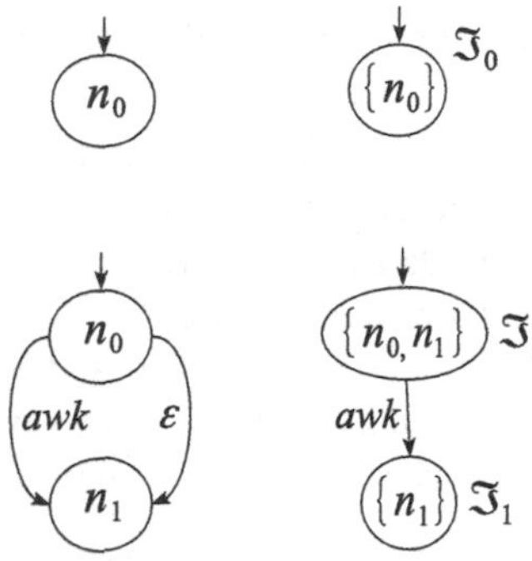

Fig. 10.6 With reference to temporal observation $\mathcal{O}'_2$ displayed on the left-hand side of Fig. 10.5, the initial nondeterministic index space $\mathcal{N}$ (top left), the corresponding index space $\mathcal{I}$ (top right), the expanded nondeterministic index space $\mathcal{N} \cup \Delta\mathcal{N}$ (bottom left), and the index space equivalent to the latter (bottom right)

Example 10.5. With reference to the temporal observation $\mathcal{O}'_2$ displayed on the left-hand side of Fig. 10.5, the top of Fig. 10.6 shows the initial nondeterministic index space $\mathcal{N}$ (left) and the equivalent index space $\mathcal{I}$ (right), corresponding to level 0 in the partition hierarchy. Displayed in the bottom left of the figure is the expanded nondeterministic index space $\mathcal{N} \cup \Delta\mathcal{N}$, where $\Delta\mathcal{N}$ consists of the state n_1 and transitions $n_0 \xrightarrow{awk} n_1$ and $n_0 \xrightarrow{\varepsilon} n_1$.

Based on line 30 of *LISCA*, the set $\bar{\mathbb{N}}$ of states in $\mathcal{N}$ which are exited by transitions in $\Delta\mathcal{N}$ is the singleton $\{n_0\}$. The stack $\mathcal{B}$ is initialized in line 32 by buds $(\Im_0, awk, \{n_1\})$ and $(\Im_0, \varepsilon, \{n_1\})$.

The processing of these buds in the main loop (lines 33–60) gives rise to the following actions:

1. Bud $(\Im_0, \varepsilon, \{n_1\})$, rule $\mathcal{R}_1$: the extension of $\Im_0$ is enlarged by the auxiliary procedure *Expand* to $\|\Im_0\| = \{n_0, n_1\}$, with the update $Ext(\Im_0)$ being inserted into $\mathcal{U}$.
2. Bud $(\Im_0, awk, \{n_1\})$, rule $\mathcal{R}_3$: both the state $\Im_1$ and the transition $\Im_0 \xrightarrow{awk} \Im_1$ are created (line 41), with $\|\Im_1\|$ becoming $\{n_1\}$ (line 42), while the update $New(\Im_0 \xrightarrow{awk} \Im_1)$ is appended to $\mathcal{U}$ (line 44).

The index space resulting from the processing of the two buds is displayed in the bottom right of Fig. 10.6. The update sequence returned by *LISCA* in line 61 is

$$\mathcal{U} = [Ext(\Im_0), New(\Im_0 \xrightarrow{awk} \Im_1)] . \tag{10.8}$$

These actions are exploited in the technique of circular pruning introduced in the next section.

10.4 Circular Pruning

Circular pruning amounts to intertwining the generation of the index space and the reconstruction of the condensed behavior in order to perform pruning on these graphs at each *layering step*, with the latter consisting of the following sequence of *layering actions*:

1. Generation of the next layer of $Nsp(\mathcal{O})$ (states at the same next level along with relevant transitions),
2. Update of the corresponding $Isp(\mathcal{O})$ by means of the *LISCA* algorithm,
3. Extension of the condensed behavior based on the updates of $Isp(\mathcal{O})$,
4. Pruning of the condensed behavior,
5. Pruning of $Isp(\mathcal{O})$ based on the updated condensed behavior,
6. Backward propagation of the pruning of $Isp(\mathcal{O})$ to $Nsp(\mathcal{O})$.

Each of the six actions in a layering step is detailed below.

10.4.1 Generation of the Next Layer of $Nsp(\mathcal{O})$

The first layering action exploits the topological nature of $Nsp(\mathcal{O})$, based on the hierarchical partition of states, where each part embodies a set of states sharing the same level. As such, $Nsp(\mathcal{O})$ is composed of $n+1$ layers, where n is the number of nodes in the temporal observation $\mathcal{O}$. The first layer includes only the initial state $n_0 = \emptyset$. Each successive layer consists of the states reached by transitions exiting states in the previous layer.

Starting from the initial state n_0, we require that $Nsp(\mathcal{O})$ be generated one layer at a time. We call this the *layered growth* of the prefix space.

Example 10.6. With reference to the temporal observation $\mathcal{O}'_2$ displayed on the left-hand side of Fig. 10.5, Fig. 10.7 shows the nondeterministic index space $Nsp(\mathcal{O}'_2)$

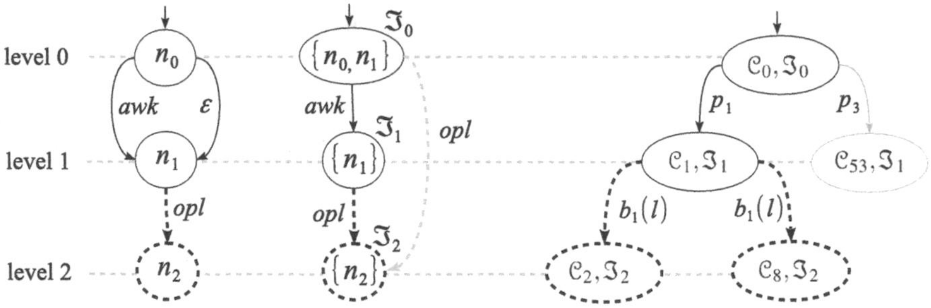

Fig. 10.7 With reference to the temporal observation $\mathcal{O}'_2$ displayed on the left-hand side of Fig. 10.5, $Nsp(\mathcal{O}'_2)$ (left) and $Isp(\mathcal{O}'_2)$ (center) up to level 2, along with the corresponding portion of $\mathbf{Bhv}(\wp(\mathcal{A}_2))$ (right), where $\wp(\mathcal{A}_2)$ is defined in Example 10.3

(left), which is extended to the second layer (level = 2), giving rise to the new state n_2 and the new transition $n_1 \xrightarrow{opl} n_2$ (both represented by dashed lines).

10.4.2 Update of $Isp(\mathcal{O})$

The second layering action consists in updating $Isp(\mathcal{O})$ based on the set of transitions involved in the generation of the new layer of $Nsp(\mathcal{O})$. This can be accomplished by the algorithm *LISCA*, where the updates of $Isp(\mathcal{O})$ relevant to the subsequent extension of the condensed behavior (layering action 3) based on the updates of $Isp(\mathcal{O})$ are returned in $\mathcal{U}$. Five relevant updates are considered:

1. $Ext(\mathfrak{I})$: extension of an existing state $\mathfrak{I}$ (line 15),
2. $Mrg(\mathfrak{I}, \mathfrak{I}')$: merging of states $\mathfrak{I}$ and $\mathfrak{I}'$ (lines 18–22),
3. $New(\mathfrak{I} \xrightarrow{\ell} \mathfrak{I}')$: creation of a new transition, possibly toward a newly created state (lines 39 and 41),
4. $Red(\mathfrak{I} \xrightarrow{\ell} \mathfrak{I}', \mathfrak{I}'')$: redirection of the transition towards $\mathfrak{I}''$ (line 50),
5. $Dup(\mathfrak{I} \xrightarrow{\ell} \mathfrak{I}', \mathfrak{I}'')$: duplication of state $\mathfrak{I}'$ and redirection of the transition toward $\mathfrak{I}''$ (lines 53 and 54).

It is important to keep the ordering of updates as they were generated by *LISCA*.[8] Besides, notice how, in $New(\mathfrak{I} \xrightarrow{\ell} \mathfrak{I}')$, state $\mathfrak{I}'$ is new in $Isp(\mathcal{O})$ (not necessarily created by this transition, though), since $\|\mathfrak{I}'\|$ includes (at least) one state in the new layer of $Nsp(\mathcal{O})$.

Example 10.7. With reference to Fig. 10.7, consider the generation of the second layer of $Nsp(\mathcal{O}'_2)$, which gives rise to the state n_2 and the transition $n_1 \xrightarrow{opl} n_2$. According to *LISCA*, the index space $Isp(\mathcal{O}'_2)$ is updated by the new state $\mathfrak{I}_2$ and the new transitions $\mathfrak{I}_0 \xrightarrow{opl} \mathfrak{I}_2$ and $\mathfrak{I}_1 \xrightarrow{opl} \mathfrak{I}_2$, with the following updates:

$$\mathcal{U} = [New(\mathfrak{I}_0 \xrightarrow{opl} \mathfrak{I}_2), New(\mathfrak{I}_1 \xrightarrow{opl} \mathfrak{I}_2)] . \tag{10.9}$$

10.4.3 Extension of the Condensed Behavior

The third layering action performs an extension of $\mathbf{Bhv}(\wp(\mathcal{A}))$ based on the extensions of $Isp(\mathcal{O})$ carried out by *LISCA* and recorded in $\mathcal{U}$. Specifically, the updates are considered in the order in which they have been stored in $\mathcal{U}$, and processed as follows:

[8] This is why $\mathcal{U}$ is a sequence rather than a set.

1. $Ext(\mathfrak{I})$: Each state $(\mathcal{C}, \mathfrak{I})$ in $\mathbf{Bhv}(\wp(\mathcal{A}))$ is qualified as belonging to the new frontier of $\mathbf{Bhv}(\wp(\mathcal{A}))$. This information is exploited for subsequent pruning of the latter.
2. $Mrg(\mathfrak{I}, \mathfrak{I}')$: Each state $(\mathcal{C}, \mathfrak{I})$ in $\mathbf{Bhv}(\wp(\mathcal{A}))$ is transformed into $(\mathcal{C}, \mathfrak{I}')$, where $\mathfrak{I}$ is replaced by $\mathfrak{I}'$. Since this transformation may possibly generate state duplication, duplicate states $(\mathcal{C}, \mathfrak{I}')$ are merged into a single state, with redirection of entering/exiting transitions of removed duplicates to/from the remaining single state $(\mathcal{C}, \mathfrak{I}')$.
3. $New(\mathfrak{I} \xrightarrow{\ell} \mathfrak{I}')$: Each state $(\mathcal{C}, \mathfrak{I})$ in $\mathbf{Bhv}(\wp(\mathcal{A}))$ is considered. Let T_ℓ be the set of transitions $a \xrightarrow{t(c)} a'$ exiting a leaving state of $\mathcal{C}$ such that $t(c)$ is an observable transition via the label ℓ. Then, for each $a \xrightarrow{t(c)} a' \in T_\ell$, $\mathbf{Bhv}(\wp(\mathcal{A}))$ is extended by $(\mathcal{C}, \mathfrak{I}) \xrightarrow{t(c)} (\mathcal{C}', \mathfrak{I}')$, where $\mathcal{C}' = Cond(a', \mathcal{V})$. If there is a T_ℓ which is not empty, then $\mathfrak{I} \xrightarrow{\ell} \mathfrak{I}'$ is marked as consistent.
4. $Red(\mathfrak{I} \xrightarrow{\ell} \mathfrak{I}', \mathfrak{I}'')$: The redirection of $\mathfrak{I} \xrightarrow{\ell} \mathfrak{I}'$ toward $\mathfrak{I}''$ (line 50) is mimicked in $\mathbf{Bhv}(\wp(\mathcal{A}))$ as follows. For each transition $(\mathcal{C}, \mathfrak{I}) \xrightarrow{t(c)} (\mathcal{C}', \mathfrak{I}')$ in $\mathbf{Bhv}(\wp(\mathcal{A}))$, two cases are possible, depending on whether or not there is another transition entering $(\mathcal{C}', \mathfrak{I}')$. In the case of another transition entering $(\mathcal{C}', \mathfrak{I}')$, if there is no state $(\mathcal{C}', \mathfrak{I}'')$ then the latter is created; in any case, $(\mathcal{C}, \mathfrak{I}) \xrightarrow{t(c)} (\mathcal{C}', \mathfrak{I}')$ is eventually redirected toward $(\mathcal{C}', \mathfrak{I}'')$. If, instead, no other transition enters $(\mathcal{C}', \mathfrak{I}')$, then the state $(\mathcal{C}', \mathfrak{I}')$ is transformed into $(\mathcal{C}', \mathfrak{I}'')$, where $\mathfrak{I}'$ is replaced by $\mathfrak{I}''$. If this transformation causes state duplication, then duplicate states $(\mathcal{C}', \mathfrak{I}'')$ are merged into a single state, with possible transition redirections.
5. $Dup(\mathfrak{I} \xrightarrow{\ell} \mathfrak{I}', \mathfrak{I}'')$: The subsequent redirection of $\mathfrak{I} \xrightarrow{\ell} \mathfrak{I}'$ toward $\mathfrak{I}''$ (line 54) is mimicked in $\mathbf{Bhv}(\wp(\mathcal{A}))$ as follows. For each transition $(\mathcal{C}, \mathfrak{I}) \xrightarrow{t(c)} (\mathcal{C}', \mathfrak{I}')$ in $\mathbf{Bhv}(\wp(\mathcal{A}))$, two cases are possible, depending on whether or not there is another transition entering $(\mathcal{C}', \mathfrak{I}')$. In the case of another transition entering $(\mathcal{C}', \mathfrak{I}')$, if there is no state $(\mathcal{C}', \mathfrak{I}'')$, then a state $(\mathcal{C}', \mathfrak{I}'')$ is created and the transition $(\mathcal{C}, \mathfrak{I}) \xrightarrow{t(c)} (\mathcal{C}', \mathfrak{I}')$ is redirected toward $(\mathcal{C}', \mathfrak{I}'')$. Otherwise, the state $(\mathcal{C}', \mathfrak{I}')$ is transformed into $(\mathcal{C}', \mathfrak{I}'')$, where $\mathfrak{I}'$ is replaced by $\mathfrak{I}''$. If this transformation causes state duplication, then duplicate states $(\mathcal{C}', \mathfrak{I}'')$ are merged into a single state, with possible transition redirections.[9]

Example 10.8. With reference to Fig. 10.7, consider the update of $Isp(\mathcal{O}'_2)$ analyzed in Example 10.7, which gives rise to the update sequence $\mathcal{U}$ in (10.9). According to the portion of the condensed behavior space $\mathbf{Bsp}(\mathcal{A}_2, \mathcal{V})$ displayed in Fig. 10.3, the processing of the three updates causes the extension of the condensed behavior $\mathbf{Bhv}(\wp'(\mathcal{A}_2))$ as follows:

[9] Unlike $Red(\mathfrak{I} \xrightarrow{\ell} \mathfrak{I}', \mathfrak{I}'')$, in the case of $Dup(\mathfrak{I} \xrightarrow{\ell} \mathfrak{I}', \mathfrak{I}'')$, the state $(\mathcal{C}', \mathfrak{I}'')$ cannot exist already, because $\mathfrak{I}''$ has been newly created in $Isp(O)$.

1. $New(\mathfrak{I}_0 \xrightarrow{opl} \mathfrak{I}_2)$: the set T_{opl} of transitions exiting a leaving state in $\mathcal{C}_0$ is empty; hence, no transition exiting the state $(\mathcal{C}_0, \mathfrak{I}_0)$ of $\mathbf{Bhv}(\wp'(\mathcal{A}_2))$ is created. Consequently, the transition $\mathfrak{I}_0 \xrightarrow{opl} \mathfrak{I}_2$ is *not* marked as consistent in $Isp(\mathcal{O}'_2)$.
2. $New(\mathfrak{I}_1 \xrightarrow{opl} \mathfrak{I}_2)$: we have $T_{opl} = \{1 \xrightarrow{b_1(l)} 2, 5 \xrightarrow{b_1(l)} 8\}$; hence, two transitions are created in $\mathbf{Bhv}(\wp'(\mathcal{A}_2))$, $(\mathcal{C}_1, \mathfrak{I}_1) \xrightarrow{b_1(l)} (\mathcal{C}_2, \mathfrak{I}_2)$ and $(\mathcal{C}_1, \mathfrak{I}_1) \xrightarrow{b_1(l)} (\mathcal{C}_8, \mathfrak{I}_2)$, with the transition $\mathfrak{I}_1 \xrightarrow{opl} \mathfrak{I}_2$ being marked as consistent in $Isp(\mathcal{O}'_2)$.

10.4.4 Pruning of the Condensed Behavior

Once the condensed behavior $\mathbf{Bhv}(\wp(\mathcal{A}))$ has been extended based on the updates stored in $\mathcal{U}$ by *LISCA*, pruning of $\mathbf{Bhv}(\wp(\mathcal{A}))$ is enabled. Specifically, a state $(\mathcal{C}, \mathfrak{I})$ is removable from $\mathbf{Bhv}(\wp(\mathcal{A}))$ when the following three conditions hold:

1. No transition exits $(\mathcal{C}, \mathfrak{I})$,
2. No transition will exit $(\mathcal{C}, \mathfrak{I})$ in future extensions of $\mathbf{Bhv}(\wp(\mathcal{A}))$,
3. $\mathfrak{I}$ is not final in $Isp(\mathcal{O})$.

While conditions 1 and 3 can easily be checked, the second one requires some reasoning. The statement that no transition will exit $(\mathcal{C}, \mathfrak{I})$ in subsequent extensions of $\mathbf{Bhv}(\wp(\mathcal{A}))$ amounts to saying that no transition will exit $\mathfrak{I}$ in subsequent extensions of $Isp(\mathcal{O})$. A transition can possibly exit $\mathfrak{I}$ only if $\mathfrak{I}$ is either a new state or an old state whose extension has been enlarged when a bud relevant to label ε was processed (line 36 of *LISCA*).

Thus, based on $\mathcal{U}$, it is possible to determine the new frontier of $\mathbf{Bhv}(\wp(\mathcal{A}))$, namely those states whose $\mathfrak{I}$'s are new or have been expanded. The difference between the frontier of the previous layer and the frontier of the current layer is precisely the set of states of $\mathbf{Bhv}(\wp(\mathcal{A}))$ from which no future exiting transition will be created. Besides, the removal of a state from $\mathbf{Bhv}(\wp(\mathcal{A}))$ may cause upward cascade pruning of its ancestors too.

Example 10.9. Consider Example 10.8, where the condensed behavior $\mathbf{Bhv}(\wp'(\mathcal{A}_2))$ displayed on the right-hand side of Fig. 10.7 is extended by two states and two transitions (represented by dashed lines). Based on the three conditions above for the pruning of a state, we are driven to the conclusion that the state $(\mathcal{C}_{53}, \mathfrak{I}_1)$ can be removed from $\mathbf{Bhv}(\wp'(\mathcal{A}_2))$. In fact, this state does not belong to the new frontier of $\mathbf{Bhv}(\wp'(\mathcal{A}_2))$, nor is it exited by a transition. Moreover, the index $\mathfrak{I}_1$ is not final in $Isp(\mathcal{O}'_2)$. The removal of the state $(\mathcal{C}_{53}, \mathfrak{I}_1)$ provokes the removal of its entering transition $(\mathcal{C}_0, \mathfrak{I}_0) \xrightarrow{p_3} (\mathcal{C}_{53}, \mathfrak{I}_1)$ too.

10.4.5 Pruning of $Isp(\mathcal{O})$

Once the condensed behavior has been extended based on $\mathcal{U}$, the index space $Isp(\mathcal{O})$ can be pruned based on the unmarked (inconsistent) transitions. To this end, each transition $\Im \xrightarrow{\ell} \Im'$ *not* marked as consistent in $\mathcal{U}$ is removed from $Isp(\mathcal{O})$. Furthermore, if $\Im'$ becomes isolated (no entering transition), then $\Im'$ too is removed from $Isp(\mathcal{O})$. Contextually, $\Im$ is marked by the inconsistent label ℓ; this information is exploited by *LISCA* in order to initialize the bud set with consistent buds only, thereby discarding buds whose label belongs to the set of inconsistent labels marking the relevant state $\Im$, namely $Inc(\Im)$ (line 32).

In principle, the removal of an isolated state in $Isp(\mathcal{O})$ may create dangling transitions (those exiting such a state) and, consequently, disconnected states. Nevertheless, we can be sure that the removal of an inconsistent transition from $Isp(\mathcal{O})$ will not cause such a downward cascade of dangling states. In other words, if we remove $\Im \xrightarrow{\ell} \Im'$, at most $\Im'$ will be isolated. In fact, if $\Im'$ becomes isolated after the removal of the inconsistent entering transition, then it is necessarily a new state created by *LISCA*. As such, $\Im'$ cannot be exited by any transition.

Example 10.10. Consider Example 10.8, where the condensed behavior displayed on the right-hand side of Fig. 10.7 is extended by two states and two transitions (represented by dashed lines). Since the extension of $Isp(\mathcal{O}'_2)$ by the transition $\Im_0 \xrightarrow{opl} \Im_2$ is not followed by any new transition exiting the state $(\mathcal{C}_0, \Im_0)$ and marked by *opl* in $\mathbf{Bhv}(\wp(\mathcal{A}_2))$, the transition $\Im_0 \xrightarrow{opl} \Im_2$ is not marked as consistent in $Isp(\mathcal{O}'_2)$; in other words, this transition is inconsistent with $\mathbf{Bhv}(\wp(\mathcal{A}_2))$. Consequently, it can be pruned from $Isp(\mathcal{O}'_2)$.

10.4.6 Pruning of $Nsp(\mathcal{O})$

The pruning of the index space is eventually propagated as a pruning of the nondeterministic index space $Nsp(\mathcal{O})$. Specifically, let $\mathbb{I}$ be the set of nodes in $Isp(\mathcal{O})$ which have been either created or extended by *LISCA*, namely the *frontier* of $Isp(\mathcal{O})$.[10] A node n in the current layer of $Nsp(\mathcal{O})$ is *inconsistent* when no state in $\mathbb{I}$ contains n, that is,

$$n \notin \bigcup_{\Im \in \mathbb{I}} \|\Im\| \,. \qquad (10.10)$$

Inconsistent nodes are removed from $Nsp(\mathcal{O})$, along with its entering transitions.

Example 10.11. Shown in Fig. 10.8 is a continuation of the scenario displayed in Fig. 10.7, where $Nsp(\mathcal{O}'_2)$ (left) is extended to the third level, with the creation of states n_3 and n_4 and their entering transitions. The corresponding extension of the

[10] States in $Isp(\mathcal{O})$ that have been pruned are not in $\mathbb{I}$.

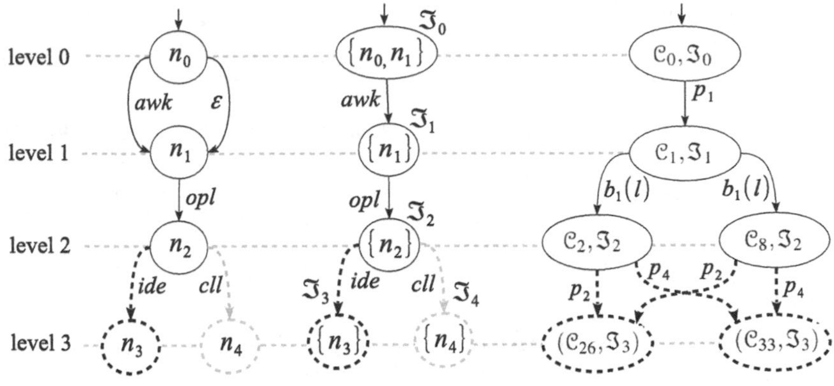

Fig. 10.8 With reference to Fig. 10.4, the portion of $Nsp(\mathcal{O}'_2)$ up to level 3 (left), the corresponding $Isp(\mathcal{O}'_2)$ (center), and the corresponding portion of $\mathbf{Bhv}(\wp(\mathcal{A}_2))$ (right)

(pruned) index space is displayed in the center of the figure, with the creation of states $\mathfrak{I}_3$ and $\mathfrak{I}_4$, along with their entering transitions.

Displayed on the right-hand side of Fig. 10.8 is the extension of the condensed behavior (dashed part), consisting of two states and four transitions. However, since none of these transitions is relevant to the label *cll*, the transition $\mathfrak{I}_2 \xrightarrow{cll} \mathfrak{I}_4$ is inconsistent, and therefore it is pruned from the index space.

According to the pruning condition (10.10) for the nondeterministic index space, the frontier of the index space is $\mathbb{I} = \{\mathfrak{I}_3, \mathfrak{I}_4\}$. Hence, based on condition (10.10), state n_4 of the nondeterministic index space is inconsistent and, as such, is removed, along with its entering transition $n_2 \xrightarrow{cll} n_4$.

10.5 Two-Phase Decoration

To generate the solution of a diagnosis problem, we need to decorate each system state a of each condensation $\mathcal{C}$ of the reconstructed condensed behavior $\mathbf{Bhv}(\wp(\mathcal{A}))$ with the sets of diagnoses associated with all the trajectories ending at a.

Based on Requirement 10.3, such a decoration is a two-phase process. First, a relative decoration is performed on the internal states of each condensation. Then, an absolute decoration is achieved by superimposing on the relative decoration the faults relevant to the prefixes of the trajectories from the initial state of $\mathbf{Bhv}(\wp(\mathcal{A}))$ to each internal state of each condensation.

Definition 10.4 (Relative Decoration). Let $\mathcal{C} = (\Sigma, \bar{A}, \bar{\tau}, \bar{a}, A_l)$ be a condensation relevant to a system $\mathcal{A}$, and $\mathcal{R}$ a ruler for $\mathcal{A}$. The *relative decoration* of $\mathcal{C}$ based on $\mathcal{R}$, written $Rdec(\mathcal{C}, \mathcal{R})$, is the automaton obtained from $\mathcal{C}$ by extending each state $a \in \bar{A}$ with a set $\mathcal{D}_r$ of diagnoses defined as follows:

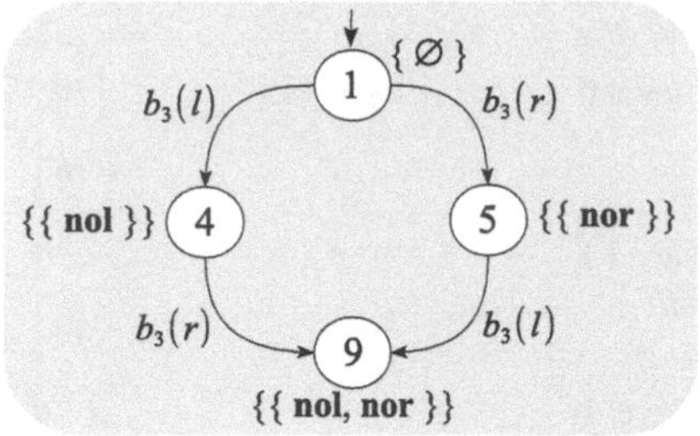

Fig. 10.9 Relative decoration of condensation $\mathcal{C}_1$

$$\mathcal{D}_{\rm r} = \{\, \delta \mid \delta = h_{[\mathcal{R}]}, h \text{ is a segment of trajectory in } \mathcal{C} \text{ from } \bar{a} \text{ to } a \,\} . \qquad (10.11)$$

Based on Def. 10.4, we conclude that the notion of a relative decoration equals the notion of a diagnosis closure, as defined in Def. 7.6, namely,

$$Rdec(\mathcal{C}, \mathcal{R}) = Dcl(\bar{a}, \mathcal{V}, \mathcal{R}) . \qquad (10.12)$$

Example 10.12. Consider the condensed behavior displayed on the right-hand side of Fig. 10.4. With reference to the condensed behavior space displayed in Fig. 10.3, Fig. 10.9 shows the relative decoration of the condensation $\mathcal{C}_1$ (based on the ruler $\mathcal{R}$ specified in Table 4.1).

Definition 10.5 (Absolute Decoration). Let $\beta = (\mathcal{C}, \mathfrak{I})$ be a state of a condensed behavior $\mathbf{Bhv}(\wp(\mathcal{A}))$, and $\mathcal{R}$ a ruler for $\mathcal{A}$. The *absolute decoration* of $\mathcal{C}$ based on $\mathcal{R}$, written $Adec(\mathcal{C}, \mathcal{R})$, is inductively defined as follows. If β is the initial state, then $Adec(\mathcal{C}, \mathcal{R}) = Rdec(\mathcal{C}, \mathcal{R})$. If β is not the initial state, then $Adec(\mathcal{C}, \mathcal{R})$ is obtained from $\mathcal{C}$ by extending each state a in $\mathcal{C}$ with a set $\mathcal{D}_{\rm a}$ of diagnoses, defined as follows:

$$\begin{aligned} \mathcal{D}_{\rm a} = \mathcal{D}_{\rm r} \bowtie \{\, \delta' \mid\ & \delta' \in (\mathcal{D}' \oplus f), \\ & (a', \mathcal{D}') \text{ is a leaving state in } Adec(\mathcal{C}', \mathcal{R}), \\ & (\mathcal{C}', \mathfrak{I}') \xrightarrow{t(c)} (\mathcal{C}, \mathfrak{I}) \in \mathbf{Bhv}(\wp(\mathcal{A})), \text{ either } (t(c), f) \in \mathcal{R} \text{ or } f = \varepsilon \,\}, \end{aligned} \qquad (10.13)$$

where $(a, \mathcal{D}_{\rm r})$ is a state in $Rdec(\mathcal{C}, \mathcal{R})$ (and $\oplus$ and $\bowtie$ are the operators of the diagnosis union (Def. 7.7) and diagnosis join (Def. 7.10), respectively).

Example 10.13. The absolute decoration of the condensed behavior $\mathbf{Bhv}(\wp'(\mathcal{A}_2))$ displayed on the right-hand side of Fig. 10.4 is displayed in Fig. 10.10.

The solution Δ of the diagnosis problem $p(\mathcal{A})$ is distilled from the absolute decoration of the condensed behavior as the union of the set of diagnoses marking states of the condensations associated with indices which are final in the index space:

$$\Delta(\wp(\mathcal{A})) = \{\, \delta \mid \delta \in \mathcal{D}, (a, \mathcal{D}) \in \mathcal{C}, (\mathcal{C}, \mathfrak{I}) \in \mathbf{Bhv}(\wp(\mathcal{A})), \mathfrak{I} \text{ is final in } Isp(\mathcal{O}) \,\} . \qquad (10.14)$$

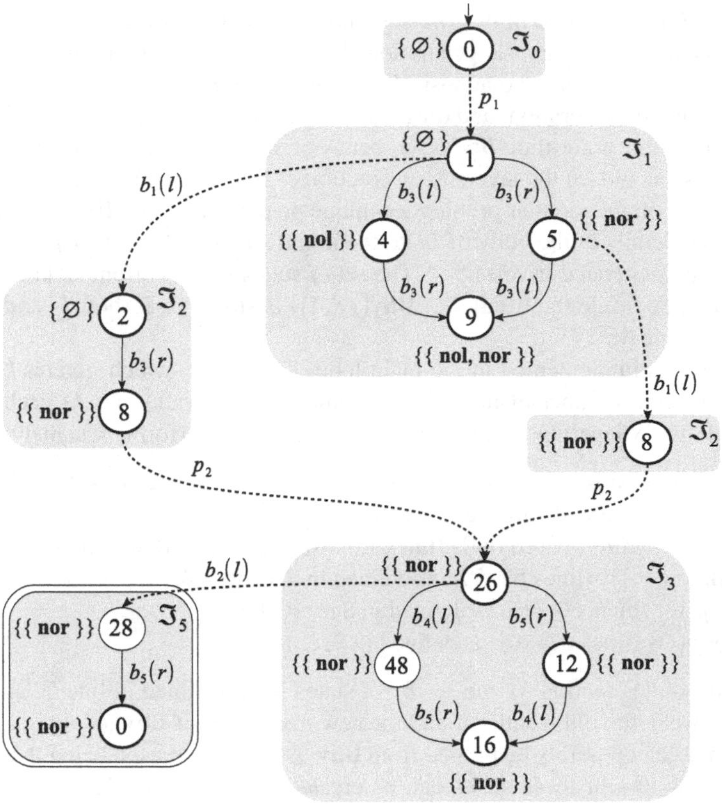

Fig. 10.10 Absolute decoration of condensed behavior $\mathbf{Bhv}(\wp'(\mathcal{A}_2))$

Example 10.14. According to the absolute decoration of the condensed behavior displayed on Fig. 10.10, and based on (10.14), the solution of the problem $\wp'(\mathcal{A}_2)$ is

$$\Delta(\wp'(\mathcal{A}_2)) = \{\{\mathbf{nor}\}\}, \tag{10.15}$$

which is obtained by the union of the set of diagnoses marking states of the condensation $\mathcal{C}_{28}$ associated with final index $\mathfrak{I}_5$. This solution includes just one candidate diagnosis, namely $\{\mathbf{nor}\}$. According to the ruler $\mathcal{R}$ defined in Example 4.10, we are driven to the conclusion that breaker r did not open.

10.6 Lazy Diagnosis Engine

Listed below is the specification of the algorithm *LDE*, which takes as input a diagnosis problem $\wp(\mathcal{A})$ and outputs its solution, namely $\Delta(\wp(\mathcal{A}))$. In doing so, it operates in lazy mode, based on the three laziness requirements defined in Sec. 10.1.

Essentially, *LDE* is composed of two parts: (condensed) behavior reconstruction (lines 5–72) and candidate diagnosis generation (lines 73–85). Implicitly, it comprises all three steps of monolithic diagnosis defined in Chapter 5, namely behavior reconstruction, behavior decoration, and diagnosis distillation.

However, unlike the algorithm *Build* for behavior reconstruction defined in Sec. 5.2.1, emphasis is put on the layered construction of both $Nsp(\mathcal{O})$ and $Isp(\mathcal{O})$, and, more generally, on the circular pruning technique presented in Sec. 10.4.

Considering the details of the body of *LDE*, the initial states of $Nsp(\mathcal{O})$, $Isp(\mathcal{O})$, and $\mathbf{Bhv}(\wp(\mathcal{A}))$ are generated in lines 5–7. The set of states in the current layer of $Nsp(\mathcal{O})$ is stored in $\mathbb{N}$, while the frontier of $\mathbf{Bhv}(\wp(\mathcal{A}))$ is stored in $\mathbb{B}$ (both $\mathbb{N}$ and $\mathbb{B}$ are initialized in line 8).

Circular pruning is implemented in the main loop, lines 9–72, which iterates k times, with k being the number of nodes in the temporal observation $\mathcal{O}$. At each iteration, each of the six layering actions defined in Sec. 10.4 is performed, namely:

1. Generation of the next layer of $Nsp(\mathcal{O})$ (line 10), as defined in Sec. 10.4.1,
2. Update of $Isp(\mathcal{O})$ by *LISCA* (line 12), as defined in Sec. 10.4.2,
3. Extension of $\mathbf{Bhv}(\wp(\mathcal{A}))$ based on $\mathcal{U}$ (lines 13–50), as defined in Sec. 10.4.3,
4. Pruning of $\mathbf{Bhv}(\wp(\mathcal{A}))$ (lines 53–59), as defined in Sec. 10.4.4,
5. Pruning of $Isp(\mathcal{O})$ (lines 60–66), as defined in Sec. 10.4.5,
6. Pruning of $Nsp(\mathcal{O})$ (lines 67–70), as defined in Sec. 10.4.6.

To prune $\mathbf{Bhv}(\wp(\mathcal{A}))$ (action 4), the set $\mathbb{B}^*$ of states is determined in line 52 as the difference between the old frontier $\mathbb{B}$ and the new frontier $\mathbb{B}'$ of $\mathbf{Bhv}(\wp(\mathcal{A}))$.

Only states in $\mathbb{B}^*$ can possibly be pruned from $\mathbf{Bhv}(\wp(\mathcal{A}))$, the reason being that no transition exiting any of these states can be created in the future. Hence, if a nonfinal state is exited by no transition, then it will inevitably be disconnected from any final state in $\mathbf{Bhv}(\wp(\mathcal{A}))$. As such, it is a spurious state that has to be pruned.

Once $\mathbf{Bhv}(\wp(\mathcal{A}))$ has been generated, based on Def. 10.4, relative decoration is performed on $\mathbf{Bhv}(\wp(\mathcal{A}))$ in line 73.[11] Then, absolute decoration is carried out in lines 74–84 based on Def. 10.5.

Eventually, based on (10.14), the solution $\Delta(\wp(\mathcal{A}))$ is generated in line 85.

The specification of the algorithm *LDE* is as follows:

1. **algorithm** LDE(**in** $\wp(\mathcal{A})$, **out** $\Delta(\wp(\mathcal{A}))$)
2. $\quad\wp(\mathcal{A}) = (a_0, \mathcal{V}, \mathcal{O}, \mathcal{R})$: a diagnostic problem for $\mathcal{A}$,
3. $\quad\Delta(\wp(\mathcal{A}))$: the solution of $\wp(\mathcal{A})$;
4. **begin** $\langle LDE \rangle$
5. $\qquad$Create the initial state of $Nsp(\mathcal{O})$, $n_0 = \emptyset$;
6. $\qquad$Create the initial state of $Isp(\mathcal{O})$, $\mathfrak{I}_0 = \{n_0\}$;
7. $\qquad$Create the initial state of $\mathbf{Bhv}(\wp(\mathcal{A}))$, $\beta_0 = (\mathcal{C}_0, \mathfrak{I}_0)$, where $\mathcal{C}_0 = Cond(a_0, \mathcal{V})$;
8. $\qquad\mathbb{N} := \{n_0\}$, $\mathbb{B} := \{\beta_0\}$;
9. $\qquad$**loop** k times, where k is the number of nodes of temporal observation $\mathcal{O}$

[11] Although not detailed in the code, once relative decoration has been performed on a state $(\mathcal{C}, \mathfrak{I})$, the same decoration can be reused for a different state $(\mathcal{C}, \mathfrak{I}')$ sharing the same condensation $\mathcal{C}$.

10.   Generate the next layer $\mathbb{N}'$ of $Nsp(\mathcal{O})$;

11.   Let $\Delta\mathcal{N}$ be the corresponding expansion of $Nsp(\mathcal{O})$;

12. $\mathcal{U} := LISCA(Nsp(\mathcal{O}), Isp(\mathcal{O}), \Delta\mathcal{N})$;

13.   **foreach** update $U \in \mathcal{U}$ performed on $Isp(\mathcal{O})$ **do**

14.   **if** $U = Mrg(\mathfrak{I}, \mathfrak{I}')$ **then**

15.   **foreach** state $(\mathcal{C}, \mathfrak{I})$ in $\mathbf{Bhv}(\wp(\mathcal{A}))$ **do** replace $\mathfrak{I}$ with $\mathfrak{I}'$ **endfor**;

16.   **if** duplicate states $(\mathcal{C}, \mathfrak{I}')$ are generated **then**

17.   Keep only one such state, by removing other duplicate states and redirecting corresponding dangling transitions to/from the kept state

18.   **endif**

19.   **elsif** $U = New(\mathfrak{I} \xrightarrow{\ell} \mathfrak{I}')$ **then**

20.   **foreach** state $(\mathcal{C}, \mathfrak{I})$ in $\mathbf{Bhv}(\wp(\mathcal{A}))$ **do**

21.   Let $T_\ell = \{ a \xrightarrow{t(c)} a' \mid a$ is a leaving state of $\mathcal{C}, t(c)$ is observable via label $\ell \}$;

22.   **if** $T_\ell \neq \emptyset$ **then**

23.   Mark U as consistent;

24.   **foreach** $a \xrightarrow{t(c)} a' \in T_\ell$ **do**

25.   Extend $\mathbf{Bhv}(\wp(\mathcal{A}))$ by $(\mathcal{C}, \mathfrak{I}) \xrightarrow{t(c)} (\mathcal{C}', \mathfrak{I}')$, where $\mathcal{C}' = Cond(a', \mathcal{V})$

26.   **endfor**

27.   **endif**

28.   **endfor**

29.   **elsif** $U = Red(\mathfrak{I} \xrightarrow{\ell} \mathfrak{I}', \mathfrak{I}'')$ **then**

30.   **foreach** transition $(\mathcal{C}, \mathfrak{I}) \xrightarrow{\ell} (\mathcal{C}', \mathfrak{I}')$ in $\mathbf{Bhv}(\wp(\mathcal{A}))$ **do**

31.   **if** there is another transition entering $(\mathcal{C}', \mathfrak{I}')$ **then**

32.   **if** there is no state $(\mathcal{C}', \mathfrak{I}'')$ **then** create state $(\mathcal{C}', \mathfrak{I}'')$ **endif**;

33.   Redirect $(\mathcal{C}, \mathfrak{I}) \xrightarrow{\ell} (\mathcal{C}', \mathfrak{I}')$ to $(\mathcal{C}', \mathfrak{I}'')$

34.   **else**

35.   Update $(\mathcal{C}', \mathfrak{I}')$ to $(\mathcal{C}', \mathfrak{I}'')$;

36.   **if** there is another state $(\mathcal{C}', \mathfrak{I}'')$ **then** merge the two states **endif**

37.   **endif**

38.   **endfor**

39.   **elsif** $U = Dup(\mathfrak{I} \xrightarrow{\ell} \mathfrak{I}', \mathfrak{I}'')$ **then**

40.   **foreach** transition $(\mathcal{C}, \mathfrak{I}) \xrightarrow{\ell} (\mathcal{C}', \mathfrak{I}')$ in $\mathbf{Bhv}(\wp(\mathcal{A}))$ **do**

41.   **if** there is another transition entering $(\mathcal{C}', \mathfrak{I}')$ **then**

42.   Create state $(\mathcal{C}', \mathfrak{I}'')$;

43.   Redirect $(\mathcal{C}, \mathfrak{I}) \xrightarrow{\ell} (\mathcal{C}', \mathfrak{I}')$ to $(\mathcal{C}', \mathfrak{I}'')$

44.   **else**

45.   Update $(\mathcal{C}', \mathfrak{I}')$ to $(\mathcal{C}', \mathfrak{I}'')$;

if there is another state $(\mathcal{C}', \mathfrak{I}'')$ **then** merge the two states **endif**
endif
endfor
endif
endfor;
$\mathbb{B}' := \{(\mathcal{C}, \mathfrak{I}) \mid (\mathcal{C}, \mathfrak{I}) \in \mathbf{Bhv}(\wp(\mathcal{A})),$
$Ext(\mathfrak{I}) \in \mathcal{U} \vee Mrg(\mathfrak{I}_1, \mathfrak{I}) \in \mathcal{U} \vee New(\mathfrak{I}_1 \xrightarrow{\ell} \mathfrak{I}) \in \mathcal{U} \vee$
$Red(\mathfrak{I}_1 \xrightarrow{\ell} \mathfrak{I}_2, \mathfrak{I}) \in \mathcal{U} \vee \; Dup(\mathfrak{I}_1 \xrightarrow{\ell} \mathfrak{I}_2, \mathfrak{I}) \in \mathcal{U}\}$;
$\mathbb{B}^* := \mathbb{B} - \mathbb{B}'$;
foreach $(\mathcal{C}, \mathfrak{I}) \in \mathbb{B}^*$ **do**
if $\mathfrak{I}$ is not final in $Isp(\mathcal{O})$ **and** there is no transition exiting $(\mathcal{C}, \mathfrak{I})$ **then**
Remove $(\mathcal{C}, \mathfrak{I})$ from both $\mathbf{Bhv}(\wp(\mathcal{A}))$ and $\mathbb{B}^*$;
Remove the dangling transitions from parents of $(\mathcal{C}, \mathfrak{I})$;
Insert into $\mathbb{B}^*$ all parents of $(\mathcal{C}, \mathfrak{I})$
endif
endfor;
foreach update $New(\mathfrak{I} \xrightarrow{\ell} \mathfrak{I}') \in \mathcal{U}$ that is *not* marked as consistent **do**
Mark $\mathfrak{I}$ with the inconsistent label ℓ;
Remove transition $\mathfrak{I} \xrightarrow{\ell} \mathfrak{I}'$ from $Isp(\mathcal{O})$;
if $\mathfrak{I}'$ is no longer entered by any transition in $Isp(\mathcal{O})$ **then**
Remove $\mathfrak{I}'$ from $Isp(\mathcal{O})$
endif
endfor;
Let $\mathbb{I}$ be the set of states in $Isp(\mathcal{O})$ either created or extended by actions in $\mathcal{U}$;
foreach state n in $Nsp(\mathcal{O})$ that is not included in any state of $\mathbb{I}$ **do**
Remove n from $Nsp(\mathcal{O})$, along with its entering transitions
endfor;
$\mathbb{N} := \mathbb{N}', \mathbb{B} := \mathbb{B}'$
endloop;
Based on ruler $\mathcal{R}$, perform relative decoration of all states in $\mathbf{Bhv}(\wp(\mathcal{A}))$;
Mark the initial state of $\mathbf{Bhv}(\wp(\mathcal{A}))$;
while $\mathbf{Bhv}(\wp(\mathcal{A}))$ includes an unmarked state **do**
Pick up an unmarked state $(\mathcal{C}, \mathfrak{I})$ in $\mathbf{Bhv}(\wp(\mathcal{A}))$ such that all its parent states are marked;
foreach state $(a, \mathcal{D})$ in the decorated condensation $\mathcal{C}$ **do**
foreach $(\mathcal{C}', \mathfrak{I}', a') \xrightarrow{t(c)} (\mathcal{C}, \mathfrak{I})$ in $\mathbf{Bhv}(\wp(\mathcal{A}))$ **do**
$\mathcal{D}'_{\mathrm{f}} := \mathcal{D}' \oplus f$, where $\mathcal{D}'$ is associated with a', either $(t(c), f) \in \mathcal{R}$ or $f = \varepsilon$;
$\mathcal{D} := \mathcal{D} \bowtie \mathcal{D}'_{\mathrm{f}}$
endfor
endfor;
Mark $(\mathcal{C}, \mathfrak{I})$

84. **endwhile**;
85. $\Delta := \{\, \delta \mid \delta \in \mathcal{D}, (a, \mathcal{D}) \in \mathcal{C}, (\mathcal{C}, \mathfrak{I}) \in \mathbf{Bhv}(\wp(\mathcal{A})), \mathfrak{I}$ is final in $Isp(\mathcal{O})\,\}$;
86. **return** Δ
87. **end** $\langle LDE \rangle$.

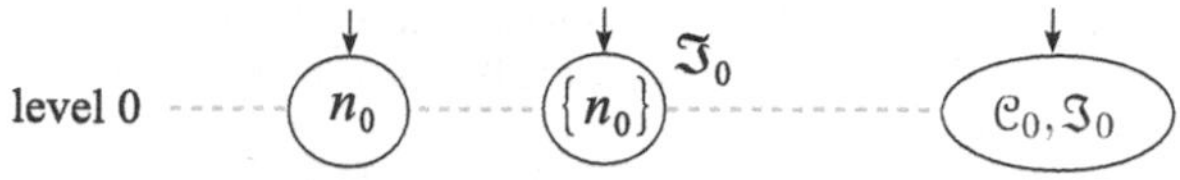

Fig. 10.11 With reference to the temporal observation $\mathcal{O}'_2$ displayed on the left-hand side of Fig. 10.5, the initial states of $Nsp(\mathcal{O}'_2)$ (left), $Isp(\mathcal{O}'_2)$ (center), and $\mathbf{Bhv}(\wp'(\mathcal{A}_2))$ (right), where $\wp'(\mathcal{A}_2)$ is defined in Example 10.3

Example 10.15. Consider the diagnosis problem $\wp'(\mathcal{A}_2) = (a_0, \mathcal{V}, \mathcal{O}'_2, \mathcal{R})$ introduced in Example 10.3. According to *LDE*, the solution of $\wp'(\mathcal{A}_2)$ is generated as follows.

First, the initial states of $Nsp(\mathcal{O}'_2)$, $Isp(\mathcal{O}'_2)$, and $\mathbf{Bhv}(\wp'(\mathcal{A}_2))$ are created, as shown in Fig. 10.11 (level 0).

Then, the loop in lines 9–72 is iterated four times (the number of nodes in $\mathcal{O}'_2$), performing the layering steps listed below:

1. As displayed in Fig. 10.12, $Nsp(\mathcal{O}'_2)$ is expanded to the first layer, and determinized into (the first layer of) $Isp(\mathcal{O}'_2)$ by *LISCA*, producing the update sequence
$$\mathcal{U} = [\,Ext(\mathfrak{I}_0), New(\mathfrak{I}_0 \xrightarrow{awk} \mathfrak{I}_1)\,]\,. \tag{10.16}$$
Updates are processed one by one (lines 13–50) in order to extend $\mathbf{Bhv}(\wp'(\mathcal{A}_2))$. Specifically, the states $(\mathcal{C}_1, \mathfrak{I}_1)$ and $(\mathcal{C}_{53}, \mathfrak{I}_1)$ are created, along with the transitions $(\mathcal{C}_0, \mathfrak{I}_0) \xrightarrow{p_1} (\mathcal{C}_1, \mathfrak{I}_1)$ and $(\mathcal{C}_0, \mathfrak{I}_0) \xrightarrow{p_1} (\mathcal{C}_{53}, \mathfrak{I}_1)$. After the processing of the updates, we have $\mathbb{B}' = \{(\mathcal{C}_0, \mathfrak{I}_0), (\mathcal{C}_1, \mathfrak{I}_1), (\mathcal{C}_{53}, \mathfrak{I}_1)\}$ and $\mathbb{B}^* = \emptyset$ (lines 51 and 52). Hence, the loop in lines 53–59 is not applicable. Since the transition $\mathfrak{I}_0 \xrightarrow{awk} \mathfrak{I}_1$ in $Isp(\mathcal{O}'_2)$ gives rise to the creation of two transitions in $\mathbf{Bhv}(\wp'(\mathcal{A}_2))$, the update $New(\mathfrak{I}_0 \xrightarrow{awk} \mathfrak{I}_1)$ is marked as consistent in line 23. Consequently, the code in lines 60–70 is not applicable either.
2. As displayed in Fig. 10.7, $Nsp(\mathcal{O}'_2)$ is expanded to the second layer, and determinized into (the second layer of) $Isp(\mathcal{O}'_2)$ by *LISCA*, producing
$$\mathcal{U} = [New(\mathfrak{I}_0 \xrightarrow{opl} \mathfrak{I}_2), New(\mathfrak{I}_1 \xrightarrow{opl} \mathfrak{I}_2)]\,. \tag{10.17}$$
The processing of $New(\mathfrak{I}_0 \xrightarrow{opl} \mathfrak{I}_2)$ results in no transitions in $\mathbf{Bhv}(\wp'(\mathcal{A}_2))$; hence $\mathfrak{I}_0 \xrightarrow{opl} \mathfrak{I}_2$ is *not* marked as consistent. Instead, $New(\mathfrak{I}_1 \xrightarrow{opl} \mathfrak{I}_2)$ results in the creation of the transitions $(\mathcal{C}_1, \mathfrak{I}_1) \xrightarrow{b_1(l)} (\mathcal{C}_1, \mathfrak{I}_2)$ and $(\mathcal{C}_1, \mathfrak{I}_1) \xrightarrow{b_1(l)} (\mathcal{C}_8, \mathfrak{I}_2)$. Therefore, $\mathfrak{I}_1 \xrightarrow{opl} \mathfrak{I}_2$ is marked as consistent in $Isp(\mathcal{O}'_2)$. After the processing of

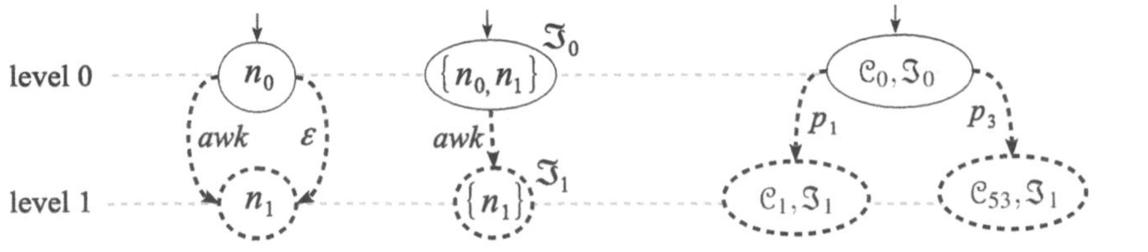

Fig. 10.12 With reference to the temporal observation $\mathcal{O}'_2$ displayed on the left-hand side of Fig. 10.5, the first extension of $Nsp(\mathcal{O}'_2)$ (left), $Isp(\mathcal{O}'_2)$ (center), and $\mathbf{Bhv}(\wp(\mathcal{A}_2))$ (right), where $\wp(\mathcal{A}_2)$ is defined in Example 10.3

the updates, we have $\mathbb{B}' = \{(\mathcal{C}_2, \mathfrak{I}_2), (\mathcal{C}_8, \mathfrak{I}_2)\}$ and $\mathbb{B}^* = \{(\mathcal{C}_0, \mathfrak{I}_0), (\mathcal{C}_{53}, \mathfrak{I}_1)\}$. In the pruning of $\mathbf{Bhv}(\wp(\mathcal{A}_2))$ (lines 53–59), the state $(\mathcal{C}_{53}, \mathfrak{I}_1)$ is removed along with its entering transition. In the pruning of $Isp(\mathcal{O}'_2)$ (lines 60–66), $\mathfrak{I}_0$ is marked by the inconsistent label *opl*, while $\mathfrak{I}_0 \xrightarrow{opl} \mathfrak{I}_2$ is removed. However, this pruning does not cause any backward pruning of $Nsp(\mathcal{O}'_2)$ (lines 67–70), because no state in $Isp(\mathcal{O}'_2)$ is removed; therefore state n_2 in $Nsp(\mathcal{O}'_2)$ still belongs to $\|\mathfrak{I}_2\|$, where $\mathbb{I} = \{\mathfrak{I}_2\}$.

3. As displayed in Fig. 10.8, $Nsp(\mathcal{O}'_2)$ is expanded to the third layer, and determinized into (the third layer of) $Isp(\mathcal{O}'_2)$ by *LISCA*, producing

$$\mathcal{U} = [New(\mathfrak{I}_2 \xrightarrow{ide} \mathfrak{I}_3), New(\mathfrak{I}_2 \xrightarrow{cll} \mathfrak{I}_4)]\,. \tag{10.18}$$

The processing of $New(\mathfrak{I}_2 \xrightarrow{ide} \mathfrak{I}_3)$ results in the creation of the states $(\mathcal{C}_{26}, \mathfrak{I}_3)$ and $(\mathcal{C}_{33}, \mathfrak{I}_3)$ in $\mathbf{Bhv}(\wp(\mathcal{A}_2))$, along with the transitions $(\mathcal{C}_2, \mathfrak{I}_2) \xrightarrow{p_2} (\mathcal{C}_{26}, \mathfrak{I}_3)$, $(\mathcal{C}_2, \mathfrak{I}_2) \xrightarrow{p_4} (\mathcal{C}_{33}, \mathfrak{I}_3)$, $(\mathcal{C}_8, \mathfrak{I}_2) \xrightarrow{p_2} (\mathcal{C}_{26}, \mathfrak{I}_3)$, and $(\mathcal{C}_8, \mathfrak{I}_2) \xrightarrow{p_4} (\mathcal{C}_{33}, \mathfrak{I}_3)$. Hence, $\mathfrak{I}_2 \xrightarrow{ide} \mathfrak{I}_3$ is marked as consistent in $Isp(\mathcal{O}'_2)$. By contrast, $New(\mathfrak{I}_2 \xrightarrow{cll} \mathfrak{I}_4)$ results in no transitions in $\mathbf{Bhv}(\wp(\mathcal{A}_2))$; hence $\mathfrak{I}_2 \xrightarrow{cll} \mathfrak{I}_4$ is *not* marked as consistent. After the processing of the updates, we have $\mathbb{B}' = \{(\mathcal{C}_{26}, \mathfrak{I}_3), (\mathcal{C}_{33}, \mathfrak{I}_3)\}$ and $\mathbb{B}^* = \{(\mathcal{C}_2, \mathfrak{I}_2), (\mathcal{C}_8, \mathfrak{I}_2)\}$. Since both states in $\mathbb{B}^*$ are exited by transitions, no pruning of $\mathbf{Bhv}(\wp(\mathcal{A}_2))$ is applicable in lines 53–59. Instead, the pruning of $Isp(\mathcal{O}'_2)$ in lines 60–66 removes the inconsistent transition $\mathfrak{I}_2 \xrightarrow{cll} \mathfrak{I}_4$. Moreover, this pruning causes backward pruning of $Nsp(\mathcal{O}'_2)$ (lines 67–70). In fact, $\mathbb{I} = \{\mathfrak{I}_3\}$ (as $\mathfrak{I}_4$ was removed). Since state n_4 in $Nsp(\mathcal{O}'_2)$ does not belong to $\|\mathfrak{I}_3\|$, both n_4 and its entering transition $n_2 \xrightarrow{cll} n_4$ are pruned from $Nsp(\mathcal{O}'_2)$.

4. As displayed in Fig. 10.13, $Nsp(\mathcal{O}'_2)$ is expanded to the fourth (and last) layer, and determinized into $Isp(\mathcal{O}'_2)$ by *LISCA*, producing

$$\mathcal{U} = [New(\mathfrak{I}_3 \xrightarrow{cll} \mathfrak{I}_5)]\,. \tag{10.19}$$

The processing of the update results in the creation of the final state $(\mathcal{C}_{28}, \mathfrak{I}_5)$, along with the transition $(\mathcal{C}_{26}, \mathfrak{I}_3) \xrightarrow{b_2(l)} (\mathcal{C}_{28}, \mathfrak{I}_5)$. Hence, $\mathfrak{I}_3 \xrightarrow{cll} \mathfrak{I}_5$ is marked as consistent in $Isp(\mathcal{O}'_2)$. After the processing of the update, we have $\mathbb{B}' =$

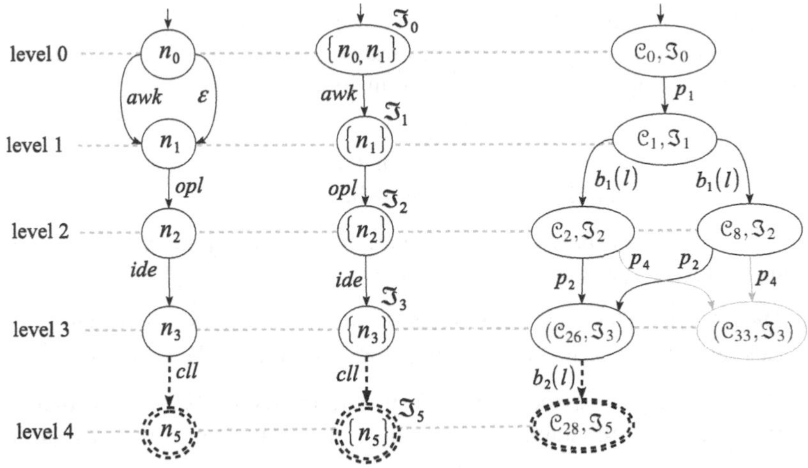

Fig. 10.13 With reference to the temporal observation $\mathcal{O}'_2$ displayed on the left-hand side of Fig. 10.5, the final extension of $Nsp(\mathcal{O}'_2)$ (left), $Isp(\mathcal{O}'_2)$ (center), and $\mathbf{Bhv}(\wp(\mathcal{A}_2))$ (right), where $\wp(\mathcal{A}_2)$ is defined in Example 10.3

$\{(\mathcal{C}_{28}, \mathfrak{I}_5)\}$ and $\mathbb{B}^* = \{(\mathcal{C}_{26}, \mathfrak{I}_3), (\mathcal{C}_{33}, \mathfrak{I}_3)\}$. Since the state $(\mathcal{C}_{33}, \mathfrak{I}_3)$ fulfills the condition in line 54, it is removed along with its two entering transitions (lines 55 and 56). However, since the transition $\mathfrak{I}_3 \xrightarrow{cll} \mathfrak{I}_5$ is consistent in $Isp(\mathcal{O}'_2)$, no pruning of $Isp(\mathcal{O}'_2)$ or $Nsp(\mathcal{O}'_2)$ is performed.

At this point, the generation of $\mathbf{Bhv}(\wp(\mathcal{A}_2))$ is complete. Once pruned, it equals the condensed behavior displayed on the right-hand side of Fig. 10.4.

The decoration of $\mathbf{Bhv}(\wp(\mathcal{A}_2))$ is performed in lines 73–84, giving rise to the decorated condensed behavior displayed in Fig. 10.10. The solution determined in line 85 is therefore $\Delta(\wp(\mathcal{A}_2)) = \{\{\mathbf{nor}\}\}$.

10.7 Bibliographic Notes

Lazy diagnosis of active systems based on pruning techniques was presented in [42, 74, 107].

10.8 Summary of Chapter 10

Condensation. Given a state $\bar{a}$ of a system $\mathcal{A}$, and a viewer $\mathcal{V}$ for $\mathcal{A}$, the subgraph of the behavior space of $\mathcal{A}$ rooted in $\bar{a}$ and encompassing all paths of unobservable transitions (and states involved) rooted in a.

Condensed Behavior Space. Given a system $\mathcal{A}$ and a viewer $\mathcal{V}$ for $\mathcal{A}$, an automaton where each state is a condensation and transitions between states (condensations) are marked by observable component transitions based on $\mathcal{V}$. It is denoted by $\mathbf{Bsp}(\mathcal{A}, \mathcal{V})$.

Lazy Index-Space Generation. The process of intertwining the generation of the index space with the generation of the system behavior.

Circular Pruning. The technique of intertwining the pruning of the index space and the pruning of the reconstructed behavior of the system.

Relative Decoration. The association of each state in a condensation $\mathcal{C}$ with a set of diagnoses relevant to the segments of trajectories starting at the root of $\mathcal{C}$.

Absolute Decoration. The association of each state in a condensation $\mathcal{C}$ with a set of diagnoses relevant to the trajectories starting at the root of the initial state of the condensed behavior.

Two-Phase Decoration. The technique of associating a set of diagnoses with each system state in a condensation in two steps: relative decoration and absolute decoration.

Lazy Diagnosis Engine. A diagnosis engine that generates the solution of a diagnosis problem by lazy techniques, including circular pruning and two-phase decoration.

Chapter 11
Sensitive Diagnosis

In previous chapters we have introduced several diagnosis techniques for active systems. Even though these techniques are based on different requirements and pursue different goals, they nonetheless share a common fundamental assumption, namely that a diagnosis is a set of faults, where *each fault is associated with a component transition*, which is considered abnormal.

Specifically, based on a set of associations $(t(c), f)$, namely a ruler, where $t(c)$ is an abnormal transition of component c and f a fault label, each trajectory h of the system gives rise to a corresponding diagnosis including the set of faults associated with the abnormal transitions of h.

So, whatever the active system may be, faults are invariably associated with component transitions. This is called the *atomic assumption*, as faults are defined at the atomic level of component transitions.

One may argue about whether the atomic assumption is satisfactory in any circumstances. In principle, faults might be associated with patterns of transitions, possibly involving different components. For example, a fault f can be associated with a sequence $[t_1(c_1), t_2(c_2), t_3(c_3)]$ of three transitions, each one relevant to a different component. As such, fault f occurs when a trajectory of the system involves that string of transitions.

More general patterns of transitions can be defined based on operators of regular expressions. For example, the regular expression $r = t_1(c_1)t_2(c_2)^*t_3(c_3)$ is matched by any string of transitions starting with $t_1(c_1)$, followed by zero or more repetitions of $t_2(c_2)$, and ending with $t_3(c_3)$. In other words, if we associate fault f with r, f occurs when a trajectory of the system incorporates a string of transitions belonging to the regular language of r, for instance, $[t_1(c_1), t_3(c_3)]$ or $[t_1(c_1), t_2(c_2), t_2(c_2), t_3(c_3)]$.

From this perspective, the atomic assumption becomes only a special case of a more general assumption, which we call the *pattern assumption*: *a fault of the system is associated with a pattern of component transitions* (rather than with a single component transition). Consequently, what is considered abnormal is not a single component transition involved in the pattern, but a specific combination of transitions matching such a pattern.

G. Lamperti et al., *Introduction to Diagnosis of Active Systems*,
https://doi.org/10.1007/978-3-319-92733-6_11

According to the pattern assumption, a fault occurs because of the contributions of several transitions, possibly relevant to different components. Therefore, generally speaking, what is faulty is not a single component, but rather the system as a whole.

A further extension to the diagnosis process can be achieved by conceiving the active system as a hierarchy of subsystems. In other words, the network of components forming the system can be recursively partitioned into subsystems. Conceptually, the system is viewed as a tree, where each internal node is a subsystem, and each leaf node is a component.

For example, a protected transmission line can be viewed as composed of two subsystems: one set of protection hardware on the left and one set of protection hardware on the right. Each set of protection hardware is in turn composed of two components: one breaker and one protection device. Thus, the protected line is the root of a tree containing two internal nodes (two instances of the protection hardware) and four leaf nodes (two breakers and two protection devices).

The hierarchical organization of the active system can be exploited to generalize the pattern assumption. Considering our example, a fault can be associated with the protection hardware (breaker and protection device) when either the protection device fails to command the breaker to open or the breaker does not open when commanded.

This way, the corresponding pattern involves transitions of both the breakers and the protection devices. Consequently, the alphabet of the regular expression defining such a pattern is the union of the transitions of the breakers and protection devices.

In principle, we can define the patterns of any nonleaf nodes in terms of component transitions. For example, we can define patterns for the protected line (the root of the tree) in terms of transitions of breakers and protection devices.

The drawback of this approach is that it provides no support for abstraction. In fact, the hierarchy represented by the tree implicitly defines different levels of abstraction, corresponding to the different levels of the tree. Defining a pattern for an internal node in terms of transitions of the leaf nodes (components) encompassed is like decomposing a software system into a hierarchy of functional modules and codifying each module in terms of built-in statements only.

In order to preserve abstraction and separation of concerns, we can define for each internal node (subsystem) not only patterns associated with faults, but also patterns associated with *interface symbols*. In particular, a fault can be an interface symbol too, but not necessarily.

The point is, for each subsystem in the hierarchy, both faults and interface symbols can be defined in terms of regular expressions.

As the name suggests, an interface symbol associated with a node is what is visible to the parent of that node in its own pattern definition. That is, if ξ is an internal node with children $\xi_1, \ldots, \xi_n$, then the alphabet for the patterns defined on ξ is the union of the interface symbols defined for $\xi_1, \ldots, \xi_n$. In particular, if a child ξ_i, $i \in [1 \cdots n]$, is a leaf node (component), then component transitions are interface symbols by definition.

This way, faults can be defined at any level of the system hierarchy, and diagnoses can incorporate faults associated with different subsystems.

11.1 Context Sensitivity

Conceptually, the topology of an active system may be organized in a hierarchy, where the root corresponds to the whole system, leaves to components, and intermediate nodes to subsystems. Since faults in an active system are defined at the level of components, there is no possibility to provide a hierarchy of diagnoses adhering to the hierarchy of the system. Trivially, a subsystem is faulty if and only if it includes (at least) one faulty component. We call this commonly used approach *context-free diagnosis*. While context-free diagnosis may be adequate for simple active systems, it is doomed to be unsatisfactory when applied to structured active systems. The hierarchical structure of an active system suggests that one should organize the diagnosis rules into a hierarchy that parallels the hierarchical structure of the active system: each subsystem has its own set of diagnosis rules, which may or may not depend on the rules of inner subsystems. We call this alternative approach *context-sensitive diagnosis*.

The idea of context sensitivity in diagnosis was inspired by the problem of (formal) language specification. The classical hierarchy proposed by Chomsky [29] incorporates four types of generative grammars of growing expressiveness. Within this hierarchy, *Type-2* and *Type-1* grammars are means to specify the syntax of context-free and context-sensitive languages, respectively. For practical reasons, Type-2 grammars in Backus-Naur Form (BNF) notation have become the standard for the specification of the syntax of programming languages. However, since, generally speaking, the syntax rules of programming languages depend on the context, production rules in BNF notation are unable to enforce all the syntax constraints of the language.[1] In compilers, the checking of these (context-sensitive) constraints is generally left to semantic analysis.

Loosely speaking, the idea of context sensitivity can be profitably injected into the diagnosis of active systems in three ways:

1. A transition of a component is not considered either normal or faulty on its own: it depends on the context in which such a transition is triggered,
2. A fault in a component is not necessarily ascribed to one transition: it can be the result of a pattern of transitions,
3. Normal or faulty behavior of a subsystem is not necessarily determined by the normal or faulty behavior of its components (or inner subsystems).

Context-sensitive diagnosis is linked to some seemingly contradictory but nonetheless possible results, called *paradoxes*.

[1] For example, stating that the list of actual parameters in a function call should match, in number and types, the list of formal parameters.

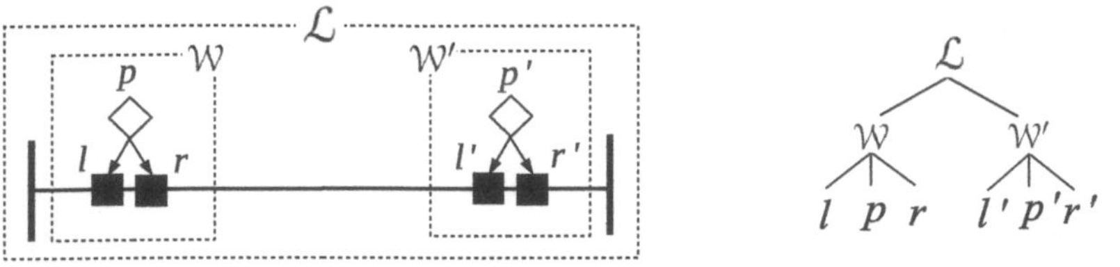

Fig. 11.1 Protected power transmission line (left) and corresponding context hierarchy (right)

11.1.1 Positive Paradox

The *positive paradox* states that a (sub)system may be normal despite the faulty behavior of a number of its components (or inner subsystems). Example 11.1 aims to clarify this assertion.

Example 11.1. Shown on the left side of Fig. 11.1 is a simplified representation of a power transmission line $\mathcal{L}$. The line is protected at both ends by redundant sets of protection hardware, called $\mathcal{W}$ and $\mathcal{W}'$, each involving one protection device and two breakers. As such, $\mathcal{W}$ and $\mathcal{W}'$ are two instances of the active system $\mathcal{A}_2$ introduced in Chapter 4 and displayed in Fig. 4.2. For instance, $\mathcal{W}$ incorporates a protection device p and breakers l and r. When lightning strikes the line, a short circuit may occur on the latter. The protection system is designed to open the breakers in order to isolate the line, which then causes the extinguishing of the short. To this end, when it detects a short circuit, a protection device is expected to trigger both breakers to open. If this causes the extinguishing of the short circuit, the protection device commands both breakers to close in order to reconnect the line to the power network. A protection device is faulty either when, upon detection of a short circuit, it commands breakers to close (instead of opening) or, after the extinguishing of the short circuit, it commands breakers to open (instead of closing). A breaker is faulty when, after receiving a command from the protection device to change state, it remains in the same state. A minimal (nonredundant) set of protection hardware would incorporate one single protection device and one single breaker at each end of the line. Thus, when one of the two devices is faulty, the line cannot be isolated. By contrast, in the redundant protection hardware shown in Fig. 11.1, the normal behavior of the protection device and of *one* breaker (for each end) suffices to guarantee the isolation of the line. Consider the following two scenarios:[2]

1. In $\mathcal{W}$, l is faulty, while p and r are normal. In $\mathcal{W}'$, l' is faulty, while p' and r' are normal. The diagnosis is $\delta_1 = \{l, l'\}$. Owing to redundancy, the behavior of the protected line is in fact normal, as the line is isolated despite the faulty behavior of the two breakers.
2. In $\mathcal{W}$, l is faulty, while p and r are normal. In $\mathcal{W}'$, p' is faulty, while l' and r' are normal. The diagnosis is $\delta_2 = \{l, p'\}$. However, owing to p', $\mathcal{W}'$ fails to open, causing the failure of the whole protected line.

[2] For simplicity, we assume that the fault labels are just the names of faulty components.

Although the two diagnoses δ_1 and δ_2 differ in one component only (l' vs. p'), the behavior of the protected line in the first scenario is normal, while it is faulty in the second. The point is, *the given diagnoses do not explicitly account for such a distinction*. More generally, we can consider several subsystems of the protected line within a hierarchy, and require a diagnosis for each of them. Such a hierarchy is displayed on the right side of Fig. 11.1, where $\mathcal{L}$ denotes the whole protected line. According to this hierarchy, the diagnosis of the first scenario is $\{l, l'\}$, while in the second scenario the diagnosis is $\{l, p', \mathcal{W}', \mathcal{L}\}$. Comparing the two diagnoses, we conclude that in the first scenario two components are faulty but their misbehavior is not propagated to higher subsystems. By contrast, in the second scenario, besides two faulty components (l and p'), $\mathcal{W}'$ and $\mathcal{L}$ are faulty too.

11.1.2 Negative Paradox

The *negative paradox* states that a (sub)system can be faulty despite the normal behavior of all its components (or inner subsystems). This is true also for systems other than active systems:

1. A software system can be faulty even if all its software components are bug-free,
2. Injustice may occur even if all laws are respected,
3. A human society can be destined to dictatorship notwithstanding all its democratic institutions.[3]

Example 11.2 instantiates the negative paradox in application domain of power transmission networks that we have been referring to.

Example 11.2. With reference to Fig. 11.1, the isolation of the line causes the extinguishing of the short circuit because the short is no longer fed by any current. This is why the protection system is designed to reconnect the line (by closing breakers) to the network once the short has been extinguished. Suppose now that, instead of a lightning strike, what causes the short is a tree falling on the line. In this case, the reconnection of the line to the network is expected to activate the short circuit again, as the tree is likely still to be on the line, thereby causing the line to be isolated anew (and, this time, permanently). Clearly, even assuming that the behavior

[3] An anecdote about the Austrian logician, mathematician, and philosopher Kurt Friedrich Gödel is told in [66]. In December 1947, Gödel went to his American citizenship hearing in New Jersey. As his witnesses, Gödel brought his two closest friends, Oskar Morgenstern and Albert Einstein. Gödel, in his usual manner, had read extensively in preparing for the hearing. In the course of his studies, Gödel decided that he had discovered a flaw in the U.S. Constitution, a contradiction which would allow the U.S. to be turned into a dictatorship. Gödel seemed to feel a need to make this known. However, his friends Morgenstern and Einstein warned Gödel that it would be a disaster to confront his citizenship examiner with visions of a Constitutional flaw leading to an American dictatorship. The examiner happened to remark how fortunate it was that the U.S. was not a dictatorship, which Gödel took as a cue to explain his discovery. This, however, did not compromise his citizenship hearing, as he eventually obtained this citizenship in April 1948.

of the protection devices and breakers is normal, the behavior of the line is actually faulty.

Coping with negative paradoxes is essential when the behavior of an active system cannot be foreseen at design time. To face this uncertainty, a set of constraints can be associated with the nodes of the system hierarchy, aimed at intercepting relevant faulty behavior. These constraints parallel the requirements of software systems, which need to be validated independently of the correct behavior of the components of the software.

11.2 Diagnosis Problem

The notion of a diagnosis problem for an active system $\mathcal{A}$ provided in Def. 4.7, namely a quadruple

$$\wp(\mathcal{A}) = (a_0, \mathcal{V}, \mathcal{O}, \mathcal{R}), \tag{11.1}$$

is still valid for a hierarchical active system. However, the notion of a ruler $\mathcal{R}$ must be extended by context sensitivity as follows.

Definition 11.1 (Context-Sensitive Ruler). Let $\mathcal{A}$ be an active system. A ruler for $\mathcal{A}$ is a triple

$$\mathcal{R} = (\Xi, \mathcal{H}, \mathcal{S}), \tag{11.2}$$

where Ξ is the *context domain*, $\mathcal{H}$ the *context hierarchy*, and $\mathcal{S}$ the *semantics*. These three elements are defined as follows:

1. The context domain $\Xi = \{\xi_1, \ldots, \xi_n\}$ is the set of subsystems of $\mathcal{A}$, called *contexts*, that are designated to be relevant to the output of the diagnosis task.
2. The context hierarchy $\mathcal{H}$ defines a partition of each context $\xi_i \in \Xi$, $i \in [1 \cdots n]$, in terms of other subcontexts (or components), namely $(\xi_i, \{\xi_{i_1}, \ldots, \xi_{i_{n_i}}\}) \in \mathcal{H}$ (where $\xi_{i_1}, \ldots, \xi_{i_{n_i}}$ are the child nodes of ξ_i in $\mathcal{H}$); as such, $\mathcal{H}$ implicitly defines a tree (more generally, a forest), where the leaf nodes are components.
3. Let $\mathbf{N}$ be a set of identifiers called the *name space*, $\mathbf{F}$ a subset of $\mathbf{N}$ called the *fault space*, and $\mathbf{I}$ a subset of $\mathbf{N}$ called the *interface space*. The semantics $\mathcal{S}$ is a set of pairs $(\xi, \mathcal{P})$, where $\xi \in \Xi$ and $\mathcal{P} = [P_1, \ldots, P_k]$ is the list of *semantic patterns*, with each semantic pattern $P_j \in \mathcal{P}$, $P_j = (N_j, r_j)$, being an association between a name in $N_j \in \mathbf{N}$ and a regular expression r_j.

The *alphabet* of the regular expression r_j is inductively defined as follows:

1. The alphabet of a component is the set of transitions of its model,
2. Let $\mathbb{N}$ be the set of names used for semantic patterns in $\bar{\mathcal{P}}$, where $(\bar{\xi}, \bar{\mathcal{P}}) \in \mathcal{S}$; let $\mathbb{I}$ be the subset of names in $\mathbb{N}$ that are also in the interface space, namely $\mathbb{I} = \mathbb{N} \cap \mathbf{I}$. Each symbol in the alphabet Σ of $\bar{\xi}$ is a subset of $\mathbb{I}$, called an *interface symbol*; in other words, $\Sigma(\bar{\xi})$ is the powerset of $\mathbb{I}$, namely $\Sigma(\bar{\xi}) = 2^{\mathbb{I}}$,

3. Let $(\xi, \{\xi_1, \dots, \xi_m\}) \in \mathcal{H}$, $(\xi, \mathcal{P}) \in \mathcal{S}$, $P_j \in \mathcal{P}$, and $P_j = (N_j, r_j)$; the alphabet Σ of r_j is the union of the alphabets of the subcontexts (possibly components) of ξ and the semantic-pattern names defined up to $P_{j-1} \in \mathcal{P}$:

$$\Sigma(r_j) = \left(\bigcup_{i=1}^{m} \Sigma(\xi_i) \right) \cup \left(\bigcup_{i=1}^{j-1} \{N_i\} \right) . \tag{11.3}$$

The syntax of the regular expression on the alphabet Σ is defined inductively as follows (assuming x and y to be regular expressions denoting languages $L(x)$ and $L(y)$, respectively):

1. ε denotes the language $\{\varepsilon\}$ (where ε is the *null* symbol),
2. If $a \in \Sigma$, then a denotes the singleton language $\{a\}$,
3. (x) denotes $L(x)$ (parentheses are allowed as usual),
4. $x^?$ denotes $L(x) \cup \{\varepsilon\}$ (optionality),
5. x^* denotes $\bigcup_{i=0}^{\infty} (L(x))^i$ (iteration zero or more times),
6. x^+ denotes $\bigcup_{i=1}^{\infty} (L(x))^i$ (iteration one or more times),
7. xy denotes $L(x)L(y)$ (concatenation),
8. $x \mid y$ denotes $L(x) \cup L(y)$ (alternative),
9. $x \& y$ is a shorthand for $(xy \mid yx)$ (free concatenation).

The *plain form* of the regular expression r is the iterated macro-substitution of each name in $\mathbf{N}$ (involved in r) by the corresponding regular expression. We assume that the plain form of each regular expression is nonempty.

Example 11.3. A context-sensitive ruler $\mathcal{R}_{\mathcal{W}}$ for the protection hardware $\mathcal{W}$ defined in Example 4.6 is specified as follows. The context domain is $\Xi = \{\xi_l, \xi_r, \xi_{\mathcal{W}}\}$, where both ξ_l and ξ_r include just the breakers l and r, respectively. The context hierarchy $\mathcal{H}$ is drawn in Fig. 11.2. $\mathbf{N} = \mathbf{F} = \{\mathbf{nol}, \mathbf{ncl}, \mathbf{nor}, \mathbf{ncr}, \mathbf{fop}, \mathbf{fcp}, \mathbf{fdw}, \mathbf{frw}\}$ (an explanation is provided in Table 11.1). $\mathbf{I} = \{\mathbf{nol}, \mathbf{ncl}, \mathbf{nor}, \mathbf{ncr}\}$. $\mathcal{S}$ is defined in Table 11.1.[4]

For the breakers l and r, two semantic patterns are defined, involving one transition, corresponding to failure either to open or to close. For the protection hardware $\mathcal{W}$, four semantic patterns are defined. The first two patterns, **fop** and **fcp**, are relevant to the misbehavior of the protection device (by sending breakers the wrong command). The semantic pattern **fdw** indicates a failure in disconnecting the left-hand end of the line: either the protection device sends the wrong command to the

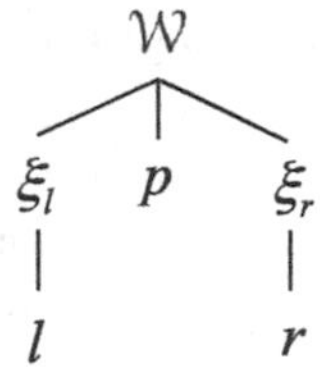

Fig. 11.2 Context hierarchy for protection hardware $\mathcal{W}$

[4] In Table 11.1, with a slight abuse of notation, singletons are represented by the single element.

Table 11.1 Semantic patterns

Context	*Semantic patterns*	*Explanation*
ξ_l	**nol** $= b_{3l}$	Breaker l does not open
	ncl $= b_{4l}$	Breaker l does not close
ξ_r	**nor** $= b_{3r}$	Breaker r does not open
	ncr $= b_{4r}$	Breaker r does not close
$\mathcal{W}$	**fop** $= p_3$	Protection device p fails to open
	fcp $= p_4$	Protection device p fails to close
	fdw $=$ **fop** $\mid p_1($**nol** & **nor**$)$	Protection hardware $\mathcal{W}$ fails to disconnect
	frw $=$ **fcp** $\mid p_2($**ncl** $\mid$ **ncr**$)$	Protection hardware $\mathcal{W}$ fails to reconnect

breakers (**fop**) or it sends the correct command (p_1), yet both breakers remain closed (**nol** & **nor**). Finally, the semantic pattern **frw** indicates a failure in reconnecting the left-hand end of the line: either the protection device sends the wrong command to the breakers (**fcp**) or it sends the correct command (p_2), yet either l or r remains open (**ncl** | **ncr**).

11.3 Problem Solution

As a consequence of context sensitivity, in order to define a diagnosis we need to introduce the notion of *trajectory projection*. Intuitively, the projection of a trajectory h on a context ξ, written $h[\xi]$, is the mode in which h is perceived by ξ in terms of interface symbols (if ξ includes other contexts) and/or component transitions (if ξ includes components).

If ξ includes components only, $h[\xi]$ is simply the subsequence of transitions in h which belong to those components. If, instead, when ξ embodies other contexts, $h[\xi]$ is bound to include interface symbols of such contexts. Each of these symbols corresponds to one or more matchings of relevant semantic patterns.

We now provide the precise definition of $h[\xi]$ in operational terms. Let $\xi_1, \ldots, \xi_n$ be the set of contexts involved (either directly or indirectly) in the context hierarchy rooted in ξ (including the latter). Let C_ξ denote the set of components involved in that hierarchy. The operational specification, called *Projection*, generates all projections $h[\xi_1], \ldots, h[\xi_n]$ in one run as detailed below:

algorithm *Projection* (**in** $\mathcal{H}$, **in** h, **in** ξ, **out** $H_\xi = \{h[\xi_1], \ldots, h[\xi_n]\}$)
  $\mathcal{H}$: the context hierarchy of a system $\mathcal{A}$,
  h: a trajectory of $\mathcal{A}$,
  ξ: a context in $\mathcal{H}$;
  $H_\xi = \{h[\xi_1], \ldots, h[\xi_n]\}$: the trajectory projections $h[\xi_1], \ldots, h[\xi_n]$ for ξ and all descendant contexts of ξ in $\mathcal{H}$;
  begin $\langle$*Projection*$\rangle$

7. Initialize each $h[\xi_i]$, $i \in [1 \cdots n]$, to the empty sequence;
8. **foreach** $t(c) \in h$ such that $c \in C_\xi$ **do**
9. Let ξ_p be the parent context of c in the hierarchy;
10. Shift $t(c)$ to $h[\xi_p]$;
11. **repeat**
12. Let $\mathbb{I}$ be the set of symbols $N \in \mathbf{I}$ (interface) such that (N, r) is a semantic pattern for ξ_p **and** a suffix of $h[\xi_p]$ matches r (in plain form);
13. **if** $\mathbb{I} \neq \emptyset$ **and** $\xi_p \neq \xi$ **and** ξ_p has a parent ξ_p^* **then**
14. Append $\mathbb{I}$ to $h[\xi_p^*]$;
15. $\xi_p := \xi_p^*$
16. **endif**
17. **until** $\mathbb{I} = \emptyset$ **or** $\xi_p = \xi$
18. **endfor**
19. **end** $\langle$*Projection*$\rangle$.

Example 11.4. Based on the behavior space $Bsp(\mathcal{A}_2)$ displayed in Fig. 4.4, consider the trajectory $h = [p_1, b_{3r}, b_{3l}, p_4, b_{3r}, b_{1l}]$. With reference to the context hierarchy displayed in Fig. 11.2, Table 11.2 traces the computation of $h[\mathcal{W}]$ by the algorithm *Projection*. For each transition considered in the main loop, identified by $i \in [1 \cdots 6]$, the configurations of h, $h[\xi_l]$, $h[\xi_r]$, and $h[\mathcal{W}]$ can be represented as follows:

0. Initialization (line 7).
1. p_1 is shifted to $h[\mathcal{W}]$; $\mathbb{I} = \emptyset$.
2. b_{3r} is shifted to $h[\xi_r]$; $\mathbb{I} = \{\mathbf{nor}\}$, **nor** is appended to $h[\mathcal{W}]$.
3. b_{3l} is shifted to $h[\xi_l]$; $\mathbb{I} = \{\mathbf{nol}\}$, **nol** is appended to $h[\mathcal{W}]$.
4. p_4 is shifted to $h[\mathcal{W}]$; $\mathbb{I} = \emptyset$.
5. b_{3r} is shifted to $h[\xi_r]$; $\mathbb{I} = \{\mathbf{nor}\}$, **nor** is appended to $h[\mathcal{W}]$.
6. b_{1l} is shifted to $h[\xi_l]$; $\mathbb{I} = \emptyset$.

The configurations in the last line are the actual projections of the trajectory h on the contexts ξ_l, ξ_r, and $\mathcal{W}$. In particular, we have $h[\mathcal{W}] = [p_1, \mathbf{nor}, \mathbf{nol}, p_4, \mathbf{nor}]$. In contrast, as expected, the projections on the contexts ξ_l and ξ_r include the subsequences of transitions relevant to components l and r, respectively.

Table 11.2 Tracing of *Projection* algorithm for trajectory h

i	h	$h[\xi_l]$	$h[\xi_r]$	$h[\mathcal{W}]$
0	$[p_1, b_{3r}, b_{3l}, p_4, b_{3r}, b_{1l}]$	$[]$	$[]$	
1	$[b_{3r}, b_{3l}, p_4, b_{3r}, b_{1l}]$	$[]$	$[]$	$[p_1]$
2	$[b_{3l}, p_4, b_{3r}, b_{1l}]$	$[]$	$[b_{3r}]$	$[p_1, \mathbf{nor}]$
3	$[p_4, b_{3r}, b_{1l}]$	$[b_{3l}]$	$[b_{3r}]$	$[p_1, \mathbf{nor}, \mathbf{nol}]$
4	$[b_{3r}, b_{1l}]$	$[b_{3l}]$	$[b_{3r}]$	$[p_1, \mathbf{nor}, \mathbf{nol}, p_4]$
5	$[b_{1l}]$	$[b_{3l}]$	$[b_{3r}, b_{3r}]$	$[p_1, \mathbf{nor}, \mathbf{nol}, p_4, \mathbf{nor}]$
6	$[]$	$[b_{3l}, b_{1l}]$	$[b_{3r}, b_{3r}]$	$[p_1, \mathbf{nor}, \mathbf{nol}, p_4, \mathbf{nor}]$

Definition 11.2 (Context-Sensitive Diagnosis). Let h be a trajectory of an active system $\mathcal{A}$, and $\mathcal{R}$ a ruler for $\mathcal{A}$. Let $\|r\|$ denote the language of the plain form of a regular expression r. The *diagnosis* of h based on the ruler $\mathcal{R} = (\Xi, \mathcal{H}, \mathcal{S})$, written $h_{[\mathcal{R}]}$, is the set of fault labels:

$$h_{[\mathcal{R}]} = \{\, f \mid f \in \mathbf{F}, (f,r) \in \mathcal{P}, (\xi, \mathcal{P}) \in \mathcal{S}, e \in \|r\|, e \sqsubseteq h[\xi] \,\} \,. \tag{11.4}$$

In other words, a diagnosis $h_{[\mathcal{R}]}$ is the set of faults f involved in the semantic-pattern list $\mathcal{P}$ of a context ξ, such that there exists a string e, in the language of the (plain form of the) regular expression r relevant to f, that is a (contiguous) substring ($\sqsubseteq$) of the projection of trajectory h on ξ.

Example 11.5. Consider the protection hardware $\mathcal{W}$ defined in Example 4.6 (called system $\mathcal{A}_2$ in Chapter 4). The behavior space of $\mathcal{W}$ is displayed in Fig. 4.4. The context-sensitive ruler $\mathcal{R}_\mathcal{W} = (\Xi, \mathcal{H}, \mathcal{S})$ for $\mathcal{W}$ is specified in Example 11.3.

Consider the trajectory $h = [p_1, b_{3r}, b_{3l}, p_4, b_{3r}, b_{1l}]$ defined in Example 11.4. According to (11.4) and based on the semantic patterns specified in Table 11.1, to determine the diagnosis of h based on $\mathcal{R}_\mathcal{W}$, we need to perform pattern matching on the projections of h on the contexts ξ_l, ξ_r, and $\mathcal{W}$.

Considering $h[\xi_l] = [b_{3l}, b_{1l}]$, since $[b_{3l}] \sqsubseteq h[\xi_l]$, we have $\mathbf{nol} \in h_{[\mathcal{R}_\mathcal{W}]}$. For $h[\xi_r] = [b_{3r}, b_{3r}]$, since $[b_{3r}] \sqsubseteq h[\xi_r]$, we have $\mathbf{nor} \in h_{[\mathcal{R}_\mathcal{W}]}$. Finally, for $h[\mathcal{W}] = [p_1, \mathbf{nor}, \mathbf{nol}, p_4, \mathbf{nor}]$, since $[p_4] \sqsubseteq h[\mathcal{W}]$ and $[p_1, \mathbf{nor}, \mathbf{nol}] \sqsubseteq h[\mathcal{W}]$ (with $[p_1, \mathbf{nor}, \mathbf{nol}]$ being a string in the language of $p_1(\mathbf{nol} \,\&\, \mathbf{nor})$), both of the fault labels **fcp** and **fdw** are in $h_{[\mathcal{R}_\mathcal{W}]}$.

In summary, $h_{[\mathcal{R}_\mathcal{W}]} = \{\mathbf{nol}, \mathbf{nor}, \mathbf{fcp}, \mathbf{fdw}\}$, meaning that both breakers fail to open, the protection device sends the wrong command to the breakers (open rather than close), and the protection hardware fails to disconnect the left-hand end of the line.

After we have revisited the notion of a diagnosis according to Def. 11.2, the notion of the solution of a diagnosis problem $\wp(\mathcal{A}) = (a_0, \mathcal{V}, \mathcal{O}, \mathcal{R})$ remains that defined in Def. 4.8, namely,

$$\Delta(\wp(\mathcal{A})) = \{\, \delta \mid \delta = h_{[\mathcal{R}]}, h \in Bsp(\mathcal{A}), h_{[\mathcal{V}]} \in \|\mathcal{O}\| \,\} \,. \tag{11.5}$$

That is, the solution of $\wp(\mathcal{A})$ is the set of diagnoses of the trajectories of $\mathcal{A}$ based on the ruler $\mathcal{R}$ such that their trace on the viewer $\mathcal{V}$ is a candidate trace of the temporal observation $\mathcal{O}$.

Example 11.6. Consider the diagnosis problem $\wp(\mathcal{A}_2) = (a_0, \mathcal{V}, \mathcal{O}_2, \mathcal{R})$ defined in Example 4.12. Here, system $\mathcal{A}_2$ is the protection hardware $\mathcal{W}$, with the context hierarchy displayed in Fig. 11.2.

Let $\wp(\mathcal{W}) = (a_0, \mathcal{V}, \mathcal{O}_2, \mathcal{R}_\mathcal{W})$ be a diagnosis problem for the protection hardware $\mathcal{W}$, where the ruler $\mathcal{R}_\mathcal{W}$ is specified in Example 11.3, while the behavior space $Bsp(\mathcal{W}) = Bsp(\mathcal{A}_2)$ is displayed in Fig. 4.4.

The solution of $\wp(\mathcal{W})$ can be determined based on (11.5). First, we have to select the trajectories in $Bsp(\mathcal{W})$ whose trace is in $\|\mathcal{O}_2\|$. These trajectories are listed in Table 4.4, namely $h_1, \ldots, h_{26}$.

Table 11.3 Trajectories, projections, and candidate diagnoses

h	*Trajectory*	$h[\xi_l]$	$h[\xi_r]$	$h[\mathcal{W}]$	$h_{[\mathcal{R}_\mathcal{W}]}$
h_1	$[p_1, b_1(l), b_3(r), p_4]$	$[b_1(l)]$	$[b_3(r)]$	$[p_1, \mathbf{nor}, p_4]$	$\{\mathbf{nor}, \mathbf{fcp}, \mathbf{frw}\}$
h_2	$[p_1, b_1(l), b_3(r), p_4, b_3(r)]$	$[b_1(l)]$	$[b_3(r), b_3(r)]$	$[p_1, \mathbf{nor}, p_4, \mathbf{nor}]$	$\{\mathbf{nor}, \mathbf{fcp}, \mathbf{frw}\}$
h_3	$[p_1, b_1(l), b_3(r), p_4, b_3(r), b_6(l)]$	$[b_1(l), b_6(l)]$	$[b_3(r), b_3(r)]$	$[p_1, \mathbf{nor}, p_4, \mathbf{nor}]$	$\{\mathbf{nor}, \mathbf{fcp}, \mathbf{frw}\}$
h_4	$[p_1, b_1(l), b_3(r), p_4, b_6(l)]$	$[b_1(l), b_6(l)]$	$[b_3(r)]$	$[p_1, \mathbf{nor}, p_4]$	$\{\mathbf{nor}, \mathbf{fcp}, \mathbf{frw}\}$
h_5	$[p_1, b_1(l), b_3(r), p_4, b_6(l), b_3(r)]$	$[b_1(l), b_6(l)]$	$[b_3(r), b_3(r)]$	$[p_1, \mathbf{nor}, p_4, \mathbf{nor}]$	$\{\mathbf{nor}, \mathbf{fcp}, \mathbf{frw}\}$
h_6	$[p_1, b_1(l), b_3(r), p_2]$	$[b_1(l)]$	$[b_3(r)]$	$[p_1, \mathbf{nor}, p_2]$	$\{\mathbf{nor}\}$
h_7	$[p_1, b_1(l), b_3(r), p_2, b_2(l)]$	$[b_1(l), b_2(l)]$	$[b_3(r)]$	$[p_1, \mathbf{nor}, p_2]$	$\{\mathbf{nor}\}$
h_8	$[p_1, b_1(l), b_3(r), p_2, b_2(l), b_5(r)]$	$[b_1(l), b_2(l)]$	$[b_3(r), b_5(r)]$	$[p_1, \mathbf{nor}, p_2]$	$\{\mathbf{nor}\}$
h_9	$[p_1, b_1(l), b_3(r), p_2, b_4(l)]$	$[b_1(l), b_4(l)]$	$[b_3(r)]$	$[p_1, \mathbf{nor}, p_2, \mathbf{ncl}]$	$\{\mathbf{ncl}, \mathbf{nor}, \mathbf{frw}\}$
h_{10}	$[p_1, b_1(l), b_3(r), p_2, b_4(l), b_5(r)]$	$[b_1(l), b_4(l)]$	$[b_3(r), b_5(r)]$	$[p_1, \mathbf{nor}, p_2, \mathbf{ncl}]$	$\{\mathbf{ncl}, \mathbf{nor}, \mathbf{frw}\}$
h_{11}	$[p_1, b_1(l), b_3(r), p_2, b_5(r)]$	$[b_1(l)]$	$[b_3(r), b_5(r)]$	$[p_1, \mathbf{nor}, p_2]$	$\{\mathbf{nor}\}$
h_{12}	$[p_1, b_1(l), b_3(r), p_2, b_5(r), b_4(l)]$	$[b_1(l), b_4(l)]$	$[b_3(r), b_5(r)]$	$[p_1, \mathbf{nor}, p_2, \mathbf{ncl}]$	$\{\mathbf{ncl}, \mathbf{nor}, \mathbf{frw}\}$
h_{13}	$[p_1, b_1(l), b_3(r), p_2, b_5(r), b_2(l)]$	$[b_1(l), b_2(l)]$	$[b_3(r), b_5(r)]$	$[p_1, \mathbf{nor}, p_2]$	$\{\mathbf{nor}\}$
h_{14}	$[p_1, b_3(r), b_1(l), p_4]$	$[b_1(l)]$	$[b_3(r)]$	$[p_1, \mathbf{nor}, p_4]$	$\{\mathbf{nor}, \mathbf{fcp}, \mathbf{frw}\}$
h_{15}	$[p_1, b_3(r), b_1(l), p_4, b_3(r)]$	$[b_1(l)]$	$[b_3(r), b_3(r)]$	$[p_1, \mathbf{nor}, p_4, \mathbf{nor}]$	$\{\mathbf{nor}, \mathbf{fcp}, \mathbf{frw}\}$
h_{16}	$[p_1, b_3(r), b_1(l), p_4, b_3(r), b_6(l)]$	$[b_1(l), b_6(l)]$	$[b_3(r), b_3(r)]$	$[p_1, \mathbf{nor}, p_4, \mathbf{nor}]$	$\{\mathbf{nor}, \mathbf{fcp}, \mathbf{frw}\}$
h_{17}	$[p_1, b_3(r), b_1(l), p_4, b_6(l)]$	$[b_1(l), b_6(l)]$	$[b_3(r)]$	$[p_1, \mathbf{nor}, p_4]$	$\{\mathbf{nor}, \mathbf{fcp}, \mathbf{frw}\}$
h_{18}	$[p_1, b_3(r), b_1(l), p_4, b_6(l), b_3(r)]$	$[b_1(l), b_6(l)]$	$[b_3(r), b_3(r)]$	$[p_1, \mathbf{nor}, p_4, \mathbf{nor}]$	$\{\mathbf{nor}, \mathbf{fcp}, \mathbf{frw}\}$
h_{19}	$[p_1, b_3(r), b_1(l), p_2]$	$[b_1(l)]$	$[b_3(r)]$	$[p_1, \mathbf{nor}, p_2]$	$\{\mathbf{nor}\}$
h_{20}	$[p_1, b_3(r), b_1(l), p_2, b_2(l)]$	$[b_1(l), b_2(l)]$	$[b_3(r)]$	$[p_1, \mathbf{nor}, p_2]$	$\{\mathbf{nor}\}$
h_{21}	$[p_1, b_3(r), b_1(l), p_2, b_2(l), b_5(r)]$	$[b_1(l), b_2(l)]$	$[b_3(r), b_5(r)]$	$[p_1, \mathbf{nor}, p_2]$	$\{\mathbf{nor}\}$
h_{22}	$[p_1, b_3(r), b_1(l), p_2, b_4(l)]$	$[b_1(l), b_4(l)]$	$[b_3(r)]$	$[p_1, \mathbf{nor}, p_2, \mathbf{ncl}]$	$\{\mathbf{ncl}, \mathbf{nor}, \mathbf{frw}\}$
h_{23}	$[p_1, b_3(r), b_1(l), p_2, b_4(l), b_5(r)]$	$[b_1(l), b_4(l)]$	$[b_3(r), b_5(r)]$	$[p_1, \mathbf{nor}, p_2, \mathbf{ncl}]$	$\{\mathbf{ncl}, \mathbf{nor}, \mathbf{frw}\}$
h_{24}	$[p_1, b_3(r), b_1(l), p_2, b_5(r)]$	$[b_1(l)]$	$[b_3(r), b_5(r)]$	$[p_1, \mathbf{nor}, p_2]$	$\{\mathbf{nor}\}$
h_{25}	$[p_1, b_3(r), b_1(l), p_2, b_5(r), b_4(l)]$	$[b_1(l), b_4(l)]$	$[b_3(r), b_5(r)]$	$[p_1, \mathbf{nor}, p_2, \mathbf{ncl}]$	$\{\mathbf{ncl}, \mathbf{nor}, \mathbf{frw}\}$
h_{26}	$[p_1, b_3(r), b_1(l), p_2, b_5(r), b_2(l)]$	$[b_1(l), b_2(l)]$	$[b_3(r), b_5(r)]$	$[p_1, \mathbf{nor}, p_2]$	$\{\mathbf{nor}\}$

According to the context hierarchy drawn in Fig. 11.2, the projections of these trajectories on the contexts ξ_l, ξ_r, and $\xi_\mathcal{W}$ can be computed as in Example 11.4, giving rise to the results displayed in Table 11.3. Finally, the corresponding diagnoses can be computed as in Example 11.5, giving rise to the results listed in the last column of Table 11.3.

Hence, the solution of the diagnosis problem includes three candidate diagnoses, namely $\delta_r = \{\mathbf{nor}\}$, $\delta_{lrw} = \{\mathbf{ncl}, \mathbf{nor}, \mathbf{frw}\}$, and $\delta_{rpw} = \{\mathbf{nor}, \mathbf{fcp}, \mathbf{frw}\}$.

In summary, according to (11.5), the solution of $\wp(\mathcal{W})$ is

$$\Delta(\wp(\mathcal{W})) = \{\, \delta_r, \delta_{lrw}, \delta_{rpw} \,\} . \tag{11.6}$$

In other words, one of the following three abnormal scenarios has occurred:

1. $\delta_r = \{\mathbf{nor}\}$: Breaker r did not open,
2. $\delta_{lrw} = \{\mathbf{ncl}, \mathbf{nor}, \mathbf{frw}\}$: Breaker r did not open, breaker l did not close, and the protection hardware $\mathcal{W}$ did not reconnect the line,
3. $\delta_{rpw} = \{\mathbf{nor}, \mathbf{fcp}, \mathbf{frw}\}$: Breaker r did not open, the protection device p failed to close the breakers, and the protection hardware $\mathcal{W}$ did not reconnect the line.

Comparing this solution with the (context-free) solution in Example 4.13, we notice that δ_{lrw} and δ_{rpw} differ from δ_{lr} and δ_{rp}, respectively, in that they include the additional fault **frw**, which is relevant to the protection hardware $\mathcal{W}$ as a whole. This extra information allows the operator in the control room to know that the protection hardware $\mathcal{W}$ failed to reconnect the line once the short circuit was extinguished.

11.4 Semantic Space

In the diagnosis process, a distinction is made between *online* and *offline* tasks. An offline task is accomplished before the system begins operating.[5] By contrast, an online task is performed while the system is operating, possibly under stringent time constraints. For instance, the modeling of the system is an offline task, while the task performed by the diagnosis engine, which is in charge of solving the diagnosis problem, is carried out online.

Since diagnosing active systems is computationally expensive, it is convenient to maximize the amount of processing performed offline [86, 94]. This provides two advantages:

1. Specific properties of the system, such as diagnosability, can be checked (and possibly changed) before the system begins operating,
2. The speed of the diagnosis engine increases, as what is performed offline does not need to be performed online.

Typically, in the quadruple defining a diagnosis problem, only the temporal observation $\mathcal{O}$ is available online, while the viewer and the ruler are defined at system-modeling time (offline). In particular, the semantic patterns can be analyzed and compiled into data structures which allow for better (online) performance of the diagnosis engine.

As is known, any regular expression can be transformed into an equivalent DFA [1].

The process of compiling semantic patterns involves the following steps:

1. For each semantic pattern (F_i, r_i) such that $F_i \in \mathbf{F}$, r_i is unfolded into its plain form and an equivalent DFA $\mathcal{D}_i$ is generated, where final states are marked by the label F_i.
2. For each semantic pattern (I_j, r_j) such that $I_j \in (\mathbf{I} - \mathbf{F})$, r_j is unfolded into its plain form and an equivalent DFA $\mathcal{D}_j$ is generated, where final states are marked by the label I_j.
3. The so-generated DFAs $\mathcal{D}_1, \ldots, \mathcal{D}_n$ *within the same list of semantic patterns* $\mathcal{P}$ (and, as such, relevant to the same context ξ) are merged to yield a new DFA called the *semantic space* of ξ, $Sem(\xi)$, where each final state S is marked by a pair $(\mathbb{I}, \mathbb{F})$, where $\mathbb{I}$ and $\mathbb{F}$ are the set of interface labels and the set of fault labels, respectively, that are associated with the states identifying S.[6]

[5] As in compilers, where tasks such as type checking are performed before the software begins operating (at *compile time* rather than at *run time*).

[6] A state of the determinized automaton is identified by a subset of the states of the corresponding DFA (see Chapter 2).

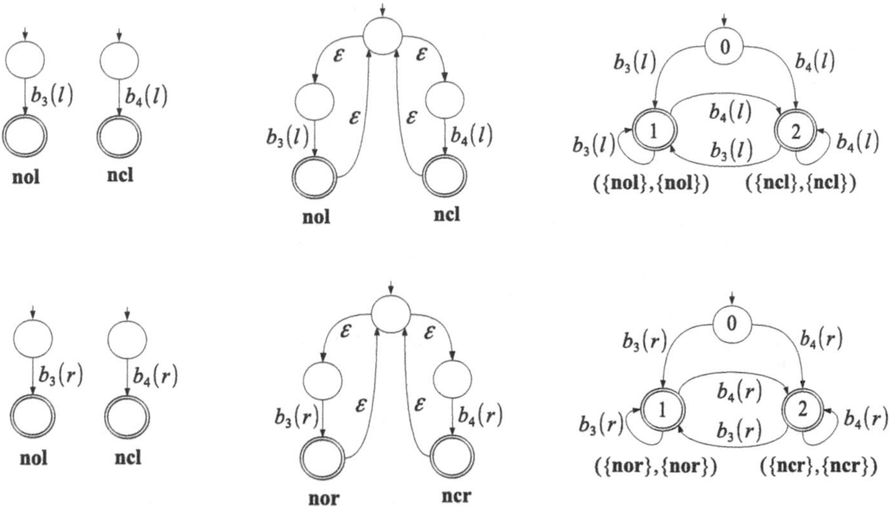

Fig. 11.3 Generation of semantic spaces $Sem(\xi_l)$ (top) and $Sem(\xi_r)$ (bottom): DFAs (left), NFA (center), and semantic space (right)

The merging of automata in step 3 is performed as follows:

3a. An NFA $\mathcal{N}$ is created by generating its initial state S_0 and one empty transition from S_0 to each initial state of $\mathcal{D}_i$, $i \in [1 \cdots n]$,

3b. In each $\mathcal{D}_i$, $i \in [1 \cdots n]$, an empty transition from each noninitial state to S_0 is inserted,

3c. $\mathcal{N}$ is determinized, thereby providing the semantic space $Sem(\xi)$, with each final state S of $Sem(\xi)$ being marked by a pair $(\mathbb{I}, \mathbb{F})$, where $\mathbb{I}$ and $\mathbb{F}$ are the interface labels and the fault labels, respectively, obtained from the union of the interface labels and fault labels, respectively, that are associated with states in S that are final in the corresponding $\mathcal{D}_i$.

The rationale for the sequence of steps above is that, during the reconstruction of the system behavior, the diagnosis engine is supposed to uncover a matching of several (possibly overlapping) semantic patterns. This means that after the matching of any transition, the same semantic pattern, or even a different one, may possibly start.

Example 11.7. With reference to the ruler $\mathcal{R}_{\mathcal{W}}$ specified in Example 4.10, the top of Fig. 11.3 shows the generation of the semantic space $Sem(\xi_l)$ for the context involving breaker l. Based on Table 11.1, the semantic patterns relevant to the fault labels **nol** and **ncl** are defined by single transitions b_{3l} and b_{4l}, respectively. For step 1 of the compiling process, the corresponding DFAs are represented on the left side of Fig. 11.3, with each one being composed of two states (initial and final) and one transition, which is marked by b_{3l} and b_{4l}. Step 2 is not applicable, as the fault labels **nol** and **ncl** are also interface labels. For step 3, the NFA $\mathcal{N}$ is outlined in

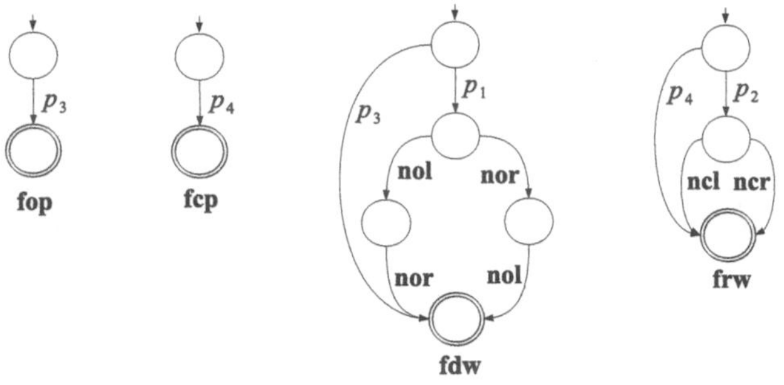

Fig. 11.4 DFAs relevant to regular expressions for the semantic patterns of the protection hardware $\mathcal{W}$ (see Table 11.1)

Table 11.4 Transition function of semantic space $Sem(\mathcal{W})$: final states 3, 4, 7, and 8 are associated with the corresponding set $\mathbb{F}$ of fault labels

State	p_1	p_2	p_3	p_4	**nol**	**nor**	**ncl**	**ncr**	$\mathbb{F}$
0	1	2	3	4					
1	1	2	3	4	5	6			
2	1	2	3	4			7	7	
3	1	2	3	4					{**fop**, **fdw**}
4	1	2	3	4					{**fcp**, **frw**}
5	1	2	3	4		8			
6	1	2	3	4	8				
7	1	2	3	4					{**frw**}
8	1	2	3	4					{**fdw**}

the center of Fig. 11.3. Finally, the actual semantic space $Sem(\xi_l)$ is displayed on the right side of the figure, where each final state is marked by a pair $(\mathbb{I}, \mathbb{F})$, with (incidentally) $\mathbb{I} = \mathbb{F}$.

The generation of the semantic space $Sem(\xi_r)$ for breaker r is outlined at the bottom of Fig. 11.3, giving rise to a similar automaton.

The generation of the semantic space for the protection hardware $\mathcal{W}$ is carried out in a similar way, based on the semantic patterns specified in the bottom section of Table 11.1. For step 1 of the compiling process, the corresponding four DFAs are outlined in Fig. 11.4. For space reasons, we omit the representation of steps 2 and 3. Instead, we report in Table 11.4 the tabular representation of the resulting automaton $Sem(\mathcal{W})$. Each row defines the transition function for states $0, 1, \ldots, 8$, where $3, 4, 7$, and 8 are final.

According to the semantic patterns defined in Table 11.1, the alphabet of $Sem(\mathcal{W})$ is $\{p_1, p_2, p_3, p_4, \mathbf{nol}, \mathbf{nor}, \mathbf{ncl}, \mathbf{ncr}\}$. If defined, the state reached by a pair (P, σ), where P is a state and σ a symbol in the alphabet, is indicated in Table 11.4 in the cell corresponding to row P and column σ; for instance, the state reached by $(2, \mathbf{ncl})$

is 7. The last column outlines for each final state the associated set of fault symbols; for instance, state 4 is marked by the set $\{\mathbf{fcp}, \mathbf{frw}\}$ of fault labels.

Proposition 11.1. *No transition enters the initial state of a semantic space.*

Proof. In the last step of the construction of a semantic space, the DFAs $\mathcal{D}_1, \ldots, \mathcal{D}_n$ are merged by creating the initial state S_0, and by inserting an empty transition from S_0 to the initial state of each $\mathcal{D}_i$, $i \in [1 \cdots n]$, and from each noninitial state of each $\mathcal{D}_i$ to S_0. Hence, S_0 is entered by empty transitions only. Based on the determinization algorithm (Chapter 2), starting from the initial state S_0^d of the DFA, obtained as the ε-closure of S_0, each new state is generated as the ε-closure of a nonempty subset of states of the NFA. Hence, to generate S_0^d, we need a nonempty transition entering S_0, which is not the case. □

Proposition 11.2. *If s is a string in the language of the semantic space $Sem(\xi)$, generated by a path from the initial state to a final state S_f, then the interface symbol associated with S_f is the set of symbols $N \in \mathbf{I}$ such that (N, r) is a semantic pattern relevant to ξ, and the language of r (in plain form) includes a string e which is a suffix of s.*

Proof. The specification of $Sem(\xi)$ requires three steps. In steps 1 and 2, a DFA $\mathcal{D}_i$, $i \in [1 \cdots n]$, is generated for each semantic pattern (N_i, r_i) relevant to ξ, where $N_i \in \mathbf{F} \cup \mathbf{I}$, with each final state of $\mathcal{D}_i$ being marked by the label N_i. This means that each $\mathcal{D}_i$ recognizes the language of r_i (in plain form). Then, in step 3, automata $\mathcal{D}_1, \ldots, \mathcal{D}_n$ are merged into an NFA $\mathcal{N}$, for which an initial state S_0 is created and connected to the initial state of each $\mathcal{D}_i$ by an empty transition. Furthermore, each noninitial state of each $\mathcal{D}_i$ is connected to S_0 by an empty transition. Based on this construction, each string s in the language of $\mathcal{N}$ is generated by a set $\boldsymbol{\pi}$ of paths from S_0 to a final state. Because of the way in which the automata $\mathcal{D}_i$ are merged into $\mathcal{N}$, each path in $\boldsymbol{\pi}$ ending at a final state marked by the symbol N_i necessarily has a suffix which is a string in the language of $\mathcal{D}_i$, and hence in the language of r_i (in plain form). Finally, $Sem(\xi)$ is obtained by determinization of $\mathcal{N}$, with the latter sharing the same language as the former, where each final state S_f is marked by a pair $(\mathbb{I}, \mathbb{F})$, with $\mathbb{F} \subseteq \mathbf{F}$ and $\mathbb{I} \subseteq \mathbf{I}$, where, in particular, $\mathbb{I}$ is the set of symbols in $\mathbf{I}$ marking the final states of $\mathcal{N}$ within S_f. Therefore, since $Sem(\xi)$ is the result of the determinization of $\mathcal{N}$, each string s in the language of $Sem(\xi)$ is generated by *one* path π, from the initial state to a final state S_f. In $\mathcal{N}$, the same string s is generated by a set $\boldsymbol{\pi}$ of paths from S_0 to the final states of $\mathcal{N}$ included in S_f. Consequently, the interface symbol associated with S_f is the set of symbols $N \in \mathbf{I}$ such that (N, r) is a semantic pattern relevant to ξ, and the language of r (in plain form) includes a string e which is a suffix of s. □

Corollary 11.1. *If s is a string in the language of the semantic space $Sem(\xi)$, generated by a path from the initial state to a final state S_f, then the set of faults associated with S_f is the set of symbols $N \in \mathbf{F}$ such that (N, r) is a semantic pattern relevant to ξ, and the language of r (in plain form) includes a string e which is a suffix of s.*

Proof. The proof is by analogy with Proposition 11.2, as the way in which the set of symbols in **F** is associated with final states of the semantic space is the same as that for the set of symbols in **I**. □

11.5 Behavior Reconstruction

The diagnosis engine takes as input a diagnosis problem $\wp(\mathcal{A}) = (a_0, \mathcal{V}, \mathcal{O}, \mathcal{R})$ and outputs the solution $\Delta(\wp(\mathcal{A}))$, as defined in (11.5). In doing so, it needs to build a context-sensitive behavior of $\mathcal{A}$ that conforms with the observation, namely $Bhv(\wp(\mathcal{A}))$. We assume that the semantic spaces generated offline by preprocessing (see Sec. 11.4) are available to the diagnosis engine. Since pattern recognition is to be performed, the states of $Bhv(\wp(\mathcal{A}))$ will incorporate information not only on the observation but also on the semantic spaces in order to maintain the state of the pattern matching.

Definition 11.3 (Context-Sensitive Behavior). Let $\boldsymbol{B}$ denote the set of states of $Bsp(\mathcal{A})$, let $\mathfrak{I}$ denote the set of states of $Isp(\mathcal{O})$, and let $\boldsymbol{P} = \boldsymbol{P}_1 \times \cdots \times \boldsymbol{P}_n$, where $\boldsymbol{P}_1, \ldots, \boldsymbol{P}_n$ are the set of states of all semantic spaces $Sem(\xi_1), \ldots, Sem(\xi_n)$, respectively. The *context-sensitive behavior* of $\wp(\mathcal{A})$,

$$Bhv(\wp(\mathcal{A})) = (\Sigma, B, \tau, \beta_0, B_f), \tag{11.7}$$

is a DFA such that:[7]

1. Σ is the alphabet, consisting of the transitions of the components in $\mathcal{A}$,
2. $B \subseteq \boldsymbol{B} \times \boldsymbol{P} \times \mathfrak{I}$ is the set of states,
3. $\beta_0 = (a_0, \mathbb{P}_0, \mathfrak{I}_0)$ is the initial state, where $\mathfrak{I}_0$ is the initial state of $Isp(\mathcal{O})$, and $\mathbb{P}_0 = (P_{10}, \ldots, P_{n0})$ is the tuple of the initial states of $Sem(\xi_1), \ldots, Sem(\xi_n)$,
4. B_f is the set of final states $(a, \mathbb{P}, \mathfrak{I}_f)$ where $\mathfrak{I}_f$ is final in $Isp(\mathcal{O})$,
5. τ is the transition function $(a, \mathbb{P}, \mathfrak{I}) \xrightarrow{t(c)} (a', \mathbb{P}', \mathfrak{I}') \in \tau$, $\mathbb{P} = (P_1, \ldots, P_n)$, $\mathbb{P}' = (P'_1, \ldots, P'_n)$, iff the following conditions hold:

   a. $a \xrightarrow{t(c)} a'$ is a transition in $Bsp(\mathcal{A})$,[8]
   b. If $t(c)$ is invisible, then $\mathfrak{I}' = \mathfrak{I}$, otherwise $\mathfrak{I}'$ equals the target state $\bar{\mathfrak{I}}$ of the transition $\mathfrak{I} \xrightarrow{\ell} \bar{\mathfrak{I}}$ in $Isp(\mathcal{O})$, where ℓ is the label associated with T in $\mathcal{V}$,
   c. $\mathbb{P}' = (P'_1, \ldots, P'_n)$ is such that, $\forall i \in [1 \cdots n]$, P'_i is defined by the following (possibly recursive) rule:

---

[7] More precisely, as in the analogous Def. 5.4, this defines the spurious behavior: the actual behavior is obtained by removing all states and transitions that are not embedded in any path from the initial state to a final state.

[8] Since $Bsp(\mathcal{A})$ is assumed to be unavailable, operationally this means that $t(c)$ is triggerable from state a.

if $t(c)$ is in the alphabet of a context ξ_i **then**
  $P'_i :=$ **if** $P_i \xrightarrow{t(c)} \bar{P}_i \in Sem(\xi_i)$ **then** $\bar{P}_i$ **else** P_{i_0} **endif**;
  if the interface symbol $\mathbb{I}$ marking P'_i is nonempty **and**
    ξ_i has a parent ξ_j in the hierarchy **then**
   Reapply the rule replacing $t(c)$ with $\mathbb{I}$ and i with j
  endif
else
  $P'_i := P_i$
endif.

As such, each state of $Bhv(\wp(\mathcal{A}))$ is a triple involving a state of the behavior space, a state of the index space of $\mathcal{O}$, and a tuple of semantic-space states. A transition marked by $t(c)$ is defined in $Bhv(\wp(\mathcal{A}))$ iff a transition marked by $t(c)$ is defined between the corresponding states of the behavior space. The index $\mathfrak{I}'$ of the new state differs from the index $\mathfrak{I}$ of the old state only if $t(c)$ is observable (according to the viewer $\mathcal{V}$). Finally, $\mathbb{P}'$ is obtained from $\mathbb{P}$ by performing a state change in the semantic space of the context ξ_i including the component relevant to $t(c)$: if the new state is marked by a nonempty interface symbol, a state change may possibly be propagated to the ancestors of ξ_i.

11.5.1 Context-Sensitive Reconstruction Algorithm

Listed below is the specification of the algorithm *Sbuild*, which reconstructs the context-sensitive behavior relevant to a given diagnosis problem $\wp(\mathcal{A})$. The algorithm takes as input a diagnosis problem $\wp(\mathcal{A})$ and the semantic spaces Π, generated (offline) based on a ruler $\mathcal{R}$.

Considering the body of *Sbuild* (lines 5–44), after the generation of the index space of $\mathcal{O}$ (line 6), the set of states B and the set of transitions τ are initialized (lines 7 and 8). Then, a loop is iterated until all states in B have been processed (lines 9–41). At each iteration, starting from the initial state, an unprocessed state $(a, \mathbb{P}, \mathfrak{I})$ is chosen from B (line 10) and its transitions are determined (lines 11–39). Each transition $t(c)$ triggerable from a is first checked against the observation, and the new index-space state $\mathfrak{I}'$ is computed (lines 12–18). Then, the new state a' is computed based on a and $t(c)$ (lines 19–21). Finally, the new tuple $\mathbb{P}'$ of semantic-space states is determined (lines 22–35). To this end, the state of the semantic space $Sem(\xi_i)$ including $t(c)$ in its alphabet may possibly be updated. Besides, if the new state P_i is marked by an interface symbol $\mathbb{I} \neq \emptyset$, then the semantic-space change may possibly be propagated to the ancestors of the context ξ_i. Eventually, the set of final states is determined (line 42), and all (spurious) states and transitions which are not in a path from the initial state to a final state are removed (line 43).

The specification of the algorithm *Sbuild* is as follows:

1. **algorithm** *Sbuild* (**in** $\wp(A)$, **in** Π, **out** $Bhv(\wp(\mathcal{A}))$)

2. $\wp(\Sigma) = (a_0, \mathcal{V}, \mathcal{O}, \mathcal{R})$: a context-sensitive diagnosis problem for $\mathcal{A}$,
3. $\Pi = (Sem(\xi_1), \ldots, Sem(\xi_n))$: the tuple of semantic spaces based on ruler $\mathcal{R}$;
4. $Bhv(\wp(\Sigma)) = (\Sigma, B, \tau, \beta_0, B_{\mathrm{f}})$: the context-sensitive behavior of $\wp(\mathcal{A})$;

5.   **begin** $\langle Sbuild \rangle$
6.   Generate the index space $Isp(\mathcal{O})$;
7. $\beta_0 := (a_0, \mathbb{P}_0, \mathfrak{I}_0)$, where $\mathfrak{I}_0$ is the initial state of $Isp(\mathcal{O})$ and $\mathbb{P}_0$ is
                                      the tuple of initial states of semantic spaces in Π;
8. $B := \{\beta_0\}$; $\tau := \emptyset$;
9.   **repeat**
10.   Choose an unmarked state $\beta = (a, \mathbb{P}, \mathfrak{I})$ in B;
11.   **foreach** component transition $t(c)$ which is triggerable in state a **do**
12.   **if** $t(c)$ is unobservable in $\mathcal{V}$ **then**
13. $\mathfrak{I}' := \mathfrak{I}$
14.   **elsif** $t(c)$ is observable in $\mathcal{V}$ via label ℓ **and** $\mathfrak{I} \xrightarrow{\ell} \bar{\mathfrak{I}}$ is in $Isp(\mathcal{O})$ **then**
15. $\mathfrak{I}' := \bar{\mathfrak{I}}$
16.   **else**
17.   **continue**
18.   **endif**;
19. $a' := a$;
20.   Set the state in a' relevant to component c with the target state of
                    $t(c)$;
21.   Insert into the links of a' the output events of $t(c)$;
22. $\mathbb{P}' := \mathbb{P}$;
23.   **if** component c is within the context hierarchy $\mathcal{H}$ **then**
24.   Let ξ_i be the context whose alphabet includes symbol $t(c)$;
25. $\mathbb{I} := t(c)$;
26.   **repeat**
27.   Let P_i be the state in $\mathbb{P}'$ corresponding to $Sem(\xi_i)$;
28.   Let P_{i_0} be the initial state of semantic space $Sem(\xi_i)$;
29. $P_i := $ **if** $P_i \xrightarrow{\mathbb{I}} \bar{P}_i \in Sem(\xi_i)$ **then** $\bar{P}_i$ **else** P_{i_0} **endif**;
30.   Let $\mathbb{I}'$ be the interface symbol marking P_i in $Sem(\xi_i)$;
31.   **if** $\mathbb{I}' \neq \emptyset$ **and** ξ_i has a parent ξ_j in $\mathcal{H}$ **then**
32. $\xi_i := \xi_j$; $\mathbb{I} := \mathbb{I}'$
33.   **endif**
34.   **until** $\mathbb{I}' = \emptyset$ **or** ξ_i has no parent in $\mathcal{H}$
35.   **endif**;
36. $\beta' := (a', \mathbb{P}', \mathfrak{I}')$;
37.   **if** $\beta' \notin B$ **then** Insert β' into B **endif**;
38.   Insert transition $\beta \xrightarrow{t(c)} \beta'$ into τ;
39.   **endfor**;
40.   Mark β
41.   **until** all states in B are marked;
42. $B_{\mathrm{f}} := \{\beta_{\mathrm{f}} \mid \beta_{\mathrm{f}} \in B, \beta_{\mathrm{f}} = (a, \mathbb{P}, \mathfrak{I}_{\mathrm{f}}), \mathfrak{I}_{\mathrm{f}} \text{ is final in } Isp(\mathcal{O})\}$;

Remove from B and τ states and transitions not connected to any state in B_f

end $\langle$*Sbuild*$\rangle$.

Example 11.8. Consider the diagnosis problem $\wp(\mathcal{W}) = (a_0, \mathcal{V}, \mathcal{O}_2, \mathcal{R}_\mathcal{W})$ defined in Example 11.6. Depicted in Fig. 11.5 is the relevant context-sensitive behavior $Bhv(\wp(\mathcal{W}))$. According to the definition, the initial state is $\beta_0 = (a_0, \mathbb{P}_0, \Im_0)$, where $\Im_0$ is the initial state of $Isp(\mathcal{O}_2)$ (displayed on the right of Fig. 5.12), while $\mathbb{P}_0 = (0,0,0)$ is the tuple of initial states of the semantic spaces $Sem(\xi_l)$, $Sem(\xi_r)$, and $Sem(\mathcal{W})$, respectively. Since the following conditions hold,

- $Bsp(\wp(\mathcal{A}))$ includes a transition $a_0 \xrightarrow{p_1} a_1$ which is visible via the label *awk* of viewer $\mathcal{V}$,
- a transition $\Im_0 \xrightarrow{awk} \Im_1$ is included in $Isp(\mathcal{O}_2)$,
- the component transition p_1 is in the alphabet of $Sem(\mathcal{W})$,
- the transition $0 \xrightarrow{p_1} 1$ is included in $Sem(\mathcal{W})$,

a new transition $(a_0, 000, \Im_0) \xrightarrow{p_1} (a_1, 001, \Im_1)$ is created in $Bhv(\wp(\mathcal{W}))$. Consider the transition $\beta_6 \xrightarrow{b_{4l}} \beta_{11}$, where $\mathbb{P} = (0,1,2)$ and $\mathbb{P}' = (2,1,7)$. In fact, based on Fig. 11.3, $Sem(\xi_l)$ includes the transition $0 \xrightarrow{b_{4l}} 2$, whose target state 2 is marked

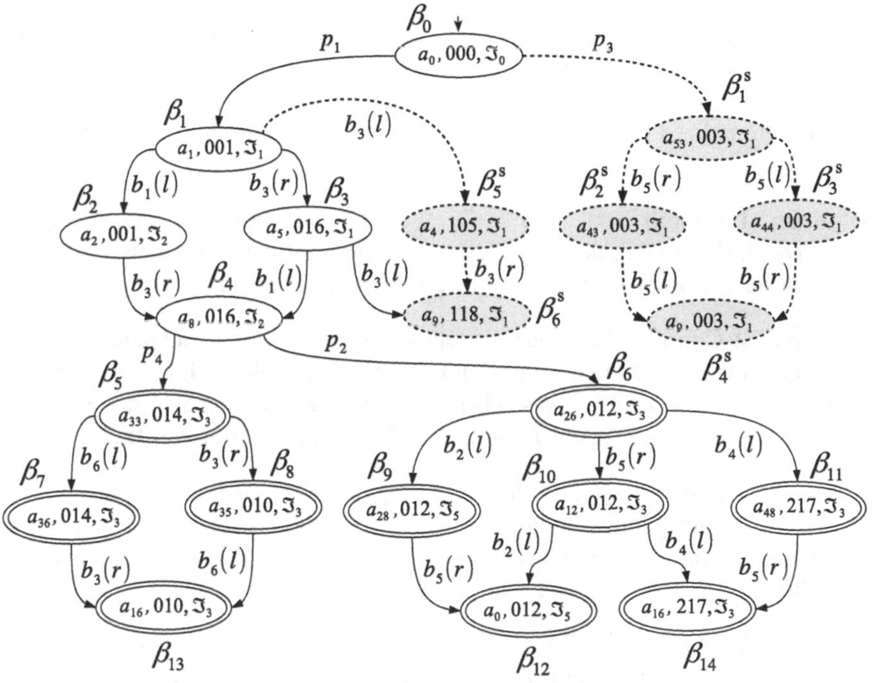

Fig. 11.5 Reconstructed context-sensitive behavior $Bhv(\wp(\mathcal{W}))$

by the interface symbol **ncl**. Since, based on Table 11.4, **ncl** is in the alphabet of $Sem(\mathcal{W})$ and a transition $2 \xrightarrow{\mathbf{ncl}} 7$ exists in the latter, the semantic-space state relevant to $\mathcal{W}$ is updated too.

The construction of the context-sensitive behavior continues until no new node is generated by the application of the transition function. In Fig. 11.5, the final states are β_5 and β_{14}, as both $\mathfrak{I}_3$ and $\mathfrak{I}_5$ are final in $Isp(\mathcal{O}_2)$. The gray part of the graph is inconsistent because it is not encompassed by any path from the initial state to a final state. Hence, it is removed, leaving $Bhv(\wp(\mathcal{W}))$ with 15 states, namely $\beta_0 \cdots \beta_{14}$.

If we compare the context-sensitive behavior in Fig. 11.5 with the behavior $Bhv(\wp(\mathcal{A}_2))$ displayed in Fig. 5.13, where $\wp(\mathcal{A}_2) = (a_0, \mathcal{V}, \mathcal{O}_2, \mathcal{R})$ differs from $\wp(\mathcal{W})$ in the ruler only, we notice that $\wp(\mathcal{W})$ includes one more state. The same applies to the spurious part of the behavior. What makes $p(\mathcal{W})$ larger than $\wp(\mathcal{A}_2)$ is the additional field $\mathbb{P}$ in the triple $(a, \mathbb{P}, \mathfrak{I})$ identifying each state of $p(\mathcal{W})$, where a is in fact a pair (S, Q), with S and Q being the tuples of component states and link configurations, respectively. Note how the states $\beta_{13} = (a_{16}, (0,1,0), \mathfrak{I}_3)$ and $\beta_{14} = (a_{16}, (2,1,7), \mathfrak{I}_3)$ differ in the $\mathbb{P}$ value only, which is the cause of the additional state. The same applies to the spurious states $\beta_4^s = (a_9, (0,0,3), \mathfrak{I}_1)$ and $\beta_6^s = (a_9, (1,1,8), \mathfrak{I}_1)$.

11.6 Behavior Decoration and Diagnosis Distillation

Once the context-sensitive behavior $Bhv(\wp(\mathcal{A}))$ has been reconstructed, the diagnosis engine is expected to generate the solution of $\wp(\mathcal{A})$ in terms of the sound and complete set of candidate diagnoses, in accordance with the notion of a candidate diagnosis formalized in Def. 11.2. To this end, we first introduce the notion of a fault set.

Definition 11.4 (Fault Set). Let $\mathbb{P} = (P_1, \ldots, P_n)$ be a tuple of semantic-space states. The *fault set* of $\mathbb{P}$, denoted $\mathcal{F}(\mathbb{P})$, is the union of the fault labels associated with each state P_i in $\mathbb{P}$, $i \in [1 \cdots n]$, in the corresponding semantic space.

Example 11.9. Consider the semantic spaces $Sem(\xi_l)$, $Sem(\xi_r)$, and $Sem(\mathcal{W})$ computed in Example 11.7, and displayed in Fig. 11.3 and Table 11.4. Let $\mathbb{P} = (0,1,4)$ be a triple of such semantic spaces, where 0, 1, and 4 are states in $Sem(\xi_l)$, $Sem(\xi_r)$, and $Sem(\mathcal{W})$, respectively. According to Def. 11.4, $\mathcal{F}(\mathbb{P}) = \{\mathbf{nor}, \mathbf{fcp}, \mathbf{frw}\}$.

Based on Def. 11.4, the technique for generating the set of candidate diagnoses is an extension of the technique presented in Sec. 5.3, based on which the reconstructed behavior is decorated by sets of diagnoses. The set of candidate diagnoses is eventually distilled from the final states of the decorated behavior.

Definition 11.5 (Decorated Context-Sensitive Behavior). Let $Bhv(\wp(\mathcal{A}))$ be the context-sensitive behavior of a diagnosis problem $\wp(\mathcal{A})$. The *decorated context-sensitive behavior* $Bhv^*(\wp(\mathcal{A}))$ is the DFA obtained from $Bhv(\wp(\mathcal{A}))$ by marking each state β of the latter with a *diagnosis set* $\Delta(\beta)$ based on the application of the following two rules:

1. For the initial state $\beta_0 = (a_0, \mathbb{P}_0, \Im_0)$, $\Delta(\beta_0) = \{\emptyset\}$,
2. For a transition $\beta \xrightarrow{t(c)} \beta'$ in $Bhv(\wp(\mathcal{A}))$, with $\beta = (a, \mathbb{P}, \Im)$ and $\beta' = (a', \mathbb{P}', \Im')$, if $\delta \in \Delta(\beta)$ then $(\delta \cup \mathcal{F}(\mathbb{P}')) \in \Delta(\beta')$.

The rationale for the above rules stems from the definition of a semantic space, where each state keeps track of the matching of semantic patterns. Assuming that the regular expression of a semantic pattern cannot be empty, the first rule initializes the diagnosis set of the initial state to empty. In the second rule, each diagnosis $\delta \in \Delta(\beta)$ is extended by the set of fault labels associated with (final) states of $\mathbb{P}'$, which certify the matching of relevant regular expressions by projections of trajectories up to the component transition $t(c)$.

Unlike the first rule, which represents the base case and, as such, is applied only once, the second rule is inductive in nature. This means that, for the sake of completeness of the decoration, the second rule should be continuously applied until the decoration of the context-sensitive behavior cannot be changed.

11.6.1 Context-Sensitive Decoration Algorithm

Listed below is the specification of the algorithm *Sdecorate*, which, based on Def. 11.5, decorates the context-sensitive behavior relevant to a context-sensitive diagnosis problem $\wp(\mathcal{A})$ by means of the recursive auxiliary procedure *Sdec*.

Considering the body of *Sdecorate* (lines 18–22), first the initial state β_0 is marked by the singleton $\{\emptyset\}$ (line 19), while all other states are marked by the empty set (line 20). Then, *Sdec* is called with parameters β_0 and $\{\emptyset\}$.

Considering the specification of *Sdec* (lines 5–17), which takes as input a state β and a set of diagnoses $\mathcal{D}$, the latter being an extension of the decoration of β, the aim of *Sdec* is to propagate $\mathcal{D}$ to each neighboring state β' of β based on the second decoration rule in Def. 11.5. To this end, if the decoration of β' is extended by a nonempty set $\mathcal{D}^+$ of diagnoses, then *Sdec* is recursively called on β' and $\mathcal{D}^+$. *Sdec* stops when no further diagnosis can be generated.

The specification of the algorithm *Sdecorate* is as follows:

1. **algorithm** *Sdecorate* (**in** $Bhv(\wp(A))$)
2. $Bhv(\wp(A))$: the context-sensitive behavior of $\wp(\mathcal{A}) = (a_0, \mathcal{V}, \mathcal{O}, \mathcal{R})$;

3. **side effects**
4. $Bhv(\wp(\mathcal{A}))$ is decorated, thereby becoming $Bhv^*(\wp(\mathcal{A}))$;

5. **auxiliary procedure** *Sdec* $(\beta, \mathcal{D})$
6. β: a state of $Bhv(\wp(\mathcal{A}))$,
7. $\mathcal{D}$: a set of diagnoses;

```
      begin ⟨Sdec⟩
        foreach transition β --t(c)--> β' do
          D+ := ∅;
          foreach δ ∈ D do
            δ' := δ ∪ F(P');
            if δ' ∉ Δ(β') then insert δ' into both Δ(β') and D+ endif
          endfor;
          if D+ ≠ ∅ then Sdec (β', D+) endif
        endfor
      end ⟨Sdec⟩;

    begin ⟨Sdecorate⟩
      Mark the initial state β0 with the singleton {∅};
      Mark all other (noninitial) states with the empty set ∅;
      Sdec (β0, {∅})
    end ⟨Sdecorate⟩.
```

Example 11.10. Consider the context-sensitive behavior $Bhv(\wp(\mathcal{W}))$ displayed in Fig. 11.5. Reported in Fig. 11.6 is the same behavior (with the spurious part missing), where states are marked by corresponding fault sets (empty fault sets are omit-

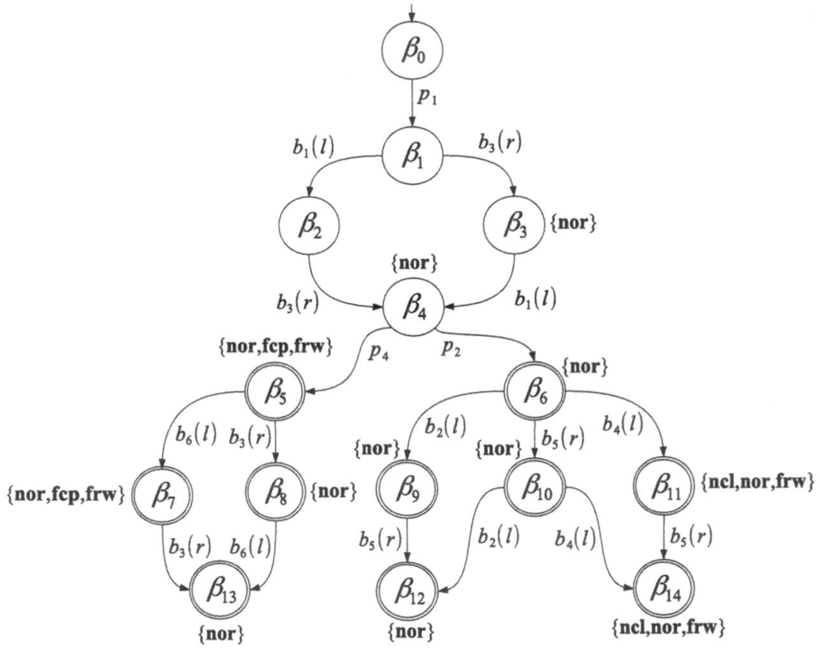

Fig. 11.6 Consistent part of the reconstructed context-sensitive behavior $Bhv(\wp(\mathcal{W}))$ displayed in Fig. 11.5, with each state $(a, \mathbb{P}, \mathfrak{I})$ being marked by a fault set $\mathcal{F}(\mathbb{P})$, as defined in Def. 11.4

Table 11.5 Trace of *Sdecorate* applied to context-sensitive behavior $Bhv(\wp(\mathcal{W}))$ in Fig. 11.6

Call	$\Delta(\beta_0)$	$\Delta(\beta_1)$	$\Delta(\beta_2)$	$\Delta(\beta_3)$	$\Delta(\beta_4)$	$\Delta(\beta_5)$	$\Delta(\beta_6)$	$\Delta(\beta_7)$	$\Delta(\beta_8)$	$\Delta(\beta_9)$	$\Delta(\beta_{10})$	$\Delta(\beta_{11})$	$\Delta(\beta_{12})$	$\Delta(\beta_{13})$	$\Delta(\beta_{14})$
$(\beta_0,\{\emptyset\})$	$\{\emptyset\}$	$\emptyset$	$\emptyset$	$\emptyset$	$\emptyset$	$\emptyset$	$\emptyset$	$\emptyset$	$\emptyset$	$\emptyset$	$\emptyset$	$\emptyset$	$\emptyset$	$\emptyset$	$\emptyset$
$(\beta_1,\{\emptyset\})$	$\{\emptyset\}$	$\{\emptyset\}$	$\emptyset$	$\emptyset$	$\emptyset$	$\emptyset$	$\emptyset$	$\emptyset$	$\emptyset$	$\emptyset$	$\emptyset$	$\emptyset$	$\emptyset$	$\emptyset$	$\emptyset$
$(\beta_2,\{\emptyset\})$	$\{\emptyset\}$	$\{\emptyset\}$	$\{\emptyset\}$	$\emptyset$	$\emptyset$	$\emptyset$	$\emptyset$	$\emptyset$	$\emptyset$	$\emptyset$	$\emptyset$	$\emptyset$	$\emptyset$	$\emptyset$	$\emptyset$
$(\beta_3,\{\emptyset\})$	$\{\emptyset\}$	$\{\emptyset\}$	$\{\emptyset\}$	$\{\delta_r\}$	$\emptyset$	$\emptyset$	$\emptyset$	$\emptyset$	$\emptyset$	$\emptyset$	$\emptyset$	$\emptyset$	$\emptyset$	$\emptyset$	$\emptyset$
$(\beta_4,\{\delta_r\})$	$\{\emptyset\}$	$\{\emptyset\}$	$\{\emptyset\}$	$\{\delta_r\}$	$\{\delta_r\}$	$\emptyset$	$\emptyset$	$\emptyset$	$\emptyset$	$\emptyset$	$\emptyset$	$\emptyset$	$\emptyset$	$\emptyset$	$\emptyset$
$(\beta_5,\{\delta_r\})$	$\{\emptyset\}$	$\{\emptyset\}$	$\{\emptyset\}$	$\{\delta_r\}$	$\{\delta_r\}$	$\{\delta_{rpw}\}$	$\emptyset$	$\emptyset$	$\emptyset$	$\emptyset$	$\emptyset$	$\emptyset$	$\emptyset$	$\emptyset$	$\emptyset$
$(\beta_7,\{\delta_{rpw}\})$	$\{\emptyset\}$	$\{\emptyset\}$	$\{\emptyset\}$	$\{\delta_r\}$	$\{\delta_r\}$	$\{\delta_{rpw}\}$	$\emptyset$	$\{\delta_{rpw}\}$	$\emptyset$	$\emptyset$	$\emptyset$	$\emptyset$	$\emptyset$	$\emptyset$	$\emptyset$
$(\beta_{13},\{\delta_{rpw}\})$	$\{\emptyset\}$	$\{\emptyset\}$	$\{\emptyset\}$	$\{\delta_r\}$	$\{\delta_r\}$	$\{\delta_{rpw}\}$	$\emptyset$	$\{\delta_{rpw}\}$	$\emptyset$	$\emptyset$	$\emptyset$	$\emptyset$	$\emptyset$	$\{\delta_{rpw}\}$	$\emptyset$
$(\beta_8,\{\delta_{rpw}\})$	$\{\emptyset\}$	$\{\emptyset\}$	$\{\emptyset\}$	$\{\delta_r\}$	$\{\delta_r\}$	$\{\delta_{rpw}\}$	$\emptyset$	$\{\delta_{rpw}\}$	$\{\delta_{rpw}\}$	$\emptyset$	$\emptyset$	$\emptyset$	$\emptyset$	$\{\delta_{rpw}\}$	$\emptyset$
$(\beta_{13},\{\delta_{rpw}\})$	$\{\emptyset\}$	$\{\emptyset\}$	$\{\emptyset\}$	$\{\delta_r\}$	$\{\delta_r\}$	$\{\delta_{rpw}\}$	$\emptyset$	$\{\delta_{rpw}\}$	$\{\delta_{rpw}\}$	$\emptyset$	$\emptyset$	$\emptyset$	$\emptyset$	$\{\delta_{rpw}\}$	$\emptyset$
$(\beta_6,\{\delta_r\})$	$\{\emptyset\}$	$\{\emptyset\}$	$\{\emptyset\}$	$\{\delta_r\}$	$\{\delta_r\}$	$\{\delta_{rpw}\}$	$\{\delta_r\}$	$\{\delta_{rpw}\}$	$\{\delta_{rpw}\}$	$\emptyset$	$\emptyset$	$\emptyset$	$\emptyset$	$\{\delta_{rpw}\}$	$\emptyset$
$(\beta_9,\{\delta_r\})$	$\{\emptyset\}$	$\{\emptyset\}$	$\{\emptyset\}$	$\{\delta_r\}$	$\{\delta_r\}$	$\{\delta_{rpw}\}$	$\{\delta_r\}$	$\{\delta_{rpw}\}$	$\{\delta_{rpw}\}$	$\{\delta_r\}$	$\emptyset$	$\emptyset$	$\emptyset$	$\{\delta_{rpw}\}$	$\emptyset$
$(\beta_{12},\{\delta_r\})$	$\{\emptyset\}$	$\{\emptyset\}$	$\{\emptyset\}$	$\{\delta_r\}$	$\{\delta_r\}$	$\{\delta_{rpw}\}$	$\{\delta_r\}$	$\{\delta_{rpw}\}$	$\{\delta_{rpw}\}$	$\{\delta_r\}$	$\emptyset$	$\emptyset$	$\{\delta_r\}$	$\{\delta_{rpw}\}$	$\emptyset$
$(\beta_{10},\{\delta_r\})$	$\{\emptyset\}$	$\{\emptyset\}$	$\{\emptyset\}$	$\{\delta_r\}$	$\{\delta_r\}$	$\{\delta_{rpw}\}$	$\{\delta_r\}$	$\{\delta_{rpw}\}$	$\{\delta_{rpw}\}$	$\{\delta_r\}$	$\{\delta_r\}$	$\emptyset$	$\{\delta_r\}$	$\{\delta_{rpw}\}$	$\emptyset$
$(\beta_{12},\{\delta_r\})$	$\{\emptyset\}$	$\{\emptyset\}$	$\{\emptyset\}$	$\{\delta_r\}$	$\{\delta_r\}$	$\{\delta_{rpw}\}$	$\{\delta_r\}$	$\{\delta_{rpw}\}$	$\{\delta_{rpw}\}$	$\{\delta_r\}$	$\{\delta_r\}$	$\emptyset$	$\{\delta_r\}$	$\{\delta_{rpw}\}$	$\emptyset$
$(\beta_{14},\{\delta_r\})$	$\{\emptyset\}$	$\{\emptyset\}$	$\{\emptyset\}$	$\{\delta_r\}$	$\{\delta_r\}$	$\{\delta_{rpw}\}$	$\{\delta_r\}$	$\{\delta_{rpw}\}$	$\{\delta_{rpw}\}$	$\{\delta_r\}$	$\{\delta_r\}$	$\emptyset$	$\{\delta_r\}$	$\{\delta_{rpw}\}$	$\{\delta_{lrw}\}$
$(\beta_{11},\{\delta_r\})$	$\{\emptyset\}$	$\{\emptyset\}$	$\{\emptyset\}$	$\{\delta_r\}$	$\{\delta_r\}$	$\{\delta_{rpw}\}$	$\{\delta_r\}$	$\{\delta_{rpw}\}$	$\{\delta_{rpw}\}$	$\{\delta_r\}$	$\{\delta_r\}$	$\{\delta_{lrw}\}$	$\{\delta_r\}$	$\{\delta_{rpw}\}$	$\{\delta_{lrw}\}$
$(\beta_{14},\{\delta_{lrw}\})$	$\{\emptyset\}$	$\{\emptyset\}$	$\{\emptyset\}$	$\{\delta_r\}$	$\{\delta_r\}$	$\{\delta_{rpw}\}$	$\{\delta_r\}$	$\{\delta_{rpw}\}$	$\{\delta_{rpw}\}$	$\{\delta_r\}$	$\{\delta_r\}$	$\{\delta_{lrw}\}$	$\{\delta_r\}$	$\{\delta_{rpw}\}$	$\{\delta_{lrw}\}$

ted). Based on the specification of the *Sdecorate* algorithm, Table 11.5 traces the sequence of calls to the *Sdec* auxiliary procedure. Specifically, each line indicates the call to *Sdec* with the relevant parameters $(\beta, \mathcal{D}^+)$, along with the diagnosis sets $\Delta(\beta_i)$, $i \in [0 \cdots 14]$. For space reasons, diagnoses are indicated as in Example 11.6, namely $\delta_r = \{\mathbf{nor}\}$, $\delta_{lrw} = \{\mathbf{nol}, \mathbf{nor}, \mathbf{frw}\}$, and $\delta_{rpw} = \{\mathbf{nor}, \mathbf{fcp}, \mathbf{frw}\}$. The first call to $Sdec(\beta_0, \{\emptyset\})$ by *Sdecorate* (line 21) is activated with a decoration in which all states but β_0 are marked by the empty diagnosis set, while $\Delta(\beta_0) = \{\emptyset\}$. The subsequent calls are generated assuming that the transitions within the loop at line 9 are considered from left to right in Fig. 5.14.

Once the reconstructed context-sensitive behavior $Bhv(\wp(\mathcal{A}))$ has been decorated, the solution of the diagnosis problem $\wp(\mathcal{A})$ can be determined as indicated in (5.12), that is, as the union of the diagnostic sets associated with the final states:

$$\Delta(\wp(\mathcal{A})) = \{\, \delta \mid \delta \in \Delta(\beta_f), \beta_f \text{ is final in } Bhv^*(\wp(\mathcal{A}))\,\} \,. \tag{11.8}$$

Example 11.11. Consider the context-sensitive behavior $Bhv(\wp(\mathcal{W}))$ displayed in Fig. 11.6, whose decoration is listed in the last line of Table 11.5. Since the final states are $\beta_5, \ldots, \beta_{14}$, the solution $\Delta(\wp(\mathcal{W}))$ will be

$$\bigcup_{i \in [5 \cdots 14]} \Delta(\beta_i) = \{\, \delta_r, \delta_{lrw}, \delta_{rpw} \,\} = \{\, \{\mathbf{nor}\}, \{\mathbf{ncl}, \mathbf{nor}, \mathbf{frw}\}, \{\mathbf{nor}, \mathbf{fcp}, \mathbf{frw}\} \,\} \,.$$

As expected, this set of candidate diagnoses equals the solution of $\wp(\mathcal{W})$ determined in Example 11.6 based on (11.5).

The equality of the solutions to the same diagnosis problem determined in Example 11.6, based on (11.5), and Example 11.11, based on (11.8), is not a coincidence, as is formally proven in Theorem 11.1 in the next section.

11.7 Correctness

Here, we formally prove the soundness and completeness of the diagnosis technique for determining the solution of a context-sensitive diagnosis problem. Theorem 11.1 states how the actual solution of $\wp(\mathcal{A})$ can be distilled from the decorated context-sensitive behavior $Bhv^*(\wp(\mathcal{A}))$.

Theorem 11.1. *Let $Bhv^*(\wp(\mathcal{A}))$ be a decorated context-sensitive behavior. The union of the set of sets of faults decorating the final states of $Bhv^*(\wp(\mathcal{A}))$ equals the solution of $\wp(\mathcal{A})$.*

Proof. The proof is grounded on Lemmas 10.1–10.7, where $\mathcal{B}^s$ and $\mathcal{B}^v$ denote $Bsp(\mathcal{A})$ and $Bhv^*(\wp(\mathcal{A}))$, respectively.

Lemma 11.1. *If a trajectory $h \in \mathcal{B}^v$, then $h \in \mathcal{B}^s$.*

Proof. This derives from the fact that B^v differs from $\mathcal{B}^s$ in the additional fields $\mathfrak{I}$ and $\mathbb{P}$, which are irrelevant to the triggering of transitions. By induction on h, starting from the initial state, each new transition applicable to $\mathcal{B}^v$ is applicable to $\mathcal{B}^s$ too.

Lemma 11.2. *If a trajectory $h \in \mathcal{B}^v$, then $h_{[\mathcal{V}]} \in \|\mathcal{O}\|$.*

Proof. Recall that $h_{[\mathcal{V}]}$ is the sequence of observable labels associated with observable transitions in the viewer $\mathcal{V}$. Based on the definition of $\mathcal{B}^v$, $h_{[\mathcal{V}]}$ belongs to the language of $Isp(\mathcal{O})$, which equals $\|\mathcal{O}\|$. Thus, $h_{[\mathcal{V}]} \in \|\mathcal{O}\|$.

Lemma 11.3. *If a trajectory $h \in \mathcal{B}^s$ and $h_{[\mathcal{V}]} \in \|\mathcal{O}\|$, then $h \in \mathcal{B}^v$.*

Proof. By induction on h, starting from the initial state, each new component transition $t(c)$ applicable to $\mathcal{B}^s$ is applicable to $\mathcal{B}^v$ too. In fact, if $t(c)$ is unobservable, no further condition is required. If $t(c)$ is observable, based on the assumption $h_{[\mathcal{V}]} \in \|\mathcal{O}\|$ and on the fact that $\|Isp(\mathcal{O})\| = \|\mathcal{O}\|$, the label associated with $t(c)$ in the viewer $\mathcal{V}$ matches a transition in $Isp(\mathcal{O})$.

Lemma 11.4. *Let π be a path in $\mathcal{B}^v$, from the initial state to a state $\beta = (a, \mathbb{P}_\beta, \mathfrak{I})$, with $\mathbb{P}_\beta = (P_1, \ldots, P_n)$. Let h be the sequence of transitions marking arcs in π. Let $\beta_i = (a_i, \mathbb{P}_{\beta_i}, \mathfrak{I}_i)$, $i \in [1 \cdots n]$, be the state in $\mathcal{B}^v$ corresponding to the nearest ancestor of β in π such that the i-th element of $\mathbb{P}_{\beta_i}$ is the initial state of $Sem(\xi_i)$. Let h_i be the prefix of h ending at β_i. Then, each P_i in $\mathbb{P}_\beta$ is the recognition state of $h[\xi_i] - h_i[\xi_i]$, where the sequence difference denotes the suffix of $h[\xi_i]$ obtained by removing from the latter its prefix $h_i[\xi]$.*

Proof. The proof is by induction on h. In the following, for $k \geq 0$, $\pi(k)$ denotes the prefix of π up to the k-th arc, while $h(k)$ denotes the prefix of h composed of k transitions.[9]

[9] In other words, $h(k)$ is the prefix of h corresponding to $\pi(k)$.

(*Basis*) For $h(0) = [\,]$, Lemma 11.4 is trivially fulfilled.

(*Induction*) *If Lemma 11.4 holds for $h(k)$, then it holds for $h(k+1)$ too.* Let t_{k+1} be the last component transition in $h(k+1)$. Two cases are possible, depending on whether or not there exists a context ξ whose alphabet includes t_{k+1}. If not, $\mathbb{P}'$ equals $\mathbb{P}$, thereby making Lemma 11.4 still true. If, instead, ξ does exist (this being the parent of the component performing the transition t_{k+1}), then, based on the recursive rule introduced in the definition of $Bhv(\wp(\mathcal{A}))$ (specifically, for the computation of $\mathbb{P}'$ in Def. 11.3, point c), the new state P' of $Sem(\xi)$ is either $\bar{P}$, if $Sem(\xi)$ includes $P \xrightarrow{t_{k+1}} \bar{P}$, or the initial state of $Sem(\xi)$, if such a transition does not exist.[10] In either case, $\bar{P}$ is the recognition state of $h(k+1)[\xi_i] - h_i[\xi_i]$. Afterwards, if the interface symbol $\mathbb{I}$ marking the new state P' is not empty and ξ has a parent ξ_p in the hierarchy, then the rule is reapplied, with t_{k+1} and ξ being substituted by $\mathbb{I}$ and ξ_p, respectively. In fact, if $\mathbb{I}$ is not empty, P' is a final state of $Sem(\xi)$. Consequently, based on Proposition 11.2, the induction assumption, and line 12 of the *Projection* specification (Sec. 11.3), $\mathbb{I}$ is the set of symbols $N \in \mathbf{I}$ such that (N, r) is a semantic pattern for ξ_p, and a suffix of $h(k+1)[\xi_p] - h_i[\xi_p]$ matches r (in plain form). The possible re-iteration of the rule mimics the operational definition of *Projection* (Sec. 11.3), specifically, the loop in lines 11–17. Finally, the condition stated in Lemma 11.4 is still true when h is replaced by $h(k+1)$.

Lemma 11.5. *Let π be a path in $\mathcal{B}^v$ ending at a final state. Let h be the sequence of component transitions marking arcs in π. Let $\pi_{[\mathcal{R}]}$ denote the set of faults relevant to π, namely,*

$$\pi_{[\mathcal{R}]} = \bigcup_{(a,\mathbb{P},\mathfrak{I}) \in \pi} \mathcal{F}(\mathbb{P}) \,. \tag{11.9}$$

Then, $h_{[\mathcal{R}]} = \pi_{[\mathcal{R}]}$.

Proof. According to Sec. 11.6, $\mathcal{F}(\mathbb{P})$ is the union of the faults associated with each state P_i of the tuple $\mathbb{P}$ in $Sem(\xi_i)$, $i \in [1 \cdots n]$. We denote the subset of $\mathcal{F}(\mathbb{P})$ relevant to the i-th state of $\mathbb{P}$ by $\mathcal{F}(P_i)$. We denote the prefix of h up to the k-th transition by $h(k)$. Hence, $h(k)_{[\mathcal{R}]}$ denotes the diagnosis of $h(k)$ based on $\mathcal{R}$. Likewise, $\pi(k)$ denotes the prefix of the path π up to the k-th arc. Hence, $\pi(k)_{[\mathcal{R}]}$ denotes the set of faults in $\pi(k)$. The proof is by induction on h.

(*Basis*) $h(0)_{[\mathcal{R}]} = \pi(0)_{[\mathcal{R}]} = \emptyset$. In fact, $h(0) = [\,]$; hence, based on (11.4), $h(0)_{[\mathcal{R}]} = \emptyset$. On the other hand, $\pi(0)$ is composed only of the initial state of $\mathcal{B}^v$, namely $(a_0, \mathbb{P}_0, \mathfrak{I}_0)$, where $\mathcal{F}(\mathbb{P}_0) = \emptyset$; hence $\pi(0)_{[\mathcal{R}]} = \emptyset$.

(*Induction*) *If Lemma 11.5 holds for $h(k)$, then it holds for $h(k+1)$ too.* In other words, we assume that if we substitute h with $h(k)$ (and hence π with $\pi(k)$), Lemma 11.5 is still true. Based on this assumption, we have to show that Lemma 11.5 is still true when h is substituted by $h(k+1)$ (and hence π by $\pi(k+1)$). To this end, let t_{k+1} be the last component transition in $h(k+1)$, and

---

[10] According to Proposition 11.1, the initial state P_0 of a semantic space cannot be entered by a transition. Thus, setting $P' = P_0$ makes the nearest ancestor of β in h, namely β_i, the actual initial recognition state of a string in the language of the corresponding regular expression.

$(a, \mathbb{P}, \mathfrak{I}) \xrightarrow{t_{k+1}} (a', \mathbb{P}', \mathfrak{I}')$ the corresponding arc in $\pi(k+1)$. Based on Lemma 11.4, each P_i in $\mathbb{P}$ is the recognition state of the suffix $h(k)[\xi_i] - h_i(k)[\xi_i]$, where $h_i(k)$ is the prefix of $h(k)$ ending at a state where the i-th semantic-space state is the initial state of $Sem(\xi_i)$. Considering each ξ_i, $i \in [1 \cdots n]$, and comparing $h(k)$ and $h(k+1)$, two cases are possible: either $h(k+1)[\xi_i] = h(k)[\xi_i]$ or $h(k+1)[\xi_i]$ is the extension of $h(k)[\xi_i]$ by a new symbol in the alphabet of ξ_i, namely $\mathbb{I}$.

If $h(k+1)[\xi_i] = h(k)[\xi_i]$ then, on the one hand, no state change is performed in $Sem(\xi_i)$; in other words, the i-th element of the tuple $\mathbb{P}'$ equals the corresponding element in $\mathbb{P}$, that is, $P_i' = P_i$. Hence, $\mathcal{F}(P_i') = \mathcal{F}(P_i)$; in other words, the contribution of P_i' to $\mathcal{F}(\mathbb{P}')$ is empty. On the other hand, since $h(k+1)[\xi_i] = h(k)[\xi_i]$, no additional match can fulfill $e \sqsubseteq h(k+1)[\xi]$ in (11.4). Hence, the contribution of the new transition t_{k+1} to $h(k+1)_{[\mathcal{R}]}$ is empty. Therefore, both in $h(k+1)_{[\mathcal{R}]}$ and in $\pi(k+1)_{[\mathcal{R}]}$, no additional fault relevant to ξ_i is generated.

If, instead, $h(k)[\xi_i]$ is extended into $h(k+1)[\xi_i]$ by a new symbol $\mathbb{I}$ (in the alphabet of ξ_i), then two cases are possible, depending on whether or not a transition $P_i \xrightarrow{\mathbb{I}} \bar{P}_i$ is matched in $Sem(\xi_i)$.

If it is matched, two cases are possible, depending on whether or not $\bar{P}_i$ is final in $Sem(\xi_i)$.

If $\bar{P}_i$ is not final then $\mathcal{F}(\bar{P}_i) = \emptyset$, and no contribution to $\mathcal{F}(\mathbb{P}')$ comes from P_i'. On the other hand, since $\bar{P}_i$ is not final, the extension of $h(k+1)[\xi_i]$ by $\mathbb{I}$ does not provoke any new match relevant to ξ_i in (11.4); in other words, no further fault relevant to ξ_i is generated.

If, instead, $\bar{P}_i$ is final, then $\mathcal{F}(\bar{P}_i)$ is the contribution in $\mathcal{F}(\mathbb{P}')$ relevant to ξ_i. That is, $\mathcal{F}(P_i') = \mathcal{F}(P_i) \cup \mathcal{F}(\bar{P}_i)$. On the other hand, based on Corollary 11.1, since $\bar{P}_i$ is final, in (11.4) additional matches for the condition $e \sqsubseteq h(k+1)[\xi]$ hold, precisely, those relevant to faults in $\mathcal{F}(\bar{P}_i)$. In other words, the contribution relevant to ξ_i to $h(k+1)_{[R]}$ equals $\mathcal{F}(\bar{P}_i)$.

If $P_i \xrightarrow{\mathbb{I}} \bar{P}_i$ is not matched in $Sem(\xi_i)$, then $P_i' = P_{i_0}$, and no contribution to $\mathcal{F}(\mathbb{P}')$ is given by $\mathcal{F}(P_i')$.[11] On the other hand, in (11.4), the mismatch in $Sem(\xi_i)$ results in the impossibility of further matches for the condition $e \sqsubseteq h(k+1)[\xi]$; in other words, no additional diagnosis relevant to ξ_i is generated.

In summary, applying this reasoning for all contexts ξ_i, $i \in [1 \cdots n]$, we conclude that $h(k+1)_{[R]} = \pi(k+1)_{[R]}$.

Lemma 11.6. *If a trajectory $h \in \mathcal{B}^\nu$ ends at a final state β_f, then $h_{[R]} \in \Delta(\beta_f)$.*

Proof. According to Lemma 11.5, $h_{[R]} = \pi_{[R]}$, where π is the path in $\mathcal{B}^\nu$ corresponding to h, with $\pi_{[R]}$ being defined in (11.9). Thus, it suffices to show that $\pi_{[R]} \in \Delta(\beta_f)$. Based on the two rules for decoration of $\mathcal{B}^\nu$ (Sec. 11.6), the proof is by induction

11 Based on Lemma 11.4, successive states in $Sem(\xi_i)$ represent the recognition of a suffix of $h[\xi_i]$ starting from a successive state in π. At first sight, this might cause a loss of completeness for $\pi_{[\mathcal{R}]}$, specifically for the matches in (11.4) where the string e is extended by $\mathbb{I}$. The fact is, however, if $\mathbb{I}$ introduces a discontinuity in the matching of e, then no string e including the symbol $\mathbb{I}$ (in that position) will match any regular expression r. Thus, completeness is actually preserved.

on h. In the following, $h(k)$ denotes the prefix of h up to the k-th transition, $\pi(k)$ the prefix of π up to the k-th transition, and β_k the last state in $\pi(k)$.

(*Basis*) Based on (11.9), $\pi(0)_{[\mathcal{R}]} = \emptyset$. According to decoration rule (1), $\emptyset \in \Delta(\beta_0)$. Hence $\pi(0)_{[\mathcal{R}]} \in \Delta(\beta_0)$.

(*Induction*) *If* $\pi(k)_{[\mathcal{R}]} \in \Delta(\beta_k)$, *then* $\pi(k+1)_{[\mathcal{R}]} \in \Delta(\beta_{k+1})$. Let t_{k+1} be the transition marking the last arc in $\pi(k+1)$. On the one hand, according to (11.9), $\pi(k+1)_{[\mathcal{R}]} = \pi(k)_{[\mathcal{R}]} \cup \mathcal{F}(\mathbb{P}_{k+1})$, with $\mathbb{P}_{k+1}$ being relevant to β_{k+1}. On the other hand, according to decoration rule (2), for $\beta_k \xrightarrow{t_{k+1}} \beta_{k+1}$, if $\delta \in \Delta(\beta_k)$ then $(\delta \cup \mathcal{F}(\mathbb{P}_{k+1})) \in \Delta(\beta_{k+1})$. As, by assumption, $\pi(k)_{[\mathcal{R}]} \in \Delta(\beta_k)$, it follows that $\pi(k+1)_{[\mathcal{R}]} \in \Delta(\beta_{k+1})$.

Lemma 11.7. *If β_f is a final state in $\mathcal{B}^v$ and $\delta \in \Delta(\beta_f)$, then there exists a trajectory $h \in \mathcal{B}^v$ ending at β_f such that $h_{[\mathcal{R}]} = \delta$.*

Proof. Based on the decoration rules for $\mathcal{B}^v$, the diagnosis δ is incrementally generated by a path π starting from the empty diagnosis initially associated with β_0, specifically,

$$\delta = \bigcup_{(a,\mathbb{P},\mathfrak{I})\in\pi} \mathcal{F}(\mathbb{P}), \tag{11.10}$$

which, based on (11.9), equals $\pi_{[\mathcal{R}]}$, which in turn, according to Lemma 11.5, equals $h_{[\mathcal{R}]}$, with the latter being the trajectory generated by π. On the one hand, in order for h to be a trajectory, π must be finite. If π is infinite, then π must include (at least) a cycle in $\mathcal{B}^v$ traversed an infinite number of times. On the other hand, once a cycle is traversed, all associated $\mathcal{F}(\mathbb{P})$ are inserted into δ: successive iterations of the cycle do not extend δ, because of duplicate removals caused by the set-theoretic union in decoration rule (2). In other words, δ can be always generated by a finite path π and, hence, h is finite.

To prove Theorem 11.1, we show that $\delta \in \Delta(\mathcal{B}^v) \Leftrightarrow \delta \in \Delta(\wp(\mathcal{A}))$. On the one hand, if $\delta \in \Delta(\mathcal{B}^v)$ then, based on Lemmas 11.1, 11.2, and 11.7, there exists a trajectory $h \in \mathcal{B}^s$ such that $h_{[\mathcal{V}]} \in \|\mathcal{O}\|$ and $h_{[\mathcal{R}]} = \delta$; in other words, based on (11.5), $\delta \in \Delta(\wp(\mathcal{A}))$. On the other hand, if $\delta \in \Delta(\wp(\mathcal{A}))$ then, according to (11.5) and based on Lemmas 11.3 and 11.6, there exists a trajectory $h \in \mathcal{B}^v$ ending at a final state β_f such that $\delta = h_{[\mathcal{R}]}$ and $\delta \in \Delta(\beta_f)$; in other words, $\delta \subset \Delta(\mathcal{B}^v)$. □

11.8 Bibliographic Notes

The idea of injecting some sort of semantics into the diagnosis of active systems was proposed in [108], where a set of *semantic rules* was specified on a *semantic domain* (the set of subsystems relevant to diagnosis). Each rule defines the faulty behavior of a subsystem, possibly depending on the behavior of other subsystems. The diagnostic output is a set of candidate diagnoses which account for the faults of every subsystem in the semantic domain. The notion of context-sensitive diagnosis was presented in [109]. Diagnosis by semantic patterns was proposed in [117].

Context-sensitive diagnosis was extended to *higher-order DESs* in [114, 115, 116], where the DES is characterized by *behavior stratification*, like that in complex systems [15, 30, 51, 71, 121, 151]. An efficient technique for diagnosis computation for *complex active systems* with plain observations was presented in [77, 119], and extended in [118] to uncertain temporal observations.

11.9 Summary of Chapter 11

Context Sensitivity. The property of a diagnosis task of associating faults not only with abnormal transitions of components, but also with patterns of component transitions, where each pattern is relevant to a certain context.

Context Domain. Given an active system $\mathcal{A}$, the set of subsystems of $\mathcal{A}$ (called contexts) which are relevant to the output of a diagnosis task.

Context Hierarchy. Given a context domain Ξ, a partition of each context in Ξ in terms of other subcontexts (or components).

Interface Symbol. Given a context ξ, a subset of the names involved in semantic patterns defined on ξ.

Semantic Pattern. Given a context ξ, a pair (N, r), where N is a name and r a regular expression on interface symbols relevant to ξ and possibly transitions of components in ξ (if ξ includes components).

Semantics. Given a context hierarchy $\mathcal{H}$, a set of pairs $(\xi, \mathcal{P})$, where ξ is a context in $\mathcal{H}$ and $\mathcal{P}$ the set of semantic patterns defined on ξ.

Context-Sensitive Ruler. Given an active system $\mathcal{A}$, a triple $(\Xi, \mathcal{H}, \mathcal{S})$, where Ξ is the context domain, $\mathcal{H}$ the context hierarchy, and $\mathcal{S}$ the semantics.

Context-Sensitive Diagnosis Problem. Given an active system $\mathcal{A}$, a quadruple $\wp(\mathcal{A}) = (a_0, \mathcal{V}, \mathcal{O}, \mathcal{R})$, where a_0 is the initial state of $\mathcal{A}$, $\mathcal{V}$ a viewer for $\mathcal{A}$, $\mathcal{O}$ a temporal observation of $\mathcal{A}$, and $\mathcal{R}$ a context-sensitive ruler for $\mathcal{A}$.

Trajectory Projection. Given a trajectory h of a system $\mathcal{A}$ and a context ξ of $\mathcal{A}$, the mode in which h is perceived by ξ in terms of interface symbols (if ξ includes other contexts) and/or component transitions (if ξ includes components).

Context-Sensitive Diagnosis. Given a trajectory h of an active system $\mathcal{A}$, the set of fault labels f such that (f, r) is a semantic pattern defined for a context ξ of $\mathcal{A}$, and e is a string that matches r and is a contiguous subsequence of the projection of h on ξ.

Semantic Space. Given a context ξ, a DFA guiding the matching of semantic patterns defined on ξ.

Context-Sensitive Behavior. Given a context-sensitive diagnosis problem $\wp(\mathcal{A}) = (a_0, \mathcal{V}, \mathcal{O}, \mathcal{R})$, a DFA whose language is the subset of trajectories in the behavior space $Bsp(\mathcal{A})$ that are consistent with a temporal observation $\mathcal{O}$, with each state being a triple $(a, \mathbb{P}, \mathfrak{I})$, where a is a state of $Bsp(\mathcal{A})$, $\mathbb{P}$ a tuple of pattern-space states, and $\mathfrak{I}$ a state of the index space of $\mathcal{O}$.

Fault Set. Given a state $(a, \mathbb{P}, \mathfrak{I})$ of the behavior of a context-sensitive diagnosis problem, the union of the fault labels associated with each state of $\mathbb{P}$ in the corresponding semantic space.

Decorated Context-Sensitive Behavior. Given the context-sensitive behavior of $\wp(\mathcal{A})$, the DFA obtained from the behavior by marking each state β with the set of context-sensitive diagnoses relevant to the trajectories of $\mathcal{A}$ ending at β.

Chapter 12
Related Work

Since the middle 1990s, model-based diagnosis of DESs has spurred research in both the artificial intelligence and the automatic control communities. In this chapter some contributions in the literature are briefly mentioned, all of which are based on untimed DES models (similar to active systems). First, the seminal *diagnoser approach*, which is chronologically the oldest, is covered. Then, some methods that address specific aspects of the task are touched on, ending with the recent *hypothesis space approach*. Finally, a short survey of some fundamental studies on diagnosability of DESs is presented.

12.1 Diagnoser Approach

According to the diagnoser approach [143, 144, 145], a system consists of several distinct physical components and is equipped with a set of sensors. The complete (i.e., both normal and abnormal) behavior of each component is represented by an NFA: each transition is triggered by an event, which is either observable, typically a command issued by the system supervisor, or unobservable, typically a failure event.

Failures are classified into disjoint classes corresponding to different failure types: the diagnosis task is required to uniquely identify not the failure event itself but only the type of failure. The global *sensor map* asserts a correspondence between each composition of the states of all components and the set of discrete outputs of all sensors.

The diagnoser approach distinguishes between online activities, which are carried out while monitoring the system, and offline ones, which are accomplished beforehand. Diagnosis is performed online, whereas the verification of the diagnosability properties of the system and the automatic generation of the data structure, called a *diagnoser*, that will be exploited online are accomplished offline.

The diagnoser is derived from the behavioral model of the whole system, hereafter called the *system model*. A systematic procedure was presented in [145] for

G. Lamperti et al., *Introduction to Diagnosis of Active Systems*,
https://doi.org/10.1007/978-3-319-92733-6_12

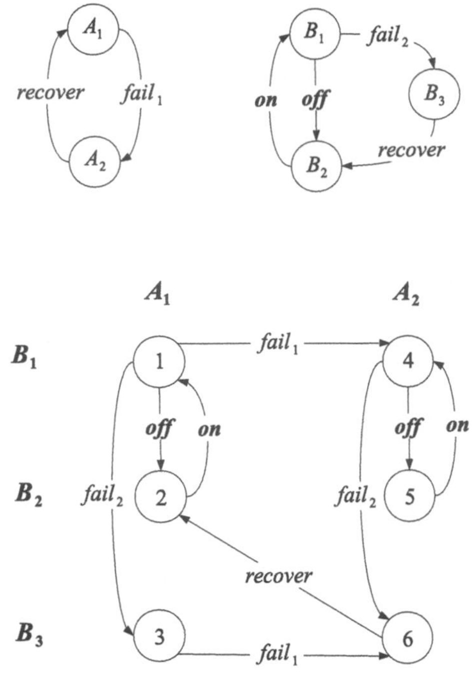

Fig. 12.1 Component models for Example 12.1 and their synchronous composition

generating the system model based on the behavioral models of the components of the system and on the sensor map. This procedure basically consists in performing the synchronous composition of the component models,[1] thus capturing the interaction among components, and then in redefining the observability of the transitions obtained, possibly introducing additional states, so as to reflect the discrete outputs of the available sensors.

In the system model, which is an NFA, unobservable events are either failures or events that cause changes in the system state which are not recorded by sensors. The system model is assumed both not to incorporate unobservable cycles and to generate a prefix-closed live language over the alphabet of all events. These two assumptions together guarantee that a failure event is, sooner or later, followed by an observable event.

Every diagnosis provided by the diagnoser is consistent with a path whose latest transition is observable, but it does not consider possible following unobservable transitions. In other words, according to the terminology introduced in Sec. 1.4.1, the diagnosis candidates produced by the diagnoser approach are refined.

Example 12.1. The top of Fig. 12.1 displays the behavioral models (observable events are shown in bold) of the two components A (left) and B (right) of the system that we are considering in this example.

[1] See [25] for a definition of the standard operation of synchronous composition, also called parallel or strict composition, on automata.

At the bottom of Fig. 12.1, the synchronous composition of the two automata is shown, where each state is the composition of the two corresponding component states in the lattice. So, for instance, state 1 is the composition (A_1, B_1).

Suppose that the system is endowed with a single sensor, which in practice distinguishes state A_1 from state A_2. The set of the discrete outputs of the sensor is $\{norm, ab\}$. Using the same formalism as in [145], the global sensor map is the following:

$$h(A_1, \bullet) = norm,$$
$$h(A_2, \bullet) = ab,$$

where h is the function providing the output of the sensor, and a bullet denotes an indifference condition, that is, the value of h does not depend on the state of component B.

Figure 12.2 portrays the construction of the system model, based on both the result of the synchronous composition above (bottom of Fig. 12.1) and the sensor map. Shaded nodes represent the states added to the synchronous composition in order to account for the discrete values given by the sensor: in this respect, the observable events are the changes of the sensor value, that is, the change from *norm* to *ab* and vice versa, indicated by *norm* $\rightsquigarrow$ *ab* and *ab* $\rightsquigarrow$ *norm*, respectively. The states numbered from 1 to 6 are the same as in the synchronous composition.

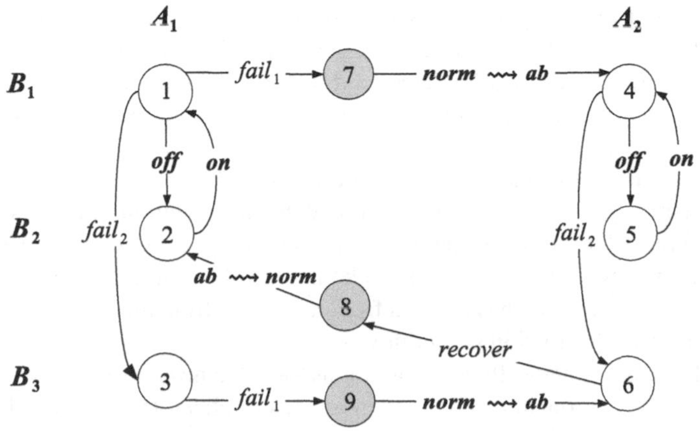

Fig. 12.2 System model for Example 12.1

12.1.1 Diagnosability

Two related notions of diagnosability were introduced in [144], with the first one being stronger than the second, which is called *I-diagnosability* and is better detailed in [145]. Such notions are defined with respect to the set of observable events of the system considered and the partition of failure events (that is, the types of failures that have to be detected). In addition, in order to define I-diagnosability, some observable *indicator* event(s) is/are associated with each class of failure events.

Roughly, a system is *diagnosable* or *I-diagnosable* if, given any occurrence of any failure event, it is possible to detect and isolate that failure within a finite number of events (i.e., a finite number of transitions of the system model) following it or following one of the indicator events associated with that kind of failure, respectively. In other words, the system is diagnosable if it is always possible to detect within a finite delay the occurrence of a failure event (and uniquely identify its type) by using the record of observed events.

A system is said to be I-diagnosable if it is possible to detect (and identify) failures not always but whenever failure events are followed by certain observable indicator events that are associated with them. Hence, a system that is diagnosable is also I-diagnosable; in particular, it is possible to identify a nonempty set of indicator events for each type of failure event such that the occurrence of any failure event is always followed by the occurrence of an indicator event and the failure can be detected and isolated within a finite delay after it.

If, instead, the system is not diagnosable, it may be I-diagnosable given certain indicator events and not I-diagnosable given some other indicator events. If the system is I-diagnosable given certain indicator events, this means that the occurrence of a failure of a certain type can certainly be detected if that occurrence is followed by that of any indicator event associated with this type, while it may not be detectable otherwise.

Checking necessary and sufficient conditions for diagnosability requires accounting for both the system model and the diagnoser if no dynamic evolution of the system can include several failures of the same type. If this condition does not hold, the check requires us to process both the system model and another automaton, say a diagnoser for multiple failures, which is derived from it (and not from the diagnoser) by means of a systematic construction procedure.

In order to check necessary and sufficient conditions for I-diagnosability, both the system model and another automaton, say a diagnoser with indicators, are needed, where the latter is derived from the former by means of another systematic construction procedure.

Notice that, when several failure events of the same failure type occur along any trajectory of the system, the definition of diagnosability does not require that all of these events be detected, it only requires that we are able to conclude within a finite delay after the first occurrence of a failure that a failure event of that type occurred.

Checking diagnosability and I-diagnosability amounts to cycle detection in the (various) diagnosers and in the system model: any of the standard cycle detection algorithms (whose complexity is polynomial) can be used. However, in order to prove that a system is diagnosable, the global system model is needed along with the diagnoser or one of its variations, which, in general, are larger than it.

The diagnoser and its variations are drawn from the global system model. In the worst case, the state space of the diagnoser is exponential in the state space of the system model [144]. As underlined in [132, 141], building the system model, although it can be done via a well-known operation of synchronization (which, however, is not enough in general, since the sensor map has to be considered also), is unrealistic owing to its intractable size for large systems. Hence, also, the diagnoser and its variations cannot be produced. Thus, in general, the diagnosability of a system is an intractable problem (as well as its diagnosis by means of the diagnoser approach).

Example 12.2. The (already defined) set of observable events of the system in Example 12.1 is

$$\{on, off, norm \rightsquigarrow ab, ab \rightsquigarrow norm\}$$

and all such events are displayed in bold in the system model (Fig. 12.2). The set of all failure events of the system is $\{fail_1, fail_2\}$; now we partition it into two distinct sets, that is, we assume that each of the two failure events has to be detected in a distinct way.

Let 1 be the initial state of the system. Given the above partition of failure events, the system is diagnosable. We omit a proof of this claim based on the method embodied by the diagnoser approach, as it would require the generation of a diagnoser for multiple failures since there are infinite trajectories including multiple instances of failures of the same type (all trajectories cycling either through states $1,3,6,2$ or through $1,4,6,2$ include multiple instances of both $fail_1$ and $fail_2$).

However, thanks to the small size and simplicity of the system considered, we can see that the system is diagnosable based on the definition of diagnosability: in fact, starting from state 1, the first occurrence of the failure $fail_1$ is detected in one transition following it ($norm \rightsquigarrow ab$), the first occurrence of the failure $fail_2$ is detected in four transitions at most, i.e., $fail_1$, $norm \rightsquigarrow ab$, *recover*, $ab \rightsquigarrow norm$ (otherwise in two transitions, i.e., *recover*, $ab \rightsquigarrow norm$), and further occurrences of the two failures need not be detected. Therefore the definition of a diagnosable system is fulfilled.

It goes without saying that the system is also I-diagnosable, since diagnosability implies I-diagnosability. Clearly, for instance, if $norm \rightsquigarrow ab$ and $ab \rightsquigarrow norm$ are chosen as indicator events for $fail_1$ and $fail_2$, respectively, then either fault can be detected with no delay (i.e., zero transitions) after the corresponding indicator event.

In [143], a generalized definition of diagnosability and a procedure to check the diagnosability of systems generating nonlive languages over the alphabet of all events were presented. The definition is the same as that for live languages except that the diagnosability condition is required to hold for terminating trajectories as well.

12.1.2 Diagnoser

Besides being essential offline in order to establish the diagnosability properties of a DES whose evolution over time cannot be affected by several failures of the same type at the same time, the diagnoser is essential online too for diagnosis. However, as remarked in the previous section, the construction of the diagnoser is intractable.

The generation of the diagnoser, which is a deterministic, completely observable automaton, starts from a known initial state, which is assumed to be normal. This means that the initial state of every evolution to be considered is known in advance. The diagnoser can be regarded as a graph.

The purpose of the diagnoser while the system is being monitored is to infer past occurrences of failures, based on observed events. Monitoring a DES amounts to following online a path in the diagnoser, progressing along a transition each time an event is observed.

If a sequence of events has already been observed, then, following the (only) path marked by this sequence of events, a node of the diagnoser is reached that contains an estimate of all the possible current states of the DES, each of which is associated with a set of faults.

Each state of the DES contained in this node is reachable in the system model by means of a sequence of transitions producing the given observation, while its corresponding set of faults contains all the faults that have occurred during that sequence, that is, the set of faults is a candidate diagnosis.

Thus, each node of the diagnoser contains all the candidate diagnoses relevant to the sequence of events observed so far. A node of a diagnoser detects a fault if all of its candidate diagnoses include that fault.

Example 12.3. Figure 12.3 displays the diagnoser of the diagnosable system of Example 12.1, assuming, as already done, that state 1 is the initial state of the system.

The label N stands for *normal* (no fault), while F_1 and F_2 are shorthand for the faults $fail_1$ and $fail_2$, respectively. The initial node of the diagnoser is $1N$, where 1 is the given initial state of the system and N corresponds to the assumption that the system is initially free of faults.

Suppose that, while the system is being monitored, $norm \rightsquigarrow ab$ is the first event that is observed. By following the (only) transition exiting the initial node that is labeled by this observable event, node $4F_1\ 6F_1F_2$ is reached, this meaning that the current state of the system, according to the model in Fig. 12.1, can be either 4 or 6, in the former case the system being affected by the fault $fail_1$, and in the latter also by the fault $fail_2$. Thus, after $norm \rightsquigarrow ab$ has been observed, the fault $fail_1$ is certain, whereas fault $fail_2$ is uncertain. This uncertainty, however, is removed once the next event has been observed. In fact, if the event *off* is observed, we can conclude that the system is affected by the fault $fail_1$ only; otherwise (that is, if the event $ab \rightsquigarrow norm$ is observed), it is affected by both faults.

This is another facet of diagnosability: a system is diagnosable if any uncertainty about the diagnosis (i.e., the set of faults affecting the system) is resolved in a finite

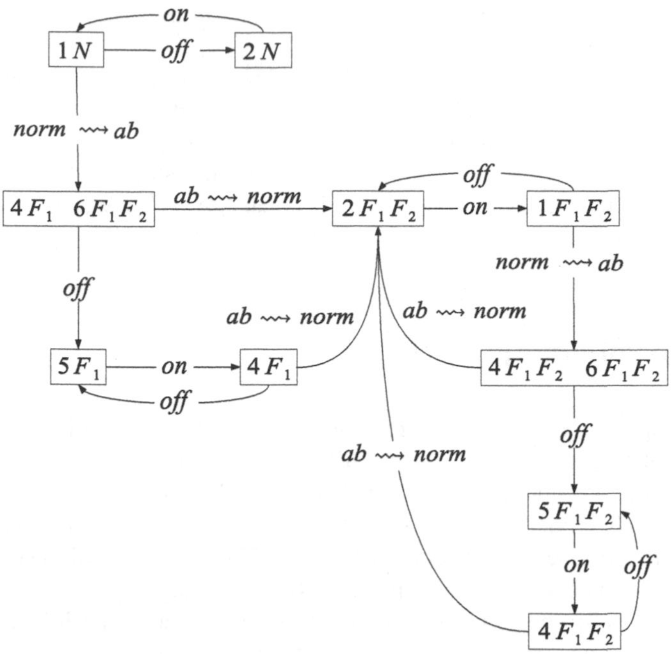

Fig. 12.3 Diagnoser for Example 12.3

number of transitions of the system model (and, consequently, in a finite number of observed events during monitoring).

If the same state in the system model can be reached by means of several distinct paths, producing the same observation but each characterized by a distinct set of faults, then the set of faults associated with that state within the (only) node of the diagnoser reachable through that observation is the intersection of all such sets.

In this case, an A is added to denote ambiguity. While the diagnoser is being constructed, this ambiguity is propagated to the successor states of the ambiguous state in the next nodes as long as no failure events are encountered. Once a fault has been encountered, the set of faults is updated and the A is dropped.

A theorem, proven in [144], states that if the DES is diagnosable or I-diagnosable, the diagnoser is capable of detecting and isolating any failure event in a bounded number of events after its occurrence or the occurrence of one of its indicator events, respectively. In addition, a bound on the detection delay corresponding to each failure type is given based on the diagnoser only. Thus, the diagnoser has been proven to be a structure capable of providing the needed diagnoses of diagnosable (as well as I-diagnosable) systems.

Nothing is added about the diagnosis of systems that are neither diagnosable nor I-diagnosable. What can be inferred, however, is that a diagnoser can be exploited for such systems as well and, in general, each time an event is observed, a set of candidate diagnoses is output, where that set is not complete in the case where a node including an A (ambiguity) has been encountered.

Diagnosis of systems generating nonlive languages [143] is performed based on a diagnoser derived from a system model that has been modified with respect to the real one so as to generate a live language. In order to extend the system model, a transition is added for each state having no successors: such a state is both the source and the target of the added transition, which is labeled by a *Stop* event, this being a fictitious observable event.

12.2 Supervision Pattern Approach

The interesting concept of a *fault supervision pattern* was introduced in [68] in order both to perform DES diagnosis and to check DES diagnosability. Such a notion, which (as is the case with all approaches mentioned in this book) relies on the representation of the behavioral model of a DES as a finite automaton, is a generalization of the notion of a fault, and resembles that of a chronicle [37] in that it describes not a single faulty transition but a whole paradigmatic faulty evolution.

Notice that a single faulty transition is a particular case of a faulty evolution; hence the approach based on fault patterns is broader than the usual diagnosis approaches. In other words, while the usual diagnosis approaches look for the occurrence(s) of all faulty transitions, the supervision pattern approach can look for the occurrence(s) of any specific fault pattern, this concept being general enough to be used to define and compute different kinds of diagnosis results, such as single-fault, multiple-fault, ordered-fault, multioccurrence-fault, intermittent-fault diagnosis, and so on.

12.2.1 Supervision Patterns

Formally, a supervision pattern is a deterministic automaton Ω, having a set of states Q_Ω, including an initial state and a set $Q_F \subseteq Q_\Omega$ of final states. State transitions are triggered by events over an alphabet Σ, a subset of which includes *faulty events* (or, simply, *faults*). The set of transitions between states is such that, for each state $q \in Q_\Omega$ and for each event $\sigma \in \Sigma$, there is a transition exiting q that is triggered by σ. In addition, Q_F is assumed to be *stable*, which means that, if a trajectory reaches a final state, then all the states that can be reached by any next transition are final. Hence, the language $\mathcal{L}_{Q_F}(\Omega)$ accepted by Ω is "extension-closed", namely

$$\mathcal{L}_{Q_F}(\Omega) = \mathcal{L}_{Q_F}(\Omega).\Sigma^* .$$

In the following, several kinds of supervision patterns are described and pictorially represented (the initial state of each pattern is identified by an entering arrow that does not exit from any state).

- *Single-fault pattern.* Figure 12.4 depicts the automaton representing the supervision pattern Ω_f for the occurrence of a fault $f \in \Sigma$, where F (double circled) is the only final state. Any trajectory $s \in \Sigma^*$ including fault f, i.e., $s \in \Sigma^*.f.\Sigma^*$, is singled out as faulty since $\mathcal{L}(\Omega_f) = \Sigma^*.f.\Sigma^*$ is the language recognized by the supervision pattern.

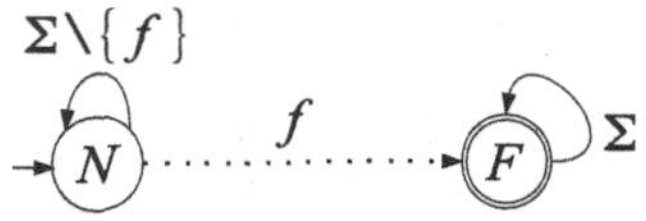

Fig. 12.4 Supervision pattern for a single fault

- *Fault-set pattern.* Figure 12.5 depicts the supervision pattern for the occurrence of a pair of faults $f_1, f_2 \in \Sigma$, independently of their order. Any trajectory including both faults, i.e., $s \in \Sigma^*.f_1.\Sigma^* \cap \Sigma^*.f_2.\Sigma^*$, belongs to the language $\mathcal{L}_{F_1}(\Omega_{f_1}) \cap \mathcal{L}_{F_2}(\Omega_{f_2})$ recognized by the supervision pattern in Fig. 12.5, which is just the synchronous product of the supervision pattern for a single fault f_1 and the supervision pattern for a single fault f_2, i.e., $\Omega_{f_1} \times \Omega_{f_2}$. More generally, the automaton resulting from the synchronous product $\times_{i=1,\cdots,l}\Omega_{f_i}$, having $\times_{i=1,\cdots,l}F_i$ as its only final state, denotes the supervision pattern recognizing the occurrence of the set of faults $\{f_1, \cdots, f_l\}$. If the set is a singleton, the pattern collapses into the single-fault pattern above.

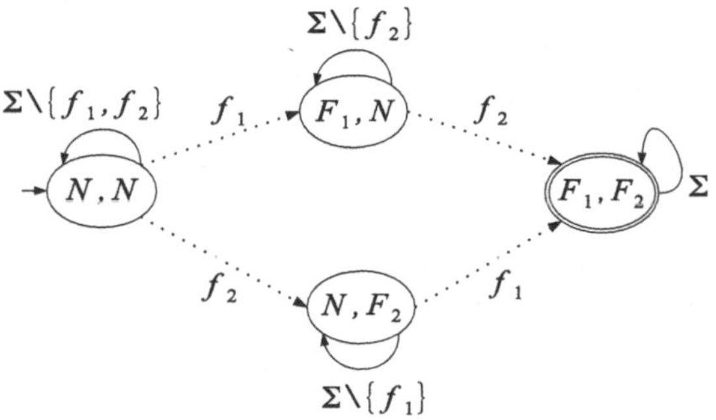

Fig. 12.5 Supervision pattern for a set of faults

- *Ordered-fault pattern.* Figure 12.6 shows the supervision pattern for the occurrence of a pair of faults $f_1, f_2 \in \Sigma$ in a given order (e.g., f_2 after f_1). Any trajectory $s \in \Sigma^*.f_1.\Sigma^*.f_2.\Sigma^*$, which includes the two ordered faults f_1 and f_2, belongs to the concatenation language $\mathcal{L}_{F_1}(\Omega_{f_1}).\mathcal{L}_{F_2}(\Omega_{f_2})$ recognized by the supervision pattern.
- *Multioccurrence-fault pattern.* Figure 12.7 depicts the supervision pattern for multiple occurrences (k times here) of the same fault $f \in \Sigma$. Any trajectory

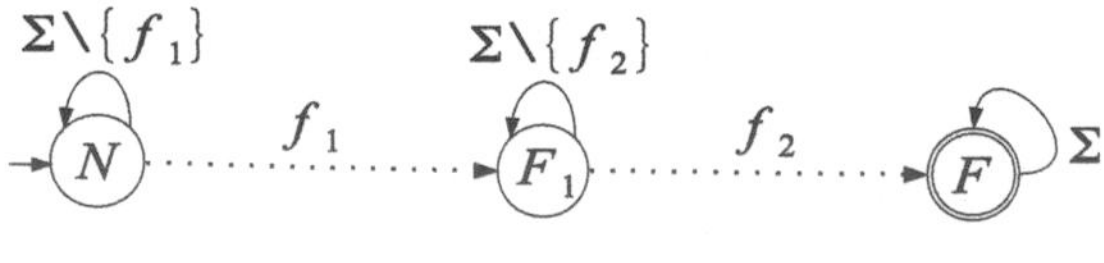

Fig. 12.6 Supervision pattern for ordered faults

$s \in \Sigma^*$ where the same fault f occurs k times (at least) belongs to the language $\mathcal{L}_F(\Omega_f)^k$ (which is the concatenation of k instances of the language $\mathcal{L}_F(\Omega_f)$) recognized by the supervision pattern.

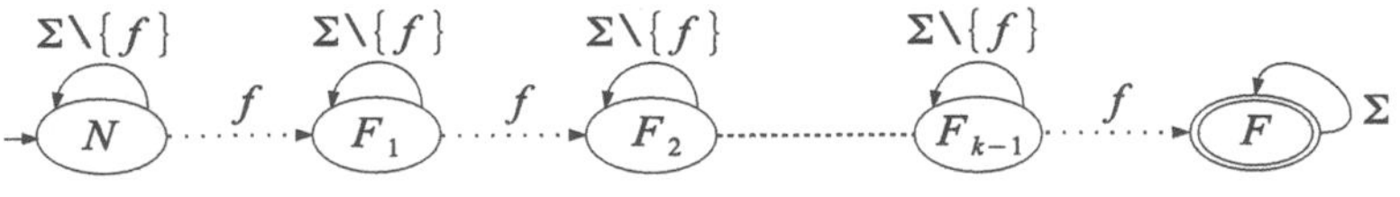

Fig. 12.7 Supervision pattern for a number of occurrences of the same fault

- *Intermittent-fault pattern.* The supervision pattern for an intermittent fault $f \in \Sigma$ is displayed in Fig. 12.8. Any trajectory $s \in \Sigma^*$ where a fault f has occurred twice (at least) without any repair (event r) belongs to the language recognized by the supervision pattern. This case is different from the previous one since it is focused on intermittent faults only, where a fault is considered as intermittent if it has occurred previously and then, without having being repaired, it has occurred (at least once) again. The authors of the supervision pattern approach implicitly mean that the latter occurrence is necessarily a manifestation of the same fault as in the former occurrence. The multioccurrence supervision pattern, instead, is focused on a specific number of occurrences of the same (kind of) fault, independently of whether such occurrences are due to intermittency or not (that is, the fault f may have been repaired between two successive occurrences).

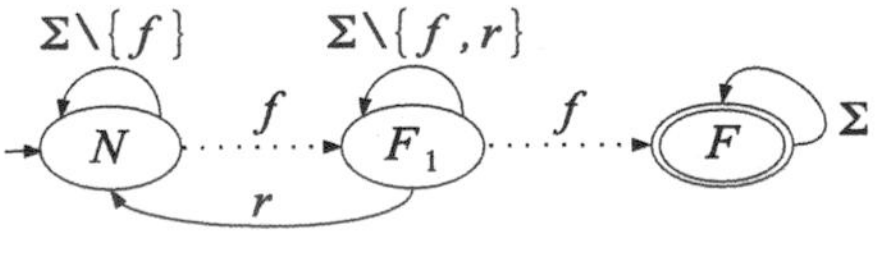

Fig. 12.8 Supervision pattern for an intermittent fault

Those listed above are just a sample of the supervision patterns that can be built. If the user is interested in a specific arrangement of fault occurrences that is meaningful for the DES at hand, s/he can build a corresponding pattern. For instance, the notion of intermittency adopted by the authors of the supervision pattern approach can be changed by building a supervision pattern that requires the same fault to occur (at least) three times without any repair.

12.2.2 Diagnosis with Patterns

As pointed out in Sec. 1.4.1, a diagnosis problem for a DES consists in a DES whose behavior is modeled as a partially observable NFA and an observation. In the supervision pattern approach, the behavioral model of the DES, denoted by G, is assumed to be an automaton where each state is such that, among its exiting transitions, there is (at least) one transition that does not belong to an unobservable cycle. The observation, denoted by μ, is assumed to be certain, that is, a sequence of observable events that have taken place in the DES.

In order to exploit supervision patterns to carry out a diagnosis task, the supervision pattern Ω applicable to the desired diagnosis has to be built. Then, that pattern has to be synchronized with G, thus producing an automaton G_Ω. Each state in G_Ω is the composition of a state of G and a state of Ω.

The set of trajectories of G accepted by Ω equals the set of trajectories accepted by G_Ω, where a state in G_Ω is final if it is composed of a final state of Ω.

The automaton G_Ω is then determinized into $Det(G_\Omega)$, whose states are called *macrostates* since each of them is a set of states of G_Ω, that is, it is a set of pairs (s_G, s_Ω), where s_G is a state of G and s_Ω is a state of Ω (trajectories are preserved by the determinization).

Notice that $Det(G_\Omega)$ is completely observable, as all unobservable transitions in G_Ω have been handled as ε-transitions by the determinization process. The automaton $Det(G_\Omega)$ can be seen as some kind of diagnoser or diagnosis checker. In fact, it has been proven that all trajectories compatible with the observation μ are accepted by Ω if and only if the only path corresponding to μ in $Det(G_\Omega)$ leads (from the initial macrostate) to a macrostate composed only of final states in G_Ω.

In other words, if in each pair (s_G, s_Ω) in the macrostate reached s_Ω is a final state in Ω, then the arrangement of faults according to the pattern has definitely occurred in G. Dually, all trajectories compatible with the observation μ are not accepted by Ω if and only if μ leads to a macrostate composed only of nonfinal states in G_Ω.

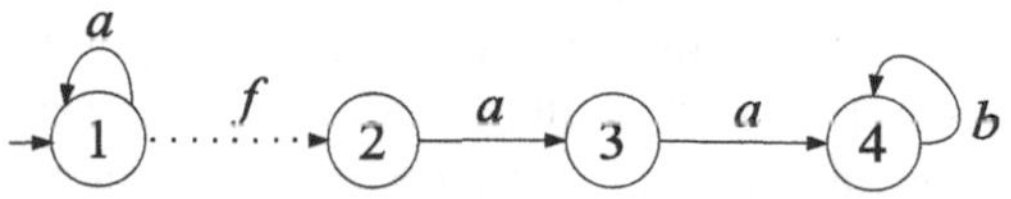

Fig. 12.9 DES model G for Example 12.4

Example 12.4. Consider a DES whose behavioral model G is shown in Fig. 12.9, where the only faulty event is f, which is not observable, and events a and b are observable. Figure 12.10 displays the NFA G_Ω resulting from the synchronous composition of G with the automaton representing the single-fault pattern (Fig. 12.4):

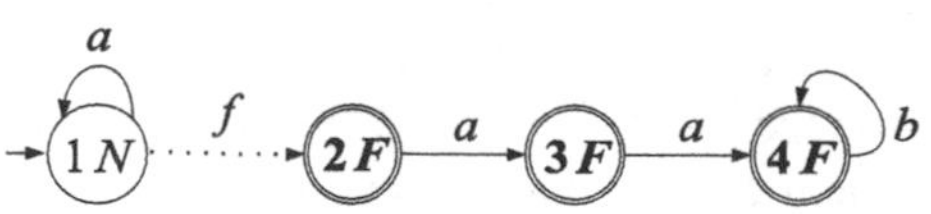

Fig. 12.10 Synchronous product G_Ω for Example 12.4

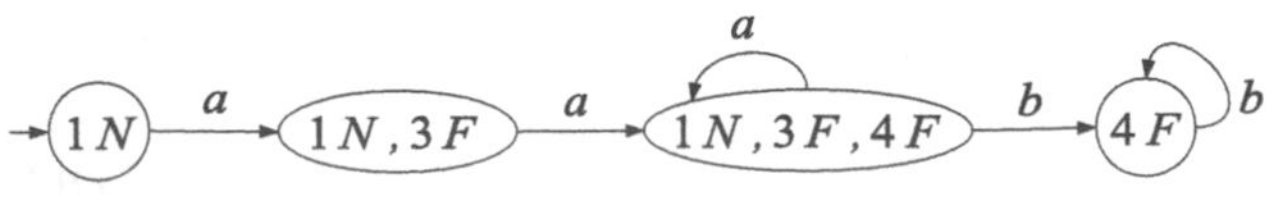

Fig. 12.11 Determinization of G_Ω

each final state of G_Ω is double circled. In this figure and in the remainder of this example, the shorthand for each pair (s_G, s_Ω) is $s_G s_\Omega$ (no comma in between). Figure 12.11 depicts the automaton $Det(G_\Omega)$ resulting from the determinization of G_Ω.

Suppose that now we get the (certain) observation $\mu = [a]$, and that we perform the diagnosis task based on the single-fault supervision pattern. Starting from the initial state in $Det(G_\Omega)$ and following the trajectory marked by the observable events in the observation, we reach the state $\{1N, 3F\}$, which is ambiguous as it means that, once the observable event a has taken place, either the DES is in state 1 (of model G) and no fault has occurred, or it is in state 3 and fault f has occurred. Since not all elements in $\{1N, 3F\}$ are final, we cannot definitely know whether fault f has occurred.

However, if the observer has gathered the observation $\mu = [a, a, b]$, we can conclude that fault f has definitely occurred, as the state reached in $Det(G_\Omega)$ is $\{4F\}$, which is the only final one.

12.3 SCL-D-Holon Approach

The scalability of model-based reasoning about DESs has been considered a challenge since the early 1990s. A contribution [17] dating back to 1993 faced this challenge by exploiting a hierarchical representation of the behavior of a DES. In particular, an algorithm that took advantage of such a hierarchical structure in order to perform a reachability analysis was provided. A decade later, the exploitation of the same structure in order to perform model-based diagnostic problem solving was proposed [65]. The idea was to adopt a distinct diagnoser for each node in the hierarchy. Later on, diagnosis closures, as proposed in the active system approach (originally called *silent closures*), were interpreted [169] as a special case of the elements defined in hierarchical diagnosis.

12.3.1 Hierarchical State Machines

Following [17], we consider *hierarchical state machines* (HSMs), which are a simplified version of Harel's statecharts [62]. As such, an HSM can be used to represent the behavior of a class of DESs. In an HSM, several states, called *substates*, can be grouped into a *superstate*. Since each substate can in turn be a superstate, a tree structure is obtained, where each node is a state. In particular, a superstate corre-

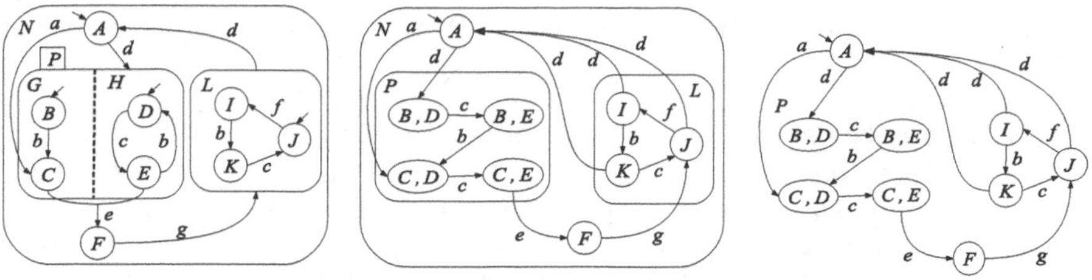

Fig. 12.12 An HSM (left), its equivalent basic HSM (center), and its equivalent flat HSM (right)

sponds to a parent node, and its own children are its *immediate substates*. States that have no substates (i.e., the leaves of the tree) are called *basic*.

A *superstate* is either an *OR-state* or an *AND-state*. If the current state of the DES represented is an OR-state, then the actual state of the DES is one of its immediate substates, just one of them (and not more than one at a time).

The immediate substates of an AND-state are partitioned into several parts (two at least), each of which can be regarded as an OR-state, that is, the states in each part are the states of an FA. If the current state of the DES is an AND-state, then the actual state of the DES is one in the FA resulting from the synchronous composition of all the FAs corresponding to the parts. The internal transitions of an AND-state are those relevant to the synchronous product of those FAs. An HSM with no AND-states is a *basic* HSM. An HSM including just basic states is *flat*. Any HSM can be transformed into an equivalent basic HSM, and the latter into an equivalent flat HSM.

Example 12.5. Consider the HSM shown on the left of Fig. 12.12: N, P, and L are superstates, as each of them includes other states; in contrast, A and F are basic states, which are immediate substates (or *children*) of N. Another child of N is L, which is an OR-state: if the current state of the DES is L, the actual state of the DES is either I or J or K (just one at any time). The only remaining immediate substate of N is P, which is an AND-state: if the DES is in state P, this is equivalent to being in both G and H.

In the center of Fig. 12.12, G and H have been replaced by their synchronous product, and hence state P is now an OR-state. Since the HSM in the center of Fig. 12.12 has no AND-state, this HSM is a basic one. If we now remove P and L, by replacing each of them with its content, and then we remove superstate N, leaving just its content, we obtain an equivalent HSM, shown on the right of Fig. 12.12, where no state is a superstate; hence this HSM is flat.

A sequence of transitions between states is a *basic path* if both its source and its destination are basic states. If the destination state of a basic path in an HSM is different from the destination of the same path in the corresponding basic HSM, the target state in the basic HSM is called the *explicit destination* of that path.

Example 12.6. In the HSM on the left of Fig. 12.12, the path consisting of just the transition (A, a, C) is a basic one. This path brings the system from state A to state

(C,D) in the equivalent basic HSM. Thus, (C,D) is the explicit destination of the path.

The work described in [17] was extended in [65], where HSMs were renamed *hierarchical finite state machines* (HFSMs). The basic states of a basic HFSM are referred to as *simple states*. A basic HFSM has a *father–child-connected* (FC-connected) form if a superstate has only transitions among its simple states and/or from the states of its children to the states of its parent and/or from the states of its parent to the states of its children. A basic HFSM that is not FC-connected can be transformed into a basic FC-connected HFSM: every transition (x,σ,y) that violates FC-connectivity can be replaced with a sequence of transitions compliant with FC-connectivity from x to a dummy state (that is, a state purposefully added) in a common ancestor of x and y, followed by another transition (or sequence of transitions, interleaved with dummy states) to y.

A basic HFSM is *reachable* if every simple state of it is reachable starting from the initial state of the system. A basic HFSM is called a *standard HFSM* if it is both reachable and FC-connected.

Example 12.7. The basic HFSM in the center of Fig. 12.12 is a standard HFSM, as can be proven by showing that all its transitions are compliant with the FC-connected form. Its simple states are A and F, where A is the initial state of the HFSM and F is reachable from it. The superstates within the hierarchical structure of this HFSM are N, P, and L. Superstate N is the root of the hierarchy: there is no transition to/from the states of its parent, since it has no parent. Its substates are A, F, P, and L, where A and F are simple; however, N does not include any transitions among its simple substates.

Superstate P has four simple substates and its parent is N: the sources of the transitions exiting P are the children of P and their targets are the children of N (in this case, there is a transition directed to F); the sources of the transitions entering P are the children of N and their targets are the children of P (in this case there are two transitions coming from A).

Superstate L has three simple substates and its parent is N: the sources of the transitions exiting L are the children of L and their targets are the children of N (in this case, there are three transitions directed to A); the sources of the transitions entering L are the children of N and their targets are the children of L (in this case there is a transition coming from F).

Notice that if the basic HFSM were to include a transition from a child of P to a child of L (e.g., from (C, E) to I), the FC-connected form requirements would be violated, as a transition exiting a superstate (e.g., P) is allowed to be directed only to a child of the parent of such a superstate (that is, to a child of N, whereas I is not a child of N).

The hierarchical structure of a standard HFSM can be represented as a tree, where each intermediate node is a superstate and each leaf is a simple state.

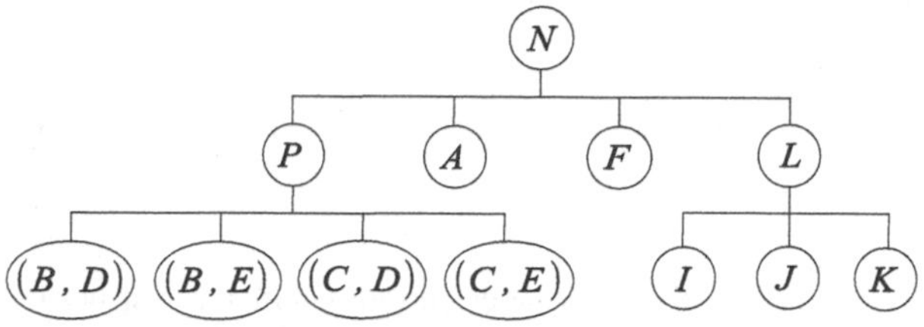

Fig. 12.13 Hierarchy of states of the HFSM in the center of Fig. 12.12

Roughly speaking, a *D-holon*[2] [65] is a possibly disconnected directed graph representing a (super)state that belongs to a standard HFSM. The set of states of the D-holon includes all the simple substates of the relevant superstate, called *internal states*. In addition, it includes the so-called *external states*, these being the simple states of immediately higher- and/or lower-level superstates that are the target of some transition(s) from an internal state. A D-holon is endowed with a set of initial states, consisting of the internal states that are the target of a transition from the simple states of immediately higher- and/or lower-level superstates. Each transition within a D-holon is marked by an event. The events marking the transitions leading to external states are called *boundary events*. Several transitions of a D-holon can be labeled by the same boundary event. An HFSM can be completely described by its D-holons and its initial state(s).

Example 12.8. The tree in Fig. 12.13 shows the hierarchical structure relevant to the standard HFSM in the center of Fig. 12.12, while Fig. 12.14 shows the D-holons associated with its superstates (N, P, and L). External states are shown in gray.

12.3.2 Diagnosis with D-Holons

The exploitation of D-holons for model-based diagnosis of DESs was proposed in [65], where it was assumed that the boundary events of any considered D-holon were observable, and an initial-state estimate at the moment the diagnosis process

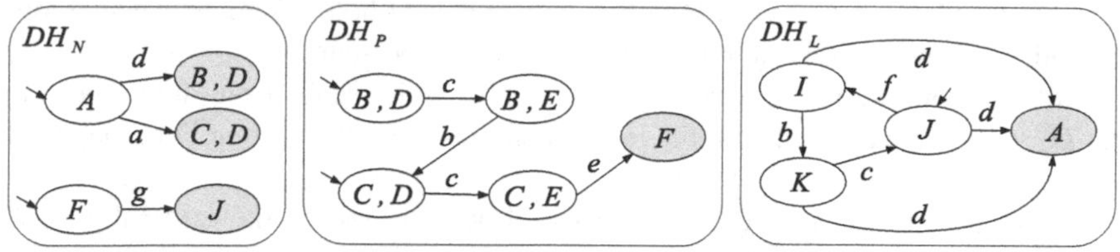

Fig. 12.14 D-holons of the HFSM in the center of Fig. 12.12

[2] The word 'holon' is borrowed from Arthur Koestler's 1967 book *The Ghost in the Machine*, where it represents a way to overcome the dichotomy between parts and wholes.

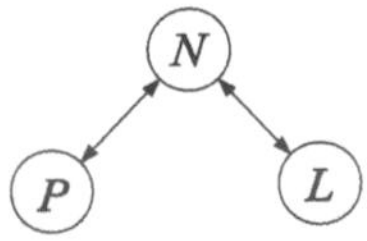

Fig. 12.15 Interactions between the D-holons in Fig. 12.14

was started was available. All the D-holons can be accommodated in a tree structure that has the same topology as the hierarchy of states, where all leaf nodes have been removed (since they represent internal states of their parents) and each intermediate state is replaced by its corresponding D-holon.

D-holons support state estimation. A D-holon is *active* if its current state estimate is not empty. At the beginning of the diagnosis process, a D-holon is active if at least one of its internal states belongs to the initial state estimate. The observation is assumed to be certain. For each observable event processed, only the currently active D-holons are used to update the state estimate. If the state estimate of a D-holon includes some external state(s), these are actually internal states of other D-holons, and then each of them has to be added to the state estimate of the D-holon it belongs to, while it has to be removed from the state estimate of the current D-holon. After this propagation has finished, each D-holon whose estimated state is not empty is active. The new set of active D-holons is used to process the next observable event, and so on.

Given a standard HFSM, in order to update the state estimate, the D-holon relevant to a superstate communicates only with the D-holon of its parent and with the D-holons of its children. In other words, a D-holon can affect only the state estimate of its parent (belonging to the next higher level in the hierarchy) and those of its children (belonging to the next lower level), while D-holons in the same level of the hierarchy have no interaction with each other. Hence, D-holons can be decoupled while working simultaneously, provided that they belong to the same level.

Example 12.9. Figure 12.15 shows the parent–child interactions between the D-holons relevant to the standard HFSM in the center of Fig. 12.12, which are the only interactions among D-holons needed.

Assume that we have to estimate the state of the DES after the sequence *acegd* has been observed. The initial state is A; hence, at the beginning, the only active D-holon is that relevant to superstate N.

At the first iteration, the observable event a is considered, which leads to the external state (C, D) of the active D-holon. Hence, this D-holon is not active anymore; instead, the new active D-holon is that relevant to superstate P.

At the second iteration, event c is considered, which leads from the initial state (C, D) of the active D-holon to the internal state (C, E); the current D-holon is still active.

At the third iteration, event e is processed, which leads to the external state F. Hence, this D-holon is not active anymore; instead, the new active D-holon is that relevant to superstate N.

At the fourth iteration, event g is considered, which leads from the initial state F of the active D-holon to the external state J; hence, the D-holon of N is deactivated and the D-holon of L is activated.

Finally, event d is processed, which leads from the (only) initial state J of the active D-holon to the external state A, which means that the D-holon of L is not active anymore, while the D-holon of N is the (only) new active D-holon. The state estimate after the given sequence has been observed includes just state A.

Notice that the D-holon activation chain is N–P–N–L–N, which is compliant with the interactions shown in Fig. 12.15.

Diagnosis can compute not just state estimates but also the sets of faults relevant to each state estimate. This requires that each D-holon be replaced with a corresponding diagnoser within the hierarchy. Since reaching an external state in a D-holon means starting from the initial state of another (parent or child) D-holon, each diagnoser relevant to a D-holon that resembles that of the diagnoser approach is self-contained and independent of the diagnosers relevant to the other D-holons.

Let the *event set* of level ℓ in the hierarchy be the set of all events labeling the transitions among the internal states within the D-holons at level ℓ and the transitions from the immediately higher or lower level that are directed to (the initial) states of the D-holons at level ℓ.

If the event sets relevant to the different levels of the hierarchy are disjoint and all the states in the initial state estimate belong to the same level, at any iteration of the diagnosis process the active D-holons/diagnosers are all located in the same level, and, since such D-holons/diagnosers cannot communicate with each other, diagnostic reasoning applicable to each active D-holon/diagnoser can be performed in parallel.

In conclusion, D-holons break up the system model into several submodels, according to a hierarchy (possibly with the help of the introduction of dummy models), thus enabling the diagnosis task to focus at any moment on the relevant submodels. This means that, at any moment, just the relevant submodels (D-holons/diagnosers) need to be stored in the central memory, thus achieving more efficient memory usage. The claim of the authors of this hierarchical diagnosis approach for DESs is that using the whole DES (flat) model in order to update the state estimate (and possibly computing candidate diagnoses) requires a large amount of memory, while the proposed approach is less expensive as far as space is concerned.

12.3.3 Diagnosis with SCL-D-Holons

An interpretation of a *diagnosis closure* [93, 110] (see Chapter 7) of the active system approach as a D-holon was offered in [169], where it was remarked that, given a DES, the hierarchical approach described in [65] does not provide any method to obtain a proper hierarchy. The authors of [169], after having suggested that diagnosis closures should be exploited as D-holons, actually exploited them in the same way as in the active system approach.

Recall that a diagnosis closure is an NFA whose initial state is the target of an observable transition in the global behavioral model of the DES considered (the *behavior space*, in the active-system-approach terminology). Such an NFA includes all the states in the global behavioral model of the DES that can be reached, starting from the initial state of the diagnosis closure, through a sequence of unobservable transitions. The states included in the diagnosis closure are interconnected with each other through all and only the unobservable transitions that interconnect them in the global behavioral model of the DES.

Every observable transition exiting a state S included in a diagnosis closure C_1 is a *leaving transition*: this is not contained in C_1; instead, it starts from S in C_1 and terminates in the initial state of a diagnosis closure C_2 (where, incidentally, C_1 may be the same diagnosis closure as C_2). Each state in a diagnosis closure is accompanied by a so-called *diagnosis attribute* [93, 110], this being a collection of sets of faults that takes into account just the evolutions of the DES within the diagnosis closure itself. Given the initial state S_0 of a diagnosis closure C, the diagnosis attribute D relevant to a state S in C includes all and only the sets of faults applicable to the paths within C leading from S_0 to S.

Basically, according to [169], every diagnosis closure can be regarded as a superstate of the standard HFSM representing the global behavior of the DES, and each superstate has its own corresponding D-holon. In the active system approach, the diagnosis closures together constitute the *monitor* of the DES. Thus, the monitor corresponds to the HFSM[3] considered and the diagnosis closures are its superstates.

In [169], the D-holon corresponding to a diagnosis closure was called an *SCL-D-holon*: it consists in the diagnosis closure considered, whose states play the role of the internal states of the SCL-D-holon, and all its leaving transitions along with the initial states of the diagnosis closures that are the targets of those leaving transitions, where those states play the role of the external states of the SCL-D-holon. Each state in an SCL-D-holon, external states included, is accompanied by its diagnosis attribute. The only (trivial) difference between diagnosis closures and SCL-D-holons is that, in their pictorial representations, the former are explicitly interconnected with each other while the latter are implicitly interconnected (in fact, they are depicted as separated from each other).

Example 12.10. Figure12.16 depicts a portion of a monitor, adapted from [110], encompassing ten diagnosis closures relevant to a DES whose behavioral model is not shown here. Transitions identifiers are omitted. Each (observable) transition between diagnosis closures is marked by the relevant observable event and, if that transition is faulty (as, in the active system approach, faulty transitions are not necessarily unobservable), by the fault itself (otherwise, if the transition is normal, it is marked by ε). In Fig. 12.16, this is the case for the transition (whose relevant fault is **fsh**) from diagnosis closure N_0 to diagnosis closure N_1.

[3] The monitor in the active system approach is not an HFSM. However, it is straightforward to transform it into an HFSM having just the root layer and the first level of the hierarchy.

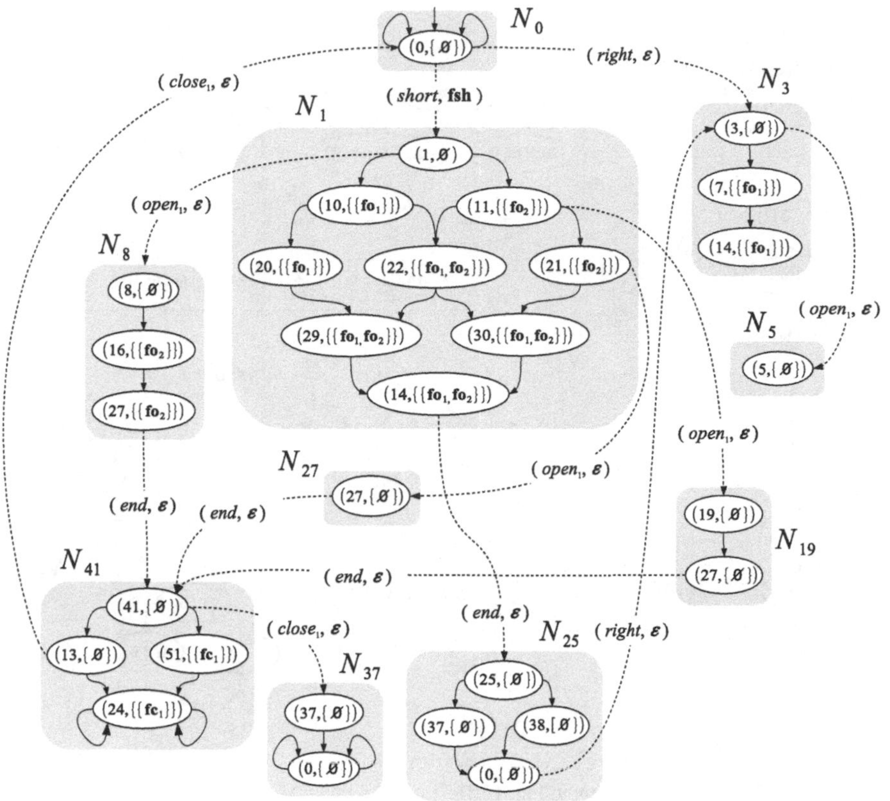

Fig. 12.16 Portion of a monitor

The ten SCL-D-holons corresponding to the diagnosis closures in Fig. 12.16 are displayed in Fig. 12.17. Notice that each SCL-D-holon has been assigned the same identifier as the corresponding diagnosis closure. External states are shaded.

The diagnosis attributes contained in the SCL-D-holons are needed to compute candidate sets of faults. Thus, the SCL-D-holons generated offline can be processed online not only in order to update the state estimate but also to compute the candidate sets of faults.

In [110], assuming that a certain observation (a plain observation) is received (this being a sequence of observable events), the set of candidate diagnoses is computed by updating the current monitoring state upon the reception of each event. A monitoring state is a set of diagnosis closures, all the diagnosis closures that can be reached starting from the initial one, given the observation received so far.

If SCL-D-holons are considered, the current monitoring state of the active system approach corresponds to the set of active SCL-D-holons. Let $\Delta_0 = \{\emptyset\}$ be the initial diagnosis, that is, it is assumed that the DES is initially free of faults, and let M_0 be the given initial set of active SCL-D-holons. Let M_{i-1} and Δ_{i-1} be the set of active

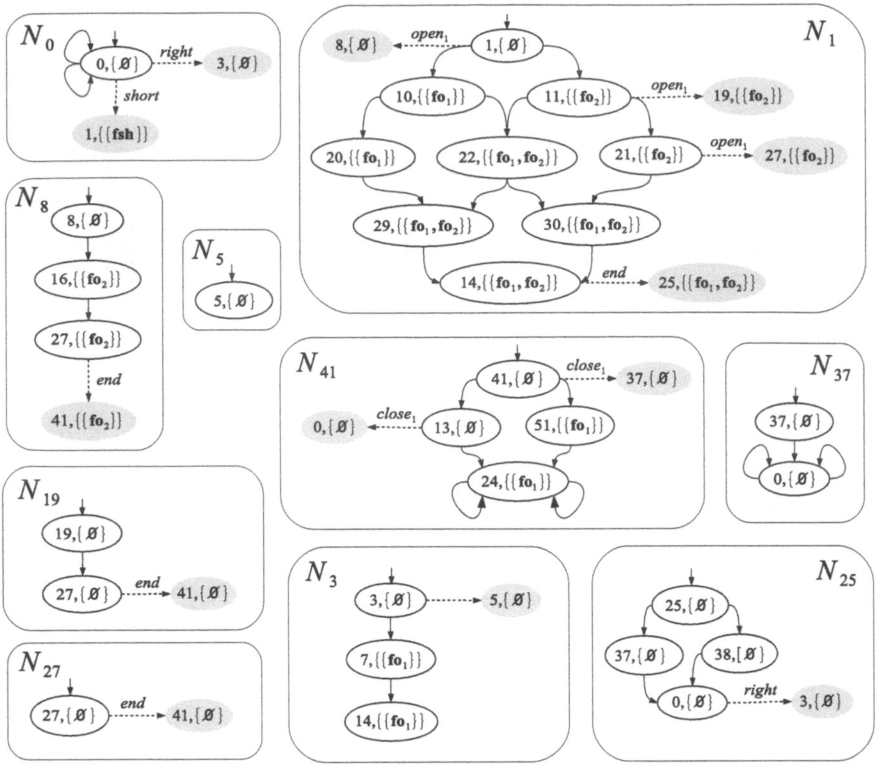

Fig. 12.17 SCL-D-holons relevant to Fig. 12.16

SCL-D-holons and the diagnosis after the $(i-1)$-th event (with $i \geq 1$) has been observed. Once the i-th event, denoted e_i, has been perceived, the new diagnosis Δ_i is obtained as follows:

$$\Delta_i = \bigcup_{N \in M_{i-1}} \left(\Delta^{\text{ext}}(N, e_i) \bowtie \Delta_{i-1}\right), \tag{12.1}$$

where $\Delta^{\text{ext}}(N, e_i)$ is the union of all the diagnosis attributes relevant to the external states in SCL-D-holon N whose entering (observable) transitions are marked by the event e_i.[4]

After the i-th event has been observed, the new set of active SCL-D-holons is

$$M_i = \{N_k \mid S_0(N_k) \in S^{\text{ext}}(N, e_i), N \in M_{i-1}\}, \tag{12.2}$$

where $S_0(N_k)$ is the initial state of SCL-D-holon N_k, and $S^{\text{ext}}(N, e_i)$ is the set of external states in SCL-D-holon N whose entering (observable) transitions are marked by the event e_i.

[4] The operator $\bowtie$ is the *diagnosis join* of two sets of diagnoses, introduced in Def. 7.10.

Example 12.11. With reference to the DES relevant to the SCL-D-holons shown in Fig. 12.17, consider the evolution, with initial state 0, relevant to the sequence of observable events $[short, open_1, end]$. As remarked, the DES is assumed to be initially free of faults; that is, $\Delta_0 = \{\emptyset\}$ is the initial diagnosis, while $M_0 = \{N_0\}$ is the initial set of active SCL-D-holons, since N_0 is the SCL-D-holon with initial state 0.

After the reception of the first observable event $e_1 = short$, the new diagnosis is $\Delta_1 = \Delta^{\text{ext}}(N_0, short) \bowtie \Delta_0 = \{\{fsh\}\} \bowtie \{\emptyset\} = \{\{fsh\}\}$. The new set of active SCL-D-holons becomes $M_1 = \{N_1\}$, as in N_0, which is the only SCL-D-holon in M_0, the only external state that is the target of an observable transition marked by *short* is 1, which is the initial state of SCL-D-holon N_1.

Once $e_2 = open_1$ has been observed, the new diagnosis is $\Delta_2 = \Delta^{\text{ext}}(N_1, open_1) \bowtie \Delta_1 = \{\emptyset, \{fo_2\}\} \bowtie \{\{fsh\}\} = \{\{fsh\}, \{fo_2, fsh\}\}$. The set of active SCL-D-holons becomes $M_2 = \{N_8, N_{19}, N_{27}\}$.

After $e_3 = end$ has been received, the diagnosis is updated as follows: $\Delta_3 = (\Delta^{\text{ext}}(N_8, end) \bowtie \Delta_2) \cup (\Delta^{\text{ext}}(N_{19}, end) \bowtie \Delta_2) \cup (\Delta^{\text{ext}}(N_{27}, end) \bowtie \Delta_2) = (\{\{fo_2\}\} \bowtie \{\{fsh\}, \{fo_2, fsh\}\}) \cup (\{\emptyset\} \bowtie \{\{fsh\}, \{fo_2, fsh\}\}) \cup (\{\emptyset\} \bowtie \{\{fsh\}, \{fo_2, fsh\}\}) = \{\{fsh\}, \{fo_2, fsh\}\}$. The new set of active SCL-D-holons is the singleton $M_3 = \{N_{41}\}$, as N_{41} is the external state that is the target of a transition marked by the observable event *end* for all the three SCL-D-holons in M_2.

In [110], a method to compute diagnoses based on the diagnosis closures in the monitor when the observation is uncertain was provided. Such a method can be adopted when SCL-D-holons replace diagnosis closures as well.

12.4 Successive Window Approach

The successive window approach [170] addresses the problem of constructing the emission sequences (also called emission orders) of observed events in such a way as to obtain sound, complete, and monotonic results in monitoring-based incremental diagnosis when the observation is temporally uncertain. In order for us to have a better grasp of the method, the notions of incremental diagnosis and monotonicity of monitoring results have first to be defined.

12.4.1 Incremental Diagnosis

By *incremental diagnosis* of DESs we mean any iterative reasoning process that, at each iteration i, does not compute from scratch the set of candidate diagnoses Δ_i relevant to the whole interval from the start of the diagnosis session up to time t_i; instead, Δ_i is obtained by updating the set of candidate diagnoses Δ_{i-1} relevant to the time interval up to time t_{i-1} (where $t_i > t_{i-1}$), computed at the previous iteration. In order to transform Δ_{i-1} into Δ_i, the incremental process takes into account the chunk of observation relevant to the time interval between t_{i-1} and t_i, where such an interval is called a *temporal window*.

Incremental diagnosis can support both diagnosis during monitoring [110, 131, 132] and a posteriori diagnosis [53, 54]. All the contributions quoted hereafter can compute sound results. However, in order to achieve additional properties of the results, they need some conditions on the observation to be fulfilled. Specifically, the reasoning methods described in [53, 54, 131, 132] yield completeness of the set of diagnosis candidates only if the observation chunk considered at each incremental step falls into a *sound temporal window*.

A temporal window is sound if every observable event emitted by the DES within the window is received within the window itself. Notice that the reception order of events (which is the only such order that is known) may not equal the emission order (which is unknown); hence, all the emission orders that are compliant with the given reception order have to be found.

The construction of emission orders can be constrained by domain knowledge (for instance, the relative reception order of all events transmitted on the same channel mirrors their emission order), or, if no (known) domain constraints are available, all permutations of the received events have to be considered. The set of all the global states in which the system could be after having produced the observation before a given temporal window is a belief state. If the windows are sound, any update of the evolutions relevant to the current window to explain the chunk of observation in a new window is based only on the states in the final belief state of the current window. Notice that it may be impossible to find an explanation for the new chunk of observation from any given final state relevant to the current diagnosis: all trajectories that have no future in the new window have to be eliminated.

As described above, the generation of all possible emission orders of observed events is simple if the temporal windows are sound. In the general case, however, there are delays between the emission of the observable events and their reception. Consequently, some observable events emitted during a temporal window may not be received within the same window. Moreover, at the end of the window, there is no guarantee that the received observation will be enough to make a diagnosis. In fact, some observable events may not have been received yet even though they were emitted before other received events. The choice of the length of the temporal window is, therefore, fundamental.

A sound temporal window is delimited by a pair of sound breakpoints, where a breakpoint t_i is sound if all observable events received before it were generated before it. Assume that an event emitted by the DES takes at most $\tau_{\max}$, the *maximum transmission delay*, before being perceived. If an observable event has been received at time t and no other event has been received since then, any time instant starting from $t + \tau_{\max}$ can be chosen as a sound breakpoint. However, if this condition never holds, we are not able to split the flow of received observable events into sound windows. In the worst case, the only sound window we can single out equals the whole observation, which results in a posteriori diagnosis, not in online diagnosis.

While slicing the whole observation can be done in a posteriori diagnosis, framing temporal windows online remains an open problem for monitoring-based diagnosis [102]. In addition, in [102] it was remarked that, if the observation is temporally uncertain and the set of candidates is updated based on an observation where some observable events that occurred before the ones we are processing are possibly missing since we have not received them yet, the criterion of soundness and completeness of monitoring-based diagnosis results is not meaningful; instead, the *monotonicity* property should be considered. The results are monotonic if every candidate (set of faults) in Δ_i either equals or is a superset of a candidate in Δ_{i-1}.

In [110] it was shown that a sufficient (not necessary) condition to obtain monotonic diagnostic results is *observation stratification*. An observation is stratified if it is a sequence of observation strata, where a stratum is a set of observation nodes (according to the terminology of the active system approach relevant to uncertain observations) such that all the nodes that are temporally unrelated with respect to the nodes in the stratum considered are in the same stratum themselves. If the chunk of observation processed at each iteration is an observation stratum, monotonicity of results is guaranteed. However, no method to isolate observation strata was provided. Moreover, in the worst case, the whole observation consists of just one stratum, which once again results in a posteriori diagnosis instead of online diagnosis.

The observation-slicing technique proposed in [53, 54] achieves monotonicity. However, such a technique can be applied to a posteriori diagnosis only, when the whole observation is known.

12.4.2 Reconstructing Emission Sequences

The successive window approach [170] has general validity and is particularly suitable when the received observable events are too dense to find sound windows or observation strata. The rationale is that, given a sequence of observable events received over two successive temporal windows, all the emission sequences of the events generated within the first temporal window can be produced (and exploited for computing a diagnosis, this being the set of all trajectories compliant with the DES model that are consistent with the observation considered) if the second window covers a time that is longer than the maximum transmission delay.

Therefore, if all the windows (possibly excluding the first one) fulfill this constraint, the emission sequences over the whole span of the monitoring process can be inferred as well. Hence, no reasoning is needed in order to split the flow of incoming observed events into windows; instead, this approach considers windows of fixed length, where this length is greater than the maximum transmission delay.

Let d_{max} be a length of time greater than the maximum transmission delay, that is, $d_{max} > \tau_{max}$. Two successive temporal windows, W_{i-1} and W_i, are *restricted* if the size of W_i is d_{max}. A sequence of temporal windows is *restricted* if every two successive windows in it are restricted. This approach assumes that, from the second window on, the size of each temporal window is equal to d_{max}; hence, if the first temporal window is denoted by W_1, the sequence of windows from W_2 on is restricted.

At each iteration of the incremental diagnosis process, a pair of restricted successive temporal windows is taken into account. In the i-th iteration, the pair of restricted windows W_iW_{i+1} is processed, and hence each window W_i is processed twice: as the second window of the pair $W_{i-1}W_i$ considered at the $(i-1)$-th iteration, and as the first one of the pair W_iW_{i+1} considered at the i-th iteration. The reasoning performed in this approach is aimed at producing all possible emission sequences of observable events by reasoning on the received sequences and taking into account domain constraints. When the pair W_iW_{i+1} is processed, only the emission order of the events emitted up to window W_i (inclusive) is taken into account.

A pair of successive temporal windows in a restricted sequence is characterized by some interesting properties:

(a) All of the events emitted in window W_{i-1} are necessarily received either in window W_{i-1} or in window W_i; hence the set of all the events received in the pair of restricted successive temporal windows $W_{i-1}W_i$ includes all the events emitted in window W_{i-1}.

(b) The emission of an event received in window W_{i-1} does not necessarily occur earlier than the emission of an event received in window W_i; hence the events received in window W_i can affect the emission order of the events received in window W_{i-1}.

(c) Each event received in window W_i was emitted either in window W_i or in W_{i-1}, if the latter window exists. Since W_1 has no previous window, all the events received in W_1 were emitted in the same window.

(d) The events received in window W_i can affect the creation of the emission sequences relevant to events received in window W_{i-1}. However, they cannot disorder any sequence relevant to W_{i-2} or earlier windows.

The approach distinguishes between the *global emission sequences*, which, at the i-th iteration, are relevant to the events emitted in the whole interval from the initial time t_0 to window W_i, and the *local emission sequences*, which are relevant to the emission order of all the events received in a pair of successive windows (W_iW_{i+1} at the i-th iteration).

The invariant feature of the incremental diagnosis process is that, when the i-th iteration starts, the set VO_{i-1} of all *valid* (i.e., consistent with the DES model) global sequences of observable events emitted from the initial time t_0 up to window W_{i-1} (inclusive) is available. At the first iteration, VO_0 is empty, since no emission sequence has been constructed yet. Notice that checking the consistency of the reconstructed sequences with the DES model ensures soundness of the results, while extending previous solutions ensures monotonicity.

The first pair of windows, W_1 and W_2, needs special treatment: the initial valid sequences are all and only the permutations of all the events received in window W_1 that fulfill the domain constraints and are consistent with the DES model.

In the i-th iteration, the pair of restricted windows W_iW_{i+1} is processed: first we find all the local emission sequences including all and only the events received in windows W_i and W_{i+1} that are consistent with the domain constraints and that can have been emitted in W_i. Let $o_{i,i-1}$ denote a generic local emission sequence relevant to W_iW_{i+1}.

At each iteration, the valid sequences built in the previous iterations are extended, and those of them that are still valid after the extension become the current set of candidates. If an extended sequence cannot be synchronized with the DES model, it is not valid, and hence it is discarded and is not considered in the next iterations. Each valid sequence for the previous iteration is extended by appending to it a subsequence, called a *valid path*. A key problem solved by the successive window approach is determining the valid start and end events of such a subsequence, given a pair of successive windows.

Within each $o_{i,i-1}$, we isolate a valid path $po_{i,i-1}$ with which to extend some valid global sequence(s) in VO_{i-1}. Such a path represents the emission sequence of all the events received in the pair of windows W_iW_{i+1} that were emitted after those already encompassed in the valid global sequences in VO_{i-1} to be extended, where such events are assumed to have been emitted in window W_i. The sequence $o_{i,i-1}$ can be regarded as the concatenation of three subsequences: the one preceding the valid path, say $\textit{before\_po}_{i,i-1}$, the valid path $po_{i,i-1}$, and the one following the valid path, say $\textit{after\_po}_{i,i-1}$.

Within the same local emission sequence $o_{i,i-1}$, several valid paths can be isolated. The end event of all such valid paths is the last (i.e., the rightmost) event in $o_{i,i-1}$ that was received in window W_i, which means that in $\textit{after\_po}_{i,i-1}$ there is no event received in W_i. A valid path implicitly assumes that its end event, besides having been received in W_i, was also emitted in W_i. However, a valid path can include events that were received in W_{i+1}, since these could have been emitted in W_i.

Hence, the method is complete if all the local sequences that intermix the events received in W_{i+1} with the events received in W_i are tried. The valid start event of a valid path depends on the previous valid sequence $s_{i-1} \in VO_{i-1}$ to be extended. In fact, no event in s_{i-1} has to belong to $po_{i,i-1}$, while all the events in $\textit{before\_po}_{i,i-1}$ have to fall in s_{i-1} and their relative order in $\textit{before\_po}_{i,i-1}$ has to be the same as in s_{i-1} (the *successive order consistency* constraint).

Once $po_{i,i-1}$ has been singled out, the new global sequence is $s_i = s_{i-1}.po_{i-1,i}$, where "." is the concatenation operator. Each global sequence s_i is endowed with a "trailer" sequence, this being $\textit{after\_po}_{i,i-1}$. The global sequence s_i assumes that the events in $\textit{after\_po}_{i,i-1}$ were emitted in W_{i+1} after all the events in s_i (which includes only events emitted up to window W_i inclusive) and that their relative order is the one exhibited in $\textit{after\_po}_{i,i-1}$. Hence, in the next iteration, s_i can in turn be extended only by appending a valid path that includes all the events in $\textit{after\_po}_{i,i-1}$, with the same relative order (the *relative order consistency* constraint).

Example 12.12. Let $o_1o_2 \ldots o_{16}$ be the incoming flow of observable events received during an online incremental diagnosis. Given $d_{\max}$, the flow is split online into temporal windows, where the length of every window is $d_{\max}$. Suppose that five windows, $W_1 \ldots W_5$, are progressively obtained, and that the five corresponding sequences are $o_1o_2o_3$, $o_4o_5o_6o_7$, $o_8o_9o_{10}$, $o_{11}o_{12}o_{13}o_{14}$, and $o_{15}o_{16}$, respectively.

At the initial iteration, the first pair of windows, that is, W_1W_2, has to be considered. The events received in those windows are all those from o_1 to o_7. Assume that there are two local sequences of all such events that are compliant with the domain constraints, these being

$$\underline{o_1}\, o_4 o_3 o_5 \underline{\underline{o_2}}\, o_6 o_7, \tag{12.3}$$

$$\underline{o_2}\, o_4 o_1 o_5 \underline{\underline{o_3}}\, o_7 o_6. \tag{12.4}$$

The first event in each sequence, which is necessarily both emitted and received in W_1, is underlined since, by construction, it is the valid start event of each valid path. The end event of each valid path is the last event that could have been both emitted and received in W_1, and is double underlined. This is necessarily the rightmost event in the local sequence that was received in W_1. Hence, we create two global sequences: $s_1 = o_1o_4o_3o_5o_2$, whose trailer is the sequence o_6o_7, and $s'_1 = o_2o_4o_1o_5o_3$, whose trailer is the sequence o_7o_6. Suppose that both of them are valid, since they can be synchronized with the DES model. Hence, $VO_1 = \{s_1, s'_1\}$.

At the second iteration, the pair of restricted temporal windows W_2W_3 is considered; this encompasses all events from o_4 to o_{10}. Assume that there is just one local sequence of all such events that is compliant with the domain constraints, this being

$$o_4 o_5 \underline{o_7}\, \underline{\underline{o_6}}\, o_8 o_9 o_{10}\ . \tag{12.5}$$

The last event of each valid path is easy to single out, since it is the rightmost event in the sequence (12.5) received in W_2. This event, which is independent of the global sequence that has to be extended, is double underlined.

Now we have to check whether each global sequence in VO_1 can be extended based on (12.5). Consider $s_1 = o_1o_4o_3o_5o_2$ first, whose trailer is o_6o_7. Since the order of the events in the sequence (12.5) is not the same as in the trailer of the global sequence, there is no valid path in (12.5) that can extend s_1.

We now consider the second global sequence of the previous step, namely $s'_1 = o_2o_4o_1o_5o_3$, whose trailer is the sequence o_7o_6, in order to single out (the start event of) a valid path in (12.5) so as to extend it. The events in the trailer of the global sequence of the previous step have to belong to the valid path singled out in the current step, and their order has to be the same as in the trailer: in this case, these conditions are fulfilled. The first two events of the local sequence (12.5), namely o_4o_5, cannot belong to the valid path, since they already occur in the global sequence s'_1.

We check whether the relative order of o_4 and o_5 in the sequence (12.5) is the same as in s'_1: this constraint is also fulfilled. Hence, the valid start of the valid path is o_7, that is, the valid path is o_7o_6 and its trailer is $o_8o_9o_{10}$. By extending s'_1 with this valid path we obtain the new global sequence $s_2 = o_2o_4o_1o_5o_3o_7o_6$, whose trailer is $o_8o_9o_{10}$. Assume that s_2 is valid, as it can be synchronized with the DES model. Hence, $VO_2 = \{s_2\}$.

At the third iteration, the pair of restricted temporal windows W_3W_4 is considered; this encompasses all events from o_8 to o_{14}. Assume that there are two local

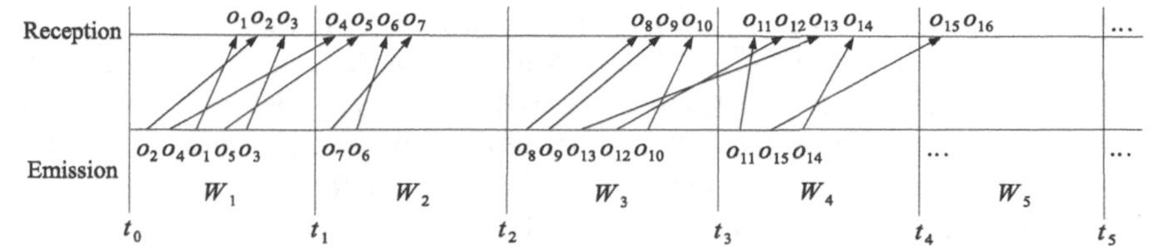

Fig. 12.18 Observable events: global emission sequence vs. reception sequence

sequences of all such events that are compliant with the domain constraints, these being

$$\underline{o_8}o_9o_{13}o_{12}\underline{\underline{o_{10}}}o_{11}o_{14}, \quad (12.6)$$

$$\underline{o_9}o_{11}o_{10}\underline{\underline{o_8}}o_{12}o_{13}o_{14}\,. \quad (12.7)$$

The end event of each valid path is easy to single out, since it is the rightmost event of each local sequence that was received in W_3. Such end events are double underlined in both of the local sequences above.

We now consider the only global valid sequence, $s_2 = o_2o_4o_1o_5o_3o_7o_6$, whose trailer is $o_8o_9o_{10}$, in order to check whether it can be extended based on the two local sequences found in the current iteration. Consider the local sequence (12.6) first. The relative order of the events in the trailer of s_2 is the same as in this local sequence, and hence the relative order consistency constraint is fulfilled. Since s_2 does not share any event with the local sequence, the successive order consistency constraint is trivially fulfilled and the *before_po*$_{34}$ subsequence of the local sequence is necessarily empty. Hence, the valid start event of the valid path is the first event of the local sequence; that is, the valid path is $o_8o_9o_{13}o_{12}o_{10}$ and its trailer is $o_{11}o_{14}$. By extending s_2 with this path we obtain the global sequence $s_3 = o_2o_4o_1o_5o_3o_7o_6o_8o_9o_{13}o_{12}o_{10}$, whose trailer is $o_{11}o_{14}$. We assume that s_3 is valid, since it can be synchronized with the DES model.

Now we consider the second local sequence, that is, (12.7). However, the relative order of the events in the trailer of s_2, which is $o_8o_9o_{10}$, is not the same as in (12.7), and hence this local sequence does not include any valid path. At the end of the third iteration, the set of valid global sequences is therefore $VO_3 = \{s_3\}$.

The last iteration takes into account the pair of restricted temporal windows W_4W_5, which encompasses all events from o_{11} to o_{16}. Assume that there is just one local sequence of all such events that is compliant with the domain constraints, this being

$$o_{13}o_{12}\underline{o_{11}}o_{15}\underline{\underline{o_{14}}}o_{16}\,. \quad (12.8)$$

We have to check how to extend the only valid global sequence in VO_3, s_3, based on (12.8). It is easy to see that both of the consistency constraints are fulfilled and that the start and end events of the valid path are those underlined and double underlined, respectively. The global sequence obtained by concatenating s_3 with this valid path is $s_4 = o_2o_4o_1o_5o_3o_7o_6o_8o_9o_{13}o_{12}o_{10}o_{11}o_{15}o_{14}$, and its trailer is o_{16}. We assume that s_4 is synchronizable with the DES model. Hence, at the end of the fourth iteration, the set of valid global sequences is $VO_4 = \{s_4\}$. The only valid global emission sequence is displayed in Fig. 12.18, along with the received sequence. The arrows show how the emission order has transformed into the reception order.

Notice that event o_{16} was necessarily emitted in window W_5.

12.5 Incomplete Model Approach

Although model-based diagnosis of DESs usually assumes the availability of a complete behavioral model of the system to be diagnosed, namely a model that encompasses all normal and faulty behaviors, completeness of such a model may be difficult (and costly) to achieve owing to the inherent complexity of the system and the consequent approximations and abstractions. Thus, the model may lack some behaviors (i.e., events and/or states in the automata representing the behavior of the components), which may lead to missing or incorrect diagnoses. In fact, some real behaviors cannot be explained by an incomplete model: in such a case, no candidate diagnosis at all is produced.

One contribution in the literature [168] is concerned with this topic. This paper defines a diagnosis as the set of all trajectories compliant with a complete behavioral model of a DES that are consistent with the (possibly uncertain) observation considered. An (uncertain) observation is represented by a so-called *observation automaton*, where any trajectory from the initial state to a final state is a sequence of observable events that may be the real sequence emitted by the DES. The diagnosis is the synchronization of the automaton representing the complete behavior of the DES with the observation automaton.

If a complete behavioral model is not available, the proposal is to adopt a new notion of diagnosis, called *P-diagnosis*, based on the new operation of *P-synchronization* (instead of synchronization) of the incomplete system model and the (complete) observation. Each P-synchronized state is a pair of states, the first relevant to the incomplete behavioral-model automaton, and the second to the observation automaton (whose transitions are all marked by observable events), and the initial state of the P-synchronized automaton is the pair of initial states of these automata.

The automaton resulting from P-synchronization is much larger than that produced by synchronization since, given a P-synchronized state $S = (s_b, s_o)$, there is an unobservable transition exiting S for each unobservable transition exiting s_b in

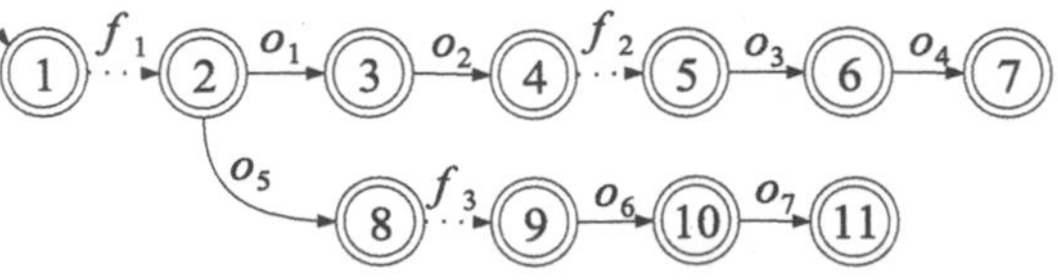

Fig. 12.19 Complete DES model

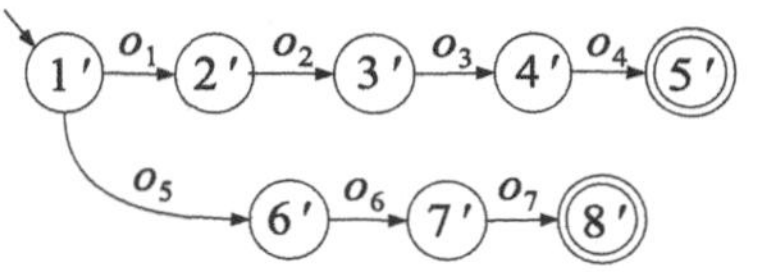

Fig. 12.20 Automaton-based uncertain observation

the behavioral model. Likewise, there is a transition exiting S for each (observable) transition exiting s_o (in the observation automaton) that is marked by an event that is not shared by any transition exiting s_b. In addition, there is an observable *synchronized* transition exiting S for each pair of observable transitions (t_b, t_o), the former exiting s_b, and the latter exiting s_o, that are marked by the same observable event.

Example 12.13. Let the automaton B shown in Fig. 12.19 be the complete behavioral model of a DES, and the automaton O shown in Fig. 12.20 be an uncertain observation relevant to it. Each state in B is a final one since a trajectory can stop in any state (in other words, the language of all the events in the DES is prefix-closed). In contrast, there are just two final states in the observation automaton O.

The observation is uncertain, since the real sequence of observable events is either $[o_1, o_2, o_3, o_4]$ or $[o_5, o_6, o_7]$, but the observer cannot single out which of them.

The automaton resulting from the synchronous composition of B and O is depicted in Fig. 12.21, where the initial state is the composition of the initial states of B and O, while each final state is the composition of a pair of final states, the first in B, and the second in O. Each trajectory from the initial state to a final state is a candidate (hence there are two candidates).

Let the automaton B' shown in Fig. 12.22 be an incomplete behavioral model relevant to the same DES (notably, states 3 and 8 are missing, along with their entering transitions). The automaton resulting from the P-synchronization of B' and O is shown in Fig. 12.23, where the only synchronized transitions are those shown by solid lines.

One can appreciate that all the candidates in the synchronized automaton relevant to the complete model are also included in the P-synchronized automaton relevant to the incomplete model. However, the P-synchronized automaton includes a large number of spurious candidates as well.

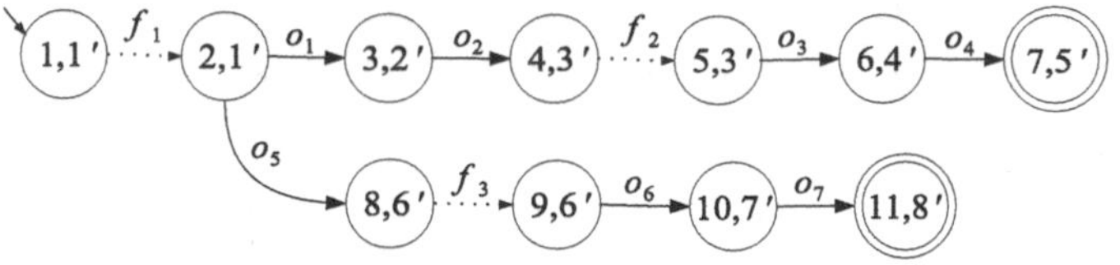

Fig. 12.21 Synchronized automaton

As highlighted by the above example, a P-synchronized automaton may include spurious candidates. A parameter P, taking a value between 0 and 1, is used for pruning

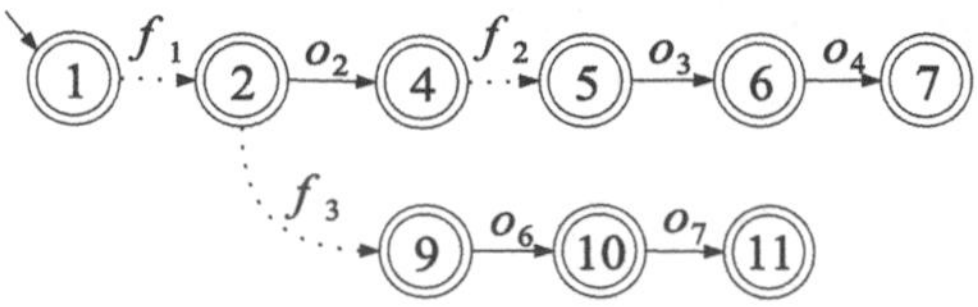

Fig. 12.22 Incomplete DES model

the trajectories in a P-synchronized automaton, based on their so-called *synchronization degree*. Given a trajectory in a P-synchronized automaton, the synchronization degree of that trajectory is the ratio of the number of (observable) synchronized transitions it includes to the number of all observable transitions it includes.

As such, the synchronization degree of a trajectory in a P-synchronized automaton is the matching ratio between a possibly incomplete trajectory of the incomplete system model and a trajectory of observable events in the observation (where both of the trajectories considered have to end in a final state).

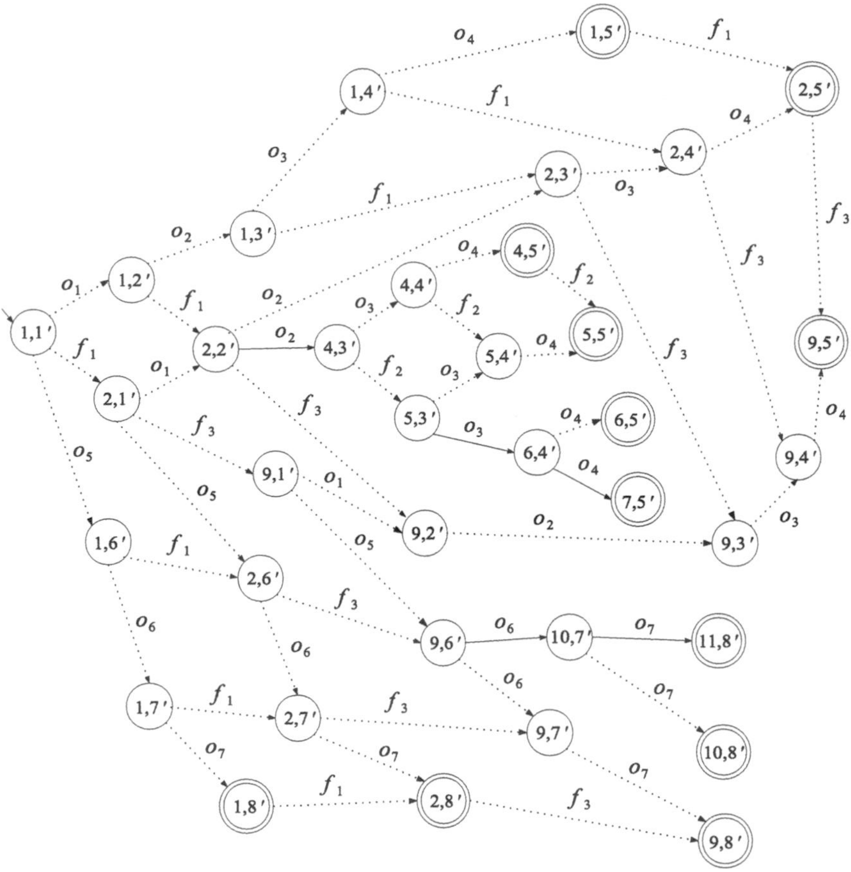

Fig. 12.23 P-synchronized automaton ($P = 0$)

If P is set to 0, no pruning is done; in this case, the diagnosis is complete (no candidate is missing) but it is not sound, since some candidates may be spurious. If $P = 1$, all the trajectories in the P-synchronized automaton whose synchronization degree is less than 1 are pruned, which means that all the trajectories that include an observable nonsynchronized transition are pruned; hence, a P-synchronized product with $P = 1$ is the same as a synchronous product. In this case, all candidates are sound; however, since the DES is incomplete, some candidates may be missing, that is, diagnosis with $P = 1$ is sound but not complete.

Example 12.14. Consider the P-synchronized automaton of the previous example, displayed in Fig. 12.23, where $P = 0$. Such an automaton can be pruned by setting the parameter P to a value greater than 0. The automaton resulting after pruning with $P = 2/3$ (that is, after pruning all the trajectories whose synchronization degree is less than 2/3) is shown in Fig. 12.24. It is easy to see that, in this automaton, the synchronization degree of every trajectory ending in state $(7,5')$ is 3/4 (which is greater than 2/3) since it includes three (observable) synchronized transitions (marked by o_2, o_3, and o_4) out of a total of four observable transitions (since a transition marked by o_1 is also included). The synchronization degree of every trajectory ending in state $(11,8')$ is, instead, 2/3 since it includes two (observable) synchronized transitions (marked by o_6, and o_7) out of a total of three observable transitions (the additional one being marked by o_5).

Moreover, an automaton resulting from P-synchronization and pruning based on the value of P can be heuristically pruned further by also removing every trajectory τ in that automaton whose synchronization degree is less than the *completeness degree* of the trajectory τ' obtained by projecting τ on the incomplete system model.

The completeness degree of a trajectory τ' in the incomplete system model is intended to be the matching ratio between that trajectory and its corresponding trajectories in the complete system model. In fact, τ' can be an abstraction of several trajectories, say the set of trajectories $trajset(\tau')$, in the complete system model, each of which has its own number of observable transitions.

The completeness degree of τ' is the minimum value of the ratio of the number of observable transitions in τ' to the number of observable transitions in a trajectory in $trajset(\tau')$. However, $trajset(\tau')$ is unknown since the complete model is not given; hence the completeness degree of τ' is a guess, an approximation whose value is chosen based on experience and domain knowledge. The value of the completeness degree of a trajectory in an incomplete DES model ranges from 0 to 1.

Example 12.15. Consider the complete system model shown in Fig. 12.19 and the incomplete system model relevant to it shown in Fig. 12.22. If the complete model is known, the completeness degree of any trajectory of the incomplete model can be computed exactly. The trajectory in the automaton in Fig. 12.22 ending in state 7 has a completeness degree equal to 3/4, since it encompasses three observable transitions (marked by o_2, o_3, and o_4) while the only corresponding trajectory in the complete model in Fig. 12.19 (the one ending in state 7) encompasses four observable transitions.

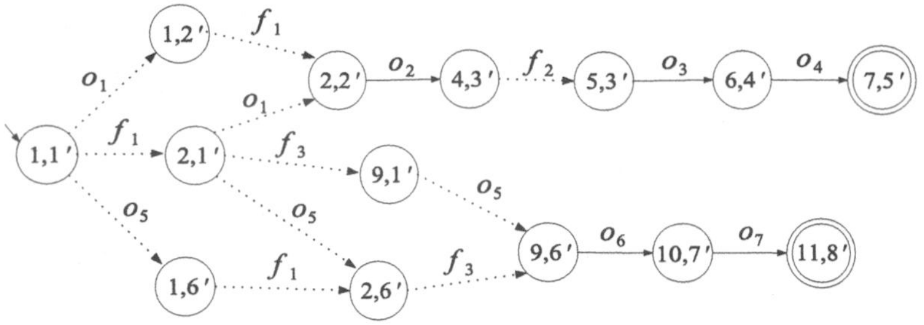

Fig. 12.24 P-synchronization with $P = 2/3$

The trajectory in the automaton in Fig. 12.22 ending in state 11 has, instead, a completeness degree equal to 2/3, since it encompasses two observable transitions (marked by o_6, and o_7) while the only corresponding trajectory in the complete model in Fig. 12.19 (the one ending in state 11) encompasses three observable transitions. However, these values are unknown when the complete model is unknown.

Consider the automaton resulting from P-synchronization and pruning discussed in Example 12.14 (with $P = 2/3$). As explained, both trajectories ending in $(7,5')$ have a synchronization degree equal to 3/4. Moreover, the projection of these trajectories on the incomplete DES model is the only trajectory ending in state 7 in Fig. 12.22, whose completeness degree (as computed above) is 3/4. Since the value of the synchronization degree is the same as that of the completeness degree, no further pruning has to be done as far as the trajectories ending in $(7,5')$ are concerned. The same applies to the trajectories in Fig. 12.24 ending in $(11,8')$: their synchronization degree is 2/3 (see Example 12.14) and their relevant completeness degree is 2/3 (see above). Hence, no further pruning of the automaton in Fig. 12.24 is carried out based on the heuristic criterion presented.

The rationale behind the heuristic criterion of pruning all the trajectories in the P-synchronized automaton whose synchronization degree is less than the completeness degree is intuitive: given a trajectory τ in the P-synchronization, its completeness degree is a (pessimistic) estimate of the fraction of observable transitions that are not missing, that is, that actually belong to the model. If the synchronization degree, which is the fraction of observable events that match with the observation, is less than the completeness degree, this means that not all of the actual observable transitions have been synchronized with the observation, and hence τ is meaningless.

12.6 Hypothesis Space Approach

Based on all the approaches to model-based diagnosis of DESs mentioned so far, one might think that, in order to solve an a posteriori diagnosis problem completely online, that is, in order to compute all the candidates (or refined candidates; see

Def. 1.2 in Chapter 1) at the desired abstraction level (where abstraction levels were introduced in Sec. 1.4.2), first all the trajectories in the behavior space of the DES that are consistent with the given observation have to be computed, and then a candidate has to be drawn from each of them.

Likewise, if only refined candidates are needed, first all the trajectories that both are consistent with the given observation and end with an observable event have to be computed, and then a refined candidate has to be drawn from each of them. This is the classical way of performing the diagnosis task.

Recent research [56, 57] has reversed this perspective: first a candidate is assumed as a hypothesis, and then it is checked whether that hypothesis is relevant to a trajectory that is consistent with the given observation. Each abstraction level corresponds to a different domain of all the hypotheses we are interested in. Such a domain, which is a set, is called a *hypothesis space*, and its elements are called *hypotheses*.

The extension of the theory of the hypothesis space to refined candidates and refined diagnosis is straightforward, although the authors of the hypothesis space approach [56, 57] do not deal with it.

12.6.1 Hypothesis Space

By definition [56], a hypothesis space H is the domain of all behavior types of a DES at a certain abstraction level. Using the same formalization as in Sec. 1.4.1, each trajectory $u \in L$, which is a behavior type in L, is associated by a *mapping* function δ with a single hypothesis $\delta(u)$ in H. The hypothesis space may be Σ^*, in which case $\delta(u) = u$: this is the least abstract hypothesis space, denoted here by H_{id}, for its mapping is the identity function. The space H_{id} is not finite, since a sequence of events may include any number of iterations of any (cyclic) sequence of events.

At the other end of the spectrum, the most abstract (and the smallest) hypothesis space may simply have two elements, *normal* and *abnormal*, that is, we are only interested in fault detection: we call this the *detection space*, denoted H_{norm}.

More common hypothesis spaces are the so-called *set space*, H_{set}, which considers each set of faults that may have occurred as a distinct hypothesis, the *multiset space*, H_{ms}, where each hypothesis records the exact number of occurrences of each fault, and the *sequence space*, H_{seq}, whose mapping associates a trajectory with the sequence of faulty events included in it, thus preserving the order of faults in addition to their type and number. While H_{norm} and H_{set} are finite, both H_{ms} and H_{seq} are not.

12.6.2 Preferred Hypotheses

In most cases, not every hypothesis in a space H is equally interesting: therefore, it is assumed that H is ordered by a reflexive *preference relation*, denoted by $\preceq$, with $h \preceq h'$ meaning that hypothesis h is either more preferable than or as preferable as h'. The set of *preferred hypotheses* in a generic set $S \subseteq H$, denoted by $min_{\preceq}(S)$, is defined as follows:

$$\{h \in S \mid \forall h' \in S, h' \preceq h \Rightarrow h' = h\} .$$

In particular, the set $min_{\preceq}(H)$ includes all and only the so-called *most preferred* hypotheses in H. In [56], some preference relations were introduced, namely:

1. In H_{set}, *subset minimality*, denoted by $\preceq_{\text{set}}$, which is defined as follows: $h \preceq_{\text{set}} h'$ if and only if $h \subseteq h'$.
2. In H_{ms}, *multiset minimality*, denoted by $\preceq_{\text{ms}}$, which prefers the hypothesis that has fewer occurrences of every fault. Formally, if we denote by $h(f)$ the number of occurrences of fault f in hypothesis h, $h \preceq_{\text{ms}} h'$ if and only if $\forall f \in \Sigma_f$, $h(f) \leq h'(f)$.
3. In H_{seq}, *subsequence minimality*, $\preceq_{\text{seq}}$, according to which $h \preceq_{\text{seq}} h'$ if and only if h is a subsequence of h'.

Here, we define a preference relation $\preceq_{\text{norm}}$ for H_{norm} also, according to which $h \preceq_{\text{norm}} h'$ if and only if $h = normal$.

Notice that several distinct preference relations can be defined in any given hypothesis space.

12.6.3 Computing Candidates

The hypothesis space approach provides a framework where a quite general definition of a candidate that holds for whichever specific domain H we are considering can be given, instead of providing different notions of a candidate for different domains (as known, a domain corresponds to an abstraction level).

Basically, a hypothesis is a candidate in a space H if it corresponds to a trajectory in the behavior space L that can *explain* the observation.[5] Formally, given a hypothesis space H whose mapping is δ, a hypothesis $h \in H$ is a *candidate* in H if $\exists u \in \Delta_{\text{id}}$ such that $\delta(u) = h$. A *diagnosis* Δ in H is the set of all the candidates in H.

Based on these definitions, we can conclude that the solution of a diagnostic problem Δ_{id}, as defined in Sec. 1.4.1, is indeed the diagnosis in H_{id}, that is, it is the solution within a specific hypothesis space (which corresponds to a specific abstraction level). However, other solutions can be provided within other hypothesis spaces. In fact, a diagnosis depends on the hypothesis space adopted, that is, $\Delta = \Delta(D, O, H)$ (or, equivalently, $\Delta = \Delta(D, O, \delta)$, where δ is the mapping from L to H). Notice that Δ_{id} is actually $\Delta(D, O, H_{\text{id}})$.

[5] The mapping function can map several trajectories in L to the same hypothesis in H.

While the traditional approaches are aimed at computing all candidates, the task tackled in [56] is aimed at computing only the *preferred* (also called *minimal*) ones. In fact, two degrees of freedom affect the desired solution of a diagnosis problem: the abstraction level of the candidates, and the preference criterion among candidates.

The notions of a preferred candidate and a preferred diagnosis can be defined in quite general terms. Given a diagnosis problem, let Δ be its diagnosis in a hypothesis space H. Given a preference relation $\preceq$ applicable to H, a hypothesis in H is a *preferred candidate* if it is an element of $min_{\preceq}(\Delta)$, and the *preferred diagnosis* in H according to $\preceq$, denoted by $\Delta_{\preceq}$, is the subset of Δ that includes all and only the preferred candidates according to the relation $\preceq$, that is, $\Delta_{\preceq} = min_{\preceq}(\Delta)$.

The preferred diagnosis $\Delta_{\preceq}$ depends on Δ, which uniquely identifies the chosen H, and on the preference relation $\preceq$ adopted in H, as several distinct preference relations can be defined for the same hypothesis space.

The hypothesis space approach can also be exploited in order to compute all candidates at a given abstraction level, provided the hypothesis space is finite.

Some algorithms to compute preferred candidates have been proposed in the literature. In summary, they perform a search in the selected hypothesis space: so, the selected H is the search space.

The set $\Delta_{\preceq}$ of preferred candidates is initialized (it is initially empty), and then a cycle is repeated until the search space is empty (sometimes this loop may be infinite). Within the loop, first a hypothesis h is extracted from the search space, and then it is checked whether h is a candidate; if it is and no candidate in $\Delta_{\preceq}$ is preferable to it, then h is added to the set of preferred candidates, any candidate that is less preferable than h is removed from $\Delta_{\preceq}$, and possibly some (ideally all) hypotheses that are less preferable than h are removed from the search space (pruning).

Different algorithms differ in the strategies they adopt for extracting each hypothesis from the search space: typically, preferred-first strategies are better. These algorithms differ from each other in the solver used to check whether a hypothesis is a candidate: several solvers can be adopted, such as planners and SAT solvers. The algorithms may also differ in the techniques used to prune the search space.

A hypothesis may correspond to several trajectories in L: an advantage of the approach is that it is enough to find a single trajectory corresponding to a hypothesis to accept that hypothesis as a candidate.

On the other hand, a disadvantage is that if the hypothesis is unfeasible, a useless search is performed. The hypothesis space approach checks each hypothesis for consistency with the given DES model and observation by use of an efficient solver, such as a planner or a SAT solver. The traditional approaches, instead, rely on ad hoc programs that find all (and only) the paths in the behavior space that are compliant with the observation.

Basically, both approaches build some paths. In the traditional approaches, this process is driven by the observation only, while in the hypothesis space approach it is both driven by the observation and constrained by a (possibly unfeasible) diagnosis hypothesis.

In the traditional approaches all the candidates are found as a result of one search, and any portion of the search space is visited just once. In the hypothesis space approach, instead, several calls of the solver are performed (one for each checked hypothesis) and the same portion of the search space may possibly be visited more than once.

Example 12.16. Referring back to the diagnosis problem in Example 1.3, where the model of the DES D is provided in Fig. 1.1 and the observation is $O = [o2, o1, o1, o1]$, recall that an incomplete set of refined candidates for this problem at several abstraction levels is provided in the example in Sec. 1.4.2.

Notice that Δ^+_{seq} is indeed the refined diagnosis in the hypothesis space H_{seq}, while Δ^+_{ms} is the refined diagnosis in H_{ms}, and Δ^+_{norm} is the refined diagnosis in H_{norm}.

Hence, the preferred refined diagnosis in H_{seq} under subsequence minimality is $min_{\preceq_{\text{seq}}}(\Delta^+_{\text{seq}}) = \{[\,]\}$. Similarly, the preferred refined diagnosis in H_{ms} under multiset minimality is $min_{\preceq_{\text{ms}}}(\Delta^+_{\text{ms}}) = \{\langle 0,0,0 \rangle\}$. The preferred refined diagnosis in H_{set} under subset minimality is $min_{\preceq_{\text{set}}}(\Delta^+_{\text{set}}) = \{\emptyset\}$, while the preferred refined diagnosis in H_{norm} is $min_{\preceq_{\text{norm}}}(\Delta^+_{\text{norm}}) = \{normal\}$.

12.7 Milestones in Diagnosability

The notions of diagnosability and I-diagnosability of partially observable DESs, defined in the diagnoser approach [144] (and briefly introduced in Sec. 12.1.1), and the intractability of the method for testing diagnosability and I-diagnosability, as proposed in the same contribution, aroused great interest in the scientific community.

Producing diagnosable devices is quite important, but this requires checking diagnosability at design time. Two groups of researchers, independently of each other, investigated the topic over the same period of time. Each group proposed a new method to verify diagnosability, as defined in [144], operating in polynomial time in the product of the size of the state space of the system considered and the number of failure types. However, both methods process the whole system model, and, if the system is distributed, the creation of that model is the result of a synchronous composition, whose complexity is not taken into account.

The two methods, which make the same assumptions as in the diagnoser approach (that is, the prefix-closed language of the events of the DES is live and there does not exist any cycle of unobservable events), are quite similar. In both of them, as in the diagnoser approach, the observation is certain, and failure events are unobservable and partitioned into (disjoint) sets, each corresponding to a different failure type.

These approaches were presented in the same journal, *IEEE Transactions on Automatic Control*, the first [69] in 2001 and the second [166] a year later. In 2003, another fundamental contribution [31] was produced.

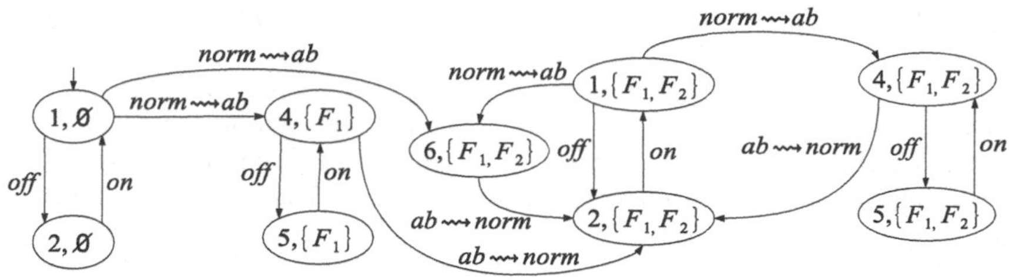

Fig. 12.25 Automaton G_o for Example 12.17

It is now usual to mention the *twin plant* approach, in this context, mainly when referring to [69]; however, there is no approach that was called by this name by its authors, although the notion of a "coupled twin model of the plant" was explicitly defined in [31]. Basically, the name refers to the product of a system model (or of some FA drawn from it) with itself and applies to every one of the approaches described below, including the extension of the diagnosability check to supervision patterns [68].

None of the methods presented here exploits any distributed compositional approach to diagnosability verification, which became a research topic in later years [120, 130, 163, 164, 165].

12.7.1 Observable Nondeterministic Automaton Method

Given an NFA G representing the model of a DES, what we call here the *observable nondeterministic automaton* method for testing diagnosability does not require the construction of a diagnoser for G; instead, it draws from G an NFA G_o, whose set of events consists of all and only the observable events of G. Each state of G_o is a pair $(x; f)$, where x is either the initial state x_0 of G or a state in G that is the target of an observable transition, while f is a set of failure types. If $x = x_0$, then $f = \emptyset$, that is, it is assumed that G is initially free of faults, as in the diagnoser approach.

Each transition from a pair $(x; f)$ to a pair $(x_1; f_1)$ in G_o represents a path in G from state x to state x_1, where the only observable transition in that path is the last one. The set f_1 is the set-theoretic union of f with all the failure types corresponding to the transitions on the path from x to x_1 in G. Thus, the constraint holds that $f_1 \supseteq f$.

Intuitively, G_o is an NFA recognizing the observable language of G; hence each state $(x; f)$ in G_o includes the set f of all the failure types that manifest themselves along a path (at least) in G that produces the same sequence of observable events as a path in G_o from the initial state $(x_0; \emptyset)$ to the state $(x; f)$.

Example 12.17. Figure 12.25 shows the automaton G_o relevant to the DES whose model G is shown in Fig. 12.2. The same partition of the set of failure events as in Example 12.2 is adopted; that is, we assume that $fail_1$ and $fail_2$ have to be detected separately. The failure type F_1 is assigned to event *fail*$_1$, while F_2 is assigned to event *fail*$_2$.

Once G_o is available, according to the observable nondeterministic automaton method for checking the diagnosability of system G, the synchronous composition of G_o with itself ($G_o \| G_o$) is computed, and denoted G_d. Thus, each state in G_d is a pair of pairs, namely $((x_1; f_1); (x_2; f_2))$.

Example 12.18. Figure 12.26 shows the automaton G_d resulting from the synchronous composition of G_o (shown in Fig. 12.25) with itself.

Finally, an algorithm checks whether there exists in G_d a cycle that includes a state $((x_1; f_1); (x_2; f_2))$ such that f_1 does not equal f_2: if this condition holds, G is not diagnosable. This check can be performed by first identifying all the states in G_d for which f_1 does not equal f_2, then deleting all the other states and their incoming and exiting transitions, and finally checking whether the remaining automaton contains a cycle. Notice that, in this last step, the observable events associated with transitions are irrelevant, and therefore G_d can be considered as a graph.

Now we aim to explain the rationale behind this method. Each pair of states in G contained in the same state in G_d can be reached by producing the same sequence of observable events. If the sets of failure events associated with these states are different, such as in the case of $((4; \{F_1\}); (6; \{F_1, F_2\}))$ in Fig. 12.26, then, based on the sequence of events the observer has received so far (in this example, only the event *norm* $\rightsquigarrow$ *ab* has been received), we cannot find out with certainty which failure events have occurred. In this example, we cannot find out whether only some failure of type F_1 has occurred, or failures of both type F_1 and type F_2.

The system is diagnosable only if we can isolate the type of failure that has occurred within a finite delay after its occurrence, that is, after a finite sequence of events has been observed after the ambiguity has manifested itself. As already remarked, given a transition $(x; f) \rightarrow (x_1; f_1)$ in G_o, the constraint holds that $f_1 \supseteq f$. Hence, in G_o, the set of failure types is the same for all states belonging to the same cycle. Consequently, in G_d, if a cycle includes a state $((x_1; f_1); (x_2; f_2))$, all the other states in the same cycle are $((\bullet; f_1); (\bullet; f_2))$, that is, they include the same sets of failure types.

Thus, if in a state in a cycle in G_d such sets are different from each other, the system is not diagnosable, since it may produce indefinitely the observable events relevant to the cycle, in which case we cannot decide within a finite delay which faults have occurred.

Example 12.19. There are several cycles in G_d (Fig. 12.26); however, none of them includes two states in G_o that are associated with distinct sets of failure types. In fact, if we remove all the states in G_d containing a pair of states in G_o associated with the same set of failure types, the only two states left are those shaded in Fig. 12.26, which are not involved in any cycle. Thus we can conclude that, as already known, the system G in Fig. 12.2 is diagnosable.

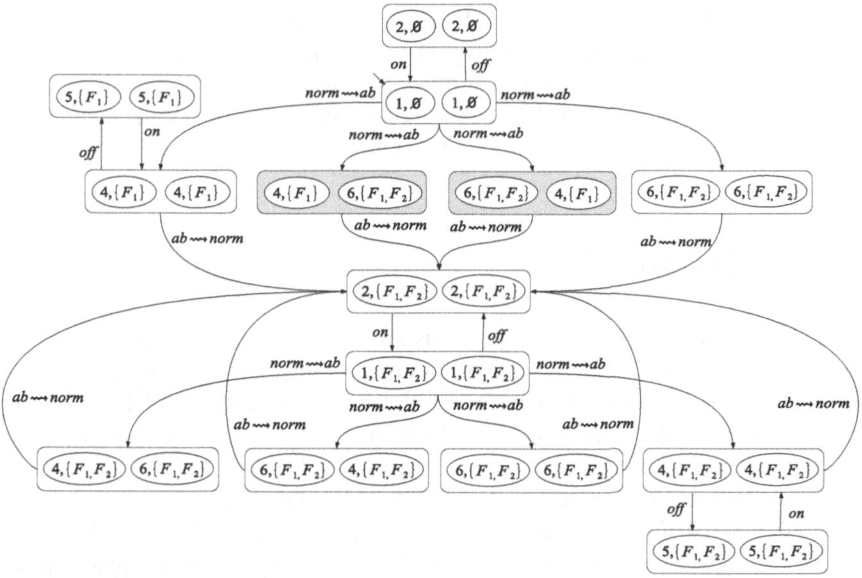

Fig. 12.26 Automaton G_d for Example 12.18

The number of states in G_o is at most $|X|2^{|F|}$, where X is the set of states in G and F is the set of failure types in G (this upper bound is obtained by assuming that, for each state in G, there is a state in G_o including it and each subset of the set of failure types).

The number of transitions in G_o is at most $|X|^2 2^{2|F|}|\Sigma_o|$, where Σ_o is the set of observable events in G (this upper bound is obtained by assuming that, for each observable event, there is a transition exiting each state and directed to each state, where the number of states considered is the maximum possible value).

The number of states in G_d is at most $|X|^2 2^{2|F|}$ (this upper bound is obtained by assuming that in G_d there is a state for each distinct pair of states in G_o), and the number of transitions in G_d is at most $|X|^4 2^{4|F|}|\Sigma_o|$ (this upper bound is obtained by assuming complete connectivity of the states in G_d).

The complexity of the construction of G_o is thus $O(|X|^2 2^{2|F|}|\Sigma_o|)$, while the complexity of the construction of G_d is $O(|X|^4 2^{4|F|}|\Sigma_o|)$ (in both cases, the complexity is given by the maximum number of transitions, since the growth of that number is asymptotically higher than that of the number of states). The complexity of detecting the presence of a certain "offending" cycle in an appropriately pruned subgraph of G_d is linear in the number of states and transitions of the subgraph.

Hence, the complexity of the whole method is $O(|X|^4 2^{4|F|}|\Sigma_o|)$. One might object that this complexity is exponential in the number of failure types. However, it can be reduced to polynomial in the number of failure types by noting that a system is diagnosable with respect to all failure types if and only if it is diagnosable with respect to each individual failure type.

In other words, one can apply the algorithm iteratively (a number of times that equals the number of distinct failure types) to test the diagnosability with respect to each individual failure type set. It follows that the complexity of each of these tests is $O(|X|^4 2^{4|1|}|\Sigma_o|) = O(|X|^4|\Sigma_o|)$. So, the overall complexity of the method for testing diagnosability is $O(|X|^4|\Sigma_o||F|)$, which is polynomial.

12.7.2 Verifier Method

While the method described in the previous section [69] deals with diagnosability only, the contribution in [166] deals with I-diagnosability as well. However, the untimed DES whose diagnosability has to be checked is modeled as a DFA G, while both in the original diagnoser approach [144] and in [69] the system model is nondeterministic.

What we call here the *verifier* method for checking diagnosability is based on the construction of a distinct NFA, called the F_i-*verifier*, for each failure type F_i.

Each state of such an automaton is a 4-tuple, which can be seen as a pair of pairs (q,l), where q is a state of G and l is a label (either N or F_i). The initial state of the F_i-verifier is (q_0,N,q_0,N), where q_0 is the initial state of G, that is, it is assumed that the system is initially free of faults, as in the diagnoser approach and in the observable nondeterministic automaton method.

Let $x = (q_1,l_1,q_2,l_2)$ be a state of the F_i-verifier. Let σ be an event that triggers in G a transition $q_1 \rightarrow q'_1$, as well as a transition $q_2 \rightarrow q'_2$. If σ is observable, then a transition, marked by σ, from x to the state $x' = (q'_1,l_1,q'_2,l_2)$, is inserted into the F_i-verifier. If, instead, σ is not observable and it is not a fault event of failure type F_i, then in the F_i-verifier there are three transitions exiting x, which are marked by σ, leading to (q'_1,l_1,q_2,l_2), (q_1,l_1,q'_2,l_2), and (q'_1,l_1,q'_2,l_2). Finally, if σ is a fault event of failure type F_i, then in the F_i-verifier there are three transitions exiting x, which are marked by σ, leading to (q'_1,F_i,q_2,l_2), (q_1,l_1,q'_2,F_i), and (q'_1,F_i,q'_2,F_i).

A (proven) theorem states that the language of the system G is diagnosable with respect to the set of observable events in G and on the partition on the set of failure events if and only if the F_i-verifier is F_i-*confusion free*, for i ranging over all parts in the partition of all the failure events of G. The F_i-verifier is F_i-confusion free if it does not include any cycle such that, for all the (4-tuple) states in the cycle, one label is N and the other is F_i.

Thus, checking diagnosability amounts to detecting the existence of a so-called F_i-confused cycle in the F_i-verifier. Notice that, in performing this check, the transition labels (that is, the events triggering transitions) are irrelevant. Therefore, the F_i-verifier can be considered as a directed graph rather than an automaton.

The F_i-verifier is F_i-confusion free if and only if the directed graph obtained by removing from the F_i-verifier all the states where the two labels are identical is acyclic. Therefore, checking diagnosability just translates into detecting the existence of a cycle in such a directed graph.

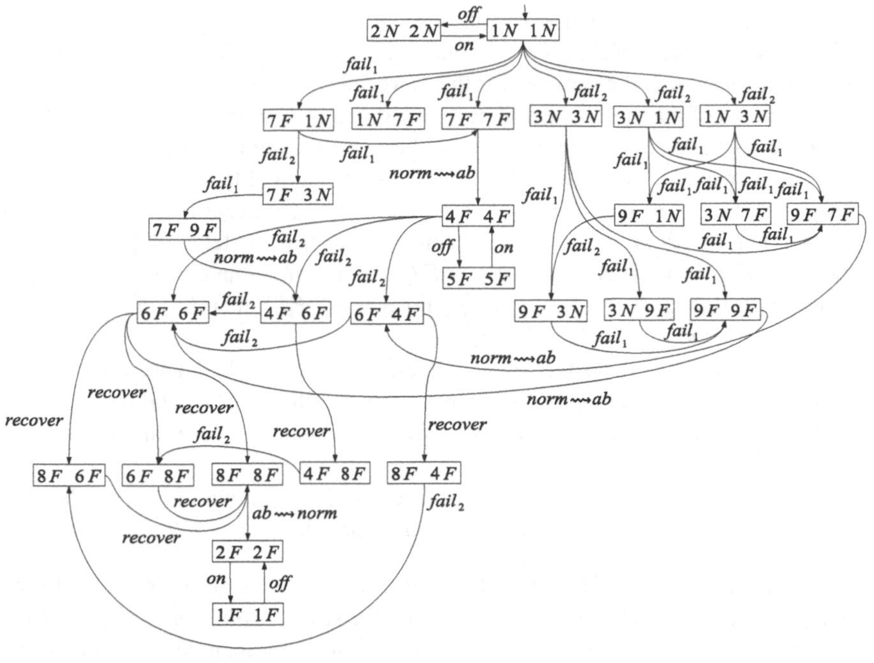

Fig. 12.27 F_1-verifier for Example 12.20

Example 12.20. Given the system G in Fig. 12.2, consider the same partition of failure events as in Example 12.17, that is, failure type F_1 is assigned to the event $fail_1$, while F_2 is assigned to the event $fail_2$.

The F_1-verifier relevant to system G is depicted in Fig. 12.27, where the label F is used for F_1. Since it does not include any F_1-confused cycle, we can conclude that this system is diagnosable as far as the part F_1 of the partition of all failure events is concerned. If the F_2-verifier is constructed, it can be concluded that G is diagnosable for part F_2 also. Hence, we conclude that G is diagnosable with respect to the given set of observable events in G and on the given partition of the set of failure events.

A proposition in [166] provides an upper bound on the number of transitions following the occurrence of a faulty transition within which the fault can be identified with certainty in a system G whose language is diagnosable. According to this proposition, any failure occurrence can be detected within a number of transitions that equals the square of the number of states in G.

The F_i-verifier can be constructed in time $O(n_1^2 n_2)$, where n_1 is the number of states in G and n_2 is the number of events in G. Cycle detection in the F_i-verifier can be done in a time $O(n_1^2 n_2)$. Hence, checking diagnosability requires a time $O(n_1^2 n_2 n_3)$, where n_3 is the number of failure types.

Similar results hold for I-diagnosability checking. As already known, if I-diagnosability is considered, an indicator map assigns to each failure type F_i a corresponding set of indicator events I_i. In order to check I-diagnosability, an $(F_i;I_i)$-

verifier has to be constructed (instead of an F_i-verifier) for each failure type F_i, where the set of observable indicator events I_i is taken into account. The initial state of the $(F_i;I_i)$-verifier is the same as for the F_i-verifier.

The creation of the (nondeterministic) transitions is the same as for the construction of the F_i-verifier; what changes is the label propagation, since the domain of labels has been extended to $\{N;F_i;F_iI_i\}$, while the domain of labels of the F_i-verifier is $\{N;F_i\}$. Let $q_1 \rightarrow q_1'$ be a transition in G, marked by the event σ, and $x \rightarrow x'$ be a corresponding transition in the $(F_i;I_i)$-verifier, where l_1 is the label of q_1 in x and l_1' is the label of q_1' in x'. Then $l_1' = l_1$ if $l_1 = F_iI_i$ or σ is neither an event of type F_i nor an event in I_i. If σ is an event of type F_i (and l_1 is not F_iI_i), then $l_1' = F_i$, while, if σ is an event in I_i and $l_1 = F_i$, then $l_1' = F_iI_i$.

A (proven) theorem states that the language of a DES G is I-diagnosable with respect to the set of observable events, the partition of failure events, and the indicator map if and only if the $(F_i;I_i)$-verifier is $(F_i;I_i)$-confusion free, for i ranging over all parts in the partition of all the failure events of G.

The $(F_i;I_i)$-verifier is $(F_i;I_i)$-confusion free if it does not include any cycle such that, in all states in the cycle, one label is N and the other one is F_iI_i.

Example 12.21. Figure 12.28 shows the verifier for I-diagnosability relevant to the system G in Fig. 12.2, when failure type F_2 and the set $I_2 = \{ab \rightsquigarrow norm\}$ of observable indicator events are considered. This verifier does not include any state where one label is N and the other is F_2I_2; hence, it is confusion free. As already known, G is I-diagnosable as far as failure F_2 and event I_2 are concerned.

According to another theorem in [166], I-diagnosability can be decided in a time $O(n_1^2 n_2 n_3)$, the same asymptotic upper bound as for diagnosability checking.

12.7.3 Coupled Plant Method

The contributions considered in Secs. 12.7.1 and 12.7.2 proposed a theoretical framework for the definition of diagnosability-checking methods; in other words, they provided operational specifications of such methods. In the same years, the contribution in [31], while basically sharing the same theoretical views as [69, 166], aimed at the definition of an effective platform for diagnosability analysis that could be applied practically in the development process of artifacts.

To achieve this goal, the diagnosability-checking problem is reduced to a model-checking problem. This way, standard model-checking tools can be reused to perform the task, without any need to reimplement the construction described in [69, 166].

The name by which the authors of the current method call the structure that is generated and processed in order to verify diagnosability, the *coupled twin plant* method, is expressive and evocative: this name is very well known and is still universally used to refer to the conceptual scaffolding, independently of whether or not symbolic model-checking techniques are actually exploited. In fact, the authors of

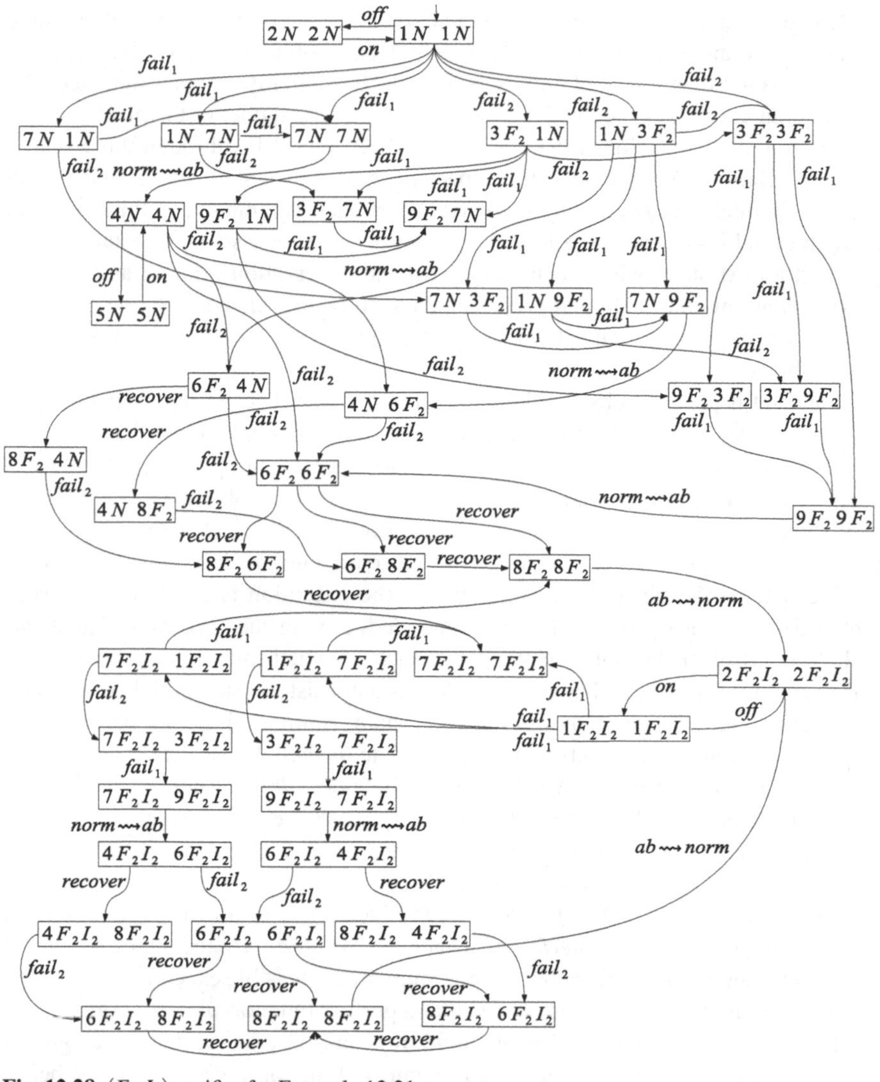

Fig. 12.28 $(F_2;I_2)$-verifier for Example 12.21

the approach provide not only the tools to perform the check but also a quite general theory of diagnosability, although in a very specific scenario.

This scenario, besides including, as usual, a system for carrying out the diagnosis task, called the *diagnosis system*, and a system to be diagnosed, called the *plant*, assumes that there is a *controller* that is connected to the plant through a feedback control loop. The controller takes as input some measurements coming from the plant, and produces as output some commands to be sent to the plant. The diagnosis system is assumed to observe both the inputs and the outputs of the controller, and to generate an estimate of the state of the plant.

This estimate, which is called a *diagnosis value*, consists in a set of possible states of the plant, referred to as a *belief state*. Hence, quite unusually, the diagnosis output does not mention any fault (as the approach implicitly assumes that pieces of information about faults are included in the states of the model of the plant).

The complete behavior of the plant is modeled as an NFA through three spaces and two relations. Each of the three spaces (input space, state space, and output space) is defined through a finite set of variables, each ranging over a finite discrete domain of values. Each element of the transition relation associates a state, an input, and a next state, while each element of the observation relation associates a state with an output. Each (and every) state has to be associated with one output at least, but it may be associated with several outputs, so as to express an uncertain observability.

The plant is partially observable in that only the sequences of inputs and outputs are observable, while the state, including the initial one, is "hidden." Every path in the NFA is a *feasible execution*, while what is perceived by the observer about such an execution is called an *observable trace*. The observable trace of a feasible execution is a sequence beginning with the output of the initial state, followed by the first input, followed by the next output, and so on, until the last output is reached.

The state of the plant at the beginning of the execution is a belief state itself, which depends on the available initial knowledge, with the constraint that it has to be a (sub)set of the states of the plant, each of which may have produced the observed initial output. A diagnosis problem is a consistent pair (initial belief state, observable trace) as, given the initial belief state, runtime diagnosis observes the sequence of inputs and outputs and tries to update the belief state, i.e., the diagnosis value. Such a value is *correct* if and only if it includes *all* the states reachable from each state in the initial belief state through every feasible execution having the given observable trace. The diagnosis function is *correct* if and only if the diagnosis value it produces is correct for any diagnosis problem.

The diagnosis function is *perfect* if it produces perfect diagnosis values only, where a diagnosis value is *perfect* if it includes *all and only* the states reachable from each state in the initial belief state through every feasible execution having the given observable trace. It is easy to see that a perfect function is also a correct one, while the converse in general does not hold. It is easy also to see that the diagnosis value produced by a correct nonperfect diagnosis function for a pair (initial belief state, observable trace) is a superset of the diagnosis value produced by the perfect function for the same pair.

Example 12.22. Since the setting addressed by the coupled plant method is different from that addressed by the approaches presented in the previous sections, we cannot reuse the DES models that we have used with those approaches; instead, we use the model in Fig. 12.29, where each state transition is labeled with an (observable) input value. The set of (observable) outputs associated with the states on the left of the figure is assumed to be a singleton including the sensor value *norm*, while that associated with the states on the right is a singleton including the sensor value *ab*.

Basically, the normal behavior of this system is represented by states 1 and 2 and by the transitions between them. All the other states denote either one or two faults.

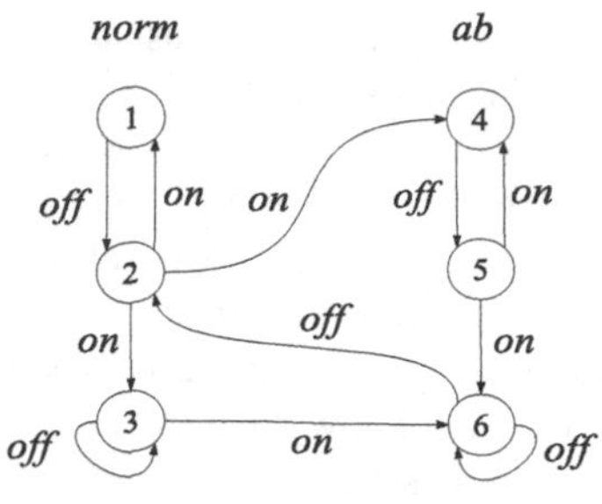

Fig. 12.29 Plant model for Example 12.22

Since such faults are intermittent, once both of them have manifested themselves, they may disappear. However, observability is partial since, while the output *ab* shows that the state is faulty, output *norm* is relevant to both the normal states and one faulty state.

A possible observable trace is the sequence $[norm, on, norm]$. Since the initial output is *norm*, the initial belief state may be any (sub)set of all the states that generate such an output, that is $\{1,2,3\}$. If we assume that no specific knowledge about the initial situation is given, the initial belief state is $\{1,2,3\}$. Hence, the diagnosis problem is $(\{1,2,3\}; [norm, on, norm])$. The perfect diagnosis relevant to this problem is $\{1,3\}$, since the corresponding feasible executions are

- *norm*, 2, *on*, 1, *norm*,
- *norm*, 2, *on*, 3, *norm*.

A correct diagnosis is any superset of the perfect diagnosis.

Diagnosability is associated with detecting (at design time) if (parts of) the hidden state of the plant can be tracked accurately (at run time) by the diagnosis system based on the observable traces.

However, according to the authors of the coupled plant method, it would be unrealistic to require that diagnosability is a property holding at all times and under all circumstances. Rather, diagnosability has to require the diagnosis function to be able to decide between alternative *conditions* on the state of the system only in a *context* where the distinction becomes critical. This is an important notion of diagnosability that accounts for the fact that we may be interested in different properties that we wish to hold at run time: diagnosability can be "customized" so as to represent each property we would like to check. Each pair (condition, context) is a customization or, in other words, it is an instance of (class) diagnosability.

A diagnosis condition is used to specify the pieces of information about the states we are interested in distinguishing. A condition c partitions the set of all the states of the plant into two parts, say $c1$ and $c2$, one including the states that fulfill the condition, the other including the states that do not. For instance, if we would like a diagnosis value not to include both faulty and nonfaulty states, $c1$ is the set of all nonfaulty states, while $c2$ is the set of faulty ones. If, instead, we do not want to mix up states that are consistent with one kind of fault $f1$ with other states, $c1$ is the set of all states that are consistent with $f1$, while $c2$ is the set of all states that are not consistent with $f1$.

By definition, a diagnosis value satisfies a given condition if it intersects just one of the two sets of states identified by the condition itself. This means that, when we are given a diagnosis value, we know whether the current state of the plant falls into one set or the other. For instance, if the condition we are taking into account is fault detection, if the current belief state includes just states falling into $c1$, although we do not know what the current state of the plant is, we know for sure that it is a nonfaulty state.

A diagnosis context (a) divides the states of the plant according to an equivalence relation, thus requiring that diagnosability be investigated only if all the states in the initial belief state fall into the same equivalence class (we say that the initial belief state *satisfies* the equivalence relation of the context); and (b) identifies a set of pairs of relevant feasible executions, where both of the initial states of such executions fall into a belief state that satisfies the equivalence relation of the context, and both executions correspond to the same observable trace.

A diagnostic problem (initial belief state, observable trace) satisfies a given context if the initial belief state satisfies the equivalence relation of the context, and there exists a pair of feasible executions, each starting from a state included in the initial belief state, and each corresponding to the observable trace, that are included in the set of pairs of feasible executions identified by the context. Basically, a diagnosis problem satisfies a context if it is reckoned as a diagnostic problem of interest by the context itself.

By definition, a diagnosis function satisfies a given condition over a given context if and only if the diagnosis value produced for each diagnosis problem (initial belief state, observable trace) that satisfies that context also satisfies the given condition.

This means that, if the initial belief state satisfies the equivalence relation of the context, and there exists some pair(s) of executions (each starting from an initial state that is included in the initial belief state) that produce the given observable trace and fall into the set of pairs identified by the context, then the diagnosis function satisfies the given condition over the given context if, when such a problem is being solved in the domain of all and only such executions, it generates an estimate of the state of the plant based on which we know whether the state of the plant falls into one set or the other defined by the condition.

By definition, a given condition is diagnosable in the plant over a given context if and only if there exists a correct diagnosis function that satisfies it. This definition has a very important meaning: it points out that diagnosability depends not only on the plant but also on the diagnosis function. Such a function has to be correct.

However, no diagnosis function can provide better discrimination than the perfect function, which, necessarily makes all the final states of all the feasible executions corresponding to the same observable trace collapse into the same final belief state. If a given condition is not diagnosable in the plant over a given context when the computation of the diagnosis value is performed by a perfect diagnosis function, then there exists no correct diagnosis function that can make it diagnosable.

In other words, by appealing to the perfect diagnosis function, we define what we may call *plant diagnosability*, as this notion depends only on the plant model, not on the diagnosis system. Diagnosability implies plant diagnosability, while the converse in general does not hold. Saying that a condition is diagnosable in the plant over a given context means that, for any pair of executions belonging to the set of pairs identified by the context, the diagnosis function produces two resulting estimates of the state such that it is possible to discriminate whether the state of the plant falls in one set or the other, denoted as $c1$ and $c2$, identified by the condition.

Example 12.23. Assume that the diagnosability of a given condition has to be checked in a given context for the DES described in Example 12.22. Also, assume that the given condition c partitions the set of states of the DES into $c1 = \{1,2\}$ (normal behavior) and $c2 = \{3,4,5,6\}$ (faulty behavior), while the given context C partitions the set of states into two equivalence classes based on their observability, $\{1,2,3\}$ and $\{4,5,6\}$. Let the context identify the set of pairs such that each pair includes two feasible executions consisting of just one step.

Notice that, since any observable trace can start with either *norm* or *ab*, the corresponding belief states are $\{1,2,3\}$ and $\{4,5,6\}$, respectively, both of which satisfy the equivalence relation of the context.

Based on the definition of diagnosability of a condition over a context given above, it is easy to see that c is not diagnosable over C, as there exists some pair(s) of feasible executions that satisfy the context but cannot be distinguished by the perfect diagnosis function. This is the case, for instance, for the pair of feasible executions (*norm*, 2, *on*, 1, *norm*; *norm*, 2, *on*, 3, *norm*) corresponding to the observable trace $[norm, on, norm]$. The former execution leads to state 2, falling into $c1$, while the latter leads to state 3, falling into $c2$, and hence the condition c is not diagnosable over the context C.

In other words, a condition c is diagnosable over a context if, for any diagnosis problem that is taken into account by the context, the diagnosis value computed by the diagnosis system enables one to know whether or not the current state of the DES at hand fulfills condition c. We have proven by a counterexample that this is not the case for the system, condition, and context considered here.

Symbolic model checking enables one to exhaustively analyze the (possibly infinite) behaviors of (large) DESs and to check whether diagnosability requirements, formally expressed in terms of linear temporal logic [45], are met. The diagnosability of a condition is verified by refutation, that is, by checking that the plant has no *critical pair*, i.e., two executions with identical observable traces, one leading to a final state that falls into $c1$ and the other to a final state that falls into $c2$.

Given a perfect diagnosis system, the absence of critical pairs associated with a condition guarantees that that condition is diagnosable over any context.

A theorem has been proven that states that the absence of critical pairs for a given condition over a given context is both necessary and sufficient for that condition to be diagnosable in that context. In other words, this theorem states what plant diagnosability is, as, in fact, the required absence of critical pairs for a given condition depends only on the plant model, not on the diagnosis system.

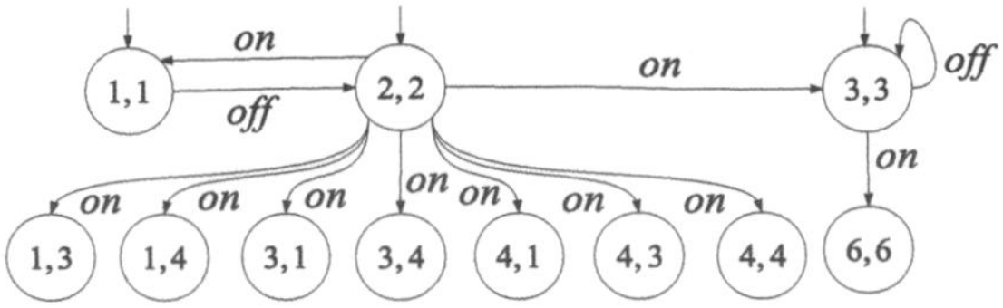

Fig. 12.30 Coupled plant for Example 12.24

In order to search for critical pairs, the so-called *coupled twin plant* is built, made of two copies of the same plant, whose inputs and outputs are forced to be identical. A state of the twin plant is a pair of states of the plant that, according to the observation relation, can produce the same output. There exists a transition from a state $(x1,x2)$ of the twin plant to a state $(x1',x2')$ if and only if there exists an input u such that both $(x1,u,x1')$ and $(x2,u,x2')$ are elements of the transition function of the plant. The pair $((x1,x2),y)$ belongs to the observation relation of the twin plant if and only if both $(x1,y)$ and $(x2,y)$ belong to the observation relation of the plant.

A theorem states that two executions producing the same observable trace are feasible in the plant if and only if there is a corresponding feasible execution in the twin plant.

Example 12.24. Considering again the same condition c and context C as in Example 12.23, we now investigate the diagnosability of c over C by using the coupled plant method. In order to check the diagnosability when the initial state falls into the equivalence class $\{1,2,3\}$, a coupled twin plant has to be built, where each state in the equivalence class can be the initial state. As the context is relevant only to one-step executions, the coupled twin plant encompasses only one-transition paths starting from each initial state.

The structure obtained is shown in Fig. 12.30. In this coupled twin plant, any final state that includes two (plant) states with one falling into $c1$ and the other into $c2$ denotes the existence of a critical pair.

In fact, any path from an initial state to a final state represents two feasible executions that are observationally identical; hence they are both considered by the diagnosis function when given the same diagnosis problem.

If the final state includes two states with one falling into $c1$ and the other into $c2$, as is the case with the final states $(3,1)$, $(1,3)$, $(4,1)$, and $(1,4)$ in Fig. 12.30, the diagnosis value is a belief state that, since it includes two such states, does not satisfy the condition. This is sufficient for us to conclude that, as already known, c is not diagnosable over C, and thus it would be useless to build a twin plant relevant to the second equivalence class, $\{4,5,6\}$.

Besides highlighting the different setting and the practical purposes of the coupled plant method, a comparison with the methods for checking diagnosability of DESs presented in the previous sections has to include a remark that its notion of diagnosability is zero-delay; that is, the property to be checked has to hold as soon as the final state relevant to the observable trace considered is reached, while the other approaches require a property to hold within a finite delay.

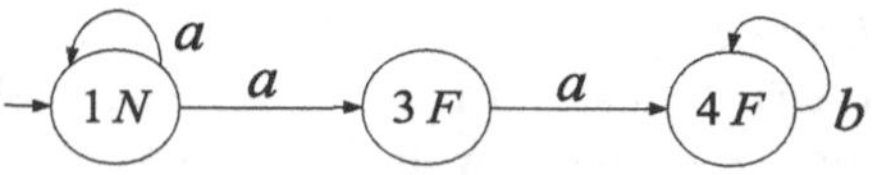

Fig. 12.31 Automaton $OBS(G_\Omega)$ for Example 12.25

Although its authors do not emphasize the point, the coupled plant method, in addition to generalizing the notion of diagnosability, which has evolved from a single absolute property into a class of relative properties, opens the way to a line of research into how diagnosability depends on the diagnosis engine at hand [154].

12.7.4 Pattern Diagnosability Method

The notion of diagnosability of a fault is extended to diagnosability of a pattern in the supervision pattern approach [68], which was briefly described in Sec. 12.2. The method for checking diagnosability of a DES by using fault supervision patterns is described in the following steps:

1. Given a DES behavioral model G, which is an FA, and a supervision pattern Ω, which is another FA, build the synchronization G_Ω for G and Ω.
2. Generate the automaton $OBS(G_\Omega)$ whose domain for the set of states is the same as for G_Ω and whose transitions are all observable. The initial state of $OBS(G_\Omega)$ is the same as for G_Ω. For each state q in $OBS(G_\Omega)$, there is an exiting transition toward each state q' that in G_Ω can be reached through a trajectory, starting from q and ending in q', whose only observable transition is the last one.
3. Similarly to the diagnosability analysis in [69, 166], produce the synchronous self-product $\Gamma = OBS(G_\Omega) \otimes OBS(G_\Omega)$.
4. Check whether there exists an *undetermined state cycle* in Γ, this being a cycle that includes both final and nonfinal states of the supervision pattern Ω; if this is so, then G is not Ω-diagnosable, otherwise G is Ω-diagnosable.

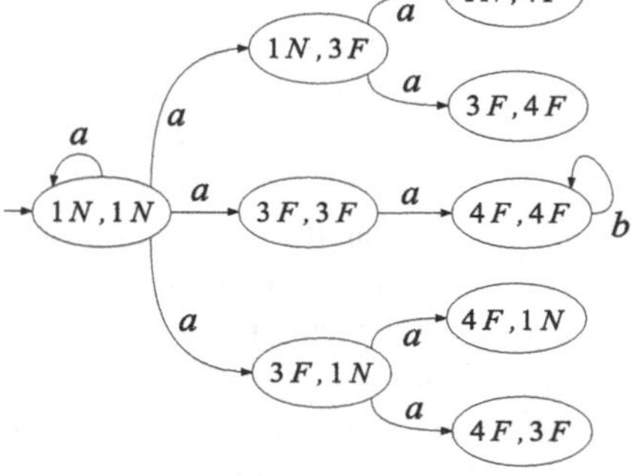

Fig. 12.32 Synchronous product Γ for Example 12.25

Example 12.25. In order to check whether the DES whose behavioral model is the FA G depicted in Fig. 12.9 is diagnosable according to the fault pattern Ω_f in Fig. 12.4, we follow the above steps.

First, we build G_Ω, as described in Example 12.4. This NFA is shown in Fig. 12.10. Second, we draw from G_Ω the corresponding completely observable automaton $OBS(G_\Omega)$, which is displayed in Fig. 12.31. Then, we produce the synchronous composition of $OBS(G_\Omega)$ with itself, i.e., $\Gamma = OBS(G_\Omega) \otimes OBS(G_\Omega)$, which is shown in Fig. 12.32. Finally, we check whether Γ includes any undetermined state cycles.

From Fig. 12.32, we can see that there is no undetermined state cycle in it, although there are some *undetermined states*, namely $(1N, 3F)$, $(1N, 4F)$, $(3F, 1N)$, and $(4F, 1N)$. Therefore, the DES whose behavioral model is G is diagnosable as far as the fault pattern Ω_f is concerned. In fact, it is easy to see that, once fault f has occurred, event b will be observed within a finite number (3) of (observable) events, where event b definitely indicates that fault f has previously occurred.

From Example 12.4, we can see that if we have not got enough observation (for instance, if the observation is $\mu = [a]$), we cannot know whether fault f has occurred. However, once we have got enough observation ($\mu = [a, a, b]$), we know that fault f has definitely occurred.

12.8 Bibliographic Notes

All of the studies quoted in the previous sections were based on untimed FAs, starting from the diagnoser approach [144, 145], where both the DES and its corresponding diagnoser were modeled as untimed FAs.

Some approaches propose to model complex DESs by means of timed FAs, which explicitly include time information. In [162], both an algorithm for checking diagnosability and a technique for constructing a diagnoser for a diagnosable timed FA were provided. In [14], a method for fault detection and isolation in real-time DESs was presented, including diagnosability checking. In [50], an approach to diagnosis of complex systems modeled as communicating timed FAs was described, where each component was modeled as a timed FA integrating different operating modes, while the communication between components was carried out by a control module. Both the formal verification of the complex system model and the construction of its diagnoser were encompassed.

A Petri net is another formalism for modeling DESs for diagnosis purposes, adopted mainly for *concurrent systems* [13, 23, 70]. Such a representation is usually more compact than the one obtained through FAs. Unlike FAs, unbounded Petri nets can represent DESs with an infinite state space.

As diagnosis during monitoring of DESs has to exhibit real-time performance, a number of offline compilation methods have been studied in order to improve the efficiency of online diagnosis processing. Information about failures is compiled offline in some formalisms, such as finite state machines, binary decision diagrams

(BDDs), ordered binary decision diagrams (OBDDs), and decomposable negation normal forms (DNNFs), and the diagnostic results are obtained online by matching the compiled model with the real observations.

The diagnoser approach proposes the compilation of the system model into an FA, called a diagnoser. Since its introduction, this approach has been extended to stochastic FAs [158] and fuzzy DESs [123] to represent DESs, with a corresponding stochastic diagnoser and fuzzy diagnoser, respectively.

In [156], an approach was presented that used OBDD-based symbolic computation [20, 21] for diagnosis of DESs. It does not rely on any diagnoser, and it exploits OBDDs to encode the system model. The set of diagnoses can be characterized by a first-order formula over such a model, and the diagnoses are obtained by manipulating the model itself with standard OBDD operators.

A compact OBDD representation of static systems was presented in [159, 161]. This representation was extended in [160] in order to support online diagnosis of dynamic systems, especially time-varying systems, by additionally encoding the system transition relations into OBDDs.

A symbolic framework for the diagnosis of DESs based on BDDs was presented in [148]. The authors focused on the BDD-level encoding of both the system and the diagnoser, and described some algorithms for synthesizing the symbolic diagnoser in terms of BDD operations. The efficiency of the diagnoser approach was improved, as BDDs are particularly suitable for compact representation of large state sets. The approach proposed in [148] concentrates mainly on compiling the diagnoser, whereas the approach proposed in [160] does not need any diagnoser and is a more general one. Adopting a diagnoser means possibly suffering from space explosion when compiling offline, whereas not adopting one means experiencing poor time performance when operating online. Therefore, in [149], a spectrum of approaches was proposed considering time and space requirements: the underlying models range from the system model without any compilation to the diagnoser model, which is the result of the compilation of the whole behavior. The diagnosis task is tackled by encoding all approaches into a symbolic framework based on BDDs.

The DNNF language is a strict superset of the OBDD language [38], and is strictly more succinct than it. This implies that (1) if a system model can be compiled into OBDD, then it can also be compiled into DNNF, and (2) a system model may have a polynomial-size compilation in DNNF, but not in OBDD. However, in [149] it was shown that BDDs are better suited for diagnosing dynamic systems, because the main operation can be performed in a *polynomial* time in the size of the BDD whereas it would require an *exponential* time in the size of the DNNF.

The hypothesis space approach (see Sec. 12.6.1) was dealt with in [26], where hypothesis spaces that are partially ordered sets (posets) were specifically addressed.

Diagnosability of DESs represented as HFSMs (see Sec. 12.3.1) was investigated in [129].

The traditional notion of diagnosability given in the diagnoser approach, which is the basis of all the diagnosability-checking methods described in Sec. 12.7, was extended in [155] to the case where the observation relevant to the diagnosis problem is temporally or logically uncertain.

12.9 Summary of Chapter 12

Diagnoser Approach. An untimed model-based approach to the diagnosis of (distributed) synchronous DESs. The complete behavior of each system component is described by a nondeterministic automaton, and all component models are subjected beforehand to a synchronous composition so as to obtain the global behavioral model of the system, from which a completely observable deterministic automaton, called the diagnoser, is drawn offline. The diagnoser is used throughout the monitoring process and provides at each step (where a new observable event is taken into account) estimates of the current system state and, for each estimated state, the corresponding candidate diagnosis, this being a set of specific faults, where faults are assumed to be persistent. The diagnoser can also be used to compute a posteriori diagnosis results. The major drawback of this approach is that it relies on the global system model, whose generation is infeasible in practice for real-sized systems owing to its computational cost.

Fault Pattern. A generalization of the notion of a fault, consisting in a faulty evolution that is meaningful for the DES at hand. It is represented by a deterministic automaton endowed with stable final states.

Supervision Pattern Approach. An untimed model-based approach to (both monitoring-based and a posteriori) diagnosis of DESs that, instead of looking for the occurrences of faults, looks for the occurrences of a fault pattern. In order to compute a diagnosis, a diagnoser (similar to that of the diagnoser approach) is built offline by determinizing the nondeterministic automaton resulting from the synchronization of the automaton representing the behavior of the DES considered and the automaton representing the relevant fault pattern. This diagnoser is used throughout the online process and provides at each step (where a new observable event is taken into account) an estimate of the current system state and, for each estimated state, whether the fault pattern has occurred.

Hierarchical Diagnosis Approach. An untimed model-based approach to the diagnosis of DESs (both monitoring-based and a posteriori). The behavior of the system is described in a hierarchical way by means of a formalism similar to statecharts, where each (AND or OR) superstate can include simple states along with superstates. This representation can be transformed into a hierarchy (tree) where transitions entering or exiting a superstate are directed either from its immediate substates to the immediate substates of its parent or vice versa. This way, the diagnosers and the diagnostic reasoning relevant to superstates belonging to the same level of the hierarchy are independent.

D-holon. A possibly disconnected directed graph that is in a one-to-one relation with a superstate in the hierarchical diagnosis approach. The purpose of D-holons is to break down the hierarchical model of the behavior of the DES, where the transitions between states are represented explicitly, into a hierarchical representation (a hierarchy of D-holons) where transitions are represented implicitly. Each node in a D-holon is a simple state (that is, it is not a superstate). The set of states within a D-holon is partitioned into internal states, which include one or several initial states, and external states. The initial states are the target of transitions entering the superstate, the other internal states are the target of transitions within the superstate, and the external states are the target of transitions exiting the superstate. Hence, external states corresponds to initial states of other D-holons, thus building an implicit link with those D-holons.

SCL-D-holon Approach. An untimed model-based approach to (both monitoring-based and a posteriori) diagnosis of DESs according to which a diagnosis closure (originally called a *silent closure*) of the active system approach is interpreted as a D-holon.

Successive Window Approach. An approach aimed at constructing all the emission sequences of observed events that are compatible with a given temporally uncertain observation of a DES and are consistent with both domain constraints and the DES model. Such reconstructed emission sequences can be exploited to obtain sound, complete, and monotonic results in monitoring-based incremental diagnosis of DESs when the observation is temporally uncertain.

Incomplete Model Approach. An approach aimed at DES diagnosis based on an incomplete behavioral model and an uncertain observation.

Hypothesis Space Approach. An untimed model-based iterative approach to the diagnosis of DESs according to which, at each iteration, a candidate, taken from a domain called the hypothesis space, is assumed as a hypothesis, and then it is checked whether that hypothesis is relevant to a trajectory (at least) that is consistent with the given observation, that is, it is checked whether the hypothesis is really a candidate. Different hypothesis spaces can be considered, corresponding to different abstraction levels of candidates. This approach relies on the availability of an engine that is capable of performing such a check.

Observable Nondeterministic Automaton Method. An untimed model-based approach to DES diagnosability analysis, usually called the *twin plant method*, according to which a completely observable NFA is drawn from the DES model, then that automaton is combined with itself, and the resulting synchronous product is checked for the presence of a cycle or cycles exhibiting a nondiagnosability condition. The complexity of the method is polynomial.

Verifier Method. An untimed model-based approach to DES diagnosability analysis, quite similar to the observable nondeterministic automaton method, but in addition providing a means to check I-diagnosability. The complexity of the method is polynomial.

Coupled Plant Method. A technique that reduces DES diagnosability checking to a model-checking problem. Most interestingly, it defines the notions of a correct diagnosis, perfect diagnosis, diagnosis context, and diagnosability condition.

Pattern Diagnosability Method. A technique similar to the observable nondeterministic automaton method to check the diagnosability of a DES as far as a fault pattern is concerned.

References

1. Aho, A., Lam, M., Sethi, R., Ullman, J.: *Compilers – Principles, Techniques, and Tools*, second edn. Addison-Wesley, Reading, MA (2006)
2. Aichernig, B., Jöbstl, E.: Efficient refinement checking for model-based mutation testing. In: *Proceedings of QSIC 2012*, pp. 21–30. Xi'an, Shaanxi, China. IEEE, Piscataway, NJ (2012)
3. Aichernig, B., Jöbstl, E., Kegele, M.: Incremental refinement checking for test case generation. In: M. Veanes, L. Viganò (eds.) *7th International Conference on Tests and Proofs (TAP 2013)*, Lecture Notes in Computer Science, vol. 7942, pp. 1–19. Springer, Berlin, Heidelberg (2013)
4. Aichernig, B., Jöbstl, E., Tiran, S.: Model-based mutation testing via symbolic refinement checking. Science of Computer Programming **97**(4), 383–404 (2015)
5. Bairoch, A., Apweiler, R.: The SWISS-PROT protein sequence database and its supplement TrEMBL in 2000. Nucleic Acids Research **28**(1), 45–48 (2000)
6. Balan, S., Lamperti, G., Scandale, M.: Incremental subset construction revisited. In: R. Neves-Silva, G. Tshirintzis, V. Uskov, R. Howlett, L. Jain (eds.) *Smart Digital Futures*, Frontiers in Artificial Intelligence and Applications, vol. 262, pp. 25–37. IOS Press, Amsterdam (2014)
7. Balan, S., Lamperti, G., Scandale, M.: Metrics-based incremental determinization of finite automata. In: S. Teufel, A.M. Tjoa, I. You, E. Weippl (eds.) *Availability, Reliability, and Security in Information Systems*, Lecture Notes in Computer Science, vol. 8708, pp. 29–44. Springer, Berlin, Heidelberg (2014)
8. Baroni, P., Lamperti, G., Pogliano, P., Tornielli, G., Zanella, M.: Automata-based reasoning for short circuit diagnosis in power transmission networks. In: *Twelfth International Conference on Applications of Artificial Intelligence in Engineering*. Capri, Italy (1997)
9. Baroni, P., Lamperti, G., Pogliano, P., Tornielli, G., Zanella, M.: A multi-interpretation approach to fault diagnosis in power transmission networks. In: *Third IFAC Symposium on Fault Detection, Supervision and Safety for Technical Processes (SAFEPROCESS'97)*, pp. 961–966. Hull, United Kingdom (1997)
10. Baroni, P., Lamperti, G., Pogliano, P., Zanella, M.: Diagnosis of active systems. In: *Thirteenth European Conference on Artificial Intelligence (ECAI'98)*, pp. 274–278. Brighton, United Kingdom (1998)
11. Baroni, P., Lamperti, G., Pogliano, P., Zanella, M.: Diagnosis of large active systems. Artificial Intelligence **110**(1), 135–183 (1999).
12. Baroni, P., Lamperti, G., Pogliano, P., Zanella, M.: Diagnosis of a class of distributed discrete-event systems. IEEE Transactions on Systems, Man, and Cybernetics – Part A: Systems and Humans **30**(6), 731–752 (2000)
13. Benveniste, A., Fabre, E., Haar, S., Jard, C.: Diagnosis of asynchronous discrete-event systems: A net unfolding approach. IEEE Transactions on Automatic Control **48**, 714–727 (2003)

G. Lamperti et al., *Introduction to Diagnosis of Active Systems*,
https://doi.org/10.1007/978-3-319-92733-6

14. Biswas, S., Sarkar, D., Bhowal, P., Mukhopadhyay, S.: Diagnosis of delay-deadline failures in real time discrete event models. ISA Transactions **46**(4), 569–582 (2007)
15. Bossomaier, T., Green, D.: *Complex Systems*. Cambridge University Press (2007)
16. Brand, D., Zafiropulo, P.: On communicating finite-state machines. Journal of the ACM **30**(2), 323–342 (1983).
17. Brave, H., Heymann, M.: Control of discrete event systems modeled as hierarchical state machines. IEEE Transactions on Automatic Control **38**(12), 1803–1819 (1993)
18. Brognoli, S., Lamperti, G., Scandale, M.: Incremental determinization of expanding automata. The Computer Journal **59**(12), 1872–1899 (2016)
19. Brusoni, V., Console, L., Terenziani, P., Dupré, D.T.: A spectrum of definitions for temporal model-based diagnosis. Artificial Intelligence **102**(1), 39–80 (1998)
20. Bryant, R.E.: Graph-based algorithms for Boolean function manipulation. IEEE Transactions on Computers **35**, 677–691 (1986)
21. Bryant, R.E.: Symbolic Boolean manipulation with ordered binary-decision diagrams. ACM Computing Surveys **24**, 293–318 (1992)
22. Bylander, T., Allemang, D., Tanner, M., Josephson, J.: The computational complexity of abduction. Artificial Intelligence **49**(1–3), 25–60 (1991)
23. Cabasino, M.P., Giua, A., Seatzu, C.: Fault detection for discrete event systems using Petri nets with unobservable transitions. Automatica **46**, 1531–1539 (2010)
24. Cassandras, C., Lafortune, S.: *Introduction to Discrete Event Systems*, Kluwer International Series in Discrete Event Dynamic Systems, vol. 11. Kluwer Academic, Boston, MA (1999)
25. Cassandras, C., Lafortune, S.: *Introduction to Discrete Event Systems*, second edn. Springer Science+Business Media, New York, (2008)
26. Ceriani, L., Zanella, M.: Model-based diagnosis and generation of hypothesis space via AI planning. In: *25th International Workshop on Principles of Diagnosis (DX-14)*. Graz, Austria (2014)
27. Cerutti, S., Lamperti, G., Scaroni, M., Zanella, M., Zanni, D.: A diagnostic environment for automaton networks. Software: Practice and Experience **37**(4), 365–415 (2007).
28. Chen, Y., Provan, G.: Modeling and diagnosis of timed discrete event systems – A factory automation example. In: *American Control Conference*, pp. 31–36. Albuquerque, NM (1997)
29. Chomsky, N.: Three models for the description of language. Transactions on Information Theory **2**(3), 113–124 (1956)
30. Chu, D.: Complexity: Against systems. Theory in Bioscience **130**(3), 229–245 (2011)
31. Cimatti, A., Pecheur, C., Cavada, R.: Formal verification of diagnosability via symbolic model checking. In: *Eighteenth International Joint Conference on Artificial Intelligence (IJCAI'03)*, pp. 363–369 (2003)
32. Console, L., Picardi, C., Ribaudo, M.: Diagnosis and diagnosability analysis using process algebras. In: *Eleventh International Workshop on Principles of Diagnosis (DX'00)*, pp. 25–32. Morelia, Mexico (2000)
33. Console, L., Picardi, C., Ribaudo, M.: Process algebras for systems diagnosis. Artificial Intelligence **142**(1), 19–51 (2002)
34. Console, L., Portinale, L., Dupré, D.T., Torasso, P.: Diagnostic reasoning across different time points. In: *Eleventh European Conference on Artificial Intelligence (ECAI'92)*, pp. 369–373. Vienna (1992)
35. Console, L., Torasso, P.: On the co-operation between abductive and temporal reasoning in medical diagnosis. Artificial Intelligence in Medicine **3**(6), 291–311 (1991)
36. Console, L., Torasso, P.: A spectrum of logical definitions of model-based diagnosis. Computational Intelligence **7**(3), 133–141 (1991)
37. Cordier, M., Dousson, C.: Alarm driven monitoring based on chronicles. In: *Fourth IFAC Symposium on Fault Detection, Supervision and Safety for Technical Processes (SAFEPROCESS 2000)*, pp. 286–291. Budapest (2000)
38. Darwiche, A., Marquis, P.: A knowledge compilation map. Journal of Artificial Intelligence Research **17**, 229–264 (2002)

39. Debouk, R., Lafortune, S., Teneketzis, D.: Coordinated decentralized protocols for failure diagnosis of discrete-event systems. Journal of Discrete Event Dynamic Systems: Theory and Applications **10**(1–2), 33–86 (2000)
40. Dressler, O., Freitag, H.: Prediction sharing across time and contexts. In: *Twelfth National Conference on Artificial Intelligence (AAAI'94)*, pp. 1136–1141. Seattle, WA (1994)
41. Ducoli, A., Lamperti, G., Piantoni, E., Zanella, M.: Coverage techniques for checking temporal-observation subsumption. In: *Eighteenth International Workshop on Principles of Diagnosis (DX'07)*, pp. 59–66. Nashville, TN (2007)
42. Ducoli, A., Lamperti, G., Piantoni, E., Zanella, M.: Lazy diagnosis of active systems via pruning techniques. In: *Seventh IFAC Symposium on Fault Detection, Supervision and Safety for Technical Processes (SAFEPROCESS 2009)*, pp. 1336–1341. Barcelona, Spain (2009)
43. Dvorak, D.: Monitoring and diagnosis of continuous dynamic systems using semiquantitative simulation. Ph.D. thesis, University of Texas at Austin, TX (1992)
44. Dvorak, D., Kuipers, B.: Model-based monitoring of dynamic systems. In: W. Hamscher, L. Console, J. de Kleer (eds.) *Readings in Model-Based Diagnosis*. Morgan Kaufmann (1992)
45. Emerson, E.: Temporal and modal logic. In: J. van Leeuwen (ed.) *Handbook of Theoretical Computer Science*, Volume B: *Formal Models and Semantics*, Chap. 16, pp. 995–1072. MIT Press (1990)
46. Fattah, Y.E., Provan, G.: Modeling temporal behavior in the model-based diagnosis of discrete-event systems (a preliminary note). In: *Eighth International Workshop on Principles of Diagnosis (DX'97)*. Mont St. Michel, France (1997)
47. Feldman, A., de Castro, H.V., van Gemund, A., Provan, G.: Model-based diagnostic decision support system for satellites. In: *IEEE Aerospace Conference (AEROCONF 2013)* (2013)
48. Friedl, J.: *Mastering Regular Expressions*, third edn. O'Reilly Media, Sebastopol, CA (2006)
49. Friedrich, G., Lackinger, F.: Diagnosing temporal misbehavior. In: *Twelfth International Joint Conference on Artificial Intelligence (IJCAI'91)*, pp. 1116–1122. Sydney, Australia (1991)
50. Gascard, E., Simeu-Abazi, Z.: Modular modeling for the diagnostic of complex discrete-event systems. IEEE Transactions on Automation Science and Engineering **10**(4), 1101–1123 (2013)
51. Goles, E., Martinez, S. (eds.): Complex Systems. Springer, Dordrecht, Netherlands (2001)
52. Grastien, A., Anbulagan, A., Rintanen, J., Kelareva, E.: Modeling and solving diagnosis of discrete-event systems via satisfiability. In: *Eighteenth International Workshop on Principles of Diagnosis (DX'07)*, pp. 114–121. Nashville, TN (2007)
53. Grastien, A., Cordier, M., Largouët, C.: Incremental diagnosis of discrete-event systems. In: *Sixteenth International Workshop on Principles of Diagnosis (DX'05)*, pp. 119–124. Monterey, CA (2005)
54. Grastien, A., Cordier, M., Largouët, C.: Incremental diagnosis of discrete-event systems. In: *Nineteenth International Joint Conference on Artificial Intelligence (IJCAI'05)*, pp. 1564–1565. Edinburgh, UK (2005)
55. Grastien, A., Haslum, P.: Diagnosis as planning: Two case studies. In: *Scheduling and Planning Applications Workshop (SPARK'11)*, pp. 37–44. Freiburg, Germany (2011)
56. Grastien, A., Haslum, P., Thiébaux, S.: Exhaustive diagnosis of discrete event systems through exploration of the hypothesis space. In: *22nd International Workshop on Principles of Diagnosis (DX'11)*, pp. 60–67. Murnau, Germany (2011)
57. Grastien, A., Haslum, P., Thiébaux, S.: Conflict-based diagnosis of discrete event systems: Theory and practice. In: *Thirteenth International Conference on Knowledge Representation and Reasoning (KR 2012)*, pp. 489–499. Rome. Association for the Advancement of Artificial Intelligence (2012)
58. Guckenbiehl, T., Schäfer-Richter, G.: Sidia: Extending prediction based diagnosis to dynamic models. In: *First International Workshop on Principles of Diagnosis*, pp. 74–82. Stanford, CA (1990)
59. Hamscher, W.: Modeling digital circuits for troubleshooting. Artificial Intelligence **51**(1–3), 223–271 (1991)

60. Hamscher, W., Console, L., de Kleer, J. (eds.): *Readings in Model-Based Diagnosis*. Morgan Kaufmann, San Mateo, CA (1992)
61. Hamscher, W., Davis, R.: Diagnosing circuits with state: An inherently underconstrained problem. In: *Fourth National Conference on Artificial Intelligence (AAAI'84)*, pp. 142–147. Austin, TX (1984)
62. Harel, D.: Statecharts: a visual formalism for complex systems. Science of Computer Programming **8**(3), 231–274 (1987)
63. Hopcroft, J.: An $n \log n$ algorithm for minimizing states in a finite automaton. In: Z. Kohave (ed.) *The Theory of Machines and Computations*, pp. 189–196. Academic Press, New York (1971)
64. Hopcroft, J., Motwani, R., Ullman, J.: *Introduction to Automata Theory, Languages, and Computation*, third edn. Addison-Wesley, Reading, MA (2006)
65. Idghamishi, A.M.: Fault diagnosis in hierarchical discrete event systems. M.Sc. thesis, Concordia University, Montreal, QC, Canada (2004)
66. Isaacson, W.: *Einstein: His Life and Universe*. Simon & Schuster, New York (2008)
67. Jéron, T., Marchand, H., Pinchinat, S., Cordier, M.: Supervision patterns in discrete event systems diagnosis. In: *Seventeenth International Workshop on Principles of Diagnosis (DX'06)*, pp. 117–124. Peñaranda de Duero, Spain (2006)
68. Jéron, T., Marchand, H., Pinchinat, S., Cordier, M.: Supervision patterns in discrete event systems diagnosis. In: *Workshop on Discrete Event Systems (WODES'06)*, pp. 262–268. Ann Arbor, MI. IEEE Computer Society (2006)
69. Jiang, S., Huang, Z., Chandra, V., Kumar, R.: A polynomial algorithm for testing diagnosability of discrete event systems. IEEE Transactions on Automatic Control **46**(8), 1318–1321 (2001)
70. Jiroveanu, G., Boel, R., Bordbar, B.: On-line monitoring of large Petri net models under partial observation. Journal of Discrete Event Dynamic Systems **18**, 323–354 (2008)
71. Kaneko, K., Tsuda, I.: *Complex Systems: Chaos and Beyond: A Constructive Approach with Applications in Life*. Springer, Berlin, Heidelberg (2013)
72. de Kleer, J., Williams, B.: Diagnosing multiple faults. Artificial Intelligence **32**(1), 97–130 (1987)
73. Lackinger, F., Nejdl, W.: Integrating model-based monitoring and diagnosis of complex dynamic systems. In: *Twelfth International Joint Conference on Artificial Intelligence (IJCAI'91)*, pp. 2893–2898. Sydney, Australia (1991)
74. Lamperti, G., Ducoli, A., Piantoni, E., Zanella, M.: Circular pruning for lazy diagnosis of active systems. In: *20th International Workshop on Principles of Diagnosis (DX'09)*, pp. 251–258. Stockholm (2009)
75. Lamperti, G., Pogliano, P.: Event-based reasoning for short circuit diagnosis in power transmission networks. In: *Fifteenth International Joint Conference on Artificial Intelligence (IJCAI'97)*, pp. 446–451. Nagoya, Japan (1997)
76. Lamperti, G., Pogliano, P., Zanella, M.: Diagnosis of active systems by automata-based reasoning techniques. Applied Intelligence **12**(3), 217–237 (2000)
77. Lamperti, G., Quarenghi, G.: Intelligent monitoring of complex discrete-event systems. In: I. Czarnowski, A. Caballero, R. Howlett, L. Jain (eds.) *Intelligent Decision Technologies 2016*, Smart Innovation, Systems and Technologies, vol. 56, pp. 215–229. Springer International Publishing Switzerland (2016)
78. Lamperti, G., Scandale, M.: From diagnosis of active systems to incremental determinization of finite acyclic automata. AI Communications **26**(4), 373–393 (2013)
79. Lamperti, G., Scandale, M.: Incremental determinization and minimization of finite acyclic automata. In: *IEEE International Conference on Systems, Man, and Cybernetics (SMC 2013)*, pp. 2250–2257. Manchester, United Kingdom (2013)
80. Lamperti, G., Scandale, M., Zanella, M.: Determinization and minimization of finite acyclic automata by incremental techniques. Software: Practice and Experience **46**(4), 513–549 (2016).

81. Lamperti, G., Vivenzi, F., Zanella, M.: On checking temporal-observation subsumption in similarity-based diagnosis of active systems. In: *Tenth International Conference on Enterprise Information Systems (ICEIS 2008)*, pp. 44–53. Barcelona, Spain (2008)
82. Lamperti, G., Vivenzi, F., Zanella, M.: Relaxation of temporal observations in model-based diagnosis of discrete-event systems. In: *ECAI 2008 Workshop on Model-Based Systems (MBS 2008)*, pp. 25–29. Patras, Greece (2008)
83. Lamperti, G., Vivenzi, F., Zanella, M.: On subsumption, coverage, and relaxation of temporal observations in reuse-based diagnosis of discrete-event systems: A unifying perspective. In: *20th International Workshop on Principles of Diagnosis (DX'09)*, pp. 353–360. Stockholm (2009)
84. Lamperti, G., Zanella, M.: Diagnosis of discrete-event systems integrating synchronous and asynchronous behavior. In: *Tenth International Workshop on Principles of Diagnosis (DX'99)*, pp. 129–139. Loch Awe, United Kingdom (1999)
85. Lamperti, G., Zanella, M.: Diagnosis of discrete-event systems with uncertain observations. In: *Fourth IFAC Symposium on Fault Detection, Supervision and Safety for Technical Processes (SAFEPROCESS 2000)*, pp. 1109–1114. Budapest (2000)
86. Lamperti, G., Zanella, M.: Generation of diagnostic knowledge by discrete-event model compilation. In: *Seventh International Conference on Knowledge Representation and Reasoning (KR 2000)*, pp. 333–344. Breckenridge, CO (2000)
87. Lamperti, G., Zanella, M.: Uncertain discrete-event observations. In: *Eleventh International Workshop on Principles of Diagnosis (DX'00)*, pp. 101–108. Morelia, Mexico (2000)
88. Lamperti, G., Zanella, M.: Uncertain temporal observations in diagnosis. In: *Fourteenth European Conference on Artificial Intelligence (ECAI 2000)*, pp. 151–155. Berlin. IOS Press, Amsterdam (2000)
89. Lamperti, G., Zanella, M.: Diagnosis of discrete-event systems from uncertain temporal observations. Artificial Intelligence **137**(1–2), 91–163 (2002). DOI 10.1016/S0004-3702(02)00123-6
90. Lamperti, G., Zanella, M.: Continuous diagnosis of discrete-event systems. In: *Fourteenth International Workshop on Principles of Diagnosis (DX'03)*, pp. 105–111. Washington, DC (2003)
91. Lamperti, G., Zanella, M.: *Diagnosis of Active Systems: Principles and Techniques*, Springer International Series in Engineering and Computer Science, vol. 741. Springer, Dordrecht, Netherlands (2003).
92. Lamperti, G., Zanella, M.: EDEN: An intelligent software environment for diagnosis of discrete-event systems. Applied Intelligence **18**, 55–77 (2003)
93. Lamperti, G., Zanella, M.: A bridged diagnostic method for the monitoring of polymorphic discrete-event systems. IEEE Transactions on Systems, Man, and Cybernetics – Part B: Cybernetics **34**(5), 2222–2244 (2004)
94. Lamperti, G., Zanella, M.: Diagnosis of discrete-event systems by separation of concerns, knowledge compilation, and reuse. In: *Sixteenth European Conference on Artificial Intelligence (ECAI 2004)* pp. 838–842. Valencia, Spain (2004)
95. Lamperti, G., Zanella, M.: Dynamic diagnosis of active systems with fragmented observations. In: *Sixth International Conference on Enterprise Information Systems (ICEIS 2004)*, pp. 249–261. Porto, Portugal (2004)
96. Lamperti, G., Zanella, M.: Monitoring and diagnosis of discrete-event systems with uncertain symptoms. In: *Sixteenth International Workshop on Principles of Diagnosis (DX'05)*, pp. 145–150. Monterey, CA (2005)
97. Lamperti, G., Zanella, M.: Monitoring-based diagnosis of discrete-event systems with uncertain observations. In: *Nineteenth International Workshop on Qualitative Reasoning (QR'05)*, pp. 97–102. Graz, Austria (2005)
98. Lamperti, G., Zanella, M.: On processing temporal observations in model-based reasoning. In: *Second MONET Workshop on Model-Based Systems (MONET'05)*, pp. 42–47. Edinburgh, United Kingdom (2005)
99. Lamperti, G., Zanella, M.: Flexible diagnosis of discrete-event systems by similarity-based reasoning techniques. Artificial Intelligence **170**(3), 232–297 (2006)

100. Lamperti, G., Zanella, M.: Incremental indexing of temporal observations in diagnosis of active systems. In: *Seventeenth International Workshop on Principles of Diagnosis (DX'06)*, pp. 147–154. Peñaranda de Duero, Spain (2006)
101. Lamperti, G., Zanella, M.: Incremental processing of temporal observations in supervision and diagnosis of discrete-event systems. In: *Eighth International Conference on Enterprise Information Systems (ICEIS 2006)*, pp. 47–57. Paphos, Cyprus (2006)
102. Lamperti, G., Zanella, M.: On monotonic monitoring of discrete-event systems. In: *Eighteenth International Workshop on Principles of Diagnosis (DX'07)*, pp. 130–137. Nashville, TN (2007)
103. Lamperti, G., Zanella, M.: Dependable monitoring of discrete-event systems with uncertain temporal observations. In: *Eighteenth European Conference on Artificial Intelligence (ECAI 2008)*, pp. 793–794. Patras, Greece. IOS Press, Amsterdam (2008)
104. Lamperti, G., Zanella, M.: Observation-subsumption checking in similarity-based diagnosis of discrete-event systems. In: *Eighteenth European Conference on Artificial Intelligence (ECAI 2008)*, pp. 204–208. Patras, Greece. IOS Press, Amsterdam (2008)
105. Lamperti, G., Zanella, M.: On processing temporal observations in monitoring of discrete-event systems. In: Y. Manolopoulos, J. Filipe, P. Constantopoulos, J. Cordeiro (eds.) *Enterprise Information Systems VIII*, Lecture Notes in Business Information Processing, vol. 3, pp. 135–146. Springer, Berlin, Heidelberg (2008)
106. Lamperti, G., Zanella, M.: Monotonic monitoring of discrete-event systems with uncertain temporal observations. In: J. Filipe, J. Cordeiro (eds.) *Enterprise Information Systems*, Lecture Notes in Business Information Processing, vol. 24, pp. 348–362. Springer, Berlin, Heidelberg (2009)
107. Lamperti, G., Zanella, M.: Diagnosis of active systems by lazy techniques. In: *12th International Conference on Enterprise Information Systems (ICEIS 2010)*, pp. 171–180. Funchal, Madeira, Portugal (2010)
108. Lamperti, G., Zanella, M.: Injecting semantics into diagnosis of discrete-event systems. In: *21st International Workshop on Principles of Diagnosis (DX'10)*, pp. 233–240. Portland, OR (2010)
109. Lamperti, G., Zanella, M.: Context-sensitive diagnosis of discrete-event systems. In: T. Walsh (ed.) *Twenty-Second International Joint Conference on Artificial Intelligence (IJCAI'11)*, vol. 2, pp. 969–975. Barcelona, Spain. AAAI Press (2011)
110. Lamperti, G., Zanella, M.: Monitoring of active systems with stratified uncertain observations. IEEE Transactions on Systems, Man, and Cybernetics – Part A: Systems and Humans **41**(2), 356–369 (2011)
111. Lamperti, G., Zanella, M.: Preliminaries on complexity of diagnosis of discrete-event systems. In: *24th International Workshop on Principles of Diagnosis (DX'13)*, pp. 192–197. Jerusalem, Israel (2013)
112. Lamperti, G., Zanella, M., Chiodi, G., Chiodi, L.: Incremental determinization of finite automata in model-based diagnosis of active systems. In: I. Lovrek, R. Howlett, L. Jain (eds.) *Knowledge-Based Intelligent Information and Engineering Systems*, Lecture Notes in Artificial Intelligence, vol. 5177, pp. 362–374. Springer, Berlin, Heidelberg (2008)
113. Lamperti, G., Zanella, M., Zanni, D.: Incremental processing of temporal observations in model-based reasoning. AI Communications **20**(1), 27–37 (2007)
114. Lamperti, G., Zhao, X.: Diagnosis of discrete-event systems with stratified behavior. In: *24th International Workshop on Principles of Diagnosis (DX'13)*, pp. 124–129. Jerusalem, Israel (2013)
115. Lamperti, G., Zhao, X.: Diagnosis of higher-order discrete-event systems. In: A. Cuzzocrea, C. Kittl, D. Simos, E. Weippl, L. Xu (eds.) *Availability, Reliability, and Security in Information Systems and HCI*, Lecture Notes in Computer Science, vol. 8127, pp. 162–177. Springer, Berlin, Heidelberg (2013)
116. Lamperti, G., Zhao, X.: Specification and model-based diagnosis of higher-order discrete-event systems. In: *IEEE International Conference on Systems, Man, and Cybernetics (SMC 2013)*, pp. 2342–2347. Manchester, United Kingdom (2013)

117. Lamperti, G., Zhao, X.: Diagnosis of active systems by semantic patterns. IEEE Transactions on Systems, Man, and Cybernetics: Systems **44**(8), 1028–1043 (2014)
118. Lamperti, G., Zhao, X.: Diagnosis of complex active systems with uncertain temporal observations. In: F. Buccafurri, A. Holzinger, A.M. Tjoa, E. Weippl (eds.) *Availability, Reliability, and Security in Information Systems*, Lecture Notes in Computer Science, vol. 9817, pp. 45–62. Springer International Publishing AG Switzerland (2016)
119. Lamperti, G., Zhao, X.: Viable diagnosis of complex active systems. In: *IEEE International Conference on Systems, Man, and Cybernetics (SMC 2016)*, pp. 457–462. Budapest (2016)
120. de León, H.P., Bonigo, G., Briones, L.B.: Distributed analysis for diagnosability in concurrent systems. In: *24th International Workshop on Principles of Diagnosis (DX'13)*, pp. 142–147. Jerusalem, Israel (2013)
121. Licata, I., Sakaji, A.: *Physics of Emergence and Organization*. World Scientific (2008)
122. Lin, F.: Diagnosability of discrete-event systems and its applications. Discrete Event Dynamical Systems **4**, 197–212 (1994)
123. Liu, F., Qiu, D.: Diagnosability of fuzzy discrete-event systems: A fuzzy approach. IEEE Transactions on Fuzzy Systems **17**, 372–384 (2009)
124. Lunze, J.: Diagnosis of quantized systems based on a timed discrete-event model. IEEE Transactions on Systems, Man, and Cybernetics – Part A: Systems and Humans **30**(3), 322–335 (2000)
125. Moore, E.: Gedanken-experiments on sequential machines. Automata Studies, Annals of Mathematics Studies **34**, 129–153 (1956)
126. Mozetič, I.: Model-based diagnosis: An overview. In: V. Marik, O. Stepankova, R. Trappl (eds.) Advanced topics in Artificial Intelligence, pp. 419–430. Springer (1992)
127. Ng, H.: Model-based, multiple-fault diagnosis of dynamic, continuous physical devices. IEEE Expert **6**(6), 38–43 (1991)
128. Pan, J.Y.C.: Qualitative reasoning with deep-level mechanism models for diagnosis of mechanism failures. In: *First IEEE Conference on AI Applications*, pp. 295–301. Denver, CO (1984)
129. Paoli, A., Lafortune, S.: Diagnosability analysis of a class of hierarchical state machines. Journal of Discrete Event Dynamic Systems: Theory and Applications **18**(3), 385–413 (2008)
130. Pencolé, Y.: Diagnosability analysis of distributed discrete event systems. In: *Sixteenth European Conference on Artificial Intelligence (ECAI 2004)*, pp. 43–47. Valencia, Spain (2004)
131. Pencolé, Y., Cordier, M.: A formal framework for the decentralized diagnosis of large scale discrete event systems and its application to telecommunication networks. Artificial Intelligence **164**(1–2), 121–170 (2005)
132. Pencolé, Y., Cordier, M., Rozé, L.: Incremental decentralized diagnosis approach for the supervision of a telecommunication network. In: *Twelfth International Workshop on Principles of Diagnosis (DX'01)*, pp. 151–158. San Sicario, Italy (2001)
133. Pietersma, J., van Gemund, A.: Benefits and costs of model-based fault diagnosis for semiconductor manufacturing equipment. In: *17th International Symposium on Systems Engineering (INCOSE'07)*, pp. 324–335. San Diego, CA (2007)
134. Poole, D.: Representing diagnosis knowledge. Annals of Mathematics and Artificial Intelligence **11**, 33–50 (1994)
135. Portinale, L., Magro, D., Torasso, P.: Multi-modal diagnosis combining case-based and model-based reasoning: A formal and experimental analysis. Artificial Intelligence **158**(1), 109–153 (2004)
136. Qiu, W., Kumar, R.: Decentralized failure diagnosis of discrete event systems. IEEE Transactions on Systems, Man, and Cybernetics – Part A: Systems and Humans **36**(2), 384–395 (2006)
137. Rabin, M., Scott, D.: Finite automata and their decision problems. IBM Journal of Research and Development **3**(2), 114–125 (1959). DOI 10.1147/rd.32.0114
138. Reeven, B.: Model-based diagnosis in industrial context. M.Sc. thesis, Delft University of Technology, Netherlands (2011)
139. Reiter, R.: A theory of diagnosis from first principles. Artificial Intelligence **32**(1), 57–95 (1987)

140. Rintanen, J., Grastien, A.: Diagnosability testing with satisfiability algorithms. In: *20th International Joint Conference on Artificial Intelligence (IJCAI'07)*, pp. 532–537. Hyderabad, India (2007)
141. Rozé, L.: Supervision of telecommunication network: A diagnoser approach. In: *Eighth International Workshop on Principles of Diagnosis (DX'97)*, pp. 103–111. Mont St. Michel, France (1997)
142. Rozé, L., Cordier, M.: Diagnosing discrete-event systems: Extending the "diagnoser approach" to deal with telecommunication networks. Journal of Discrete Event Dynamic Systems: Theory and Application **12**, 43–81 (2002)
143. Sampath, M., Lafortune, S., Teneketzis, D.: Active diagnosis of discrete-event systems. IEEE Transactions on Automatic Control **43**(7), 908–929 (1998)
144. Sampath, M., Sengupta, R., Lafortune, S., Sinnamohideen, K., Teneketzis, D.: Diagnosability of discrete-event systems. IEEE Transactions on Automatic Control **40**(9), 1555–1575 (1995)
145. Sampath, M., Sengupta, R., Lafortune, S., Sinnamohideen, K., Teneketzis, D.: Failure diagnosis using discrete-event models. IEEE Transactions on Control Systems Technology **4**(2), 105–124 (1996)
146. Schoemaker, E.: Personal communication, ASML Netherlands (2008)
147. Schullerus, G., Krebs, V.: Diagnosis of a class of discrete-event systems based on parameter estimation of a modular algebraic model. In: *Twelfth International Workshop on Principles of Diagnosis (DX'01)*, pp. 189–196. San Sicario, Italy (2001)
148. Schumann, A., Pencolé, Y., Thiébaux, S.: Diagnosis of discrete-event systems using binary decision diagrams. In: *Fifteenth International Workshop on Principles of Diagnosis (DX'04)*, pp. 197–202. Carcassonne, France (2004)
149. Schumann, A., Pencolé, Y., Thiébaux, S.: A spectrum of symbolic on-line diagnosis approaches. In: *22nd Conference on Artificial Intelligence (AAAI-07)*, pp. 335–340. Vancouver, Canada. AAAI Press (2007)
150. Shortliffe, E.: Computer-Based Medical Consultations: MYCIN. American Elsevier, New York (1976)
151. Sibani, P., Jensen, H.: *Stochastic Dynamics of Complex Systems*, Complexity Science, vol. 2. World Scientific (2013)
152. Sohrabi, S., Baier, J., McIlraith, S.: Diagnosis as planning revisited. In: *Twelfth International Conference on Knowledge Representation and Reasoning (KR 2010)*, pp. 26–36. Toronto, Canada. Association for the Advancement of Artificial Intelligence (2010)
153. Su, R., Wonham, W.: Global and local consistencies in distributed fault diagnosis for discrete-event systems. IEEE Transactions on Automatic Control **50**(12), 1923–1935 (2005)
154. Su, X., Grastien, A.: Verifying the precision of diagnostic algorithms. In: *21st European Conference on Artificial Intelligence (ECAI 2014)*, pp. 861–866. Prague (2014)
155. Su, X., Zanella, M., Grastien, A.: Diagnosability of discrete-event systems with uncertain observations. In: *5th International Joint Conference on Artificial Intelligence (IJCAI'16)*, pp. 1265–1571. New York (2016)
156. Sztipanovits, J., Misra, A.: Diagnosis of discrete event systems using ordered binary decision diagrams. In: *Proceedings of the 7th International Workshop on Principles of Diagnosis (DX-96)*. Val Morin, Québec, Canada (1996)
157. Thompson, S.: *Haskell: The Craft of Functional Programming*, 2nd edn, ICSS Series. Pearson Education (2011)
158. Thorsley, D., Teneketzis, D.: Diagnosability of stochastic discrete-event systems. IEEE Transactions on Automatic Control **50**, 476–492 (2005)
159. Torasso, P., Torta, G.: Computing minimum-cardinality diagnoses using OBDDs. In: A. Günter, R. Kruse, B. Neumann (eds.) *KI-2003: Advances in Artificial Intelligence*, Lecture Notes in Computer Science, vol. 2821, pp. 224–238. Springer, Berlin, Heidelberg (2003)
160. Torta, G.: Compact representation of diagnoses for improving efficiency in model based diagnosis. Ph.D. thesis, Dipartimento di Informatica, Università di Torino, Italy (2005)
161. Torta, G., Torasso, P.: On the use of OBDDs in model-based diagnosis: An approach based on the partition of the model. Knowledge-Based Systems **19**(5), 316–323 (2006)

162. Tripakis, S.: Fault diagnosis for timed automata. In: W. Damm, E.R. Olderog (eds.) *Formal Techniques in Real-Time and Fault-Tolerant Systems*, Lecture Notes in Computer Science, vol. 2469, pp. 205–221. Springer, Berlin, Heidelberg (2002)
163. Yan, Y., Ye, L., Dague, P.: Diagnosability for patterns in distributed discrete event systems. In: *21st International Workshop on Principles of Diagnosis (DX'10)*, pp. 345–352. Portland, OR (2010)
164. Ye, L., Dague, P., Yan, Y.: A distributed approach for pattern diagnosability. In: *20th International Workshop on Principles of Diagnosis (DX'09)*, pp. 179–186. Stockholm (2009)
165. Ye, L., Dague, P., Yan, Y.: An incremental approach for pattern diagnosability in distributed discrete event systems. In: *21st IEEE International Conference on Tools with Artificial Intelligence (ICTAI'12)*, pp. 123–130. Newark, NJ (2009)
166. Yoo, T., Lafortune, S.: Polynomial-time verification of diagnosability of partially observed discrete-event systems. IEEE Transactions on Automatic Control **47**(9), 1491–1495 (2002)
167. Zad, S., Kwong, R., Wonham, W.: Fault diagnosis in timed discrete-event systems. In: *38th IEEE Conference on Decision and Control (CDC'99)*, pp. 1756–1761. Pheonix, AZ. IEEE, Piscataway, NJ (1999)
168. Zhao, X., Ouyang, D.: Model-based diagnosis of discrete event systems with an incomplete system model. In: *Eigtheenth European Conference on Artificial Intelligence (ECAI 2008)*, pp. 189–193. IOS Press, Amsterdam (2008)
169. Zhao, X., Ouyang, D.: On-line diagnosis of discrete-event systems: A hierarchical approach. In: *2008 IEEE Conference on Robotics, Automation and Mechatronics*, pp. 785–790. IEEE (2008)
170. Zhao, X., Ouyang, D.: On-line diagnosis of discrete-event systems with two successive temporal windows. AI Communications **21**, 249–262 (2008)

Index

G. Lamperti et al., *Introduction to Diagnosis of Active Systems*,
https://doi.org/10.1007/978-3-319-92733-6

Printed in the United States
By Bookmasters